Methods in Enzymology

Volume 321

NUMERICAL COMPUTER METHODS

Part C

METHODS IN ENZYMOLOGY

Methods in Enzymology

Volume 321

Numerical Computer Methods

Part C

EDITED BY

Michael L. Johnson

CENTER FOR BIOMATHEMATICAL TECHNOLOGY
UNIVERSITY OF VIRGINIA HEALTH SYSTEM
CHARLOTTESVILLE, VIRGINIA

Ludwig Brand

JOHNS HOPKINS UNIVERSITY
BALTIMORE, MARYLAND

ACADEMIC PRESS
San Diego London Boston New York Sydney Tokyo Toronto

This book is printed on acid-free paper. 

Academic Press
A Harcourt Science and Technology Company
525 B Street, Suite 1900, San Diego, California 92101-4495, USA
http://www.academicpress.com

Academic Press Limited
Harcourt Place, 32 Jamestown Road, London NW1 7BY, UK

International Standard Book Number: 0-12-182222-2

PRINTED IN THE UNITED STATES OF AMERICA
00 01 02 03 04 05 06 MM 9 8 7 6 5 4 3 2 1

Table of Contents

Contributors to Volume 321

Article numbers are in parentheses following the names of contributors.
Affiliations listed are current.

JOACHIM BEHLKE (3), *Max Delbrück Center for Molecular Medicine, D-13122 Berlin, Germany*

EMERY N. BROWN (17), *Statistics Research Laboratory, Department of Anesthesia and Critical Care, Massachusetts General Hospital, Harvard Medical School/MIT Division of Health Sciences and Technology, Boston, Massachusetts 02114*

OLWYN BYRON (16), *Division of Infection and Immunity, Institute of Biomedical and Life Sciences, University of Glasgow, Glasgow G12 8QQ, Scotland*

LENNY M. CARRUTHERS (4), *Department of Biochemistry, University of Texas Health Sciences Center, San Antonio, Texas 78229-3900*

JONATHAN B. CHAIRES (19), *Department of Biochemistry, University of Mississippi Medical Center, Jackson, Mississippi 39216-4505*

JOHN J. CORREIA (5), *Department of Biochemistry, University of Mississippi Medical Center, Jackson, Mississippi 39216-4505*

DANIEL J. COX (21, 22), *Center for Behavioral Medicine Research, University of Virginia Health System, Charlottesville, Virginia 22908*

JULIEN S. DAVIS (2), *Cardiology Branch, National Heart, Lung and Blood Institute, National Institutes of Health, Bethesda, Maryland 20892-1650*

BORRIES DEMELER (3, 4), *Department of Biochemistry, The University of Texas Health Sciences Center at San Antonio, San Antonio, Texas 78229-3900*

EMILIOS K. DIMITRIADIS (7), *Division of Bioengineering and Physical Science, ORS, OD, National Institutes of Health, Bethesda, Maryland 20892-5766*

LEON S. FARHY (21, 22), *Department of Internal Medicine, University of Virginia Health System, Charlottesville, Virginia 22908*

KATHLEEN B. HALL (18), *Department of Biochemistry, Washington University School of Medicine, Saint Louis, Missouri 63110*

JEFFREY C. HANSEN (4), *Department of Biochemistry, University of Texas Health Sciences Center, San Antonio, Texas 78229-3900*

MICHAEL L. JOHNSON (10, 11, 12, 14, 23, 24) *Departments of Pharmacology and Internal Medicine, Center for Biomathematical Technology, National Science Foundation Science and Technology Center for Biological Timing, General Clinical Research Center, University of Virginia Health System, Charlottesville, Virginia 22908-0735*

RICHARD P. JUNGHANS (8), *Biotherapeutics Development Laboratory, Harvard Institute of Human Genetics, Harvard Medical School and Division of Hematology-Oncology, Beth Israel Deaconess Medical Center, Boston, Massachusetts 02115*

BORIS P. KOVATCHEV (21, 22), *Center for Behavioral Medicine Research, University of Virginia Health System, Charlottesville, Virginia 22908*

MICHELLE LAMPL (10, 11), *Departments of Anthropology and Pediatrics, Emory University, Atlanta, Georgia 30322*

WILLIAM R. LAWS (13), *Department of Biochemistry, Mount Sinai School of Medicine, New York, New York 10029*

MARC S. LEWIS (7, 8), *Division of Bioengineering and Physical Science, National Institutes of Health, Bethesda, Maryland 20892-5766*

LESLIE M. LOEW (1), *Department of Physiology, Center for Biomedical Imaging Technology, University of Connecticut Health Center, Framingham, Connecticut 06030*

OLE G. MOURITSEN (15), *Department of Chemistry, Technical University of Denmark, DK-2800 Lyngby, Denmark*

TIMOTHY G. MYERS (20), *Information Technology Branch, National Cancer Institute, National Institutes of Health, Bethesda, Maryland 20892*

JOHN S. PHILO (6), *Alliance Protein Laboratories, Inc., Thousand Oaks, California 91360-2823*

STEVEN M. PINCUS (9), *Guilford, Connecticut 06437*

XIAOGANG QU (19), *Department of Biochemistry, University of Mississippi Medical Center, Jackson, Mississippi 39216-4505*

EDWARD L. RACHOFSKY (13), *Department of Biochemistry, Mount Sinai School of Medicine, New York, New York 10029*

OTTO RISTAU (3), *Max Delbrück Center for Molecular Medicine, Berlin D-13122, Germany*

MADS C. SABRA (15), *Department of Chemistry, Technical University of Denmark, DK-2800 Lyngby, Denmark*

JAMES C. SCHAFF (1), *Department of Physiology, Center for Biomedical Imaging Technology, University of Connecticut Health Center, Framingham, Connecticut 06030*

VIRGIL R. SCHIRF (4), *Department of Biochemistry, University of Texas Health Sciences Center, San Antonio, Texas 78229-3900*

CHRISTOPHER H. SCHMID (17), *Biostatistics Research Center, Division of Clinical Care Research, New England Medical Center, Boston, Massachusetts 02111*

BORIS M. SLEPCHENKO (1), *Department of Physiology, Center for Biomedical Imaging Technology, University of Connecticut Health Center, Framingham, Connecticut 06030*

MARTIN STRAUME (11, 12, 21, 22), *Division of Endocrinology and Metabolism, Department of Internal Medicine, Center for Biomathematical Technology, National Science Foundation Center for Biological Timing, University of Virginia Health System, Charlottesville, Virginia 22908-0735*

WILLIAM W. VAN OSDOL (20), *ALZA Corporation, Mountain View, California 94043*

JOHANNES D. VELDHUIS (14), *General Clinical Research Center, University of Virginia Health Sciences Center, Charlottesville, Virginia 22908-0202*

JOHN N. WEINSTEIN (20), *Laboratory for Molecular Pharmacology, National Cancer Institute, National Institutes of Health, Bethesda, Maryland 20892*

DERYCK JEREMY WILLIAMS (18), *Department of Biochemistry, Washington University School of Medicine, Saint Louis, Missouri 63110*

Preface

In the five years that have elapsed since the appearance of Part B of "Numerical Computer Methods" (*Methods in Enzymology*, Volume 240), there have been substantial enhancements in computer speed, memory size, and disk space, and decreases in computer costs. There have been parallel improvements in the numerical methodologies that can now be applied to the analysis of biological data. Procedures that previously required the power of a supercomputer are now commonly run on personal computers and workstations.

Computers are not oracles! But many biomedical researchers have learned to revere computers as oracles. They assume that if a computer analyzed their data the results must be correct. Data analysis algorithms make assumptions about the nature of the experimental data and the best method of analysis based on assumptions about the experimental data. They also make compromises to save computer time and space in the memory of the computer. Computer programmers also make mistakes. Consequently, computer programs can include unwarranted assumptions and can make mistakes! Biomedical researchers cannot simply insert data into a computer and accept the results as gospel. Researchers must be aware of the assumptions and methods used by their data analysis programs. They should always question the results of a computer analysis, i.e., do they have physical meaning? The purpose of these volumes is to help biomedical researchers meet these needs.

It is clear that new methods of analyzing experimental data have enhanced the type of conclusions that can be drawn and have changed the way in which experiments are conceived and done. The biochemical community must be aware of these new developments. The aim of this volume, and of Parts A and B (Volumes 210 and 240), is to inform biomedical researchers of the modern data analysis methods that have developed concomitantly with computer hardware.

MICHAEL L. JOHNSON
LUDWIG BRAND

METHODS IN ENZYMOLOGY

VOLUME I. Preparation and Assay of Enzymes
Edited by SIDNEY P. COLOWICK AND NATHAN O. KAPLAN

VOLUME II. Preparation and Assay of Enzymes
Edited by SIDNEY P. COLOWICK AND NATHAN O. KAPLAN

VOLUME III. Preparation and Assay of Substrates
Edited by SIDNEY P. COLOWICK AND NATHAN O. KAPLAN

VOLUME IV. Special Techniques for the Enzymologist
Edited by SIDNEY P. COLOWICK AND NATHAN O. KAPLAN

VOLUME V. Preparation and Assay of Enzymes
Edited by SIDNEY P. COLOWICK AND NATHAN O. KAPLAN

VOLUME VI. Preparation and Assay of Enzymes (*Continued*)
Preparation and Assay of Substrates
Special Techniques
Edited by SIDNEY P. COLOWICK AND NATHAN O. KAPLAN

VOLUME VII. Cumulative Subject Index
Edited by SIDNEY P. COLOWICK AND NATHAN O. KAPLAN

VOLUME VIII. Complex Carbohydrates
Edited by ELIZABETH F. NEUFELD AND VICTOR GINSBURG

VOLUME IX. Carbohydrate Metabolism
Edited by WILLIS A. WOOD

VOLUME X. Oxidation and Phosphorylation
Edited by RONALD W. ESTABROOK AND MAYNARD E. PULLMAN

VOLUME XI. Enzyme Structure
Edited by C. H. W. HIRS

VOLUME XII. Nucleic Acids (Parts A and B)
Edited by LAWRENCE GROSSMAN AND KIVIE MOLDAVE

VOLUME XIII. Citric Acid Cycle
Edited by J. M. LOWENSTEIN

VOLUME XIV. Lipids
Edited by J. M. LOWENSTEIN

VOLUME XV. Steroids and Terpenoids
Edited by RAYMOND B. CLAYTON

VOLUME XVI. Fast Reactions
Edited by KENNETH KUSTIN

Volume XVII. Metabolism of Amino Acids and Amines (Parts A and B)
Edited by Herbert Tabor and Celia White Tabor

Volume XVIII. Vitamins and Coenzymes (Parts A, B, and C)
Edited by Donald B. McCormick and Lemuel D. Wright

Volume XIX. Proteolytic Enzymes
Edited by Gertrude E. Perlmann and Laszlo Lorand

Volume XX. Nucleic Acids and Protein Synthesis (Part C)
Edited by Kivie Moldave and Lawrence Grossman

Volume XXI. Nucleic Acids (Part D)
Edited by Lawrence Grossman and Kivie Moldave

Volume XXII. Enzyme Purification and Related Techniques
Edited by William B. Jakoby

Volume XXIII. Photosynthesis (Part A)
Edited by Anthony San Pietro

Volume XXIV. Photosynthesis and Nitrogen Fixation (Part B)
Edited by Anthony San Pietro

Volume XXV. Enzyme Structure (Part B)
Edited by C. H. W. Hirs and Serge N. Timasheff

Volume XXVI. Enzyme Structure (Part C)
Edited by C. H. W. Hirs and Serge N. Timasheff

Volume XXVII. Enzyme Structure (Part D)
Edited by C. H. W. Hirs and Serge N. Timasheff

Volume XXVIII. Complex Carbohydrates (Part B)
Edited by Victor Ginsburg

Volume XXIX. Nucleic Acids and Protein Synthesis (Part E)
Edited by Lawrence Grossman and Kivie Moldave

Volume XXX. Nucleic Acids and Protein Synthesis (Part F)
Edited by Kivie Moldave and Lawrence Grossman

Volume XXXI. Biomembranes (Part A)
Edited by Sidney Fleischer and Lester Packer

Volume XXXII. Biomembranes (Part B)
Edited by Sidney Fleischer and Lester Packer

Volume XXXIII. Cumulative Subject Index Volumes I–XXX
Edited by Martha G. Dennis and Edward A. Dennis

Volume XXXIV. Affinity Techniques (Enzyme Purification: Part B)
Edited by William B. Jakoby and Meir Wilchek

Volume XXXV. Lipids (Part B)
Edited by John M. Lowenstein

VOLUME XXXVI. Hormone Action (Part A: Steroid Hormones)
Edited by BERT W. O'MALLEY AND JOEL G. HARDMAN

VOLUME XXXVII. Hormone Action (Part B: Peptide Hormones)
Edited by BERT W. O'MALLEY AND JOEL G. HARDMAN

VOLUME XXXVIII. Hormone Action (Part C: Cyclic Nucleotides)
Edited by JOEL G. HARDMAN AND BERT W. O'MALLEY

VOLUME XXXIX. Hormone Action (Part D: Isolated Cells, Tissues, and Organ Systems)
Edited by JOEL G. HARDMAN AND BERT W. O'MALLEY

VOLUME XL. Hormone Action (Part E: Nuclear Structure and Function)
Edited by BERT W. O'MALLEY AND JOEL G. HARDMAN

VOLUME XLI. Carbohydrate Metabolism (Part B)
Edited by W. A. WOOD

VOLUME XLII. Carbohydrate Metabolism (Part C)
Edited by W. A. WOOD

VOLUME XLIII. Antibiotics
Edited by JOHN H. HASH

VOLUME XLIV. Immobilized Enzymes
Edited by KLAUS MOSBACH

VOLUME XLV. Proteolytic Enzymes (Part B)
Edited by LASZLO LORAND

VOLUME XLVI. Affinity Labeling
Edited by WILLIAM B. JAKOBY AND MEIR WILCHEK

VOLUME XLVII. Enzyme Structure (Part E)
Edited by C. H. W. HIRS AND SERGE N. TIMASHEFF

VOLUME XLVIII. Enzyme Structure (Part F)
Edited by C. H. W. HIRS AND SERGE N. TIMASHEFF

VOLUME XLIX. Enzyme Structure (Part G)
Edited by C. H. W. HIRS AND SERGE N. TIMASHEFF

VOLUME L. Complex Carbohydrates (Part C)
Edited by VICTOR GINSBURG

VOLUME LI. Purine and Pyrimidine Nucleotide Metabolism
Edited by PATRICIA A. HOFFEE AND MARY ELLEN JONES

VOLUME LII. Biomembranes (Part C: Biological Oxidations)
Edited by SIDNEY FLEISCHER AND LESTER PACKER

VOLUME LIII. Biomembranes (Part D: Biological Oxidations)
Edited by SIDNEY FLEISCHER AND LESTER PACKER

VOLUME LIV. Biomembranes (Part E: Biological Oxidations)
Edited by SIDNEY FLEISCHER AND LESTER PACKER

VOLUME LV. Biomembranes (Part F: Bioenergetics)
Edited by SIDNEY FLEISCHER AND LESTER PACKER

VOLUME LVI. Biomembranes (Part G: Bioenergetics)
Edited by SIDNEY FLEISCHER AND LESTER PACKER

VOLUME LVII. Bioluminescence and Chemiluminescence
Edited by MARLENE A. DELUCA

VOLUME LVIII. Cell Culture
Edited by WILLIAM B. JAKOBY AND IRA PASTAN

VOLUME LIX. Nucleic Acids and Protein Synthesis (Part G)
Edited by KIVIE MOLDAVE AND LAWRENCE GROSSMAN

VOLUME LX. Nucleic Acids and Protein Synthesis (Part H)
Edited by KIVIE MOLDAVE AND LAWRENCE GROSSMAN

VOLUME 61. Enzyme Structure (Part H)
Edited by C. H. W. HIRS AND SERGE N. TIMASHEFF

VOLUME 62. Vitamins and Coenzymes (Part D)
Edited by DONALD B. MCCORMICK AND LEMUEL D. WRIGHT

VOLUME 63. Enzyme Kinetics and Mechanism (Part A: Initial Rate and Inhibitor Methods)
Edited by DANIEL L. PURICH

VOLUME 64. Enzyme Kinetics and Mechanism (Part B: Isotopic Probes and Complex Enzyme Systems)
Edited by DANIEL L. PURICH

VOLUME 65. Nucleic Acids (Part I)
Edited by LAWRENCE GROSSMAN AND KIVIE MOLDAVE

VOLUME 66. Vitamins and Coenzymes (Part E)
Edited by DONALD B. MCCORMICK AND LEMUEL D. WRIGHT

VOLUME 67. Vitamins and Coenzymes (Part F)
Edited by DONALD B. MCCORMICK AND LEMUEL D. WRIGHT

VOLUME 68. Recombinant DNA
Edited by RAY WU

VOLUME 69. Photosynthesis and Nitrogen Fixation (Part C)
Edited by ANTHONY SAN PIETRO

VOLUME 70. Immunochemical Techniques (Part A)
Edited by HELEN VAN VUNAKIS AND JOHN J. LANGONE

VOLUME 71. Lipids (Part C)
Edited by JOHN M. LOWENSTEIN

VOLUME 72. Lipids (Part D)
Edited by JOHN M. LOWENSTEIN

VOLUME 73. Immunochemical Techniques (Part B)
Edited by JOHN J. LANGONE AND HELEN VAN VUNAKIS

VOLUME 74. Immunochemical Techniques (Part C)
Edited by JOHN J. LANGONE AND HELEN VAN VUNAKIS

VOLUME 75. Cumulative Subject Index Volumes XXXI, XXXII, XXXIV–LX
Edited by EDWARD A. DENNIS AND MARTHA G. DENNIS

VOLUME 76. Hemoglobins
Edited by ERALDO ANTONINI, LUIGI ROSSI-BERNARDI, AND EMILIA CHIANCONE

VOLUME 77. Detoxication and Drug Metabolism
Edited by WILLIAM B. JAKOBY

VOLUME 78. Interferons (Part A)
Edited by SIDNEY PESTKA

VOLUME 79. Interferons (Part B)
Edited by SIDNEY PESTKA

VOLUME 80. Proteolytic Enzymes (Part C)
Edited by LASZLO LORAND

VOLUME 81. Biomembranes (Part H: Visual Pigments and Purple Membranes, I)
Edited by LESTER PACKER

VOLUME 82. Structural and Contractile Proteins (Part A: Extracellular Matrix)
Edited by LEON W. CUNNINGHAM AND DIXIE W. FREDERIKSEN

VOLUME 83. Complex Carbohydrates (Part D)
Edited by VICTOR GINSBURG

VOLUME 84. Immunochemical Techniques (Part D: Selected Immunoassays)
Edited by JOHN J. LANGONE AND HELEN VAN VUNAKIS

VOLUME 85. Structural and Contractile Proteins (Part B: The Contractile Apparatus and the Cytoskeleton)
Edited by DIXIE W. FREDERIKSEN AND LEON W. CUNNINGHAM

VOLUME 86. Prostaglandins and Arachidonate Metabolites
Edited by WILLIAM E. M. LANDS AND WILLIAM L. SMITH

VOLUME 87. Enzyme Kinetics and Mechanism (Part C: Intermediates, Stereochemistry, and Rate Studies)
Edited by DANIEL L. PURICH

VOLUME 88. Biomembranes (Part I: Visual Pigments and Purple Membranes, II)
Edited by LESTER PACKER

VOLUME 89. Carbohydrate Metabolism (Part D)
Edited by WILLIS A. WOOD

VOLUME 90. Carbohydrate Metabolism (Part E)
Edited by WILLIS A. WOOD

VOLUME 108. Immunochemical Techniques (Part G: Separation and Characterization of Lymphoid Cells)
Edited by GIOVANNI DI SABATO, JOHN J. LANGONE, AND HELEN VAN VUNAKIS

VOLUME 109. Hormone Action (Part I: Peptide Hormones)
Edited by LUTZ BIRNBAUMER AND BERT W. O'MALLEY

VOLUME 110. Steroids and Isoprenoids (Part A)
Edited by JOHN H. LAW AND HANS C. RILLING

VOLUME 111. Steroids and Isoprenoids (Part B)
Edited by JOHN H. LAW AND HANS C. RILLING

VOLUME 112. Drug and Enzyme Targeting (Part A)
Edited by KENNETH J. WIDDER AND RALPH GREEN

VOLUME 113. Glutamate, Glutamine, Glutathione, and Related Compounds
Edited by ALTON MEISTER

VOLUME 114. Diffraction Methods for Biological Macromolecules (Part A)
Edited by HAROLD W. WYCKOFF, C. H. W. HIRS, AND SERGE N. TIMASHEFF

VOLUME 115. Diffraction Methods for Biological Macromolecules (Part B)
Edited by HAROLD W. WYCKOFF, C. H. W. HIRS, AND SERGE N. TIMASHEFF

VOLUME 116. Immunochemical Techniques (Part H: Effectors and Mediators of Lymphoid Cell Functions)
Edited by GIOVANNI DI SABATO, JOHN J. LANGONE, AND HELEN VAN VUNAKIS

VOLUME 117. Enzyme Structure (Part J)
Edited by C. H. W. HIRS AND SERGE N. TIMASHEFF

VOLUME 118. Plant Molecular Biology
Edited by ARTHUR WEISSBACH AND HERBERT WEISSBACH

VOLUME 119. Interferons (Part C)
Edited by SIDNEY PESTKA

VOLUME 120. Cumulative Subject Index Volumes 81–94, 96–101

VOLUME 121. Immunochemical Techniques (Part I: Hybridoma Technology and Monoclonal Antibodies)
Edited by JOHN J. LANGONE AND HELEN VAN VUNAKIS

VOLUME 122. Vitamins and Coenzymes (Part G)
Edited by FRANK CHYTIL AND DONALD B. MCCORMICK

VOLUME 123. Vitamins and Coenzymes (Part H)
Edited by FRANK CHYTIL AND DONALD B. MCCORMICK

VOLUME 124. Hormone Action (Part J: Neuroendocrine Peptides)
Edited by P. MICHAEL CONN

VOLUME 125. Biomembranes (Part M: Transport in Bacteria, Mitochondria, and Chloroplasts: General Approaches and Transport Systems)
Edited by SIDNEY FLEISCHER AND BECCA FLEISCHER

VOLUME 126. Biomembranes (Part N: Transport in Bacteria, Mitochondria, and Chloroplasts: Protonmotive Force)
Edited by SIDNEY FLEISCHER AND BECCA FLEISCHER

VOLUME 127. Biomembranes (Part O: Protons and Water: Structure and Translocation)
Edited by LESTER PACKER

VOLUME 128. Plasma Lipoproteins (Part A: Preparation, Structure, and Molecular Biology)
Edited by JERE P. SEGREST AND JOHN J. ALBERS

VOLUME 129. Plasma Lipoproteins (Part B: Characterization, Cell Biology, and Metabolism)
Edited by JOHN J. ALBERS AND JERE P. SEGREST

VOLUME 130. Enzyme Structure (Part K)
Edited by C. H. W. HIRS AND SERGE N. TIMASHEFF

VOLUME 131. Enzyme Structure (Part L)
Edited by C. H. W. HIRS AND SERGE N. TIMASHEFF

VOLUME 132. Immunochemical Techniques (Part J: Phagocytosis and Cell-Mediated Cytotoxicity)
Edited by GIOVANNI DI SABATO AND JOHANNES EVERSE

VOLUME 133. Bioluminescence and Chemiluminescence (Part B)
Edited by MARLENE DELUCA AND WILLIAM D. MCELROY

VOLUME 134. Structural and Contractile Proteins (Part C: The Contractile Apparatus and the Cytoskeleton)
Edited by RICHARD B. VALLEE

VOLUME 135. Immobilized Enzymes and Cells (Part B)
Edited by KLAUS MOSBACH

VOLUME 136. Immobilized Enzymes and Cells (Part C)
Edited by KLAUS MOSBACH

VOLUME 137. Immobilized Enzymes and Cells (Part D)
Edited by KLAUS MOSBACH

VOLUME 138. Complex Carbohydrates (Part E)
Edited by VICTOR GINSBURG

VOLUME 139. Cellular Regulators (Part A: Calcium- and Calmodulin-Binding Proteins)
Edited by ANTHONY R. MEANS AND P. MICHAEL CONN

VOLUME 140. Cumulative Subject Index Volumes 102–119, 121–134

VOLUME 141. Cellular Regulators (Part B: Calcium and Lipids)
Edited by P. MICHAEL CONN AND ANTHONY R. MEANS

VOLUME 142. Metabolism of Aromatic Amino Acids and Amines
Edited by SEYMOUR KAUFMAN

VOLUME 143. Sulfur and Sulfur Amino Acids
Edited by WILLIAM B. JAKOBY AND OWEN GRIFFITH

VOLUME 144. Structural and Contractile Proteins (Part D: Extracellular Matrix)
Edited by LEON W. CUNNINGHAM

VOLUME 145. Structural and Contractile Proteins (Part E: Extracellular Matrix)
Edited by LEON W. CUNNINGHAM

VOLUME 146. Peptide Growth Factors (Part A)
Edited by DAVID BARNES AND DAVID A. SIRBASKU

VOLUME 147. Peptide Growth Factors (Part B)
Edited by DAVID BARNES AND DAVID A. SIRBASKU

VOLUME 148. Plant Cell Membranes
Edited by LESTER PACKER AND ROLAND DOUCE

VOLUME 149. Drug and Enzyme Targeting (Part B)
Edited by RALPH GREEN AND KENNETH J. WIDDER

VOLUME 150. Immunochemical Techniques (Part K: *In Vitro* Models of B and T Cell Functions and Lymphoid Cell Receptors)
Edited by GIOVANNI DI SABATO

VOLUME 151. Molecular Genetics of Mammalian Cells
Edited by MICHAEL M. GOTTESMAN

VOLUME 152. Guide to Molecular Cloning Techniques
Edited by SHELBY L. BERGER AND ALAN R. KIMMEL

VOLUME 153. Recombinant DNA (Part D)
Edited by RAY WU AND LAWRENCE GROSSMAN

VOLUME 154. Recombinant DNA (Part E)
Edited by RAY WU AND LAWRENCE GROSSMAN

VOLUME 155. Recombinant DNA (Part F)
Edited by RAY WU

VOLUME 156. Biomembranes (Part P: ATP-Driven Pumps and Related Transport: The Na,K-Pump)
Edited by SIDNEY FLEISCHER AND BECCA FLEISCHER

VOLUME 157. Biomembranes (Part Q: ATP-Driven Pumps and Related Transport: Calcium, Proton, and Potassium Pumps)
Edited by SIDNEY FLEISCHER AND BECCA FLEISCHER

VOLUME 158. Metalloproteins (Part A)
Edited by JAMES F. RIORDAN AND BERT L. VALLEE

VOLUME 159. Initiation and Termination of Cyclic Nucleotide Action
Edited by JACKIE D. CORBIN AND ROGER A. JOHNSON

VOLUME 160. Biomass (Part A: Cellulose and Hemicellulose)
Edited by WILLIS A. WOOD AND SCOTT T. KELLOGG

VOLUME 161. Biomass (Part B: Lignin, Pectin, and Chitin)
Edited by WILLIS A. WOOD AND SCOTT T. KELLOGG

VOLUME 162. Immunochemical Techniques (Part L: Chemotaxis and Inflammation)
Edited by GIOVANNI DI SABATO

VOLUME 163. Immunochemical Techniques (Part M: Chemotaxis and Inflammation)
Edited by GIOVANNI DI SABATO

VOLUME 164. Ribosomes
Edited by HARRY F. NOLLER, JR., AND KIVIE MOLDAVE

VOLUME 165. Microbial Toxins: Tools for Enzymology
Edited by SIDNEY HARSHMAN

VOLUME 166. Branched-Chain Amino Acids
Edited by ROBERT HARRIS AND JOHN R. SOKATCH

VOLUME 167. Cyanobacteria
Edited by LESTER PACKER AND ALEXANDER N. GLAZER

VOLUME 168. Hormone Action (Part K: Neuroendocrine Peptides)
Edited by P. MICHAEL CONN

VOLUME 169. Platelets: Receptors, Adhesion, Secretion (Part A)
Edited by JACEK HAWIGER

VOLUME 170. Nucleosomes
Edited by PAUL M. WASSARMAN AND ROGER D. KORNBERG

VOLUME 171. Biomembranes (Part R: Transport Theory: Cells and Model Membranes)
Edited by SIDNEY FLEISCHER AND BECCA FLEISCHER

VOLUME 172. Biomembranes (Part S: Transport: Membrane Isolation and Characterization)
Edited by SIDNEY FLEISCHER AND BECCA FLEISCHER

VOLUME 173. Biomembranes [Part T: Cellular and Subcellular Transport: Eukaryotic (Nonepithelial) Cells]
Edited by SIDNEY FLEISCHER AND BECCA FLEISCHER

VOLUME 174. Biomembranes [Part U: Cellular and Subcellular Transport: Eukaryotic (Nonepithelial) Cells]
Edited by SIDNEY FLEISCHER AND BECCA FLEISCHER

VOLUME 175. Cumulative Subject Index Volumes 135–139, 141–167

VOLUME 176. Nuclear Magnetic Resonance (Part A: Spectral Techniques and Dynamics)
Edited by NORMAN J. OPPENHEIMER AND THOMAS L. JAMES

VOLUME 177. Nuclear Magnetic Resonance (Part B: Structure and Mechanism)
Edited by NORMAN J. OPPENHEIMER AND THOMAS L. JAMES

VOLUME 178. Antibodies, Antigens, and Molecular Mimicry
Edited by JOHN J. LANGONE

VOLUME 179. Complex Carbohydrates (Part F)
Edited by VICTOR GINSBURG

VOLUME 180. RNA Processing (Part A: General Methods)
Edited by JAMES E. DAHLBERG AND JOHN N. ABELSON

VOLUME 181. RNA Processing (Part B: Specific Methods)
Edited by JAMES E. DAHLBERG AND JOHN N. ABELSON

VOLUME 182. Guide to Protein Purification
Edited by MURRAY P. DEUTSCHER

VOLUME 183. Molecular Evolution: Computer Analysis of Protein and Nucleic Acid Sequences
Edited by RUSSELL F. DOOLITTLE

VOLUME 184. Avidin–Biotin Technology
Edited by MEIR WILCHEK AND EDWARD A. BAYER

VOLUME 185. Gene Expression Technology
Edited by DAVID V. GOEDDEL

VOLUME 186. Oxygen Radicals in Biological Systems (Part B: Oxygen Radicals and Antioxidants)
Edited by LESTER PACKER AND ALEXANDER N. GLAZER

VOLUME 187. Arachidonate Related Lipid Mediators
Edited by ROBERT C. MURPHY AND FRANK A. FITZPATRICK

VOLUME 188. Hydrocarbons and Methylotrophy
Edited by MARY E. LIDSTROM

VOLUME 189. Retinoids (Part A: Molecular and Metabolic Aspects)
Edited by LESTER PACKER

VOLUME 190. Retinoids (Part B: Cell Differentiation and Clinical Applications)
Edited by LESTER PACKER

VOLUME 191. Biomembranes (Part V: Cellular and Subcellular Transport: Epithelial Cells)
Edited by SIDNEY FLEISCHER AND BECCA FLEISCHER

VOLUME 192. Biomembranes (Part W: Cellular and Subcellular Transport: Epithelial Cells)
Edited by SIDNEY FLEISCHER AND BECCA FLEISCHER

VOLUME 193. Mass Spectrometry
Edited by JAMES A. MCCLOSKEY

VOLUME 194. Guide to Yeast Genetics and Molecular Biology
Edited by CHRISTINE GUTHRIE AND GERALD R. FINK

VOLUME 195. Adenylyl Cyclase, G Proteins, and Guanylyl Cyclase
Edited by ROGER A. JOHNSON AND JACKIE D. CORBIN

VOLUME 196. Molecular Motors and the Cytoskeleton
Edited by RICHARD B. VALLEE

VOLUME 197. Phospholipases
Edited by EDWARD A. DENNIS

VOLUME 198. Peptide Growth Factors (Part C)
Edited by DAVID BARNES, J. P. MATHER, AND GORDON H. SATO

VOLUME 199. Cumulative Subject Index Volumes 168–174, 176–194

VOLUME 200. Protein Phosphorylation (Part A: Protein Kinases: Assays, Purification, Antibodies, Functional Analysis, Cloning, and Expression)
Edited by TONY HUNTER AND BARTHOLOMEW M. SEFTON

VOLUME 201. Protein Phosphorylation (Part B: Analysis of Protein Phosphorylation, Protein Kinase Inhibitors, and Protein Phosphatases)
Edited by TONY HUNTER AND BARTHOLOMEW M. SEFTON

VOLUME 202. Molecular Design and Modeling: Concepts and Applications (Part A: Proteins, Peptides, and Enzymes)
Edited by JOHN J. LANGONE

VOLUME 203. Molecular Design and Modeling: Concepts and Applications (Part B: Antibodies and Antigens, Nucleic Acids, Polysaccharides, and Drugs)
Edited by JOHN J. LANGONE

VOLUME 204. Bacterial Genetic Systems
Edited by JEFFREY H. MILLER

VOLUME 205. Metallobiochemistry (Part B: Metallothionein and Related Molecules)
Edited by JAMES F. RIORDAN AND BERT L. VALLEE

VOLUME 206. Cytochrome P450
Edited by MICHAEL R. WATERMAN AND ERIC F. JOHNSON

VOLUME 207. Ion Channels
Edited by BERNARDO RUDY AND LINDA E. IVERSON

VOLUME 208. Protein–DNA Interactions
Edited by ROBERT T. SAUER

VOLUME 209. Phospholipid Biosynthesis
Edited by EDWARD A. DENNIS AND DENNIS E. VANCE

VOLUME 210. Numerical Computer Methods
Edited by LUDWIG BRAND AND MICHAEL L. JOHNSON

Volume 211. DNA Structures (Part A: Synthesis and Physical Analysis of DNA)
Edited by David M. J. Lilley and James E. Dahlberg

Volume 212. DNA Structures (Part B: Chemical and Electrophoretic Analysis of DNA)
Edited by David M. J. Lilley and James E. Dahlberg

Volume 213. Carotenoids (Part A: Chemistry, Separation, Quantitation, and Antioxidation)
Edited by Lester Packer

Volume 214. Carotenoids (Part B: Metabolism, Genetics, and Biosynthesis)
Edited by Lester Packer

Volume 215. Platelets: Receptors, Adhesion, Secretion (Part B)
Edited by Jacek J. Hawiger

Volume 216. Recombinant DNA (Part G)
Edited by Ray Wu

Volume 217. Recombinant DNA (Part H)
Edited by Ray Wu

Volume 218. Recombinant DNA (Part I)
Edited by Ray Wu

Volume 219. Reconstitution of Intracellular Transport
Edited by James E. Rothman

Volume 220. Membrane Fusion Techniques (Part A)
Edited by Nejat Düzgüneş

Volume 221. Membrane Fusion Techniques (Part B)
Edited by Nejat Düzgüneş

Volume 222. Proteolytic Enzymes in Coagulation, Fibrinolysis, and Complement Activation (Part A: Mammalian Blood Coagulation Factors and Inhibitors)
Edited by Laszlo Lorand and Kenneth G. Mann

Volume 223. Proteolytic Enzymes in Coagulation, Fibrinolysis, and Complement Activation (Part B: Complement Activation, Fibrinolysis, and Nonmammalian Blood Coagulation Factors)
Edited by Laszlo Lorand and Kenneth G. Mann

Volume 224. Molecular Evolution: Producing the Biochemical Data
Edited by Elizabeth Anne Zimmer, Thomas J. White, Rebecca L. Cann, and Allan C. Wilson

Volume 225. Guide to Techniques in Mouse Development
Edited by Paul M. Wassarman and Melvin L. DePamphilis

Volume 226. Metallobiochemistry (Part C: Spectroscopic and Physical Methods for Probing Metal Ion Environments in Metalloenzymes and Metalloproteins)
Edited by James F. Riordan and Bert L. Vallee

VOLUME 227. Metallobiochemistry (Part D: Physical and Spectroscopic Methods for Probing Metal Ion Environments in Metalloproteins)
Edited by JAMES F. RIORDAN AND BERT L. VALLEE

VOLUME 228. Aqueous Two-Phase Systems
Edited by HARRY WALTER AND GÖTE JOHANSSON

VOLUME 229. Cumulative Subject Index Volumes 195–198, 200–227

VOLUME 230. Guide to Techniques in Glycobiology
Edited by WILLIAM J. LENNARZ AND GERALD W. HART

VOLUME 231. Hemoglobins (Part B: Biochemical and Analytical Methods)
Edited by JOHANNES EVERSE, KIM D. VANDEGRIFF, AND ROBERT M. WINSLOW

VOLUME 232. Hemoglobins (Part C: Biophysical Methods)
Edited by JOHANNES EVERSE, KIM D. VANDEGRIFF, AND ROBERT M. WINSLOW

VOLUME 233. Oxygen Radicals in Biological Systems (Part C)
Edited by LESTER PACKER

VOLUME 234. Oxygen Radicals in Biological Systems (Part D)
Edited by LESTER PACKER

VOLUME 235. Bacterial Pathogenesis (Part A: Identification and Regulation of Virulence Factors)
Edited by VIRGINIA L. CLARK AND PATRIK M. BAVOIL

VOLUME 236. Bacterial Pathogenesis (Part B: Integration of Pathogenic Bacteria with Host Cells)
Edited by VIRGINIA L. CLARK AND PATRIK M. BAVOIL

VOLUME 237. Heterotrimeric G Proteins
Edited by RAVI IYENGAR

VOLUME 238. Heterotrimeric G-Protein Effectors
Edited by RAVI IYENGAR

VOLUME 239. Nuclear Magnetic Resonance (Part C)
Edited by THOMAS L. JAMES AND NORMAN J. OPPENHEIMER

VOLUME 240. Numerical Computer Methods (Part B)
Edited by MICHAEL L. JOHNSON AND LUDWIG BRAND

VOLUME 241. Retroviral Proteases
Edited by LAWRENCE C. KUO AND JULES A. SHAFER

VOLUME 242. Neoglycoconjugates (Part A)
Edited by Y. C. LEE AND REIKO T. LEE

VOLUME 243. Inorganic Microbial Sulfur Metabolism
Edited by HARRY D. PECK, JR., AND JEAN LEGALL

VOLUME 244. Proteolytic Enzymes: Serine and Cysteine Peptidases
Edited by ALAN J. BARRETT

VOLUME 245. Extracellular Matrix Components
Edited by E. RUOSLAHTI AND E. ENGVALL

VOLUME 246. Biochemical Spectroscopy
Edited by KENNETH SAUER

VOLUME 247. Neoglycoconjugates (Part B: Biomedical Applications)
Edited by Y. C. LEE AND REIKO T. LEE

VOLUME 248. Proteolytic Enzymes: Aspartic and Metallo Peptidases
Edited by ALAN J. BARRETT

VOLUME 249. Enzyme Kinetics and Mechanism (Part D: Developments in Enzyme Dynamics)
Edited by DANIEL L. PURICH

VOLUME 250. Lipid Modifications of Proteins
Edited by PATRICK J. CASEY AND JANICE E. BUSS

VOLUME 251. Biothiols (Part A: Monothiols and Dithiols, Protein Thiols, and Thiyl Radicals)
Edited by LESTER PACKER

VOLUME 252. Biothiols (Part B: Glutathione and Thioredoxin; Thiols in Signal Transduction and Gene Regulation)
Edited by LESTER PACKER

VOLUME 253. Adhesion of Microbial Pathogens
Edited by RON J. DOYLE AND ITZHAK OFEK

VOLUME 254. Oncogene Techniques
Edited by PETER K. VOGT AND INDER M. VERMA

VOLUME 255. Small GTPases and Their Regulators (Part A: Ras Family)
Edited by W. E. BALCH, CHANNING J. DER, AND ALAN HALL

VOLUME 256. Small GTPases and Their Regulators (Part B: Rho Family)
Edited by W. E. BALCH, CHANNING J. DER, AND ALAN HALL

VOLUME 257. Small GTPases and Their Regulators (Part C: Proteins Involved in Transport)
Edited by W. E. BALCH, CHANNING J. DER, AND ALAN HALL

VOLUME 258. Redox-Active Amino Acids in Biology
Edited by JUDITH P. KLINMAN

VOLUME 259. Energetics of Biological Macromolecules
Edited by MICHAEL L. JOHNSON AND GARY K. ACKERS

VOLUME 260. Mitochondrial Biogenesis and Genetics (Part A)
Edited by GIUSEPPE M. ATTARDI AND ANNE CHOMYN

VOLUME 261. Nuclear Magnetic Resonance and Nucleic Acids
Edited by THOMAS L. JAMES

VOLUME 262. DNA Replication
Edited by JUDITH L. CAMPBELL

Volume 302. Green Fluorescent Protein
Edited by P. Michael Conn

Volume 303. cDNA Preparation and Display
Edited by Sherman M. Weissman

Volume 304. Chromatin
Edited by Paul M. Wassarman and Alan P. Wolffe

Volume 305. Bioluminescence and Chemiluminescence (Part C)
Edited by Thomas O. Baldwin and Miriam M. Ziegler

Volume 306. Expression of Recombinant Genes in Eukaryotic Systems
Edited by Joseph C. Glorioso and Martin C. Schmidt

Volume 307. Confocal Microscopy
Edited by P. Michael Conn

Volume 308. Enzyme Kinetics and Mechanism (Part E: Energetics of Enzyme Catalysis)
Edited by Daniel L. Purich and Vern L. Schramm

Volume 309. Amyloid, Prions, and Other Protein Aggregates
Edited by Ronald Wetzel

Volume 310. Biofilms
Edited by Ronald J. Doyle

Volume 311. Sphingolipid Metabolism and Cell Signaling (Part A)
Edited by Alfred H. Merrill, Jr., and Yusuf A. Hannun

Volume 312. Sphingolipid Metabolism and Cell Signaling (Part B) (in preparation)
Edited by Alfred H. Merrill, Jr., and Yusuf A. Hannun

Volume 313. Antisense Technology (Part A: General Methods, Methods of Delivery and RNA Studies)
Edited by M. Ian Phillips

Volume 314. Antisense Technology (Part B: Applications)
Edited by M. Ian Phillips

Volume 315. Vertebrate Phototransduction and the Visual Cycle (Part A)
Edited by Krzysztof Palczewski

Volume 316. Vertebrate Phototransduction and the Visual Cycle (Part B)
Edited by Krzysztof Palczewski

Volume 317. RNA–Ligand Interactions (Part A: Structural Biology Methods)
Edited by Daniel W. Celander and John N. Abelson

Volume 318. RNA–Ligand Interactions (Part B: Molecular Biology Methods)
Edited by Daniel W. Celander and John N. Abelson

Volume 319. Singlet Oxygen, UV-A, and Ozone
Edited by Lester Packer and Helmut Sies

[1] Physiological Modeling with Virtual Cell Framework

By JAMES C. SCHAFF, BORIS M. SLEPCHENKO, and LESLIE M. LOEW

This article describes a computational framework for cell biological modeling and simulation that is based on the mapping of experimental biochemical and electrophysiological data onto experimental images. The framework is designed to enable the construction of complex general models that encompass the general class of problems coupling reaction and diffusion.

Introduction

A general computational framework for modeling cell biological processes, the *Virtual Cell,* is being developed at the National Resource for Cell Analysis and Modeling at the University of Connecticut Health Center. The Virtual Cell is intended to be a tool for experimentalists as well as theorists. Models are constructed from biochemical and electrophysiological data mapped to appropriate subcellular locations in images obtained from a microscope. Chemical kinetics, membrane fluxes, and diffusion are thus coupled and the resultant equations are solved numerically. The results are again mapped to experimental images so that the cell biologist can fully utilize the familiar arsenal of image processing tools to analyze the simulations.

The philosophy driving the Virtual Cell project requires a clear operational definition of the term *model.* The idea is best understood as a restatement of the scientific method. A model, in this language, is simply a collection of hypotheses and facts that are brought together in an attempt to understand a phenomenon. The choices of which hypotheses and facts to collect and the manner in which they are assembled themselves constitute additional hypotheses. A prediction based on the model is, in one sense, most useful if it *does not* match the experimental details of the process—it then unequivocally tells us that the elements of the model are inaccurate or incomplete. Although such negative results are not always publishable, they are a tremendous aid in refining our understanding. If the prediction does match the experiment, it *never* can guarantee the truth of the model, but should suggest other experiments that can test the validity of critical elements; ideally, it should also provide new predictions that can, in turn, be verified experimentally. The Virtual Cell is itself *not* a model. It is intended to be a computational framework and tool for cell biologists to

0076-6879/00 $30.00

create models and to generate predictions from models via simulations. To ensure the reliability of such a tool, all underlying math, physics, and numerics must be thoroughly validated. To ensure the utility and accessibility of such a tool to cell biologists, the results of such simulations must be presented in a format that may be analyzed using procedures comparable to those used to analyze the results of experiments.

In this article we describe the current status of the mathematics infrastructure, design considerations for model management, and the user interface. This is followed by application to the calcium wave that follows fertilization of a frog egg. Additional details can be found in an earlier publication[1] and on our web site: http://www.nrcam.uchc.edu/.

Modeling Abstractions for Cellular Physiology

Background

Often theoreticians develop the simplest model that reproduces the phenomenon under study.[2] These may be quite elegant, but are often not very extensible to other related phenomena. Other modeling efforts characterize single physiological mechanisms,[3,4] but these are often developed *ad hoc* rather than as part of a reusable and consistent framework.

Our approach to modeling concentrates on the mechanisms as well as the phenomena. The goal of this approach is to provide a direct method of evaluating single models of individual mechanisms in the context of several experiments. This approach enables the encapsulation of sufficient complexity that, after independent validation, allows it to be used as a meaningful predictive tool. To include sufficient complexity without overwhelming the user, the models are specified in their most natural form. In the case of chemical reactions, the models are represented by a series of reactants, products, modifiers (e.g., enzymes), their stoichiometry, and their kinetic constants.

One of the obstacles to modeling is the lack of general-purpose simulation and analysis tools. Each potential modeler must have resources in software development and numerical methods at his or her disposal. Each time the model or the computational domain is altered, the program must be changed. And in practice, the modeling of a new phenomenon requires a new simulation program to be written. This is a time-consuming and

[1] J. Schaff, C. C. Fink, B. Slepchenko, J. H. Carson, and L. M. Loew, *Biophys. J.* **73,** 1135 (1997).

[2] R. Kupferman, P. P. Mitra, P. C. Hohenberg, and S. S-H. Wang, *Biophys. J.* **72,** 2430 (1997).

[3] J. Sneyd, J. Keizer, and M. J. Sanderson, *FASEB J.* **9,** 1463 (1995).

[4] Y.-X. Lit and J. Rinzel, *J. Theor. Biol.* **166,** 463 (1994).

error-prone exercise, especially when developed without a proper software methodology.

We are developing a general, well-tested framework for modeling and simulation for use by the cell biology community. The application of the underlying equations to our framework with nearly arbitrary models and geometry is rigorously investigated. The numerical approach is then properly evaluated and tuned for performance. This methodology results in a proper basis for a general-purpose framework. Our approach requires no user programming; rather the user specifies models using biologically relevant abstractions such as reactions, compartments, molecular species, and experimental geometry. This allows a very flexible description of the physiological model and arbitrary geometry. The framework accommodates arbitrary geometry and automatically generates code to implement the specified physiological model.

Another problem is the lack of a standard format for expressing those models. Even implementing published models can be a nontrivial exercise. Some of the necessary details required for implementation can be missing or buried in the references. Often the models are obscured by geometrical assumptions used to simplify the problem. A standard modeling format is required to facilitate the evaluation and integration of separate models. This standard format should separately specify physiological models and cellular geometry in an implementation-independent way.

We suggest that the abstract physiological models used with the Virtual Cell framework can form the basis of such a standard.

The current implementation of the cell model description[1] involves the manipulation of abstract modeling objects that reside in the Modeling Framework as Java objects. These modeling objects can be edited, viewed, stored in a remote database, and analyzed using the WWW-based user interface (see User Interface section). These objects are categorized as Models, Geometry, and Simulation Context objects. This corresponds to the naming convention used in the current Modeling Framework software.

There are also mature efforts in Metabolic Pathway modeling such as GEPASI.[5] This package allows a simple and intuitive interface for specifying reaction stoichiometry and kinetics. The kinetics are specified by selecting from a predefined (but extensible) list of kinetic models (such as Michaelis–Menten) describing enzyme-mediated production of metabolites. These packages are focused on biochemical pathways where the spatial aspects of the system are ignored. However, for these simplified descriptions, they provide Metabolic Control Analysis (local sensitivity analysis

[5] P. Mendes, *Comput. Applic. Biosci.* **9,** 563 (1993).

tools), structural analysis (mass conservation identification), and a local stability analysis.

To provide a simple interface to a general-purpose modeling and simulation capability, the problem must be broken up into manageable pieces. In the case of the Virtual Cell, these pieces consist of abstract physiological models defined within the context of cell structure, experimental cell geometry, and a mapping of the physiological models to the specific geometry including the conditions of that particular problem. For such an interface to be consistent and maintainable, it must map directly to the underlying software architecture.

An intuitive user interface is essential to the usability of a complex application. A prototype user interface was developed to provide an early platform for modeling and simulation, and for investigating user interface requirements. The design goal was to capture the minimum functionality required for practical use of the virtual cell.

The Virtual Cell application is built on top of a distributed, component-based software architecture (Fig. 1). The physiological interface is a WWW-

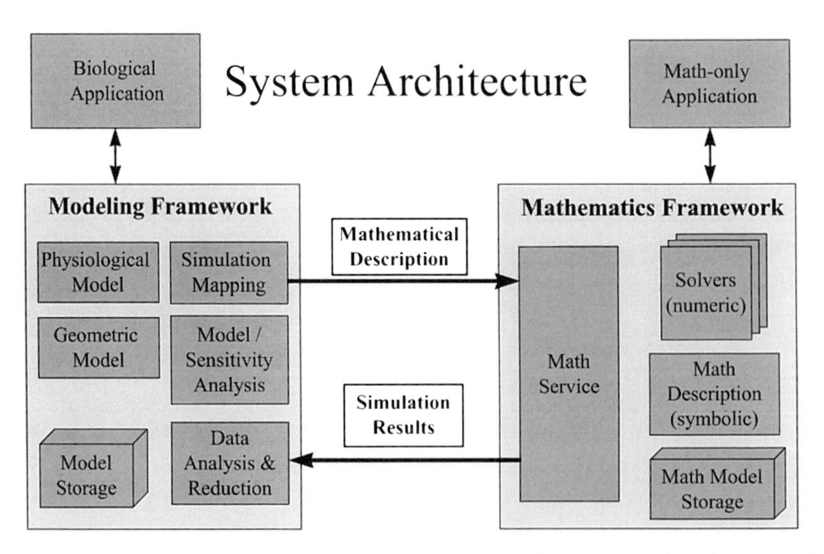

FIG. 1. The System Architecture features two distributed, component-based frameworks that create a stable, extensible, and maintainable research software platform. The Modeling Framework provides the biological abstractions necessary to model and simulate cellular physiology. The Mathematics Framework provides a general-purpose solver for the mathematical problems in the application domain of computational cellular physiology. This framework exposes a very high level system interface that allows a mathematical problem to be posed in a concise and implementation-independent mathematical description language, and for the solution to be made available in an appropriate format.

accessible Java applet that provides a graphical user interface to the capabilities of the Modeling Framework. The Mathematics Framework is automatically invoked to provide solutions to particular simulations. The architecture is designed such that the location of the user interface and the corresponding back-end services (model storage, simulation, data retrieval) are transparent to the majority of the application. The typical configuration is a Java applet running in a WWW browser, with the Database, Simulation Control, and Simulation Data services executing on a remote machine (WWW server).

Specification of Physiological Models

A physiological model of the cell system under study is defined as a collection of Cellular Structures, Molecular Species, Reactions, and Fluxes. These concepts define cellular mechanisms in the context of cell structure, and are sufficient to define nonspatial, compartmental simulations. With the addition of a specific cellular geometry (usually from a microscope image), a spatial simulation is defined. The goal is to capture the physiology independently of the geometric domain (cell shape and scale) and specific experimental context, such that the resulting physiological mechanisms are modular and can be incorporated into new models with minimum modification.

Cellular Structures. Cellular structures are abstract representations of hierarchically organized, mutually exclusive regions within cells where the general topology is defined but the specific geometric shapes and sizes are left undefined. These regions are categorized by their intrinsic topology: *compartments* represent three-dimensional volumetric regions (e.g., cytosol), *membranes* represent two-dimensional surfaces (e.g., plasma membrane) separating the *compartments,* and *filaments* represent one-dimensional contours (e.g., microtubules) lying within a single *compartment.* Using this definition, the extracellular region is separated from the cytosol (and hence the ER, nucleus, etc.) by the plasma membrane. All structures can contain *molecular species* and a collection of *reactions* that describe the biochemical behavior of those species within that structure.

These cellular structures are used for bookkeeping purposes, and must be mapped to a specific cellular geometry before any quantitative simulations or analysis can be performed. Although this introduces two parallel representations of anatomical structures (topology and shape), the separation of physiological models from specific simulation geometry allows the same model to be used in various compartmental and one-, two-, and three-dimensional geometric contexts without modification.

Molecular Species. Species are unique molecular species (e.g., Ca^{2+}, IP3) or important distinct states of molecular species (e.g., calcium-bound IP3-Receptor, unbound IP3-Receptor). These *species* can be separately defined in multiple *compartments, membranes,* or *filaments. Species* are described by linear densities when located on *filaments,* by surface densities when located in *membranes,* and by concentrations when located in *compartments. Species* diffusion is defined separately for each cellular compartment and is specified by diffusion constants.

Species can participate in reactions, fluxes, and diffusion. The behavior of *species* can be described by

1. Diffusion of *species* within *compartments, membranes,* or *filaments*
2. Directed motion of *Species* along *filaments*
3. Flux of *species* between two adjacent *compartments* through the associated *membrane*
4. Advection (e.g., binding reactions) of *species* between *compartments* and either *membranes* or *filaments*

Reactions and Fluxes. Reactions are objects that represent complete descriptions of the stoichiometry and kinetics of biochemical reactions. *Reactions* are collections of related *reaction steps* (e.g., membrane receptor binding or cytosolic calcium buffering) and *membrane fluxes* (e.g., flux through an ion channel).

Each *reaction step* is associated with a single cellular structure. The stoichiometry of a *reaction step* is expressed in terms of reactants, products, and catalysts, which are related to species in a particular cellular structure (e.g., Ca^{2+} in cytosol). This stoichiometry is depicted in the user interface as a biochemical pathway graph.

Reaction steps located in a compartment involve only those *species* that are present in the interior of that compartment. *Reaction steps* located on a *membrane* can involve *species* that are present on that *membrane* as well as *species* present in the *compartments* on either side of that *membrane.* Reaction steps located on a filament can involve species that are present on that filament as well as species present in the compartment that contains the filament. The kinetics of a *reaction step* can be specified as either mass action kinetics with forward and reverse rate coefficients or, more generally, by an arbitrary rate expression in terms of reactants, products, and modifiers.

Each *flux step* is associated with a single *membrane* and describes the flux of a single flux carrier species through that *membrane.* A flux carrier must be a species that is defined in both neighboring *compartments.* For example, a *flux step* associated with a calcium channel in the ER membrane would have calcium as the flux carrier, and thus calcium must be defined in both the ER and cytosol. A single inward flux is defined by convention

(μM μm s^{-1}) and enforces flux conservation across the *membrane.* The flux is an arbitrary function of flux carrier concentration (on either side of the *membrane*), membrane-bound modifier surface density, and of modifier concentration in the *compartments* on either side of the *membrane.*

Specification of Cellular Geometry

A particular experimental context requires a concrete description of the cellular geometry to fully describe the behavior of the cellular system. This geometric description defines the specific morphology of the cell, and any of its spatially resolvable organelles. This geometry is usually taken directly from experimental images that have been segmented into mutually exclusive regions by their pixel intensity. For example, an image may be segmented into only two classes of pixel intensities, white (pixel intensity of 255) and black (pixel intensity of 0), where white regions represent cytosol and black regions represent the extracellular milieu. The actual size of the domain is defined by specifying the scale of a pixel in microns to properly define a simulation domain.

Alternatively, the geometry may be specified analytically as an ordered list of inequalities. Each resolved region can be represented as an inequality in x, y, and z. Each point in the geometric domain can be assigned to the compartment corresponding to the first inequality that is satisfied. For example, if cytosol is represented by a sphere at the origin of radius R, and the rest of the domain is considered extracellular, then the nested compartments can be easily specified as follows.

$$\begin{array}{lll} \text{cytosol:} & x^2 + y^2 + z^2 < R & \text{(true within sphere at origin)} \\ \text{extracellular:} & 1 < 2 & \text{(always true)} \end{array} \quad (1)$$

Mapping Biology to Mathematical Description

Mapping Cellular Structures to Experimental Geometry. The resulting geometry then has to be mapped to the corresponding *cellular structures* defined in the *physiological model.* Each mutually exclusive volumetric region in the geometry is naturally mapped to a single *compartment,* and an interface (surface) separating any two regions maps to the corresponding *membrane.* A *compartment* that is not spatially resolved in the geometry may be considered continuously distributed within the geometric region of its parent *compartment.* For example, if a particular geometry specifies spatial mappings for only extracellular, cytosol, and the plasma membrane, then the endoplasmic reticulum and its *membrane* (which are interior to the cytosol *compartment*) are continuously distributed within the cytosol region of the geometry. For such continuously distributed *compartments*, the volume fraction (e.g., er volume to total cytosolic volume) and internal surface to volume ratios (e.g., er membrane surface to er volume) are required to reconcile distributed fluxes and membrane binding rates to the spatially resolved *compartments*. Note that unresolved structures, such as the endoplasmic reticulum, need not be uniformly distributed within the cytosol *compartment* even if they are described as continuous.

When spatial simulations are not required (compartmental models) specification of volume fractions for all *compartments* and surface to volume ratios for all *membranes* are sufficient to represent the geometric mapping.

Mapping of Reactions and Species to Equations and Variables. After the *compartments* and *membranes* have been mapped to concrete geometry, the mapping of *reactions* to systems of equations is well defined. One or more *compartments* and *membranes* are mapped onto a region in the computational domain as discussed in the previous section. All of the *reaction steps* and *membrane fluxes* that are associated with this set of *compartments* and *membranes* are collected.

Within a single computational subdomain, *species* dynamics can be represented as a system of differential equations [Eq. (2)] expressing the time rate of change of each *species* C_i concentration as the sum of their reaction kinetics and their diffusion term (Laplacian scaled by diffusion rate D_i). The reaction kinetics can be expressed in terms of a linear combination of reaction rates v_j (and fluxes through distributed *membranes*) that are scaled by their corresponding stoichiometry c_{ij}:

$$\begin{bmatrix} \frac{dC_1}{dt} \\ \frac{dC_2}{dt} \\ \vdots \\ \frac{dC_n}{dt} \end{bmatrix} = \mathbf{S} \cdot \mathbf{v} = \begin{bmatrix} c_{11} & \cdots & c_{1m} & D_1 & 0 & \cdots & 0 \\ c_{21} & \cdots & c_{2m} & 0 & D_2 & \cdots & 0 \\ \vdots & \ddots & \vdots & \vdots & \vdots & \ddots & \vdots \\ c_{n1} & \cdots & c_{nm} & 0 & 0 & \cdots & D_n \end{bmatrix} * \begin{bmatrix} v_1 \\ \vdots \\ v_m \\ \nabla^2 C_1 \\ \nabla^2 C_2 \\ \vdots \\ \nabla^2 C_n \end{bmatrix} \quad (2)$$

As a result of conservation of mass in biochemical systems, the number of independent differential equations is often less than the number of *species*. The application of a stoichiometry (or structural) analysis[6–8] allows the

$$\begin{bmatrix} C_{r+1} \\ \vdots \\ C_n \end{bmatrix} = \begin{bmatrix} L_1 \\ \vdots \\ L_{n-r} \end{bmatrix} \cdot \begin{bmatrix} C_1 \\ C_2 \\ \vdots \\ C_r \end{bmatrix} + \begin{bmatrix} K_1 \\ \vdots \\ K_{n-r} \end{bmatrix} \tag{3}$$

automatic extraction of a minimum set of independent variables ($C_1 \cdots C_r$ where r is the rank of **S**), and a set of dependent variables ($C_{r+1} \cdots C_n$) described by conservation relationships L_i [Eq. (3)] (derived from the left null space of S) that allow expression of all dependent *species* concentrations as linear combinations of independent *species* concentrations. Then only the equations associated with the independent *species* are used in the mathematical description.

Fast Kinetics. Interrelated cellular processes often occur on a wide range of timescales causing a problem that is commonly encountered when computing numerical solutions of reaction/diffusion problems in biological applications. It manifests itself as a set of equations that is said to be stiff in time. The direct approach to solving this type of a system would result in a numerical algorithm taking prohibitively small time steps dictated by the fastest kinetics. The typical techniques used to avoid this are employing stiff solvers[9] or analytic pseudo-steady approximations.[10] We have developed a general approach based on separation of fast and slow kinetics.[11] We then perform the pseudo-steady approximation on fast processes. Within this approximation, fast reactions are considered to be at rapid equilibrium. This assumption results in a system of nonlinear algebraic equations. To treat it correctly, the set of independent variables within the fast subsystem has to be determined. This can be done by means of the structural analysis described earlier. The difference is that now these identified invariants (mass conservation relationships) are only invariant with respect to the fast kinetics. They therefore have to be updated at each time step via the slow

[6] C. M. Villa and T. W. Chapman, *Ind. Eng. Chem. Res.* **34,** 3445 (1995).

[7] C. Reder, *J. Theor. Biol.* **135,** 175 (1988).

[8] J. P. Sorensen and W. E. Stewart, *AIChE J.* **26,** 99 (1980).

[9] C. Gear, "Numerical Initial Value Problems in Ordinary Differential Equations." Prentice-Hall, Englewood Cliffs, New Jersey, 1971.

[10] J. Wagner and J. Keizer, *Biophys. J.* **67,** 447 (1994).

[11] G. Strang, *SIAM J. Numer. Anal.* **5,** 506 (1968).

kinetics and diffusion. Our approach does not require any preliminary analytic treatment and can be performed automatically. A user must only specify the subset of reactions that has fast kinetics, and the appropriate "fast system" is automatically generated and integrated into the mathematical description. Currently we are using this capability to investigate the influence of mobile buffers on the properties of calcium waves and on the conditions of their initiation.

Stochastic Formulation for Reaction/Diffusion Systems. When the number of particles involved in the cell processes of interest is not sufficiently large and fluctuations become important, the replacement of the continuous description by a stochastic treatment is required. This might be used to simulate the motion and interactions of granules in intracellular trafficking.[12] The stochastic approach is also required for the description of spontaneous local increases in the concentration of intracellular calcium from discrete ER calcium channels (calcium sparks or puffs) observed in cardiac myocytes.[13] In the case of discrete particles, the problem is formulated in terms of locations (the state variables) and random walks instead of concentration distributions and fluxes in the continuous description, while using the same physical constant—the diffusion coefficient D. The chemical reactions between discrete particles and structures (capture) or between discrete particles and continuously distributed species are described in terms of transition probabilities (reaction rates). In the latter case we have to incorporate the stochastic formulation into a continuous reaction/diffusion framework. This can be done because the numeric treatment of the corresponding partial differential equations requires their discretization. In fact, we deal with discrete numbers that characterize each elementary computational volume. Thus for the case of a species treated continuously, the source term of the partial differential equation will contain the random contribution due to chemical reactions with discrete particles.

Monte Carlo techniques have been employed to simulate both the reaction events and Brownian motion of the discrete particles. Since the displacements at time t in unbiased random walks are normally distributed with mean zero and standard deviation $(4Dt)^{1/2}$, one can model the increment of each coordinate within a time step Δt by $(4D\Delta t)^{1/2}\, r$ where r is a random number described by the standard normal distribution with zero mean and a unity standard deviation.[14] To reproduce the standard normal distribution for the Brownian movement simulation, we use the Box–Muller

[12] K. Ainger, D. Avossa, A. S. Diana, C. Barry, E. Barbarese, and J. H. Carson, *J. Cell Biol.* **138,** 1077 (1997).

[13] H. Cheng, W. J. Lederer, and M. B. Cannell, *Science* **262,** 740 (1993).

[14] C. E. Schmidt, T. Chen, and D. A. Lauffenburger, *Biophys. J.* **67,** 461 (1994).

method.[15] In this method the variable $\lambda = (-2 \ln \xi)^{1/2} \cos 2\pi\eta$ proves to be normally distributed with the standard deviation $\sigma = 1$ provided the random variables ξ and η are uniformly distributed on the interval [0,1]. Methods based on the central limit theorem[16] might be less computationally intensive, but validation is required. The interaction of particles with the membrane has been described in terms of elastic collisions.

The selection–rejection type method has been used for the stochastic simulation of reaction dynamics. The corresponding transition between particle states is assumed to occur if a generated random number, uniformly distributed on [0,1], is less than the reaction probability. In our case, particles interact with the continuously distributed species. Thus, the reaction probability for small time steps Δt is $k\Delta t[C]$, where k is the reaction on rate, and $[C]$ stands for the concentration of a dispersed species. Clearly, for the reaction to occur, $[C]$ should satisfy the condition $[C]Sd \geq n$, where S is the particle surface, d is the characteristic interaction distance, and n is the stoichiometry number. Hence the reaction is ruled out if $[C] < n/V_0$, provided the control volume is bigger than Sd. This condition is necessary to eliminate the possibility of negative concentrations.

We have implemented a simple model representing the initial step in RNA trafficking: RNA granule assembly, where discrete particles interact with the continuously distributed species (RNA). Thus the two approaches, stochastic and deterministic, have been combined.

Compartmental Simulations

For simulation of compartmental models (single-point approximations), the ordinary differential equations (ODEs) representing the reaction kinetics are generated and passed to an interpreted ODE solver (within the client applet).

This system of equations is solved using a simple explicit integration scheme (forward difference) that is first-order accurate in time. A higher order numerical scheme will be integrated in the future.

The Compartmental Simulation (Preview) component executes a compartmental (single-point) simulation based on the defined physiological model and the geometric assumptions entered in the Feature Editor (surface to volume ratios and volume fractions). This results in a set of nonlinear ordinary differential equations that typically are solved in seconds. This allows an interactive, though manual, modification of parameters and a

[15] W. H. Press, S. A. Teukolsky, W. T. Vetterling, and B. P. Flannery, "Numerical Recipes in C: The Art of Scientific Computing." Cambridge University Press, New York, 1992.

[16] L. Devroye, "Non-Uniform Random Variate Generation." Springer-Verlag, New York, 1986.

quick determination of the effect over time. Once the simulation is complete, each species can be viewed easily.

The Equation Viewer displays the equations generated as a result of mapping the physiological model to either a cellular geometry model (spatial simulation) or a single-point approximation (compartmental model). The parameter values may be substituted (and the expression simplified) or left in their symbolic representation.

Model Analysis. It is important to determine the sensitivity of model behavior (i.e., simulation results) to the choice of which physiological mechanisms are incorporated and their parameter values. For a given model structure, the selection of parameter values is constrained, but often not completely determined, by direct empirical measurements and physical limitations, as well as inferred from the steady and dynamic behavior and stability of the composite system.

It is informative to determine the relative change in model behavior due to a relative change in parameter value. For nonspatial, compartmental models, the software computes the sensitivity of any species concentration to any parameter as a function of time evaluated at the nominal solution.

The current implementation lacks a direct steady-state solver. This must either be performed by letting the dynamic system run until it converges to a steady state or by manually doing the analytic calculations (setting rates to zero and solving the simultaneous equations).

Spatial Simulations

For the solution of a complete spatial simulation, the partial differential equations (PDEs) that correspond to diffusive species, and ODEs for nondiffusive species, are generated. These equations are sent to the remote Simulation Server where the corresponding C++ code is automatically generated, compiled, and linked with the Simulation Library. The resulting executable is then run and the results are collected and stored on the server. The Simulation Data Server then coordinates client access to the server-side simulation data for display and analysis.

The system of PDEs is mapped to a rectangular grid using a finite difference scheme based on a finite volume approach.[1] The nonlinear source terms representing reaction kinetics are evaluated explicitly (forward difference) and the resulting linearized PDE is solved implicitly (backward difference). Those membranes separating spatially resolved compartments are treated as discontinuities in the solution of the reaction/diffusion equations. These discontinuities are defined by flux jump conditions that incorporate transmembrane fluxes, binding to membrane-bound species, and conservation of mass. Each boundary condition is defined in terms of a known flux (Neumann condition) or a known concentration (Dirichlet condition).

Storage

The Database Access Form presents a rudimentary model and simulation storage capability. The current implementation allows whole physiological models, geometric models, and simulation contexts to be stored and retrieved. The simulation context is stored in a way that includes the physiological and geometric models such that it encapsulates all of the information to reproduce and describe a particular spatial simulation. There is, however, currently no ability to query the stored models for specific attributes.

Application to Existing Model

This discussion describes the application of our formalism to a physiological reaction/diffusion system within a spatial context. An existing model of fertilization calcium waves in *Xenopus laevis* eggs[17] is used as an example. This is based on the experimental observation that fertilization results in a wave of calcium that spreads throughout the egg from the point of sperm–egg fusion. The calcium wave depends on the elevation of intracellular inositol 1,4,5-trisphosphate (IP_3) and propagates with a concave shape.[18] Our discussion is limited to the mechanics of the model representation without an in-depth analysis of the model itself.

Background

In the case under consideration, the calcium dynamics is essentially determined by the calcium release from the internal endoplasmic reticulum stores through IP_3 sensitive channels and the uptake through sarcoplasmic–endoplasmic reticulum calcium ATPase (SERCA) pumps. These processes are modulated by calcium diffusion and binding to calcium buffers that are always present in the cytosol. Additionally, a small constant calcium leak from ER to the cytosol ensures that the flux balance is at the initial steady state. A simplified version[19] of the De Young–Keizer model[20] for the calcium channel is used. This model implies the independent kinetics of IP_3-binding, calcium activation, and calcium inhibition sites. A Hill equation is employed to describe calcium flux through the pumps.

As has been shown,[17] the fertilization calcium wave generation can be explained by the bistability of a matured egg. The system bistability means that the system can maintain two stable steady states with different calcium

[17] J. Wagner, Y.-X. Li, J. Pearson, and J. Keizer, *Biophys. J.* **75,** 2088 (1998).
[18] R. A. Fontanilla and R. Nuccitelli, *Biophys. J.* **75,** 2079 (1998).
[19] Y.-X. Li and J. Rinzel, *J. Theor. Biol.* **166,** 461 (1994).
[20] G. De Young and J. Keizer, *Proc. Natl. Acad. Sci. U.S.A.* **89,** 9895 (1992).

concentrations, and a wavefront that corresponds to a threshold between the regions with different steady states. Thus, in the bistable system the calcium wave can be initiated even at a steady IP_3 concentration, although a spatially heterogeneous IP_3 distribution is required to reproduce the details of the wavefront shape. In this article, we map the model proposed previously[17] onto the three-dimensional geometry and explicitly introduce buffers as participating species. We treat them as a subsystem with fast kinetics rather than describe buffering with one scaling parameter. For simplicity, we consider only immobile buffers, although our general approach allows us to also treat mobile buffers (for example, fluorescent indicators) as well as any other "fast" kinetics.

Extracting Physiological Model

A physiological model first consists of a set of physiological assumptions regarding the mechanisms important to the phenomenon under study. This defines the physiological structure of the model and includes the list of physiological mechanisms as well as the way in which they interact.

The framework application starts with outlining model compartments and participating species (see Fig. 2). In our problem three compartments—extracellular, cytosol, and ER—are separated by the plasma membrane and the ER membrane, respectively. We have no species of interest in extracellular as well as no processes that are associated with the plasma membrane in our approximation. Four species are present in the cytosol: calcium (Ca), IP_3, and the free (B) and Ca-bound (CaB) states of fixed calcium buffers. We also have calcium in the ER (Ca_ER). Four species are associated with the ER membrane: the free (Ract, Rinh) and Ca-bound (RactCa, RinhCa) states of the activation and inhibition monomer sites that regulate calcium flux through ER membrane channels.

In our application the concentration of Ca_ER is considered to be constant, and the constant IP_3 distribution is described by a static, spatial function. Thus, we end up with seven variables, corresponding to the seven identified specified species, of which only cytosolic calcium is diffusive while the other six are treated as spatially fixed.

The initial concentration and diffusion rates are specified for each species present within a compartment. The initial surface densities are specified for each species present within a membrane.

We then turn to the species interaction through the chemical reactions. One can edit the processes associated with the ER membrane in the Biochemical Pathway Editor (see Fig. 3). This Editor permits the user to define reaction models as a series of membrane fluxes and reaction steps associated with either compartments or membranes.

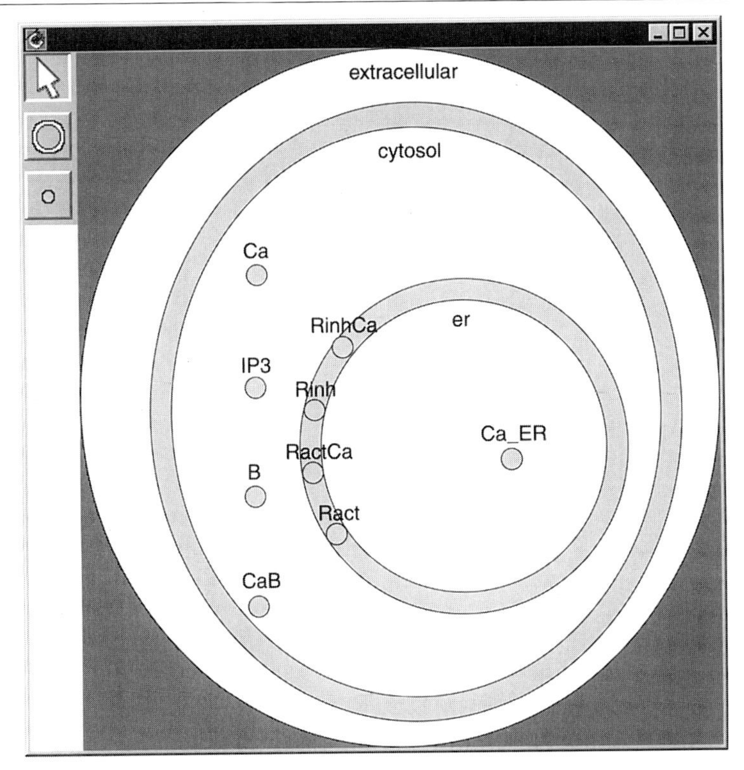

FIG. 2. The Cell Structure Editor allows specification of cellular structure and location of molecular species using a drag-and-drop user interface. In this example, there are three compartments: extracellular, cytosol, and ER, which are separated by the plasma membrane and the ER membrane, respectively. The location of a species within the editor display indicates its association with a cellular structure.

Our model contains three calcium fluxes through the ER membrane due to channels, pumps, and intrinsic leak, respectively. The channel flux is enzymatically regulated by the states of the activation and inhibition sites that participate in two reactions:

$$\mathrm{Ract} + \mathrm{Ca} \leftrightarrow \mathrm{RactCa} \tag{4}$$

$$\mathrm{Rinh} + \mathrm{Ca} \leftrightarrow \mathrm{RinhCa} \tag{5}$$

Similarly, there are editing windows for compartments. Figure 4 shows the only reaction that takes place in the cytosol:

$$\mathrm{Ca} + \mathrm{B} \leftrightarrow \mathrm{CaB} \tag{6}$$

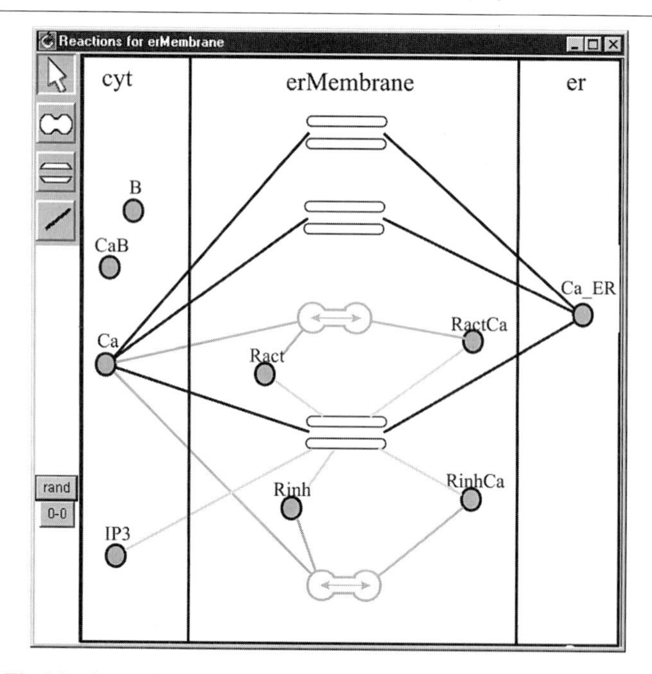

FIG. 3. The Membrane Reaction Editor allows specification of *reaction steps* (represented by dumbbell-shaped objects) and *membrane fluxes* (represented by hollow tube objects). The *reaction steps* are treated as in the Compartment Reaction Editor. The *membrane fluxes* are defined as the transmembrane flux of a single molecular species (in this case Ca^{2+}) from the exterior compartment (cytosol) to the interior compartment (ER lumen). In this example, the three membrane flux shapes represent the behavior of (from top to bottom) the SERCA pump, the membrane leak permeability, and the IP_3-receptor calcium channel. The channel flux is a function of the calcium channel activating binding site states Ract/RactCa and the inhibitory binding site states Rinh/RinhCaRI, which are modeled as independent. As in the reaction steps, flux kinetics are hyperlinked to the corresponding flux kinetics.

To complete the reaction description, we have to specify the "fast" subsystem if it exists (as described earlier this subsystem will be treated within the pseudo-steady-state approximation). As in the original published model (17), we assume reactions [Eqs. (4) and (6)] to be in a rapid equilibrium while reaction [Eq. (5)] and fluxes through the ER membrane comprise the slow system dynamics. The complete physiological model is described in Appendix 1.

After specifying the entire physiological model within the user interface, the Modeling Framework then translates the physiological description into a mathematical description (see Appendix 2). While doing this, it automatically creates the system of equations, analyzes it, and determines the mini-

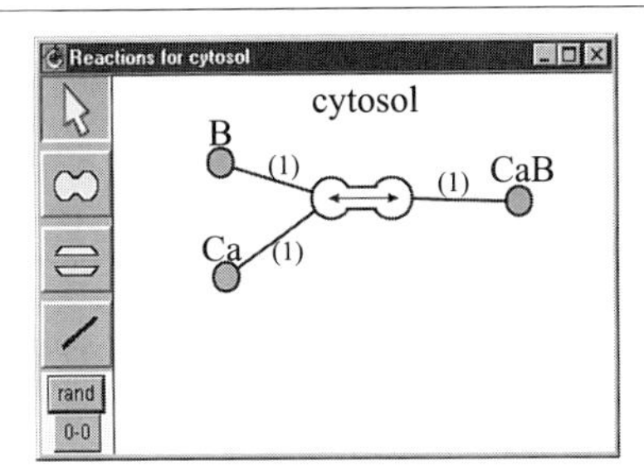

FIG. 4. Compartment Reaction Editor allows specification of *reaction steps* (represented by dumbbell-shaped objects), which are defined within the interior of a particular *compartment* and involve only those *species* that are defined within that compartment. A *species* participating as a reactant or a product has a line connecting itself to the left or right side of the reaction step shape, respectively, where the number in parentheses specifies the stoichiometry. Reaction kinetics are hyperlinked to the corresponding reaction step shape.

mal set of independent variables and equations that are sufficient for a complete solution. Thus, in our case, we finally end up with four independent variables, Ca, CaB, RactCa, and RinhCa, and the temporally constant IP3, which was explicitly defined as a variable to facilitate visualization.

It is worth mentioning that modelers have direct access to the Mathematical Framework interface, which allows them to skip the physiological description and create their own mathematical description based on a known system of equations. This Framework is a problem-solving environment (PSE) designed for the class of reaction/diffusion problems involving multiple spatially resolved compartments. Thus, one can easily test published mathematical models by translating them into the Mathematical Description Language (see Appendix 2).

Mapping Compartments to Simulation Geometry

Finally, we specify the geometry of the problem and boundary conditions. The egg cell geometry approximates a sphere, so the geometry was defined analytically as a sphere of radius 500 μm. In our case, one can reduce the computational domain by taking into account the rotational symmetry of the problem. Thus, it is sufficient to simulate the calcium distribution within a quarter of a sphere only. This geometry is specified using the Geometry Builder, a stand-alone Java applet that permits the construction of two- or three-dimensional cellular geometric models based on a series of image files or analytic geometry.

As in the previous model,[17] we do not spatially resolve ER. Instead, we treat it as continuously and uniformly distributed within the cytosol.

Correspondingly, species associated with the ER membrane are effectively described as volume variables. The information on the surface-to-volume ratio and the density of ER distribution is incorporated in the parameters LAMBDA1, LAMBDA2, and LAMBDA3 (Appendixes 1 and 2), these factors convert mass flux into rate of change of concentration in cytosol.

Spatial Simulation of Generated Equations

Finally, the procedure of automatic code generation is invoked: the mathematical description is read and the corresponding C++ files are automatically created and executed. Figure 5 shows the 3-D reconstruction (using VoxelView) of the simulation results for the calcium concentration distribution at time 70 sec after the wave initiation. We have also tested

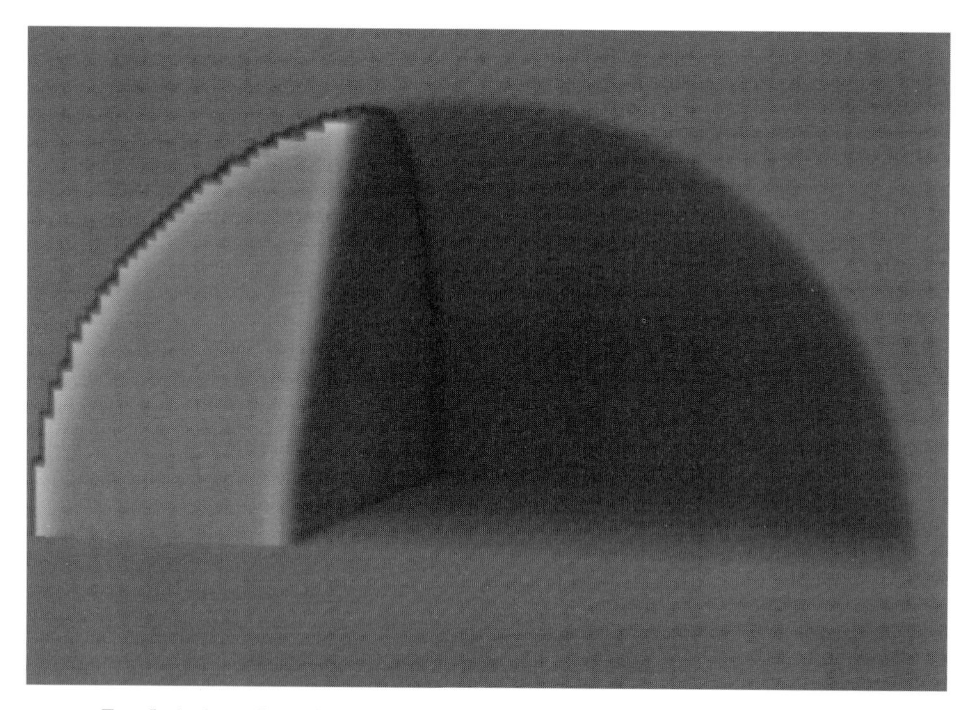

FIG. 5. A three-dimensional simulation of the fertilization calcium wave in a *Xenopus laevis* egg. The original published mathematical model[17] was directly implemented using the math interface in two and three dimensions and achieved essentially identical results. The underlying physiological model was then entered into our physiological framework and the corresponding mathematical model was automatically generated. The simulation results were in good agreement with the previous results.[17] This is a quarter of the spherical egg with a 500-μm radius. This is a volume rendering of the simulated intracellular concentration 70 sec after wave initiation (using the automatically generated mathematical model).

the two-dimensional version of the model. Our results are in good agreement with the previous results[17] for the corresponding value of the buffer scaling coefficient.

The Geometry/Mesh Editor allows participation in the choice of spatial resolution, permitting a balance of computational costs and the goodness of geometric representation. This interface directs the binding of regions of the segmented geometry to the corresponding features within the physiological model. An orthogonal mesh is specified and displayed interactively.

The Initial/Boundary Condition Editor allows the specification of initial conditions and boundary conditions for each of the species for each feature. To afford maximum flexibility, the boundary conditions for each simulation border may be specified independently for each feature. For example, the concentrations may be specified at simulation boundaries in the extracellular space to indicate a sink. A zero molecular flux may be specified at a simulation boundary belonging to cytosol to ensure the symmetry of function with the missing portion of cytosol (the implied mirror image).

The Simulation Data Viewer displays the results of the current spatial simulation. The species concentrations are displayed superimposed on the mesh. The analysis capability includes graphing the spatial distribution of a species as a line scan and graphing a time series at a single point.

Conclusions

The need for the Virtual Cell arises because the very complexity of cell biological processes severely impedes the application of the scientific method. A pair of separate factors that contribute to this problem are addressed by the Virtual Cell.

First, the large number of interdependent chemical reactions and structural components that combine to affect and regulate a typical cell biological process forces one to seek the help of a computer to build a model. This issue is the subject of an eloquent essay by Dennis Bray.[21] We are now faced with an overwhelming body of data describing the details of individual molecular events occurring inside cells. As Bray puts it, "What are we to do with the enormous cornucopia of genes and molecules we have found in living cells? How can we see the wood for the trees and understand complex cellular processes?" Brays solution: "Although we poor mortals have difficulty manipulating seven things in our head at the same time, our silicon protégés do not suffer this limitation. ... The data are accumulating and the computers are humming. What we lack are the words, the grammar and the syntax of the new language."

[21] D. Bray, *TIBS* **22,** 325 (1997).

The second factor recognizes that scientists trained in experimental cell biology are not typically equipped with sufficient mathematical, physical, or computational expertise to generate quantitative predictions from models. Conversely, theoretical biologists are often trained in the physical sciences and have difficulty communicating with experimentalists (bifurcation diagrams, for example, will not serve as a basis for a common language). By maintaining the physical laws and numerical methods in separate modular layers, the Virtual Cell is at the same time accessible to the experimental biologist and a powerful tool for the theorist. Also, by maintaining a direct mapping to experimental biochemical, electrophysiological, and/or image data, it ensures that simulation results will be communicated in a language that can be understood and applied by all biologists.

Acknowledgments

We are pleased to acknowledge the support of NIH (GM35063) and the Critical Technologies Program of the state of Connecticut. The National Resource for Cell Analysis and Modeling is funded through NIH Grant RR13186.

Appendix 1: Physiological Model Description

```
Name oocyte3d

Species CalciumBufferUnbound
Species CalciumBufferBound
Species IP3ReceptorCalciumActivationSiteUnbound
Species IP3ReceptorCalciumActivationSiteBound
Species IP3ReceptorCalciumInhibitionSiteUnbound
Species IP3ReceptorCalciumInhibitionSiteBound
Species Calcium
Species IP3

Compartment extracellular {}
Compartment cytosol {
      Context Ca Calcium          { InitialCondition 0.1153             DiffusionRate 300.0 }
      Context IP3 IP3             { InitialCondition 0.12               DiffusionRate 0.0 }
      Context B CalciumBufferUnbound { InitialCondition 1398.3876590291393 DiffusionRate 0.0 }
      Context CaB CalciumBufferBound { InitialCondition 1.6123409708605976 DiffusionRate 0.0 }
}
Membrane plasmaMembrane cytosol extracellular {
      SurfaceToVolume 1.0
      VolumeFraction 0.5
}
Compartment er {
      Context Ca_ER Calcium {  InitialCondition 10.0   DiffusionRate 0.0     }
}
Membrane erMembrane er cytosol {
      SurfaceToVolume 1.0
      VolumeFraction 0.5
      Context Ract IP3ReceptorCalciumActivationSiteUnbound {
            InitialCondition 0.009123393902531743 DiffusionRate 0.0 }
```

```
      Context RactCa IP3ReceptorCalciumActivationSiteBound {
              InitialCondition 8.766060974682582E-4 DiffusionRate 0.0 }
      Context Rinh IP3ReceptorCalciumInhibitionSiteUnbound {
              InitialCondition 0.009286200705751254 DiffusionRate 0.0 }
      Context RinhCa IP3ReceptorCalciumInhibitionSiteBound {
              InitialCondition 7.137992942487463E-4 DiffusionRate 0.0 }
}
Reaction TestReaction {
      SimpleReaction cytosol {
            Reactant Ca Calcium cytosol 1
            Reactant B CalciumBufferUnbound cytosol 1
            Product CaB CalciumBufferBound cytosol 1
            Kinetics MassActionKinetics {
                    Fast
                    Parameter K 100.0;
                    ForwardRate 1.0;
                    ReverseRate K;
            }
      }
      SimpleReaction erMembrane {
            Reactant Ca Calcium cytosol 1
            Reactant Ract IP3ReceptorCalciumActivationSiteUnbound erMembrane 1
            Product RactCa IP3ReceptorCalciumActivationSiteBound erMembrane 1
            Kinetics MassActionKinetics {
                    Fast
                    Parameter dact 1.2;
                    ForwardRate 1.0;
                    ReverseRate dact;
            }
      }
      SimpleReaction erMembrane {
            Reactant Ca Calcium cytosol 1
            Reactant Rinh IP3ReceptorCalciumInhibitionSiteUnbound erMembrane 1
            Product RinhCa IP3ReceptorCalciumInhibitionSiteBound erMembrane 1
            Kinetics MassActionKinetics {
                    Parameter TAU 4.0;
                    Parameter dinh 1.5;
                    ForwardRate (1.0 / TAU);
                    ReverseRate (dinh / TAU);
            }
      }
      FluxStep erMembrane Calcium {
            Catalyst IP3 IP3 cytosol
            Catalyst RactCa IP3ReceptorCalciumActivationSiteBound erMembrane
            Catalyst Ract IP3ReceptorCalciumActivationSiteUnbound erMembrane
            Catalyst Rinh IP3ReceptorCalciumInhibitionSiteUnbound erMembrane
            Catalyst RinhCa IP3ReceptorCalciumInhibitionSiteBound erMembrane
            Kinetics GeneralKinetics {
                    Parameter LAMBDA1 75.0;
                    Parameter dI 0.025;
                    Rate (-LAMBDA1 * (Ca_ER-Ca) * pow(((IP3/(IP3 + dI)) * (RactCa/(RactCa +
Ract)) * (Rinh/(RinhCa + Rinh))), 3.0));
            }
      }
      FluxStep erMembrane Calcium {
            Kinetics GeneralKinetics {
                    Parameter LAMBDA2 75.0;
                    Parameter vP 0.1;
                    Parameter kP 0.4;
                    Rate (LAMBDA2 * vP * Ca * Ca/((kP * kP) + (Ca * Ca)));
            }
      }
      FluxStep erMembrane Calcium {
```

```
            Kinetics GeneralKinetics {
                    Parameter LAMBDA3 75.0;
                    Parameter vL 5.0E-4;
                    Rate (- LAMBDA3 * vL * (Ca_ER - Ca));
            }
        }
}
```

Appendix 2: Mathematical Description

```
name oocyte3d_generated

Constant K 100.0;                          Constant dact 1.2;        Constant TAU 4.0;
Constant dinh 1.5;                         Constant LAMBDA1 75.0;    Constant dI 0.025;
Constant LAMBDA2 75.0;                     Constant vP 0.1;          Constant kP 0.4;
Constant LAMBDA3 75.0;                     Constant vL 5.0E-4;       Constant Ca_init 0.1153;
Constant B_init 1398.3876590291393;        Constant CaB_init 1.6123409708605976;
Constant Ract_init 0.009123393902531743;   Constant RactCa_init 8.766060974682582E-4;
Constant Rinh_init 0.009286200705751254;   Constant RinhCa_init 7.137992942487463E-4;
Constant Ca_ER 10.0;

VolumeVariable Ca          VolumeVariable IP3        VolumeVariable CaB
VolumeVariable RactCa      VolumeVariable RinhCa

Function IP3_init      0.12 + (exp(-0.133333 * (500.0 - sqrt(x*x + y*y + z*z))) * (0.12+(0.84 *
(x < -170.0) * exp(-pow((0.0025 * sqrt(y*y + z*z)), 4.0)))));
Function K_B_total     (B_init + CaB_init);
Function K_Ract_total  (Ract_init + RactCa_init);
Function K_Rinh_total  (Rinh_init + RinhCa_init);
Function B         (K_B_total - CaB);
Function Ract      (K_Ract_total - RactCa);
Function Rinh      (K_Rinh_total - RinhCa);

CartesianDomain {
        Dimension 3
        Size 1050.0 525.0 525.0
        Origin -525.0 0.0 0.0
        Compartment   cytosol ((x*x + y*y + z*z) < (500.0*500.0));
        Compartment   extracellular     1.0;
}
CompartmentSubDomain cytosol {
      Priority 2
      BoundaryXmDirichlet      BoundaryXp Dirichlet      BoundaryYm Dirichlet
      BoundaryYp Dirichlet     BoundaryZm Dirichlet      BoundaryZp Dirichlet
      PdeEquation Ca {
                 Rate     ((vL *(Ca_ER - Ca) * LAMBDA3) - (LAMBDA2*vP*Ca*Ca/(kP*kP + Ca*Ca)) -
((Rinh*Ca/TAU) - (RinhCa*dinh/TAU)) + ((Ca_ER - Ca) * pow((IP3/(IP3 + dI)*RactCa/(RactCa + Ract)
* Rinh/(RinhCa + Rinh)), 3.0) * LAMBDA1));
            Diffusion        300.0;
            Initial    0.1153; }
      OdeEquation IP3 {
            Rate 0.0;
            Initial 0.12 + (exp(-0.13333 * (500.0 - sqrt(x*x + y*y + z*z))) * (0.12 + (0.84 *
(x < -170.0) * exp(-pow((0.0025 * sqrt(y*y + z*z)), 4.0))))); }
      OdeEquation CaB {       Rate 0.0;          Initial 1.6123409708605976; }
      OdeEquation RactCa {    Rate 0.0;          Initial 8.766060974682582E-4; }
      OdeEquation RinhCa {
            Rate ((Rinh * Ca/TAU) - (RinhCa * dinh/TAU));
            Initial 7.137992942487463E-4; }
      FastSystem {
            FastInvariant     (RactCa + Ca + CaB);
```

```
            FastInvariant     IP3;
            FastRate          ((Ca * (K_B_total - CaB)) - (K*CaB));
            FastRate          ((Ca * (K_Ract_total - RactCa)) - (dact * RactCa));
            FastInvariant     RinhCa; }
}

CompartmentSubDomain extracellular {
        Priority 1
        BoundaryXm Dirichlet            BoundaryXp Dirichlet
        BoundaryYm Dirichlet            BoundaryYp Dirichlet
        BoundaryZm Dirichlet            BoundaryZp Dirichlet
        PdeEquation Ca {       Rate    0.0;    Diffusion    300.0;    Initial    0.0;    }
        OdeEquation IP3 {      Rate 0.0;       Initial 0.0; }
        OdeEquation CaB {      Rate 0.0;       Initial 0.0; }
        OdeEquation RactCa {   Rate 0.0;       Initial 0.0; }
        OdeEquation RinhCa {   Rate 0.0;       Initial 0.0; }
}

MembraneSubDomain cytosol extracellular {
        JumpCondition Ca {
                InFlux 0.0;
                OutFlux 0.0;
        }
}

Mesh { Size 100 50 50  }
Task { Output 35.0            Unsteady 0.1 0.0 105.0 }
```

[2] Kinetic Analysis of Dynamics of Muscle Function

By JULIEN S. DAVIS

Introduction

Characterizing the dynamics of contraction is key to understanding muscle function. Apart from its own intrinsic value, this is where new insights gained from single molecule and solution studies are integrated at the level of the fully organized and functional system. This, together with the current explosion in the use of transgenic animals linked to mutational analysis to study muscle function and disease in fibers[1] prompted the writing of this chapter.

Semiquantitative methods of analyzing tension transients persist in the literature, making it difficult to contrast and compare otherwise important and insightful contributions to the field. The problem is particularly acute because of the multidisciplinary nature of muscle research. Backgrounds in research range from genetics and cell biology to biophysical chemistry and physics. Fortunately, developments in digital data collection and computer-

[1] H. L. Sweeney and E. L. F. Holzbaur, *Ann. Rev. Physiol.* **58,** 751 (1996).

0076-6879/00 $30.00

based methods of analysis have transformed this task—once arcane methods used by the few are now available to all. By using the methods outlined in this chapter, force transients elicited by changes in fiber length, temperature, pressure, and the concentrations of reactants and products of the actomyosin ATPase can be readily analyzed and fully specified as sets of observed rate constants and associated amplitudes. Subdivision of tension transients into component kinetic phases in this way is critical for the complete quantitation of the mechanical response of a particular fiber type and for the elucidation of mechanisms of energy transduction and control.

The interior of a muscle fiber is very different from the homogeneous in-solution environment in which most chemical and biochemical reactions are studied. Cyclic interactions between the contractile proteins occur in a crystal-like array of actin and myosin subunits. Movement of myosin and actin filaments past one another subjects attached cross-bridges to strain, while at the same time altering the spatial relationships between detached actin and myosin molecules. Technique-related issues like the stiffness of the attachment between fiber and force transducers and step motors may also change the kinetic response. These factors all complicate analysis of the tension transients in unique ways. Approaches have to be carefully adapted from homogeneous solution kinetics where they were developed. Skinned (membrane permeabilized) fibers are preferred, because the concentrations of diffusable components can be better controlled. Yet, as we shall see, despite all the inherent complexity of a muscle fiber, tension transients can be fully quantified as individual kinetic phases with quite remarkable precision.

Emphasis is on the analysis of the mechanical responses of contracting muscle fibers to small perturbations; it is the method of choice to reliably quantify and characterize entire tension transients. Small changes in force, in product, and in reactant concentrations allow the nonlinear (second and higher order in concentration) differential equations to be linearized (first and pseudo-first order in concentration) and the power of classical *chemical relaxation spectrometry,* a kinetic approach developed by Eigen and De Maeyer,[2] to be applied to the system. Emphasis here is on the muscle-specific aspects of the adaptation of classical small-perturbation or chemical relaxation kinetic theory to give reliable kinetic data on muscle contraction. Beyond this, the books by Bernasconi,[3] Gutfreund,[4] and Moore and Pear-

[2] M. Eigen and L. De Maeyer, *in* "Technique of Organic Chemistry" (S. L. Friess, E. S. Lewis, and A. Weissberger, eds.). Wiley, New York, 1963.

[3] C. F. Bernasconi, "Relaxation Kinetics." Academic Press, New York, 1976.

[4] H. Gutfreund, "Kinetics for the Life Sciences." Cambridge University Press, Cambridge, 1995.

son,[5] respectively, provide outstanding treatments of relaxation kinetics, of biological kinetics, and of general chemical kinetics. This and earlier volumes of the series should be consulted for definitive presentations of computerized methods of data analysis and a useful elementary introduction to kinetic analysis.[6]

Adaptation of classical small-perturbation or chemical relaxation kinetic theory to give reliable kinetic data on muscle contraction is the theme. Small-perturbation methods commonly used in muscle research include temperature jump (T-jump), length jump (L-jump), pressure jump (P-jump), and the various concentration jumps (released from precursor "caged" compounds by photolysis) that include phosphate jump (P_i-jump), adenosine diphosphate jump (ADP-jump), and adenosine triphosphate jump (ATP-jump) experiments. These changes in conditions are usually applied to fibers contracting isometrically with the actomyosin ATPase cycle functioning under steady-state conditions. The more precise the dissection of tension transients, the deeper and more reliable the insights gained. Methods for doing this, cautionary tales, and specific examples follow.

Tension Transients in Muscle Fibers

Distinct from reactions in solution, kinetic phases in muscle fibers arise from two quite different sources: (1) perturbation of the contractile cycle (tension generation and the cyclic hydrolysis of ATP by the actomyosin ATPase) and (2) perturbation of the viscoelastic properties of the muscle proteins, processes defined as largely independent of the structural changes associated with the cross-bridge cycle.

Kinetic phases associated with the cross-bridge cycle correspond most closely to the familiar world of solution chemical kinetics. In such systems, theory dictates that the total number of small-perturbation kinetic phases is fixed, determined primarily by the number of intermediate states present.[3] There are $N - 1$ kinetic phases for N intermediates states in a linear pathway (where intermediate A interconverts to B, to C, etc.). More complex pathways also have a fixed number of kinetic phases, but the relationship to the number of intermediate states is more complex. Thus the cross-bridge cycle in a muscle fiber is associated with a characteristic, finite, and unique set of kinetic phases.

All possible phases are seldom observed in any one experiment. Detec-

[5] W. J. Moore and R. G. Pearson, "Kinetics and Mechanism," 3rd ed. John Wiley and Sons, New York, 1981.

[6] J. S. Davis, *Methods Enzymol.* **210,** 374 (1992).

tion of a particular kinetic phase depends both on the method of perturbation and on whether the resultant change in tension is sufficient to be detected. Similar and overlapping rates can also reduce the number of kinetic phases observed in an experiment. It is, however, essential to bear in mind that the total number of kinetic phases associated with a particular mechanism is independent of the method of perturbation.

Purely viscoelastic phases in the 100-μs and slower time domain time appear to be exponential in form. This is not unexpected, because models composed of springs (elastic elements) and dashpots (viscous elements) generally respond in this way to perturbation—in fact, the application of Kelvin elements to muscle mechanics and the analysis of tension transients has a long history (e.g., McMahon[7]). Thus, both cross-bridge-related and viscoelastic-related kinetic phases are similar in form. This simplifies analysis because the entire tension transient in the time domain we are concerned with is fully determined as the sum of a series of exponential phases, each with an observed rate constant and amplitude—some cross-bridge cycle related, others purely elastic in origin.[8,9]

Mathematical Basis for Determining Rate Constants

Extraction of the rate constants and amplitudes for each phase (numbered 1 to n) requires that the complete time course of the tension transient be analyzed as the sum of series of exponentials. One form of the equation used to fit such a multiexponential tension transient is as follows:

$$A_t = A_0 + A_1\exp(-k_1t) + A_2\exp(-k_2t) + A_3\exp(-k_3t) + A_i\exp(-k_it) + \ldots + A_n\exp(-k_nt)$$

where A_t is the amplitude of the tension transient at time t, A_0 the tension at zero time, and A_i the amplitudes and k_i the apparent rate constants of each exponential phase. This characteristic behavior is fundamental to small-perturbation relaxation kinetics and is the direct consequence of little change in concentrations of reactants products and intermediates, and energy levels of the system. For a further discussion of what constitutes a small-perturbation relaxation consult Bernasconi.[3] The important concept is that kinetic phases obtained under such conditions are simultaneous and not sequential. Large-perturbation experiments, like the initiation of contraction by the photolysis of caged ATP, initiate a sequential, and

[7] T. A. McMahon, "Muscles, Reflexes and Locomotion." Princeton University Press, Princeton, New Jersey, 1984.

[8] J. S. Davis and W. F. Harrington, *Biophys. J.* **65,** 1886 (1993).

[9] J. S. Davis and M. E. Rodgers, *Biophys. J.* **68,** 2032 (1995).

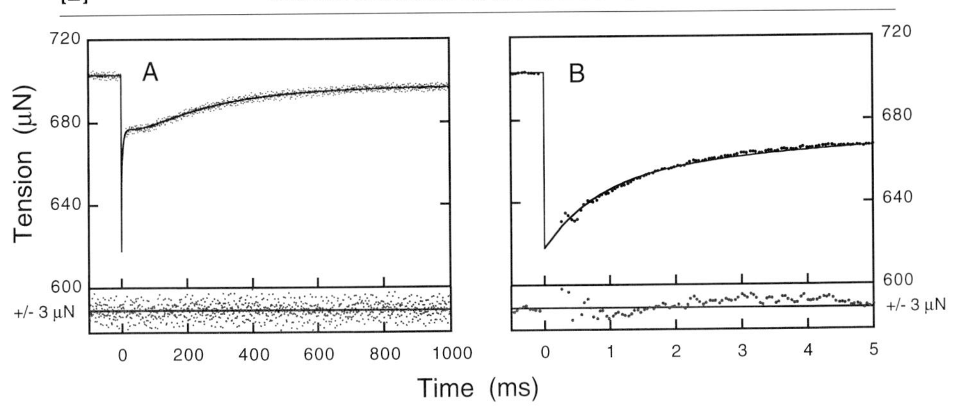

FIG. 1. Kinetics of tension generation following an L-jump. The response of a skinned, maximally Ca^{2+}-activated, rabbit psoas fiber undergoing isometric contraction at 11° to an L-jump of −1.5 nm per half sarcomere applied in 160 μs is shown on (A) slow and (B) fast time bases. Eight transients were averaged in each case. The best fit line to phases 2_{fast}, 2_{slow}, 3, and 4 is drawn through the data. Residuals are plotted below each transient. Resolved fits to these individual phases and their sum are illustrated in Fig. 2. [Data adapted from J. S. Davis and M. E. Rodgers, *Biophys. J.* **68,** 2032 (1995).]

therefore quite different series of interconversions of intermediate states (e.g., see Ref. 4).

Examples of typical tension transients recorded in the author's laboratory are described first to give the reader a clear picture of their characteristic features. Details of the methods of data selection, conditioning, and nonlinear least-squares analysis follow these examples.

Tension Transients Elicited by Two Different Methods of Perturbation

L-jump and laser T-jump tension transients from fast rabbit psoas fibers[8,9] are used as examples and are illustrated in Figs. 1 and 3, respectively. Experimental conditions are similar in both cases. Note that observed rate constants (e.g., an L-jump phase like phase 2_{slow}) have units of time^{-1}; time constants or relaxation times (e.g., a T-jump phase like τ_1) have the units of time.[6] The convention with T-jump is to use "tau" to name the phase, i.e., τ_1, τ_2, τ_3, etc. Since time^{-1} is used throughout this chapter, the symbol for a T-jump observed rate constant is $1/\tau$. One is the reciprocal of the other; the legends for Figs. 1 and 3 demonstrate their use.

L-jump tension transients are the more complex, because all states that generate tension or bear tension (either transiently or continuously) are perturbed. Only in the last decade was it realized that the entire L-jump tension transient could be analyzed as the sum of four exponential kinetic phases[8]; prior to this, the slower phases were considered nonexponential in form. The four component exponential phases resolved by nonlinear least-squares analysis are shown in Fig. 2. A description of the tension transient and its constituent L-jump or Huxley–Simmons kinetic phases as they are often called follows: The applied step change in the length causes a synchronous and immediate change in tension, termed phase 1 (Fig. 2D). Recovery of the prejump isometric tension after the abrupt change in tension of phase 1 is governed by the sum of four exponential processes. During the Huxley–Simmons phase 2 most of the initial isometric tension, lost during phase 1, is recovered. Phase 2 has two exponential subcomponents, namely, phase 2_{slow} (Fig. 2C) and phase 2_{fast} (Fig. 2D). Phase 3 (Fig. 2B) is evident next, and has a characteristic amplitude opposite in sign to

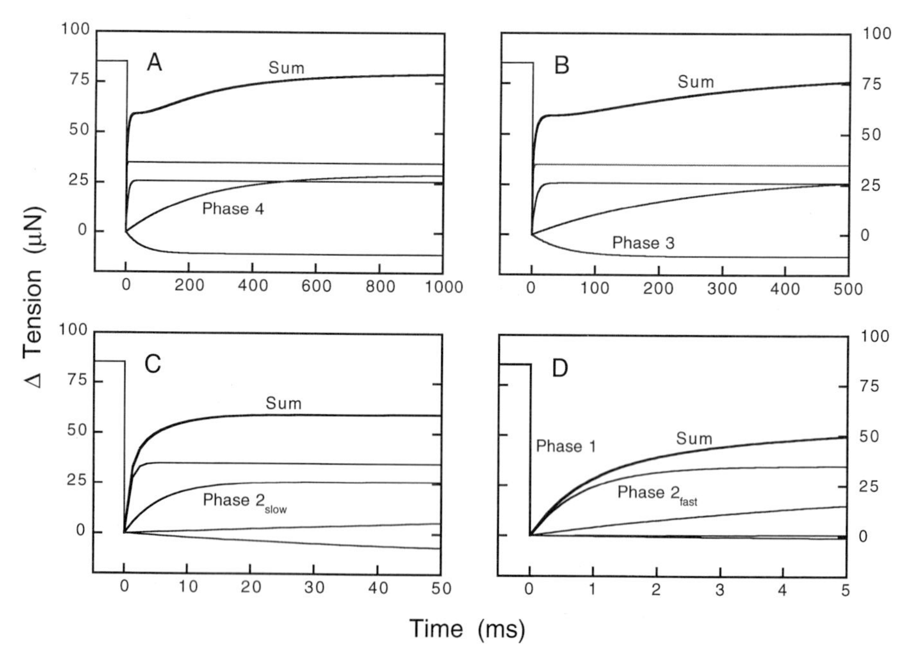

FIG. 2. The four exponential kinetic phases that sum to give the L-jump tension transient of Fig. 1. (A) Time base in which phase 4 (5.1 sec^{-1} and 34.4 μN) dominates the transient. (B) Time base in which phase 3 (20.5 sec^{-1} and -16.8 μN) dominates the transient. (C) Time base in which phase 2_{slow} (138.1 sec^{-1} and 17.7 μN) dominates the transient. Note the false endpoint of the tension transient (sum) produced by a fortuitous cancellation of the amplitudes of the slower phases. (D) Time base in which phase 2_{fast} (925 sec^{-1} and 50.9 μN) dominates the transient, the amplitude of phase 1 was -86 μN.

that of phase 2 and tension recovery. Phase 4 (Fig. 2A), the slowest of the relaxations, returns fiber tension to its prejump isometric value.

In contrast to an L-jump, a laser T-jump only perturbs temperature-sensitive steps of the cross-bridge cycle.[8] The nonlinear least-squares fit to the raw data (Fig. 3) and the simulation of the overall fit and exponential components (Fig. 4) show that a simpler three exponential fit fully accounts for the T-jump tension transient. The response is, in fact, even less complicated than this: only two of the phases (τ_2 and τ_3) arise directly from the effect of temperature on the cross-bridge cycle; the third, τ_1, is, in a sense, an artifact because it arises from a mini L-jump imposed by fiber expansion and not the T-jump. In theory, and as shown in Table I and discussed later, these T-jump kinetic phases each cross-correlate with an equivalent L-jump kinetic phase.

Note that frequently made statements to the effect that one kinetic phase follows another in small-perturbation experiments is a sure sign that the fundamental principles of the technique have not been grasped. Individual kinetic phases are, however, certainly more apparent in certain time domains than in others. This is highlighted by labeling kinetic phases dominant in particular time domains in Figs. 2A–D and Figs. 4A and B.

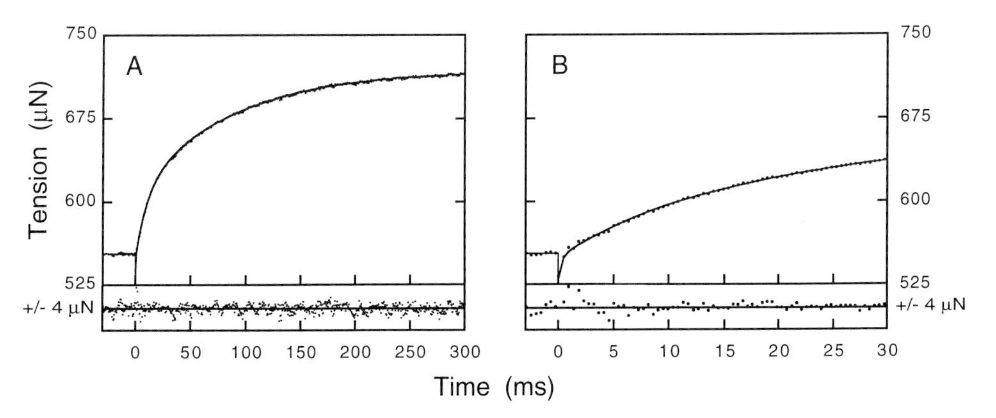

FIG. 3. Kinetics of tension generation following a laser T-jump. A skinned, maximally Ca^{2+}-activated, rabbit psoas fiber undergoing isometric contraction was heated by 5° in <1 μs to a postjump temperature of 11°. (A) Slow time base; (B) same data on a faster time base. Residuals, plotted below each transient, were obtained by subtracting the nonlinear least-squares fit to the sum of three exponentials (continuous line drawn through the data points) from the experimental data. The T-jump kinetic phases are termed, from fast to slow, tau 1 (τ_1), tau 2 (τ_2), and tau 3 (τ_3). Resolved fits to these individual phases and their sum are illustrated in Fig. 4. [Data adapted from J. S. Davis and M. E. Rodgers, *Biophys. J.* **68,** 2032 (1995).]

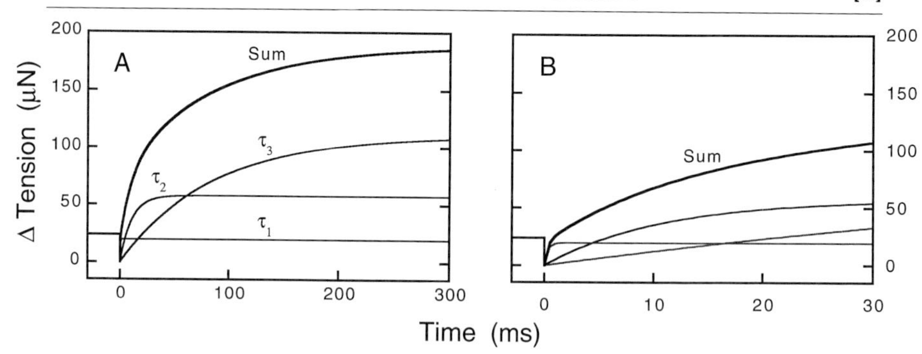

FIG. 4. The three exponential phases that sum to give the laser T-jump tension transient of Fig. 3. Kinetic components simulated from the fitted parameters are plotted on two different time bases in (A) and (B). Fitted rate constants and amplitudes were for τ_1 (3449 sec^{-1} and 19.7 μN); τ_2 (94.8 sec^{-1} and 57.6 μN) and τ_3 (11.8 sec^{-1} and 111.2 μN), respectively.

Time Span of Tension Transients

The time span of a tension transient is primarily dependent on fiber type and, to a lesser degree, on the conditions of the experiment (temperature, reactant concentrations, etc.). Times as long as 20 sec are sometimes required to capture slow fiber tension transients. However, not all rate processes track fiber type. Transitions associated with the cross-bridge cycle and therefore function are the most likely to change. General mechanical properties, such as fiber elasticity (phases 1 and 2_{fast} in L-jump experiments), change little—particularly in the same class of animal.[10]

Tension transients should be analyzed in their entirety. The reason is quite simple. All kinetic phases recorded in a small-perturbation experiment occur simultaneously and therefore overlap one another. Each time point of the tension transient will have greater or lesser contributions from all other exponential components. Errors caused by not doing this can be very serious and are discussed later.

Technical constraints limit the maximum speed at which tension transients can be recorded. The sampling rate of 20 μs per point used to collect the transients shown in Figs. 1 and 3 is typical. The rate at which the step change (perturbation) in length, temperature, pressure, or reactant and product concentrations can be applied and the response time of the tension transducer are the main limiting factors. In the two examples presented (Figs. 1 and 3), 90% of the near-linear L-jump ramp was applied in 160 μs, whereas the much faster laser T-jump was complete in 0.3 μs.[8,9] With rates

[10] J. S. Davis, *Biophys. J.* **76,** A269 (1999).

of perturbation like these and a tension transducer of 100 μs rise time, all known kinetic phases originating from the contractile cycle itself and one kinetically controlled elastic response can be quantified.[8,9,11] The fastest cross-bridge cycle related phase (phase 2_{slow}) peaks at a rate of ~1000 sec^{-1}, but is usually slower. Phase 2_{fast}, the (damped) elastic response in striated muscle fibers, occurs at ~1000 sec^{-1}.[10] These limit, high-speed phases will be reasonably well resolved in tension transients recorded in the 100-μs and longer time domains. Exploring elastic responses in the sub-100-μs time domains requires specialized methods to accommodate the rate of transmission of force and length changes along the fiber and nonlinearities in the response of the tension transducer.

Data Conditioning and Analysis

Establishing Zero Time

Establishing a correct zero time for the start of the transient is important. It is straightforward in experiments in which the perturbation is virtually instantaneous (e.g., laser T-jump). However, some thought is necessary when using techniques that impose slower perturbations (e.g., L-jump) on the fiber. Observed rate constants (first order and pseudo-first order) are unaffected by the choice of the zero time. However, amplitudes (particularly the fastest phases) are sensitive to the selected starting point. Usually, the midpoint of the applied ramp is adequate for most purposes.[8] If a more precise treatment is required, the reader is referred to Bernasconi[3] for methods to analyze transients recorded during, and just after, the completion of linear or exponential perturbations.

Selection of Artifact-Free Data

Tension transducers and step motors take time to stabilize after rapid movement and changes in tension. This ongoing movement imposes damped oscillations or ringing on the recorded signal. Because it is nonrandom, this ringing is not reduced by computer averaging a series of tension transients to improve the signal-to-noise ratio. These outlier artifacts should be excluded from the data used for the nonlinear least-squares fit. Inspection of averaged rather than single tension transients will reveal where the cutoff should be made. Remnants of a damped oscillation are evident in the 220- to 600-μs time domain of main and residual plots of Fig. 1B. Possibly the

[11] J. S. Davis and W. F. Harrington, *Adv. Exp. Med. Biol.* **332,** 513 (1993).

fit to phase 2_{fast} could have been improved by excluding these data points from the nonlinear least-squares fit to the raw data.

Data Sets with Time Bases Spanning 4 to 5 Orders of Magnitude

As a first step, it is useful to obtain the best fit possible to the entire tension transient.[8] Tension transients last a long time—an extreme example being the extended 100-μs to 20-sec transients typical of some slow muscle fibers. Analysis of data sets with linear-spaced time intervals generally fails under such conditions—too little weight is given to fast phases and too much to slower phases and the approach to the endpoint. Logarithmic spacing of these data results in a considerable improvement of fit.[8,9] The actual method used to achieve this depends on the way the data sets are recorded and the fitting routine employed.[12,13] The data set for Fig. 1, for example, was recorded as 64,000 12-bit data points at a sampling rate of 20 μs per point. A subset of 1000 log-spaced data points was selected from the 64,000 point raw data set. Then the Savitzky-Golay[14] simplified nonlinear least-squares fitting routine was used to average the selected datum point together with a set of adjacent and nonoverlapping points. The advantage of the approach is that much of the information present in the original data set is preserved for analysis. The Savitzky–Golay routine is relatively easy to program; alternatively commercially available versions of the routine do exist.

Kinetic parameters (rates and amplitudes) were then fitted to the reduced log-spaced 1000 data points with a nonlinear least-squares fitting routine—in the examples shown, Nonlin.[15] The fit should be further improved by decreasing the time base by consecutive factors of 5 to 10 and generating new log-spaced 1000-point data sets as described for the full transient. Rates and amplitudes of slower phase(s), determined outside the time domain of the new data set, are locked as constants during the fit to the faster phases. If necessary, time base selection can be optimized by noting time domains when particular kinetic phases dominate the transient (see Figs. 2 and 4). When the number of raw data points falls to a few hundred at the fastest time base, the fit can be applied directly to the original, linearly spaced data. Checking the overall fit by returning to the full data set is wise.

[12] A. R. Walmsley and C. R. Bagshaw, *Analyt. Biochem.* **176,** 313 (1989).
[13] M. Sharrock, *Rev. Sci. Instrum.* **48,** 1202 (1977).
[14] A. Savitzky and J. E. Golay, *Analyt. Chem.* **36,** 1627 (1964).
[15] M. L. Johnson and S. G. Frasier, *Methods Enzymol.* **117,** 301 (1985).

Artifacts: Real and Imaginary

Failure to Establish Steady State

A stationary state (steady state or equilibrium) is an essential requirement for small-perturbation relaxation experiments. Examples can easily be found in the current literature where these conditions have not been met. Rates and amplitudes determined under such conditions relate to ill-defined and changing states of the cross-bridge cycle and thus cannot be compared. Significant over- or undershoot of the endpoint of the transient is a characteristic artifact and clear indication that a steady state has not been not established. Always check to see that tension and sarcomere length are stable for the equivalent (no perturbation applied) of the time course of the record.

Misapplication of the Brenner protocol[16] can destabilize the steady state. This an effective method that reduces sarcomere length heterogeneity by shortening isometrically contracting fibers against a light load and then rapidly restretching them back to their original length. Unfortunately, the procedure profoundly disturbs the steady state. Allowing six half-times of the slowest component of the kinetics (either phase 4 or the ATPase rate) provides a useful first approximation for the time needed to reestablish the steady state. This calculation, though useful, does not account for time necessary to reestablish reactant and product concentration gradients across the fiber radius. It is therefore prudent to monitor tension and sarcomere length after application of the Brenner protocol to determine empirically whether steady-state conditions have been fully reestablished.

Resolving Different Phases of Similar Rate

This is a common problem in kinetics. Phases less than a factor of 3 apart in rate generally cannot be resolved from each another. An approximate fit to these too-close-in-rate phases will enable other phases in the transient to be determined. Often, changing conditions (e.g., temperature) can then be used to differentially alter the rates of the overlapping phases so that they can be determined independently. For example, in the paper by Davis and Rodgers,[9] the Arrhenius plots therein show conditions where L-jump phases 3 and 4 overlap at 1°, but allow phases 2_{slow} and 2_{fast} to be reliably determined. Likewise, in a different paper by the same authors,[17] Arrhenius plots of T-jump kinetic data show τ_2 and $\tau_{negative}$ too close in rate to be resolved at 31°; at 6°, τ_3 and $\tau_{negative}$ are now too close in rate to be resolved.

[16] B. Brenner, *Biophys. J.* **41,** 99 (1983).

[17] J. S. Davis and M. E. Rodgers, *Proc. Natl. Acad. Sci. U.S.A.* **92,** 10482 (1995).

Extrapolation of the Arrhenius data can be used to estimate the observed rate constants and amplitudes in the region of overlap.

Truncated Data Sets

The use of truncated or limited data sets for kinetic analysis is probably the prime source of error in the analysis of tension transients. Fortunately, the widespread introduction of digital data recording and computer-based methods of data analysis has eliminated the need to treat kinetic data in this way.

An historical account of the evolution of the methods to subdivide the L-jump Huxley–Simmons phase 2 into kinetic phases illustrates the pitfalls of using truncated data sets. The discovery and first analysis of the components of phase 2 were performed in the classical experiments of Huxley and Simmons[18] with tension transients much like that of Fig. 1. With the limitations of then available techniques, L-jump tension transients were truncated by drawing tangents from part of the transient where phase 2 appeared complete (linear segment) back to an intercept at zero time (see Ford *et al.*[19] for details)—at the time, an entirely reasonable approach to the analysis of an otherwise intractable data set. The L-jump data in Fig. 1, and more clearly in Fig. 2C, show just such an apparent endpoint (plateau) in the tension transient in the 20- to 50-ms time domain. The resultant truncated data sets were then analyzed as three, possibly more, exponential processes, all with the same physical properties characteristic of a force generating transition.

The above analysis using truncated data sets contrasts markedly in outcome with nonlinear least-squares analysis of the entire transient by Davis and colleagues[8,9,11] in which phase 2 subdivided into phase 2_{fast} and 2_{slow}. Inspection of Fig. 2C immediately reveals why the use of truncated data sets failed. Contributions from phases 3 and 4 are not taken into account in the truncated data set where they make significant variable, but hidden contributions to the transient in the apparent time domain of phase 2. The exponential fit is forced on the truncated data, and phase 2 splits into three apparent phases. Note that there are occasions when linear back-extrapolation of the type performed on these truncated data sets will work—the snag is that you rarely, if ever, know this beforehand.

In the above instance, improving the method of data analysis had a profound effect on our understanding of force generation. Rather than tension generation being viewed as a series of like steps declining in potential energy (thermal ratchet model), phase 2_{slow} now appears to be the sole

[18] A. F. Huxley and R. M. Simmons, *Nature* **233,** 533 (1971).
[19] L. E. Ford, A. F. Huxley, and R. M. Simmons, *J. Physiol.* **269,** 441 (1977).

tension-generating step, while phase 2_{fast}, the other component, appears to arise from the damped (slowed) elastic responses of the fiber.[8] Unfortunately, the use of truncated data sets still persists in the current literature, thus causing confusion.

Strain-Dependent Rate Constants

It is implicit that the rate of the force-generating step will be strain sensitive. Other structural changes in the cross-bridge cycle not directly associated with tension generation could also be sensitive to strain. It is easy to imagine the shape of attached cross-bridges being distorted by strain imposed by the relative movement of thick and thin filaments. In L-jump experiments, it is next to impossible to apply a sufficiently small perturbation to eliminate the strain sensitivity of phase 2_{slow}, a property consistent with its role in *de novo* tension generation. The rate recorded is proportional to fiber tension at the completion of phase 1.

Strain-neutral rates can be obtained in laser T-jump experiments, or by some other method where changes to fiber length and tension seen by the cross-bridge immediately after the jump are minimal. An alternative approach is to use sinusoidal oscillations of fiber length in place of a step function. Here, fiber strain is averaged and the rates determined are very close in value to those determined in strain neutral T-jump experiments (see Ref. 17). This otherwise useful method does not provide as much information as step methods of perturbation (see Bernasconi[3]) and has limited applicability. Of course the same result could be obtained by averaging rates obtained from same amplitude step-stretch and step-release experiments.

Series Compliance and Analysis of Kinetics of Tension Transients

It is easy to imagine how overly elastic (compliant) attachments to the ends of a fiber would communicate changes in tension and length inefficiently between fiber and apparatus. Commonly referred to as *end compliance,* it can arise from different sources. Glued or clamped ends used to attach the fiber to the apparatus may add significant end compliance. Transducers and step motors are generally designed as stiff devices and so add minimal end compliance to the system.

The effect is revealed in L-jump experiments where a motor applies a fixed-size step change in length to one end of a fiber. Monitoring sarcomere spacing reveals just how much of the applied change in length is transmitted to the fiber. Chemically cross-linking the ends of the fibers with glutaraldehyde can be used to reduce end compliance by creating a rigid well-defined link between fiber and equipment.[20]

[20] P. B. Chase and M. J. Kushmerick, *Biophys. J.* **53,** 935 (1988).

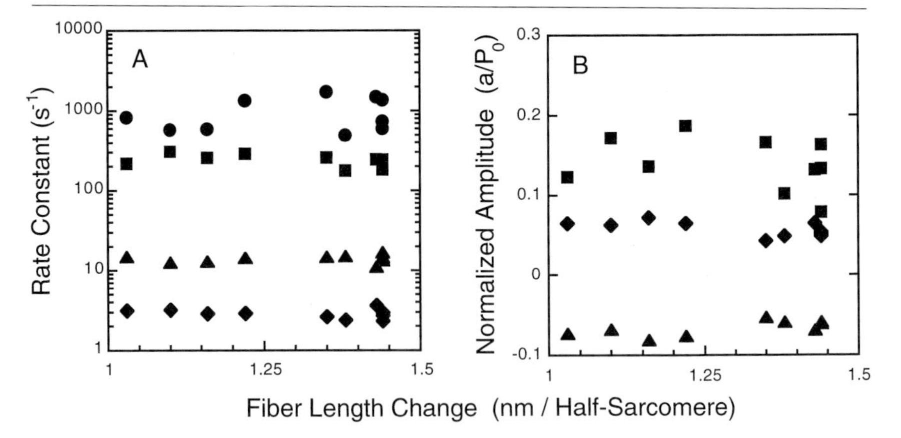

FIG. 5. The effect of end compliance on the rates and amplitudes of small-perturbation relaxation experiments. The plots show that different end compliances have little effect on the kinetics. (A) −1.5-nm per half-sarcomere L-jump was applied by the step motor to the fiber ends, and sarcomere lengths were measured from the diffraction pattern produced by a He-Ne laser set at the Bragg angle to the fiber. End compliance ranged from low for fibers with glutaraldehyde fixed ends (right) to high for poorly clamped fibers (left). (A) Effect of end compliance on the observed rate constants: circles, phase 2_{fast}; squares, phase 2_{slow}; triangles, phase 3; and diamonds, phase 4. (B) Effect of end-compliance on the normalized amplitudes of phase 2_{slow}, phase 3, and phase 4 (phase 2_{fast} is omitted because it obscured these data). Fiber temperature was 1°.

TABLE I

RELATIONSHIP BETWEEN EXPONENTIAL PHASES RECORDED IN SMALL-PERTURBATION KINETIC EXPERIMENTS ON SKINNED RABBIT PSOAS MUSCLE FIBERS CONTRACTING UNDER ISOMETRIC CONDITIONS[a]

Perturbation method	Cross-correlation of kinetic phases				
L-jump	Phase 2_{fast}	Phase 2_{slow}	Phase 3	Phase 4_a	Phase 4_b
T-jump	(τ_1)	τ_2	$\tau_{negative}$		τ_3
P-jump			Phase 2		
P_i-jump			k_{P_i}		
ADP-jump					k_{ADP}

[a] A summary of correlations proposed by Davis and colleagues between their L-jump and T-jump phases and phases recorded by others in pressure-jump experiments P_i-jump (caged phosphate) experiments and ADP-jump (caged ADP) experiments. Phase τ_1 is set parenthetically because it arises indirectly from fiber expansion and not the T-jump. Reproduced with permission from J. S. Davis, *Adv. Exp. Med. Biol.* **453,** 343 (1998).

Remarkably, end compliance appears to have a minimal effect on the kinetics of small-amplitude perturbation experiments,[8] a decided advantage of the technique. Figure 5 summarizes unpublished data obtained by Davis and colleagues that was used to reach this conclusion and to show how rates and amplitudes of all Huxley–Simmons phases are little affected by a range of fiber end compliances. Glutaraldehyde cross-linked fibers with low end compliance behave much like poorly clamped fibers with significant end compliance. It is, of course, worthwhile cultivating good technique, the lower the end compliance the better—it certainly does have an effect on rates in large-perturbation experiments where it has a well documented and long history.

Cross-Correlation of Kinetic Phases Obtained by Different Methods of Perturbation

If the method of perturbation alone is changed, under otherwise similar experimental conditions, individual kinetic phases can be cross-correlated by matching rates. The method works well for all phases—phase 2_{slow} included, once its extreme strain sensitivity, discussed earlier, is accounted for.[8] The theoretical basis for a finite and unique set of kinetic phases being associated with a particular mechanism was alluded to earlier.

A reliable, but more elaborate way of cross-correlating phases is to perform the mechanical experiments under different conditions and type each kinetic phase by its unique changes in rate.[8,9,17] Changing temperature is a convenient way to create these phase-specific "signatures" and was the method first used to cross-correlate phases. Arrhenius plots (natural log of rate versus the reciprocal of absolute temperature) are an appropriate way of presenting these phase-specific "kinetic signatures." Table I lists correlations[21] developed between the reference set of five kinetic phases seen in L-jump experiments and T-jump, P-jump,[22] and concentration jump (P_i-jump[23] and ADP-jump[24]) small-perturbation experiments. In Table I, I have used the L-jump or Huxley–Simmons kinetic phases as a reference set. Kinetic phases characteristic of a particular method of perturbation can then be cross-correlated to the reference L-jump set. Much can be learned about the primary structural changes in the fiber associated with particular kinetic phases by such an approach.

[21] J. S. Davis, *Adv. Exp. Med. Biol.* **453,** 343 (1998).

[22] N. S. Fortune, M. A. Geeves, and K. W. Ranatunga, *Proc. Natl. Acad. Sci. U.S.A.* **88,** 7323 (1991).

[23] J. A. Dantzig, Y. E. Goldman, N. C. Millar, J. Lacktis, and E. Homsher, *J. Physiol.* **451,** 247 (1992).

[24] Z. Lu, R. L. Moss, and J. W. Walker, *J. Gen. Physiol.* **101,** 867 (1993).

[3] Molecular Parameters from Sedimentation Velocity Experiments: Whole Boundary Fitting Using Approximate and Numerical Solutions of Lamm Equation

By BORRIES DEMELER, JOACHIM BEHLKE, and OTTO RISTAU

Introduction

Recent advances in analytical ultracentrifugation hardware have brought about substantial improvements in the accuracy, precision, and range of data obtained from this instrumentation. These improvements are made possible by digital data acquisition, microprocessor-controlled experimental parameters, high-quality optical components, and by the introduction of a Raleigh interference system that permits the acquisition of data at lower protein concentration with a higher signal-to-noise ratio and a higher data density than possible with absorbance optics. A fluorescence optical system is currently under development that should again increase the sensitivity by an order of at least 2 magnitudes over the absorbance optical system.[1] Computer programs for the analysis of sedimentation data from the Beckman XL-A commonly in use today rely on the graphical transformation of experimental data to obtain $G(s)$ integral distributions of the sedimentation velocity boundary,[2] calculations of $g(s)$ distributions from dC/dt profiles,[3] or simply on midpoint or second moment point determinations. Because of the improved data quality it has now become feasible to extract additional information from the sedimentation velocity boundary, such as diffusion coefficients and partial concentrations; and for certain well-conditioned systems it is even possible to determine association constants.

Although several sophisticated and computer-intensive data analysis methods have been developed, they were previously impractical to implement on all but the fastest mainframe computers. However, today it has become possible to perform these analyses on modern desktop computers, which provide sufficient computing performance and permit analyses of large data sets. These advances have given new impetus to the development of computer software employing advanced analysis methods for the routine analysis of sedimentation data.

In this article we want to focus on two types of approaches that have

[1] T. Laue, University of New Hampshire, Durham, Personal Communication (1998).
[2] K. E. van Holde and W. O. Weischet, *Biopolymers* **17,** 1387 (1978).
[3] W. Stafford, *Anal. Biochem.* **203,** 295 (1992).

0076-6879/00 $30.00

recently been implemented in software adapted for data obtained from the Beckman XL-A. Both approaches involve nonlinear least-squares fitting of the moving boundary of sedimentation velocity and approach-to-equilibrium experiments to solutions of the Lamm equation. In the first approach, the finite element method is used to solve the Lamm equation (and modifications of the Lamm equation) numerically (also called the Claverie method, after its introduction by Claverie *et al.* in 1975),[4] whereas in the second approach analytical solutions of the Lamm equation, subject to various approximations, are used. We outline the characteristics of these methods as they are used today, contrast their strengths and weaknesses, and elaborate on their range and utility.

Model

The sedimentation process of a solute k in a sector-shaped cell, subject to a centrifugal force, is modeled by the Lamm equation.[5] It is given in its general form by

$$\frac{\partial C_k}{\partial t} = -\frac{1}{r}\frac{\partial (rJ_k)}{\partial r}, \qquad J_k = D_k\frac{\partial C_k}{\partial r} - s_k\omega^2 rC_k \tag{1}$$

with these boundary conditions:

$$J(r_m, t) = J(r_b, t) = 0 \tag{2}$$

which require that no material enters or leaves the cell and that the total concentration stays constant throughout the experiment and remains between the meniscus and the bottom of the cell. The initial conditions are given by

$$0 \leq t \leq T, \qquad C(r, 0) = C_0(r) \tag{3}$$

In Eqs. (1)–(3), C is the solute concentration at time t and position r away from the center of rotation, J is the flux, t is the time, T is the elapsed time at the end of the experiment, and $r_m \leq r \leq r_b$, where r_m and r_b are the radii of the meniscus and the bottom of the cell; ω is the angular velocity, and s_k and D_k are the sedimentation and the diffusion coefficients of solute k. Clearly, only a small fraction of experimental conditions encountered in a laboratory will be adequately modeled by this equation, and modifications to the basic Lamm equation are necessary to accommodate other systems. A number of commonly encountered systems and their models incorporated in this study are listed next.

[4] J. M. Claverie, H. Dreux, and R. Cohen, *Biopolymers* **14,** 1685 (1975).
[5] O. Lamm, *Ark. Mat. Astron. Fys.* **21B,** 1 (1929).

Ideal, Noninteracting, Multicomponent System

Multiple ideal, noninteracting solutes can be modeled by summing the concentration gradients for each individual solute:

$$C_{\text{total}} = \sum_{k=1}^{i} C_k \tag{4}$$

where i is the total number of components in the system.

Concentration Dependency of s and D

In the general case, concentration-dependent nonideal solution behavior has to be considered, and s_k and D_k are not constant, but functions of concentration. As described by Claverie,[6] this can be represented in the form of Eqs. (5) and (6), where σ_k and δ_k are constant parameters that describe the variation of s_k and D_k from their value at infinite dilution, s_{k0} and D_{k0}:

$$s_k = s_{k0}\,(1 - \sigma_k C_k) \tag{5}$$
$$D_k = D_{k0}\,(1 + \delta_k C_k) \tag{6}$$

where $\sigma = \delta = 0$ corresponds to the ideal case where the sedimentation and diffusion coefficient are independent of C. The latter is generally true for small, globular macromolecules at low concentration. The models for concentration dependence of s and D as shown in Eqs. (5) and (6) represent the simplest case where interactions between different components are neglected.

Reversible Self-Association

A monomer–dimer self-associating equilibrium can be represented by

$$M + M \underset{k_2}{\overset{k_1}{\rightleftharpoons}} D \tag{7}$$

with $k_1/k_2 = K = C_D/(C_M)^2$, and $C_D + C_M = C_L$. Here K is the equilibrium constant, and k_1 and k_2 are the forward and backward rate constants, respectively; C_L is the loading concentration; C_M is the monomer concentration; and C_D is the dimer concentration. If the rate constant for the association and dissociation is sufficiently fast (diffusion controlled), the equilibrium that existed prior to sedimentation is constantly disturbed and reestablished during sedimentation and, consequently, a weight-average sedimentation coefficient is observed at each point in the cell. The weight-average is dependent on the partial concentrations of each species M and D:

[6] J.-M. Claverie, *Biopolymers* **15,** 843 (1976).

$$s_{\text{avg}} = \frac{s_M C_M + s_D C_D}{C_M + C_D} \tag{8}$$

For the diffusion coefficient, the following relationship holds:

$$D_{\text{avg}} = \frac{D_M \dfrac{\partial C_M}{\partial r} + D_D \dfrac{\partial C_D}{\partial r}}{\dfrac{\partial C_M}{\partial r} + \dfrac{\partial C_D}{\partial r}} \tag{9}$$

The monomer concentration of each species is easily computed from the total concentration and the equilibrium concentration by solving the quadratic $KC_M^2 + C_M - C_L = 0$. The dimer concentration is simply the difference between the loading concentration and the monomer concentration. While a monomer–dimer system is most commonly encountered in the laboratory, the model can theoretically be expanded to cover the general case of a monomer–dimer–$\cdots$–n–mer system, although the extraction of additional parameters beyond those of a monomer–dimer system may be difficult at best and therefore not practical. A general model for a rapidly self-associating system has been described by Fujita,[7] but the analysis of data corresponding to models of larger association states than a monomer–dimer system is questionable due to the large number of highly correlated parameters that are difficult to resolve even under optimal conditions.

Finite Element Solution

A brief description of the finite element solution is included here; a more complete derivation can be found in Claverie *et al.*[4,6] For an in-depth discussion of the finite element method, the reader is referred to Zienkiewicz.[8] The formulas presented here are sufficient for the reader to code a working simulation program. The finite element solution of the Lamm equation uses a variational formulation of Eq. (1), which is given by

$$\int_\Omega \frac{\partial C_k}{\partial t} vr\, dr + \int_\Omega \frac{\partial (rJ_k)}{\partial r} v\, dr = \int_\Omega f_k\, vr\, dr \tag{10}$$

where Ω is the $[r_m, r_b]$ domain and v is a test function. To discretize Eq. (10) in r, the distance between meniscus and the bottom of the cell is divided into $N + 1$ equally spaced increments of length $h = (r_b - r_m)/N$, where N is the number of elements in the space discretization. Integrating by parts and inserting the explicit expression of J_k [see Eq. (1)], and using

[7] H. Fujita, "Foundations of Ultracentrifugal Analysis." Wiley, New York, 1975.

[8] O. C. Zienkiewicz, "The Finite-Element Method in Engineering Science." McGraw-Hill, London, 1971.

the linear forms of S_k and D_k given in Eqs. (5) and (6), the following matrix equations are obtained:

$$
\begin{aligned}
(B + \Delta t D_0 A^1 - \Delta t s_0 \omega^2 A^2) C_{n+1} &= B C_n + \Delta t D_0 (U C^{U1} + V C^{V1} + W C^{W1}) \\
&+ \Delta t D_0 (U C^{U2} + V C^{V2} + W C^{W2}) - \Delta t s_0 \omega^2 A^2 C^{A1} - \Delta t s_0 \omega^2 A^2 C^{A2}
\end{aligned} \tag{11}
$$

where A^1, A^2, and B are tridiagonal $N + 1$ by $N + 1$ matrices whose elements are given by

$$
\begin{aligned}
&A^1_{1,1} = \frac{r_1}{h} + \frac{1}{2}, \quad A^1_{i,i} = 2\frac{r_i}{h} \qquad (i = 2, \dots, N) \\
&A^1_{N+1,N+1} = \frac{r_{N+1}}{h} - \frac{1}{2}, \quad A^1_{i,i-1} = A^1_{i-1,i} = \frac{-r_i}{h} + \frac{1}{2} \qquad (i = 2, \dots, N+1) \\
&A^2_{1,1} = -\frac{r_1^2}{2} - \frac{r_1 h}{3} - \frac{h^2}{12}, \quad A^2_{i,i} = \frac{-2 r_i h}{3} \qquad (i = 2, \dots, N) \\
&A^2_{N+1,N+1} = \frac{r_{N+1}^2}{2} - \frac{r_{N+1} h}{3} + \frac{h^2}{12}, \quad A^2_{i,i-1} = \frac{r_i^2}{2} - \frac{2 r_i h}{3} + \frac{h^2}{4} \\
&\qquad (i = 2, \dots, N+1) \\
&A^2_{i,i+1} = \frac{-r_i^2}{2} - \frac{2 r_i h}{3} - \frac{h^2}{4} \qquad (i = 2, \dots, N) \\
&B_{1,1} = \frac{r_1 h}{3} + \frac{h^2}{12}, \quad B_{i,i} = \frac{2 r_i h}{3} \qquad (i = 2, \dots, N) \\
&B_{N+1,N+1} = \frac{r_{N+1} h}{3} - \frac{h^2}{12}, \quad B_{i,i-1} = B_{i-1,i} = \frac{r_i h}{6} - \frac{h^2}{12} \qquad (i = 2, \dots, N+1)
\end{aligned} \tag{12}
$$

For the case of concentration dependency of D, the matrices U, V, and W also have to be calculated, where V is a tridiagonal matrix of dimension $N + 1$ by $N + 1$, and U and W are of dimension $N + 1$ by N:

$$
\begin{aligned}
&V_{1,1} = \frac{r_1}{2h} + \frac{1}{6}, \quad V_{N+1,N+1} = \frac{r_{N+1}}{2h} - \frac{1}{6}, \quad V_{i,i} = \frac{r_i}{h} \qquad (i = 2, \dots, N) \\
&U_{i+1,i} = -U_{i,i} = -V_{i+1,i} = \frac{r_i}{2h} + \frac{1}{6} \qquad (i = 1, \dots, N) \\
&W_{i,i} = -W_{i+1,i} = -V_{i,i+1} = \frac{r_i}{2h} + \frac{1}{3} \qquad (i = 1, \dots, N)
\end{aligned} \tag{13}
$$

All other matrix elements of A^1, A^2, B, U, V, and W not specified in Eqs. (12) and (13) are zero, leading to substantial savings in computer storage. The terms C^{U1}, C^{V1}, C^{W1}, and C^{A1} are vectors used for the calculation of the first-order concentration dependence approximations of s and D, and are calculated at each iteration as follows:

$$\begin{aligned} C_i^{U1} &= (\delta_n)_i\,(C_n)_i\,(C_n)_{i+1}, & C_i^{V1} &= (\delta_n)_i\,(C_n)_i\,(C_n)_1 \\ C_i^{W1} &= (\delta_n)_{i+1}\,(C_n)_{i+1}\,(C_n)_{i+1}, & C_i^{A1} &= (\sigma_n)_i\,(C_n)_i\,(C_n)_i \end{aligned} \tag{14}$$

For strong concentration dependence of s and D, second-order approximation vectors C^{U2}, C^{V2}, C^{W2} and C^{A2} of the concentration dependency should also be included in the model:

$$\begin{aligned} C_i^{U2} &= (\delta_n^2)_i\,(C_n)_i^2\,(C_n)_{i+1}, & C_i^{V2} &= (\delta_n^2)_i\,(C_n)_i^2\,(C_n)_i \\ C_i^{W2} &= (\delta_n^2)_{i+1}\,(C_n)_{i+1}^2\,(C_n)_{i+1}, & C_i^{A2} &= (\sigma_n^2)_i\,(C_n)_i^2\,(C_n)_i \end{aligned} \tag{15}$$

For the time discretization, the finite difference technique is used: the time coordinate t is divided into intervals of length Δt. This leads to a system of $N + 1$ linear equations in matrix form for each solute, which is given by

$$\mathbf{A}C_{n+1} = B_n \tag{16}$$

where n is an integer index for the time discretization. The matrix A is tridiagonal and can be solved with Gaussian elimination. For each solute, the concentration C is represented by a vector C_n whose elements are the values of the concentration C at radii r_m, $r_m + h, \ldots, r_b$ at time $t_0 + n\Delta t$. Starting with $n = 0$ and the initial conditions (C_0 for each solute) together with an estimate for the unknown parameters s_k and D_k, the vector C_{n+1} is recursively calculated from C_n by solving Eq. (16). An alternative method for the discretization of the Lamm equation employs a Crank–Nicholson scheme in a finite difference approach as described by Schuck.[9] A considerable improvement in the computing speed with no loss of accuracy can be obtained by implementing a moving boundary scheme as proposed by Schuck.[10] This improvement is particularly useful for slowly sedimenting species or long runs, such as approach-to-equilibrium experiments.

Nonlinear Least-Squares Parameter Estimation Using Finite Element Solutions

The inverse problem of fitting experimental data to the finite element solution of the appropriate model can be accomplished by an iterative nonlinear least-squares fitting routine. An inherent difficulty of fitting ex-

[9] P. Schuck, C. E. MacPhee, and G. J. Howlett, *Biophys. J.* **74,** 466 (1998).
[10] P. Schuck, *Biophys. J.* **75,** 1503 (1998).

perimental data with finite element solutions is the computational expense of functional evaluations. In addition, most nonlinear least-squares fitting algorithms depend on the calculation of a partial derivative for each parameter, either analytically or numerically (for a review, see Johnson and Faunt[11]). For the finite element solutions, an analytical derivative is difficult to define, and a numerical derivative is expensive to calculate, thus, a derivative-free method for the fitting process is preferred over other methods. We outline here the essential steps for the DUD method (does not use derivatives[12]), which is a derivative-free Gauss–Newton method and is preferred over other derivative-free methods, because it has a comparatively low number of functional evaluations per iteration. In the general case, five parameters are fitted for each component in the system: (1) partial concentration, (2) diffusion coefficient, (3) sedimentation coefficient, (4) concentration dependence of s, and (5) concentration dependence of D. In addition, the baseline offset, the meniscus, and the bottom of the cell position can be floated during the fitting process. However, it is advisable to float only parameters that cannot be determined otherwise. For example, in a well-calibrated instrument, the meniscus position can be determined by visual inspection of the experimental data, and the radial position at the bottom of the cell can be calculated from the centerpiece, rotor, and rotor stretching value, which is dependent on the speed (see Fig. 1). The nonlinear least-squares problem of globally fitting multiple scans of a sedimentation velocity experiment y_{ij} to a finite element solution with parameter vector $\mathbf{P} = \{p_1, p_2, \ldots, p_m\}$ is then to minimize the residual sum of squares (RSS) between the simulated and experimental data:

$$\mathrm{RSS}(\mathbf{P}) = \sum_{n=1}^{T/\Delta t} \sum_{i=1}^{N+1} [y_{n,i} - f_{n,i}(\mathbf{P})]^2 \tag{17}$$

We summarize here the jth iteration of the DUD algorithm for a system with m fitted parameters: Let $\mathbf{P}_1^j$, $\mathbf{P}_2^j$ $\mathbf{P}_3^j, \ldots, \mathbf{P}_{m+1}^j$ be a set of estimated parameters for the solution computed in the previous iteration (numbered by age with $\mathbf{P}_1^j$ being the oldest). Approximate the function $Z(\mathbf{P})$ by a linear function $\Theta^j(\mathbf{P})$, which is equal to $Z(\mathbf{P})$ at $m + 1$ points ($\mathbf{P}_i^j$; $i = 1, \ldots, m + 1$). Find $\mathbf{P}_{\mathrm{new}}^j$, the point which minimizes the distance between $\Theta^j(\mathbf{P})$ and $\mathbf{Y}$. Replace $\mathbf{P}_1^j$ with $\mathbf{P}_{\mathrm{new}}^j$ to obtain the updated parameter set. The $m + 1$ initial starting values required by DUD are generated from one user-supplied $\mathbf{P}_{m+1}$. For $i = 1, \ldots, m$, $\mathbf{P}_i$ is computed from $\mathbf{P}_{m+1}$ by displacing its ith component by a user-definable singularity factor times the

[11] M. L. Johnson and L. M. Faunt, *Methods Enzymol.* **210,** 1 (1992).
[12] M. L. Ralston and R. I. Jennrich, *Technometrics* **20,** 7 (1978).

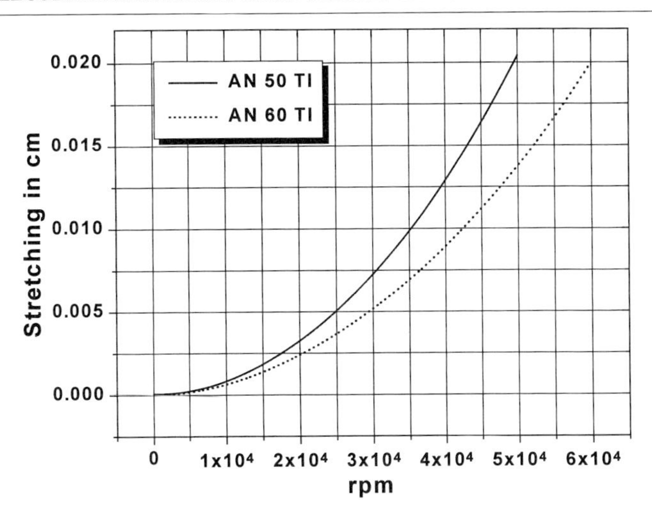

FIG. 1. Speed-dependent rotor stretching for the two rotors shipped with the Beckman XL-A/I. The stretching can be approximated with the following fourth-order polynomial: AN 50 TI: $7.754 \times 10^{-5} - 1.546 \times 10^{-8}$ rpm $+ 9.601 \times 10^{-12}$ rpm^2 $- 5.800 \times 10^{-17}$ rpm^3 $+ 6.948 \times 10^{-22}$ rpm^4; AN 60 TI: $3.128 \times 10^{-5} - 6.620 \times 10^{-9}$ rpm $+ 7.264 \times 10^{-12}$ rpm^2 $- 6.152 \times 10^{-17}$ rpm^3 $+ 5.760 \times 10^{-22}$ rpm^4; units are in centimeters. The position at the bottom of the cell is dependent on which centerpiece is used. At rest, a standard two-sector Epon/charcoal centerpiece has a position of 7.195 cm, and a standard two-sector aluminum centerpiece has a position of 7.15 cm. (Source: Beckman Instruments, Palo Alto, CA.)

corresponding component of $\mathbf{P}_{m+1}$. The algorithm can then be summarized as follows:

1. Generate an $l \times p$ matrix ΔF whose ith column is given by:

$$\Delta F_i = \mathbf{Z}(\mathbf{P}_i) - \mathbf{Z}(P_{m+1}) \tag{18}$$

where each vector $\mathbf{Z}$ is a vector of size l, generated from the simulation algorithm, where l stands for the total number of experimental observations.

2. Generate a p vector $\boldsymbol{\alpha}$ using:

$$\boldsymbol{\alpha} = (\Delta F^T \Delta F)^{-1} (\Delta F)^T [\mathbf{Y} - \mathbf{Z}(\mathbf{P}_{m+1})] \tag{19}$$

3. Compute a new parameter vector $\mathbf{P}_{\text{new}}$:

$$\mathbf{P}_{\text{new}} = \mathbf{P}_{m+1} + \boldsymbol{\Delta}\mathbf{P}\, \boldsymbol{\alpha} \tag{20}$$

where $\boldsymbol{\Delta}\mathbf{P}$ is the $m \times m$ matrix whose ith column is given by:

$$\Delta P_i = \mathbf{P}_i - \mathbf{P}_{m+1} \tag{21}$$

4. The convergence criterion is given by:

$$\mathrm{RSS}(\mathbf{P}_{\mathrm{new}}) - \mathrm{RSS}(\mathbf{P}_{\mathrm{old}}) = 0 \tag{22}$$

5. If the convergence criterion is not met, update the set of starting values $\mathbf{P}_1, \mathbf{P}_2, \ldots, \mathbf{P}_{m+1}$. Normally, the new estimate $\mathbf{P}_{\mathrm{new}}$ replaces $\mathbf{P}_1$ (the oldest number in the set), and the new starting set is formed by rearranging the set $\mathbf{P}_1, \mathbf{P}_2, \ldots, \mathbf{P}_{p+1}$ so that $\mathbf{P}_1$ is the oldest.

6. For some data sets, convergence will not occur without using a step shortening procedure that modifies $\mathbf{P}_{\mathrm{new}}$. We found the following procedure to work quite well for the cases examined to date:

$$\mathbf{P}_{\mathrm{new}} = d\,\mathbf{P}_{\mathrm{new}} + (1 - d)\mathbf{P}_{m+1} \tag{23}$$

where d is the first member of the sequence:

$$d_i = -(-10^{-3})^i, \qquad i = 1, 2, \ldots, 5 \tag{24}$$

Error Analysis of Finite Element Fitting Results

To estimate the confidence limits of the parameters obtained through a nonlinear least-squares fit, a robust error analysis is required. A good choice for such an analysis is the Monte Carlo method. This method is based on the assumption that the shape of the probability distribution $\mathbf{P}_{\mathrm{final},0} - \mathbf{P}_{\mathrm{true}}$ is similar to the probability distribution $\mathbf{P}_{\mathrm{final},i} - \mathbf{P}_{\mathrm{true}}$, where $\mathbf{P}_{\mathrm{final},0}$ is the parameter vector obtained through a nonlinear least-squares fit, and $\mathbf{P}_{\mathrm{final},i}$ is the parameter vector of a nonlinear least-squares fit to a simulated data set I. Each simulated data set is a finite element solution based on the original parameter vector $\mathbf{P}_{\mathrm{final},0}$, with the only difference that it is slightly perturbed by incorporating random errors drawn from appropriate error distributions such that the underlying instrumental errors are represented by each newly synthesized solution. Such an error distribution can be efficiently generated by the bootstrap method. To generate a simulated data set I, a fraction of the simulated points (typically about $1/e \approx 37\%$) generated from the finite element solution (based on $\mathbf{P}_{\mathrm{final},0}$ is replaced with random points drawn from the actual experimental data. Each simulated data set I is then refitted to generate the parameter vectors $\mathbf{P}_{\mathrm{final},i}$. If sufficient numbers of synthetic data sets are generated and fitted, the resulting parameter vectors $\mathbf{P}_{\mathrm{final},i}$ map out the desired probability distribution for each parameter, and the distribution of $\mathbf{P}_{\mathrm{final},i}$ will be distributed around $\mathbf{P}_{\mathrm{final},0}$ in a similar way as $\mathbf{P}_{\mathrm{final},0}$ is distributed about $\mathbf{P}_{\mathrm{true}}$. One word of caution: this procedure is computationally quite expensive. For further information about these methods, the reader is referred to Efron.[13]

[13] B. Efron, "The Jackknife, the Bootstrap, and Other Resampling Plans." Regional Conference Series in Applied Mathematics. SIAM, Philadelphia, Pennsylvania, 1982.

Analytical Solutions to the Lamm Equation

As an alternative to numerical solutions of the Lamm equation, analytical solutions incorporating various approximations can be used. The advantage of using analytical solutions generally includes improved speed of functional evaluations and the possibility of using a derivative-based nonlinear least-squares fitting algorithm, which can provide better convergence properties than nonderivative-based methods. On the downside, the boundary conditions are difficult to integrate into the solution and generally have to be approximated. In the following paragraphs, eight different approximations are considered and their applicabilities are evaluated. The first step in an analytical solution of the Lamm equation is to introduce dimensionless parameters. By substituting

$$\phi = \frac{c}{c_0}, \quad x = \left(\frac{r}{r_0}\right)^2, \quad \varepsilon = \frac{2D}{r_0^2\omega^2 s}, \quad \tau = 2s\omega^2 t \tag{25}$$

a dimensionless Lamm equation is obtained:

$$\frac{\partial\phi}{\partial\tau} = \frac{\partial\phi}{\partial x}\left[x\left(\varepsilon\frac{\partial\phi}{\partial x} - \phi\right)\right] \tag{26}$$

The explicit radius variable x in this equation complicates the integration of this equation. Thus far, only two complete solutions are known. The first solution was derived by Faxén,[14] and is valid only for infinitely long, synthetic boundary cells with $(0 < r < \infty)$. In this solution, it is assumed that the concentration buildup at the cell base has no influence on the moving boundary. Furthermore, the diffusion into the free solvent is not limited by the meniscus, because a sufficiently long column of pure solvent is supposed. To meet these assumptions, sedimentation velocity experiments must be performed with synthetic boundary cells, relatively short running times, and small boundary displacements. This solution is termed a solution of Faxén type. A second complete solution was given by Archibald.[15–17] It is valid for conventional cells and considers both the accumulation of solute at the cell base and the impermeability for solute at the meniscus. This is termed a solution of Archibald type. Because of their complexity, both of these solutions are not suitable as model functions for fitting procedures.

To generate an approximate analytical solution of the Lamm equation,

[14] H. Faxén, *Ark. Mat. Astr. Fys.* **21B,** 1 (1929).
[15] W. J. Archibald, *Phys. Rev.* **53,** 746 (1938).
[16] W. J. Archibald, *Phys. Rev.* **54,** 371 (1938).
[17] W. J. Archibald, *Ann. N.Y. Acad. Sci.* **43,** 211 (1942).

the explicit radius variable, x, has to be considered as a constant during the integration. Usually this value coincides with the starting point of sedimentation ($x = 1$). The error introduced by this procedure remains insignificant provided that the boundary migrates only a sufficiently small distance. The solution can be improved by using an adjustable value for the explicit integration variable x, which constitutes the reference point of integration.[18,19] This reference point is adjusted according to the position of the moving boundary. In practice, the value of the reference point increases with each scan in the direction of the bottom of the cell, and a different reference point is used for the analysis of each successive scan, starting with $x = 1$ for the first scan. Except for one solution under consideration here, sedimentation and diffusion coefficients are treated as constants with respect to concentration, temperature, and pressure. This is in contrast to the finite element method, which can be modeled to include concentration dependence of both s and D. The solutions presented below differ in how they consider the boundary conditions postulated by the experimental conditions [shown in Eq. (2)]. These boundary conditions describe mathematically the fact that no transport of solute is possible through the meniscus and through the bottom of the cell. The solutions are described in order of increasing complexity below.

1. The most elementary solutions do not consider either boundary condition. Nevertheless, these solutions are of value if the displacement of the boundary is small, and if a sufficiently long plateau region is present and if this plateau is not affected by the enhanced concentration at the cell bottom. Contrary to the conditions at the bottom of the cell, the situation at the meniscus depends on the cell type employed. Using synthetic boundary cells one can argue that the boundary condition is fulfilled, because the solute concentration at the meniscus remains essentially zero for the short run time employed for these experiments. However, for standard double sector cells, the application of this solution is possible only if the meniscus is completely free of solute through sedimentation. Only in that instance can a transport of solute through the meniscus be excluded, because the concentration at this position is nearly zero. On the other hand, the solute concentration at the cell base must not have completely destroyed the plateau. Larger molecules with smaller diffusion coefficients as well as faster rotor speeds favor these requirements. Under these limitations, this most simple solution can be applied successfully.

[18] L. A. Holladay, *Biophys. Chem.* **10,** 187 (1979).

[19] J. Behlke and O. Ristau, *Biophys. J.* **72,** 428 (1997).

2. The next category of solutions includes the boundary condition at the meniscus. These solutions are suitable for conventional double sector cells as long as the concentration profiles do not demonstrate an increase of concentration near the cell bottom due to back-diffusion of accumulated material at the cell bottom. If such an increase in concentration exists in the plateau region, it is important that this portion of the data not be included in the fit. In contrast to the first group of solutions, the meniscus must not be free of solute to use this type of model function, and the method is also applicable for earlier scans in the experiment.

3. The final group of solutions considers both boundary conditions. Because of its complexity, the exact solution derived by Archibald[15–17] is impractical for use in any fitting procedures, but approximate solutions with sufficient accuracy for practical application do exist.

Solutions Without Consideration of Boundary Conditions

Solutions of the Lamm equation omitting both boundary conditions are best suited to analyze the moving boundary in synthetic boundary cells. For this type of experiment, the reference point of integration has to be placed at the starting point of sedimentation between solvent and solution. For experiments displaying small displacements of the moving boundary and little diffusion of solute into the solvent, a relatively exact description of the concentration profiles can be obtained. The initial conditions for synthetic boundary cells are given by:

$$\begin{aligned} C_r &= C_0, \qquad r \geq r_0 \\ C_r &= 0, \qquad r < r_0 \end{aligned} \tag{27}$$

where r_0 is the synthetic boundary position between solution and solvent. A rigorous solution of the Lamm equation was derived by Faxén.[14] The complete solution of the Lamm equation given by Faxén includes the explicit radius variable x in the integral. After substituting,

$$\theta = \frac{C}{C_0} e^{\tau}, \quad \zeta = 2\sqrt{xe^{-\tau}}, \quad \eta = \varepsilon(1 - \varepsilon^{-\tau}) \tag{28}$$

the following Lamm equation is obtained:

$$\frac{\partial \theta}{\partial \eta} = \frac{\partial^2 \theta}{\partial \zeta^2} + \frac{1}{\zeta}\frac{\partial \theta}{\partial \zeta} \tag{29}$$

The rigorous solution of this equation according to Fujita[20] is given by

$$C = C_0 e^{-\tau}\left[1 - \int_0^{\alpha} e^{(-\beta-q)} I_0(2\sqrt{\beta q})\, dq\right] \tag{30}$$

where I_0 is the Bessel function of the first kind and zero order, and the parameters α and β are given by

$$\alpha = \frac{1}{\varepsilon(1 - e^{-\tau})}, \qquad \beta = \frac{x e^{-\tau}}{\varepsilon(1 - e^{-\tau})} \tag{31}$$

For most cases an approximation solution of Eq. (29) according to Hiester and Vermeulen[21] is sufficient. This formula is valid under conditions when $\varepsilon \sqrt{x} \sinh(\tau/2)/\sqrt{x} < 1/12$ is satisfied, and leads to

$$C_r = \frac{C_0 e^{-\tau}}{2}\left\{1 - \operatorname{erf}\left(\frac{1 - \sqrt{x e^{-\tau}}}{\sqrt{\varepsilon(1 - e^{-\tau})}}\right) + \frac{\sqrt{2}\,\varepsilon \sinh(\tau/2)}{x^{1/4}\sqrt{\pi}\,[1 + (x e^{-\tau})^{1/4}]} \exp\left[-\left(\frac{1 - \sqrt{x e^{-\tau}}}{\sqrt{\varepsilon(1 - e^{-\tau})}}\right)^2\right]\right\} \tag{32}$$

For the development of a less rigorous solution we have to substitute the variables x and Θ in Eq. (26) with $x = e^z$ and $\Theta = \phi e^{\tau}$ to obtain

$$\frac{\partial \theta}{\partial \tau} = \varepsilon e^{-z}\frac{\partial^2 \theta}{\partial z^2} - \frac{\partial \theta}{\partial z} \tag{33}$$

When the explicit radius variable z (the reference point of integration) in Eq. (33) is set to zero, we obtain the following solution[7]:

$$C_r = \frac{C_0 e^{-\tau}}{2}\operatorname{erfc}\left(\frac{z - \tau}{2\sqrt{\varepsilon\tau}}\right), \quad z = 2\ln\left(\frac{r}{r_0}\right), \quad \tau = 2\omega^2 st, \quad \varepsilon = \frac{2D}{r_0^2\omega^2 s} \tag{34}$$

where r_0 is the synthetic boundary position between solution and solvent. Fujita[20] has derived another equation, which considers the concentration dependence of the sedimentation coefficient. If a linear model for the concentration dependency of s is assumed, Eq. (26) needs to be rewritten as

$$\frac{\partial \theta}{\partial \tau} = \frac{\partial \theta}{\partial x}\left[x\left(\varepsilon\frac{\partial \theta}{\partial x} - (1 - \alpha\theta)\theta\right)\right] \tag{35}$$

[20] H. Fujita, "Mathematical Theory of Sedimentation Analysis." Academic Press, New York, 1962.

[21] N. K. Hiester and T. Vermeulen, *Chem. Eng. Progr.* **48,** 505 (1952).

The solution of this equation is given by

$$C_r = \frac{C_0 e^{-\tau}}{1-\lambda} \frac{[1-\mathrm{erf}(p)]e^{p^2}}{[1-\mathrm{erf}(p)]e^{p^2} + \sqrt{1-\lambda}\,[1+\mathrm{erf}(\zeta)]e^{\zeta^2}}$$

$$x = \left(\frac{r}{r_0}\right)^2, \quad \varepsilon = \frac{2D}{r_0^2\omega^2 s_0}, \quad \zeta = \frac{1-\sqrt{xe^{-\tau}}}{\sqrt{\varepsilon(1-e^{-\tau})}} \tag{36}$$

$$p = \frac{\zeta - \alpha\sqrt{(1-e^{-\tau})/\varepsilon}}{\sqrt{1-\lambda}}, \quad \lambda = \alpha(1-e^{-\tau}), \quad s_C = s_0\left(1-\alpha\frac{C}{C_0}\right)$$

For a more detailed derivation of this equation and the approximation, the reader is referred to Fujita (p. 93).[20]

Solutions That Consider Meniscus Boundary Condition

Solutions taking into account meniscus boundary conditions are applicable to standard double sector cells and concentration profiles with a distinct plateau. The first solution was derived by Fujita and MacCosham.[22] The initial or boundary conditions are given by

$$\begin{aligned} \theta = 1, &\quad x > 0, \quad \tau = 0 \\ \varepsilon\frac{\partial\theta}{\partial x} = \theta, &\quad x = 1, \quad \tau > 0 \end{aligned} \tag{37}$$

If the variable x in Eq. (26) is substituted with $z = \ln(x)$ the following equation is obtained:

$$\frac{\partial\theta}{\partial\tau} = \varepsilon\, e^{-z}\frac{\partial^2\theta}{\partial z^2} - \frac{\partial\theta}{\partial z} \tag{38}$$

When $e^{-z} = 1$ $(z = 0)$, and considering initial and boundary conditions given in Eq. (37), the following solution is obtained:

$$\begin{aligned} C_r = \frac{C_0 e^{-\tau}}{2}\Bigg[&\mathrm{erfc}\left(\frac{\tau - z}{2\sqrt{\varepsilon\tau}}\right) - 2\sqrt{\frac{\tau}{\pi e}}\, e^{-\left(\frac{\tau - z}{2\sqrt{\varepsilon\tau}}\right)^2} \\ &+ \left(1 + \frac{\tau + z}{\varepsilon}\right)\mathrm{erfc}\left(\frac{\tau + z}{2\sqrt{\varepsilon\tau}}\right) e^{\frac{z}{\varepsilon}}\Bigg] \end{aligned} \tag{39}$$

$$x = \left(\frac{r}{r_m}\right)^2, \quad z = \ln(x), \quad \varepsilon = \frac{2D}{r_m^2\omega^2 s}$$

An improved accuracy is attained if the radius variable z in Eq. (38) is not held constant at zero, and the product $e^{-z}\varepsilon$ is instead considered to be a

[22] H. Fujita and V. J. MacCosham, *J. Chem. Phys.* **30,** 291 (1959).

new parameter ε^* that adopts different values for each scan. Changing the definition of ε improves the accuracy of the solution considerably as judged by a comparison with data derived from the finite element method of Claverie.[4] This modification requires a displacement of the integration reference point with the moving boundary as expressed by the time τ. Only at the beginning of the experiment is the integration reference point at $z = 0$, but it changes during the experiment as described in Behlke and Ristau.[19,23]

$$z = \ln[1 + 0.5(e^{\tau} - 1)], \quad e^{-z} = \frac{1}{1 + 0.5(e^{\tau} - 1)},$$
$$\varepsilon^* = \frac{2D}{[1 + 0.5(e^{\tau} - 1)]r_m^2\omega^2 s} \tag{40}$$

Equation (39) considers only half of the boundary displacement. The boundary displacement measured at the inflection point (x_*) is described by Fujita[20]:

$$x_* - 1 = e^{\tau} - 1 \tag{41}$$

The new equation reads:

$$C_r = \frac{C_0 e^{-\tau}}{2}\left[\operatorname{erfc}\left(\frac{\tau - z}{2\sqrt{\varepsilon^*\tau}}\right) - 2\sqrt{\frac{\tau}{\pi e^*}}\, e^{-\left(\frac{\tau - z}{2\sqrt{\varepsilon^*\tau}}\right)^2} \right.$$
$$\left. + \left(1 + \frac{\tau + z}{-2\sqrt{\varepsilon^*}}\right)\operatorname{erfc}\left(\frac{\tau + z}{2\sqrt{\varepsilon\tau}}\right) e^{\frac{z}{\varepsilon}}\right] \tag{42}$$

Philo[24] reports an improvement in the accuracy of the original Fujita–MacCosham solution [Eq. (39)] by including two empirically determined parameters (α and β) that increase the accuracy of parameter estimation when compared to concentration profiles generated by the finite element method:

$$C_r = \frac{C_0 e^{-\tau}}{2}\left[\operatorname{erfc}[p(1 + \alpha\tau)] - 2\sqrt{\frac{\tau}{\pi e}}\, e^{-p^2(1 + \beta\varepsilon\tau)} \right.$$
$$\left. + \left(1 + \frac{\tau + \ln x}{-2\sqrt{\varepsilon}}\right)\operatorname{erfc}\left(\frac{\tau + \ln x}{2\sqrt{\varepsilon\tau}}\right) e^{\frac{\ln x}{\varepsilon}}\right] \tag{43}$$
$$p = \frac{\tau - \ln x}{2\sqrt{\varepsilon\tau}}, \quad \alpha = 0.2487, \quad \beta = 2, \quad \varepsilon = \frac{2D}{r_m^2\omega^2 s}$$

[23] J. Behlke and O. Ristau, *Biophys. Chem.* **70,** 133 (1998).
[24] J. S. Philo, *Biophys. J.* **72,** 435 (1997).

Another solution of the Lamm equation was developed by Holladay,[18] where, contrary to MacCosham and Fujita,[22] no logarithmic substitution for the variable x is made. Instead, the following new variables are used:

$$\phi = \theta e^{-\tau}, \qquad x = z + 1 \tag{44}$$

which lead to the following differential equation:

$$\frac{\partial \theta}{\partial \tau} = \varepsilon(z+1)\frac{\partial^2 \theta}{\partial z^2} + (\varepsilon - z - 1)\frac{\partial \theta}{\partial z} \tag{45}$$

The solution of Eq. (45) is given by:

$$\begin{aligned} C_r = \frac{C_0 e^{-\tau}}{2}\Bigg[& \operatorname{erfc}\left(\frac{\tau\alpha - z}{2\sqrt{a\varepsilon\tau}}\right) - \frac{a}{e}\exp\left(\frac{z\alpha}{a\varepsilon}\right)\operatorname{erfc}\left(\frac{\tau\alpha + z}{2\sqrt{a\varepsilon\tau}}\right) \\ & + \frac{\gamma}{\varepsilon}\exp\left(\tau + \frac{z}{\varepsilon}\right)\operatorname{erfc}\left(\frac{\tau\gamma + z}{2\sqrt{a\varepsilon\tau}}\right)\Bigg] \end{aligned} \tag{46}$$

$$\alpha = a - \varepsilon, \quad \gamma = a + \varepsilon, \quad z = x - 1, \quad a = \frac{e^{\tau} - 1}{\tau}$$

Solutions That Consider Both Boundary Conditions

Fujita[20] published a solution without derivation. Behlke and Ristau[23] communicated a derivation and additionally an improvement of this solution. This solution is based on the displacement of the reference point of integration, which depends on the movement of the boundary. This solution is given by

$$\begin{aligned} C_r = \frac{C_0 e^{-\tau}}{2}\Bigg[& \operatorname{erfc}\left(\frac{\tau\alpha_1 - z}{2\sqrt{a_1\varepsilon\tau}}\right) - \frac{a_1}{e}\exp\left(\frac{z\alpha_1}{a_1 e}\right)\operatorname{erfc}\left(\frac{\tau\alpha_1 + z}{2\sqrt{a_1 e\tau}}\right) \\ & + \frac{\gamma_1}{e}\exp\left(\frac{e\tau + z}{e}\right)\operatorname{erfc}\left(\frac{\tau\gamma_1 + z}{2\sqrt{a_1 e\tau}}\right) - \operatorname{erfc}\left(\frac{1 - z + \alpha_2\tau}{2\sqrt{a_2 e\tau}}\right) \\ & - \frac{a_2}{e}\exp\left(\frac{a_2(z-1)}{a_2 e}\right)\operatorname{erfc}\left(\frac{1 - z - \alpha_2\tau}{2\sqrt{a_2 e\tau}}\right) \\ & + \frac{\gamma_2}{e}\exp\left(\tau + \frac{z-1}{e}\right)\operatorname{erfc}\left(\frac{1 - z - \gamma_2\tau}{2\sqrt{a_2 e\tau}}\right)\Bigg] \end{aligned} \tag{47}$$

where

$$x = \left(\frac{r}{r_m}\right)^2, \quad x_b = \left(\frac{r_b}{r_m}\right)^2, \quad z = \frac{x-1}{x_b-1}, \quad e = \frac{\varepsilon}{x_b-1},$$

$$a_1 = \frac{1 + 0.5(e^{\tau} - 1)}{x_b - 1}, \tag{48}$$

$$\alpha_1 = a_1 - e, \quad \gamma_1 = a_1 + e, \quad \alpha_2 = a_2 - e, \quad \gamma_2 = a_2 + e$$

In the original solution by Fujita,[20] a_1 and a_2 were given by

$$a_1 = a_2 = \frac{1 + x_b}{2(x_b - 1)} \tag{49}$$

Since this solution incorporates both boundary conditions, the meniscus and the bottom of the cell, the accuracy of this solution allows the user not only to determine the sedimentation coefficient, *s,* and the diffusion coefficient, *D,* but it is particularly valuable for small molecules showing concentration profiles without a stable plateau.

Nonlinear Least-Squares Parameter Estimation Using Approximated Analytical Solutions

All approximate solutions of the Lamm equation contain the error function erf or the complimentary error function erfc = 1 − erf as an essential part. For its calculation, Gautschi[25] has developed a simple, but effective working procedure (an ALGOL version of this algorithm is given in the appendix). To spare computing time, it is advisable to calculate the derivatives of the model functions algebraically rather than numerically. To estimate the parameters of the analytical solutions of the Lamm equation, a nonlinear least-squares fitting method is appropriate. A short survey of different possibilities is given by M. Johnson and L. Faunt.[11] We found that a modified Gauss–Newton algorithm works well for most cases. This algorithm employs linearized model functions, which are represented by the Taylor expansion of the model function with respect to the parameters. The Taylor expansions are truncated after the first-order derivatives. Levenberg[26] and later Marquardt[27] developed a so-called "damped least-squares" method. This iteration procedure is robust enough to allow convergence even for poor starting values. A key point of the Gauss–Newton method is the Gaussian normal equation system. **A** is the so called informa-

[25] W. Gautschi, *Commun. ACM* **12,** 635 (1969).

[26] K. Levenberg, *Q. Appl. Math.* **2,** 164 (1944).

[27] D. W. Marquardt, *J. Soc. Indust. Appl. Math.* **11,** 431 (1963).

tion or Hessian matrix, an $m \times m$ square matrix. The elements of this matrix are the partial derivatives of the model function

$$\mathbf{A}\delta P = G \tag{50}$$

Next f is evaluated at the current estimate of the m parameters p, and summed up at all n data points x, y.

$$a_{i,j} = \sum \sum_{l=1}^{n} \frac{\delta f(x_l, p_1 \ldots p_m)}{\delta p_i} \cdot \frac{\delta f(x_l, p_1 \ldots p_m)}{\delta p_j} \tag{51}$$

∂P is the vector of the parameter improvements at each iteration step, the solution vector of the normal equation system. The elements of the vector G also evaluated at the current estimate of the parameters are defined by

$$g_j = \sum_{l=1}^{n} ([y_l - f(x_l, p_1 \ldots p_m)] \frac{\delta f(x_l, p_1 \ldots p_m)}{\delta p_j} \tag{52}$$

Because the information matrix is often ill conditioned, the matrix **A** may be scaled at first.

$$\mathbf{A}^* = (a_{ij}^*) = \left(\frac{a_{ij}}{\sqrt{a_{ii}}\sqrt{a_{jj}}}\right) \qquad g^* = (g_j^*) = \left(\frac{g_j}{\sqrt{a_{jj}}}\right) \tag{53}$$

Thereafter, a damping term, λ^2, is added to each diagonal element that determines the step length of the parameter change δp. In addition, if the damping values are large the method switches from the Gauss–Newton to the more slowly but rather secure steepest descent gradient procedure.[11] For the rth iteration, the normal equation system can be written as

$$(\mathbf{A}^* + \lambda_r^2 \mathbf{I})\delta^* = g^*, \qquad \delta_j = \frac{\delta_j^*}{\sqrt{a_{jj}}} \tag{54}$$

where $\mathbf{A}^*$ is a square matrix of dimension m, **I** is the unity matrix, and m is the number of floating parameters. Wynne and Wormell[28] have developed a method to optimally determine the size of the damping term. The procedure is as follows. The fitting process is started with a relatively large damping value, e.g., $\lambda_1 = 5$. After each iteration, the first improvement, D_1, of the sum of squares of the residuals (SSR) is then compared to the

[28] C. G. Wynne and P. M. J. H. Wormell, *Appl. Opt.* **2,** 1233 (1963).

second improvement, D_2, calculated with the last parameter improvements and the linearized model function.

$$D_1 = \sum_{l=1}^{n} [y_l - f(x_l, p_1^r, \ldots, p_m^r)]^2 - \sum_{l=1}^{n} [y_l - f(x_l, p_1^{r+1}, \ldots, p_m^{r+1})]^2 \tag{55}$$

The second improvement is defined by

$$D_2 = \sum_{l=1}^{n} [y_l - f(x_l, p_1^r, \ldots, p_m^r)]^2 - \sum_{l=1}^{n} \left[y_l - f(x_l, p_1^r, \ldots, p_m^r) - \sum_{i=1}^{m} \delta p_i^r \frac{f(x_l, p_1^r, \ldots, p_m^r}{\delta p_i^r} \right]^2 \tag{56}$$

Equation (56) can be expanded to:

$$D_2 = 2 \sum_{i=1}^{m} \delta p_i^r \sum_{l=1}^{n} [y_l - f(x_l, p_1^r \ldots p_m^r)] \frac{\delta f(x_l, p_1^r \ldots p_m^r)}{\delta p_i^r} - \sum_{i,j=1}^{m} \delta p_i^r \, \delta p_j^r \sum_{l=1}^{n} \frac{\delta f(x_l, p_1^r \ldots p_m^r)}{\delta p_i^r} \cdot \frac{\delta f(x_l, p_1^r \ldots p_m^r)}{\delta p_j^r} \tag{57}$$

All elements in Eq. (57), including the sums of the partial derivatives, are elements of the information matrix and therefore known at each iteration step. The ratio $R = D_1/D_2$ of both values is a measure for the linearity of the last iteration step length and is used to adjust the damping parameter in the next iteration:

$$\begin{aligned} R \geq 0.9{:} &\qquad \lambda_{(r+1)} = 0.5\,\lambda_r \\ 0.5 \geq R < 0.9{:} &\qquad \lambda_{(r+1)} = \lambda_r \\ R < 0.5{:} &\qquad \lambda_{(r+1)} = 2\,\lambda_r \end{aligned} \tag{58}$$

It is also possible to further subdivide the step sizes shown above. If the SSR value increases, the iteration is rejected and the damping term λ is increased fourfold, and the normal equation system is recalculated using the parameters from the $(r - 1)$ iteration. When solving Eq. (54), even the scaled information matrix may be ill conditioned. Therefore, it is advantageous to use the square root method. Moreover, it is possible to store the triangular matrix of the square root method and to carry out an additional iteration to solve the Gaussian normal equations. In this case, the new right-hand side (g) is now the residuals vector. This vector is obtained by introduction of the calculated parameters into the Gaussian normal equations. The additional iteration is computationally inexpensive, but increases the accuracy of the solution considerably.[29] The iteration is stopped if the following three conditions are fulfilled:

$$1: \quad \frac{\sum_{l=1}^{n} [y_l - f(x_l, p_1^r .. p_m^r)]^2 - \sum_{l=1}^{n} [y_l - f(x_l, p_1^{r-1} .. p_m^{r-1})]^2}{\sum_{l=1}^{n} y_l^2} < 3 * 10^{-7}$$

$$2: \quad \frac{\sum_{i=1}^{m} (\delta p_i^r)^2}{\sum_{i=1}^{m} (p_i^r)^2} < 10^{-9} \tag{59}$$

$$3: \quad \lambda_r < 1$$

The estimation of confidence intervals (standard deviations) of the parameters generally can be calculated from the root of the diagonal elements of the variance–covariance matrix:

$$\mathrm{VC} = \frac{\mathrm{SSR}}{(n - m)} \mathbf{A}^{-1} \tag{60}$$

The diagonal elements of the inverse of matrix **A** can be calculated without inverting the whole matrix.[29] However, strictly speaking, this procedure is only appropriate for linear regression and for model functions without correlation between parameters, which requires that all nondiagonal elements in the variance–covariance matrix be zero. The confidence intervals are generally underestimated by this method, but we have to consider that variations in the estimated parameter values between repeated experiments are often substantially larger than expected from the confidence intervals. Obviously, methodical errors in the experiment seem to play an important role. Possibilities of such errors are "window noise" described by Philo[24] or errant data points generated by the centrifuge. Additionally, the least-squares fitting algorithm requires that the noise be Gaussian distributed, but this condition is hardly assured. Even if the error of the light intensity measurement is Gaussian distributed, the logarithmic transformation of intensity data to absorbance data will also transform the presumed Gaussian error distribution to a logarithm of a Gaussian distribution.[11] If the data contain considerable noise, the estimation of the parameters and the confidence intervals may be negatively affected by this systematic error.

[29] H. R. Schwarz, H. Rutishauser, and E. Stiefel, "Numerik Symmetrischer Matrizen." B. G. Teubner, Stuttgart, 1968.

Choosing Appropriate Model

Whenever experimental data are fitted to a theoretical model it is crucially important that the correct model be chosen for the data. If the model does not describe the underlying system present in the data, any parameters obtained from the fit may be meaningless. For sedimentation data, this requirement applies regardless of which whole boundary fitting method is chosen. Unfortunately, there is no easy method for determining the proper model in each case. Under certain conditions, the correct model may not even be suitable for analysis by a whole boundary method. For example, a complex multicomponent system may have too many components and too many parameters for a reliable fit of the data. For cases where the data contain both a solvent baseline and a stable plateau (solution baseline) the data should be preprocessed with the van Holde–Weischet method.[2] This method provides a model-independent transformation of the data, which results in diffusion-corrected S-value distributions. In addition to providing information about sample composition, this method provides many useful diagnostics such as the presence of concentration dependence of S and reliably reveals the appropriate model and also provides initial guesses for the fitting parameters.[30,31] In an alternative approach, a simple model (e.g., a single ideal species) is chosen initially and a fit is attempted. Once the fit converges, the residuals are visually inspected for "runs" or by using a correlation analysis. Alternatively, subsets of early and late scans from the experiment can be analyzed separately. If the obtained parameters do not show a drift that depends on the boundary position of the subsets of scans, and the residuals display a random distribution about the mean for all scans, it can be assumed that the model is adequate.

A rigorous error analysis and cross-correlations may also suggest what aspects of the model should be changed, if any. Experience in the analysis of experimental concentration profiles demonstrates that in many cases an ideal, single component system is not present. There are many possible reasons; among the more common are nonspecific aggregates, concentration nonideality due to size, charge, or shape, and the presence of multicomponent solutes. For complex systems such as multicomponent solutes, the signal from the experimental data may not be sufficient to reliably determine the value of all parameters simultaneously. In such a case the correlation between individual parameters will be close to unity, the confidence limits will be broad, and the results will have little relevance, although the residuals could appear random, and the best model is difficult to determine.

[30] B. Demeler and H. Saber, *Biophys. J.* **74,** 444 (1998).
[31] B. Demeler, H. Saber, and J. Hansen, *Biophys. J.* **72,** 397 (1997).

Unless a valid multicomponent model can be established by a model-independent approach such as the van Holde–Weischet method, it is the authors' opinion that whole boundary fitting is generally not advisable for systems more complex than a single, ideal solute. However, for clearly separated components, such as a single, ideal species and a noninteracting, fast-moving aggregate, a partial boundary corresponding to the component under investigation may be fitted only, thus reducing the number of parameters required for the fit, and improving the covariance among parameters.[31]

When using approximated analytical solutions to model velocity data, the initial analysis should be performed with solutions of the Faxén type, since these equations require only the first part of the concentration profiles. Whenever preprocessing of the data with the van Holde–Weischet method is impractical, the following prescription may also provide reliable results. First, collect as many scans as possible, extending over the entire cell. For approximate analytical solutions, select Eq. (47) for an initial fit and then fit subsequently with Eq. (42). If the residuals appear random, and agree between both solutions, homogeneity in the solute can be assumed. All scans that are included in the fit should contain at least some stable plateau concentration. For cases of paucidispersity, approximate solutions of the Lamm equation can be used as long as the individual boundaries are clearly separated by a stable plateau region for each component. If nonspecific aggregates are present, whole boundary methods are not applicable.

Discussion

Finite element analysis can be readily employed for the analysis of ultracentrifugation data from a wide variety of experimental conditions.[9,31,32] Accurate and reliable results can be obtained for single- or two-component systems that may exhibit concentration dependency or self-association behavior. Because of the inherent integration of the boundary conditions into the finite element solution of the Lamm equation, even short column experiments and approach-to-equilibrium conditions can be successfully modeled. For such systems, the sedimentation and diffusion coefficients for each solute can be obtained. If the partial specific volume, $\bar{v}$ is known, the molecular weight can also be determined. The baseline absorbance and partial concentrations of each solute can be determined with the finite element method and used to determine stoichiometries if the molar extinction coefficients are known. For simple self-associating systems, the equilibrium constants can be determined.

[32] G. P. Todd and R. H. Haschemeyer, *Proc. Natl. Acad. Sci. U.S.A.* **78,** 6739 (1981).

While in theory the meniscus and bottom of cell radii can be fitted, it is generally not recommended to do so, since in most cases the meniscus can be determined visually with sufficient accuracy. The bottom of the cell position can be accurately obtained from values shown in Fig. 1. Deviations of the position at rest due to rotor stretching can be obtained from Fig. 1 and are added to the position at rest. When applying the finite element method to more complex systems, erroneous results are more likely produced if the information is derived from a single experiment. In such cases the signal provided by each individual component is often too small to accurately determine the hydrodynamic parameters describing it. In such a case it can be helpful to determine a subset of parameters by employing different methods or experiments and to hold the values constant throughout the fitting process.

In the experience of the authors, a single velocity experiment is generally sufficient only for the accurate determination of the parameters describing a single solute, under favorable conditions maximally a dual-component system. For more complex systems, a global approach incorporating multiple experiments employing multiple speeds and column lengths, as well as the integration of other techniques, may be necessary. Efforts are currently under way to develop fitting algorithms that utilize multiple experiments in a global fitting environment, an approach that will improve the confidence for parameters obtained from fits to more complex models incorporating more than one or two individual solutes, which may or may not be chemically interacting.

The accuracy of the finite element method is limited only by the integration step size utilized and the quality of the experimental data. Sufficiently small step sizes of less than 1.0×10^{-3} cm for the radial variable and 1 sec or less for the time variable produce errors that are smaller than the instrumental error inherent in the analytical ultracentrifuge.

In contrast to analytical solutions of approximated Lamm equations, the entire range of data over both variables of space and time can and should be utilized for the fit of the data. However, care should be taken to exclude data that exceed the linear range of the optical system. For the absorption optical system of Optima XL-A, absorbance values above 0.9 OD total absorbance are better avoided. Newer instrumentation equipped with the Raleigh interference optical system is capable of acquiring data with time intervals of only a few seconds between scans, resulting in a large number of data points for an experiment that can be utilized to improve the confidence of the fitted parameters. At this point we need to mention that the current version of interference optical system still produces significant systematic errors that need to be corrected before the data can be used directly in any whole boundary fitting method.

One approach to eliminate time-invariant baseline noise from the experimental interference data has been developed by W. Stafford (personal communication): instead of fitting $C(r, t)$ data as produced by the analytical ultracentrifuge, the differences $\Delta C/\Delta t$ are taken from the original data and fitted to the appropriate simulated differences. Let C^* be an interference scan composed from the sedimentation pattern of the solutes in the sample, $C(r, t)$, and a time-invariant error vector $\mathbf{E}$, which is superimposed onto the real data:

$$C^*(r, t_i) = C(r, t_i) + \mathbf{E} \tag{61}$$

Then the time derivative is approximated by the difference:

$$\begin{aligned}\Delta C(r, t)/\Delta t &= [C^*(r, t_i) - C^*(r, t_{i+1})]/[t_{i+1} - t_i] \\ &= [C(r, t_i) - C(r, t_{i+1})]/[t_{i+1} - t_i]\end{aligned} \tag{62}$$

and the time-invariant error vector vanishes. Although the random error will be slightly increased by this method, it can be used both for finite element fits as well as for fits to analytical solutions of approximated Lamm equations, as long as the radially invariant baseline offsets are properly adjusted.

The main disadvantage of finite element analysis is the significant computational expense involved in repeated function evaluations, as required by iterative nonlinear least-squares fitting and Monte Carlo simulations for confidence interval determinations. Approximated analytical solutions of the Lamm equation can also yield reliable information when used judiciously. To determine the accuracy of parameters derived from fits using approximated solutions, the data that are fitted need to be known *a priori.* To make this comparison, finite element simulations were used to generate data with known input values, which were then fitted with the approximated analytical solutions. Random Gaussian distributed noise with a standard deviation of 1% of the total concentration was added to the data to simulate experimental conditions. The parameters obtained from these data when fitted to Eqs. (42), (43), (46), and (47) are shown in Tables I–V. The residuals for fits to Eq. (47) are shown in Figs. 2 and 3.

Concentration profiles with stable plateaus can be analyzed by Faxén type solutions. When using standard double sector cells, Eqs. (42), (43) and (46) are most appropriate. For synthetic boundary experiments Eqs. (32), (34), and (36) can be used. Samples with larger sedimentation coefficients and smaller diffusion coefficients are adequate for this type of analysis. For concentration profiles with unstable plateaus, Eq. (47) should be used, which better models the concentration distribution at the bottom of the cell. In the solution of this equation, the entire boundary is described by the superposition of two curves based on two separate solutions, which

TABLE I
FITTING RESULTS OF SIMULATED CLAVERIE DATA CONTAINING 1% NOISE USING APPROXIMATE ANALYTICAL SOLUTIONS OF LAMM EQUATION FOR SMALL MOLECULE[a]

Eq.	C_0	s (S)	D (F)	r_m (cm)	r_b (cm)
(42)	1.0000	1.9861	9.9714	5.9000	—
(43)	0.9999	1.9893	9.9750	5.9000	—
(46)	1.0001	2.0112	10.979	5.9004	—
(47)	0.99874	2.0147	10.103	5.8991	7.2003

[a] Input parameters: C_0 = 1.0 OD, s = 2 S, D = 10 F (10^{-7} cm^2/s), speed = 50 krpm, r_m = 5.9 cm, r_b = 7.0 cm [for Eqs. (42), (43), and (46)] or 7.2 cm [for Eq. (47)]. Run times are as follows: 1 scan at 400 sec and 18 scans with a time interval of 800 sec.

TABLE II
FITTING RESULTS OF SIMULATED CLAVERIE DATA CONTAINING 1% NOISE USING APPROXIMATE ANALYTICAL SOLUTIONS OF LAMM EQUATION FOR APPROACH-TO-EQUILIBRIUM CONDITIONS[a]

Eq.	C_0	s (S)	D (F)	r_m (cm)	r_b (cm)
(42)	1.0002	0.48533	19.541	5.9015	—
(43)	0.99994	0.48953	19.629	5.9009	—
(46)	1.0002	0.51232	21.272	5.8996	—
(47)	0.99799	0.50750	20.175	5.8991	7.2023

[a] Input parameters: C_0 = 1.0 OD, s = 0.5 S, D = 20 × 10^{-7} cm^2/s, speed = 60 krpm, r_m = 5.9 cm, r_b = 6.7, noise 0.01 OD [for Eqs. (42), (43), and (46)] or 7.20 cm [for Eq. (47)]. Run times are as follows: 1 scan at 1000 sec and 18 scans with a time interval of 1000 sec. When using Eqs. (42), (43), and (46), it is advisable to consider only the early traces to improve the reliability of the results.

TABLE III
FITTING RESULTS OF SIMULATED CLAVERIE DATA CONTAINING 1% NOISE USING APPROXIMATE ANALYTICAL SOLUTIONS OF LAMM EQUATION FOR LARGE MOLECULE[a]

Eq.	C_0	s (S)	D (F)	r_m (cm)	r_b (cm)
(42)	0.99999	23.953	3.0066	5.8999	—
(43)	0.99991	23.966	2.9978	5.8999	—
(46)	0.99998	24.040	3.4052	5.9001	—
(47)	0.99856	24.003	3.0364	5.8998	7.2002

[a] Input parameters: C_0 = 1.0 OD, s = 24 S, D = 3 × 10^{-7} cm^2/s, speed = 15 krpm, r_m = 5.9 cm, r_b = 7.0 cm [for Eqs. (42), (43), and (46)] or 7.20 cm [for Eq. (47)]. Run times are as follows: 1 scan at 1000 sec and 18 scans with a time interval of 1000 sec.

TABLE IV
EFFECT OF COLUMN HEIGHT ON FITTING RESULTS OF SIMULATED CLAVERIE DATA CONTAINING 1% NOISE (DOUBLE SECTOR CELL EXPERIMENT) USING EQ. (47)[a]

Column height (cm)	Baseline	s (S)	D (F)	r_m (cm)	r_b (cm)
0.35	0.010440	0.36567	15.345	6.8915	7.1977
1.30	−0.0005412	0.50705	20.177	5.8990	7.2024

[a] Input parameters: $C_0 = 0.3$ OD, $s = 0.5$, S, $D = 20 \times 10^{-7}$ cm^2/s, speed = 60 krpm, radius range 6.85–7.20 cm or 5.9–7.2 cm. Run times are as follows: 1 scan at 1000 sec and 18 scans with a time interval of 1000 sec. Obviously, short column experiments are not suitable for the analysis with approximate analytical solutions of the Lamm equation. Results are improved when only the early concentration profiles are used for the fit.

TABLE V
EFFECT OF SPEED ON FITTING RESULTS OF SIMULATED CLAVERIE DATA CONTAINING 1% NOISE (DOUBLE SECTOR CELL EXPERIMENT) USING EQ. (47)[a]

RPM	C_0	s (S)	D (F)	r_m (cm)	r_b (cm)
60,000	0.74678	12.920	3.0422	5.9008	7.2002
15,000	0.74350	13.037	3.0217	5.8999	7.2001

[a] Input parameters: $C_0 = 0.75$, $s = 13$ S, $D = 3.0 \times 10^{-7}$ cm^2/s, $r_m = 5.9$ cm, $r_b = 7.2$ cm. Run times are as follows: 1 scan at 100 sec and 18 scans with a time interval of 200 sec (speed = 60 krpm); 1 scan at 200 sec and 18 scans with a time interval of 2400 sec (speed = 15 krpm).

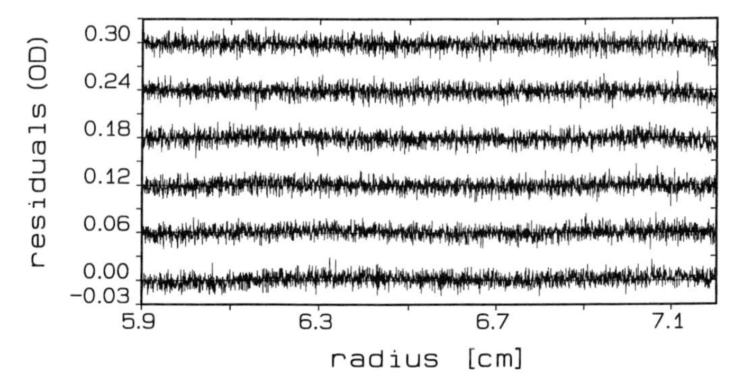

FIG. 2. Residual plots from fits using Eq. (47) to finite element generated data as shown in Table II. Parameters are $s = 0.5$ S and $D = 20 \times 10^{-7}$cm^2/s. From 18 curves considered for the parameter estimation only each third is presented. Times of curves (in seconds, from top to bottom) are 1000, 4000, 7000, 10,000, 13,000 and 16,000.

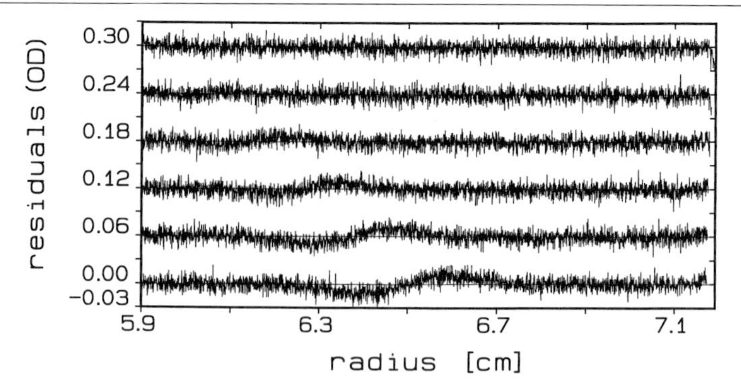

FIG. 3. Residual plots from fits using Eq. (47) to finite element generated data as shown in Table III. Parameters are $s = 24$ S and $D = 3 \times 10^{-7}$ cm^2/s. From 18 curves considered for the parameter estimation only each third is presented. Times of curves (in seconds, from top to bottom) are 1000, 4000, 7000, 10,000, 13,000, and 16,000.

are treated independently with respect to the boundary conditions. The first curve represents the change of the solute concentration in the meniscus region and the plateau, and considers the boundary condition at the meniscus (a solution of Faxén type). The second curve considers the increase of the solute concentration near the bottom and fulfills the boundary condition at the bottom of the cell.[23] Consequently, it is required that the slope of the first curve is zero near the bottom of the cell and the slope of the second curve is zero at the meniscus. Only if these conditions are satisfied can independence be guaranteed.

These requirements are generally fulfilled if long columns and moderate speed are used, and the sedimentation coefficient of the solute is not too small. Therefore, it is preferable to include earlier scans in the experiment which still display a stable plateau. As long as these conditions are observed, even small molecules with a molecular mass as low as 1 kDa can be successfully analyzed.[23] However, for experiments where these conditions cannot be fulfilled, it is recommended that finite element solutions be used instead, where these conditions do not apply. In addition to the sedimentation and diffusion coefficients, the loading concentration, baseline absorbance, and the radii at the meniscus and the bottom of the cell can be determined with Eq. (47). However, the baseline absorbance is most reliably determined only for high molecular weight solutes that clear the meniscus in late scans.

Clearly, for all methods under consideration it can be stated that the accuracy of the estimated parameters depends on the signal-to-noise ratio. Especially for poor signal-to-noise ratios, the results can be improved if additional data points are included in the fitting procedure. Additional data points can be obtained either by including additional scans and/or by decreasing the radial step size. All fits should result in residuals that are randomly distributed about the mean. If the residuals display deviations that are symmetric about the center of the boundary, it can be assumed that the accuracy of the sedimentation coefficient is only marginally affected, while the diffusion coefficient generally may be less accurate (see Table III). Systematic deviations are more frequently found in later scans where the boundary is near the bottom of the cell. Since the approximate solutions model the bottom of the cell less efficiently, deviations near the bottom of the cell are more obvious, but generally do not affect the accuracy of the solution significantly. However, if the residuals become increasingly asymmetric, the results are generally meaningless and a change in the model is required.

In contrast to the finite element method, the approximate solutions of the Lamm equation are not suited for determining concentration dependencies of the solute. The analysis of multicomponent systems is generally recommended only when the components are well separated, move in separate boundaries, and can be treated as independent fractions with a clear plateau between them. For such systems the Faxén type solutions [Eqs. (42), (43), and (46)] are suitable. Best results are obtained when only those scans are used for fitting where the midpoint of the boundary has not passed the halfway point of the solution column and where a stable plateau concentration is still visible. The main advantage for using approximate analytical solutions when compared to finite element solutions is the shorter computing time required for functional evaluations. The evaluation of the error functions is the most time-consuming part of the fitting process, but an effective method developed by Gautschi[25] can be used to reduce the computing time required for the calculation of the error function. An ALGOL version of the Gautschi algorithm is given in the appendix. Software for the analysis of sedimentation data with the methods mentioned in this article can be downloaded from the Internet at http://www.biochem.-uthscsa.edu/auc.

Appendix

The Gautschi[25] method for calculation of the complementary error function erfc simplified for real arguments exclusively.
function gautschi(x);

```
value x;
begin
integer i,k,q; real a,d,g,h,h1,h2,h3;
if x< 4.29 then
begin
```

```
h1: = 1 −x/4.29;
h: = 1.6*h1; h2: = h + h;h3: = h + x;
k: = 7 + 23*h1; I: = 9+21*h1;
a: = 1; for q: = 2 step1 to k do a: = a*h2;
d: = 0; g: = 0.5/h3;
for q: = I step −1 to k + 1 do g: = 0.5/(h3 + q*g);
for q: = k step −1 to 1 do
begin
g: = 0.5/(h3 + q*g);
d: = g*(a + d);
a: = a/h2;
end;
gautschi: = 1.12837916709551*d;
end else begin
g: = 0.5/x;
for q: = 8 step −1 to 1 do g: = 0.5/(x + q*g);
gautschi: = 1.12837916709551*g;
end;
end;
if x<0 then erfc(x): = 2-exp(−x^2)*gautschi(abs(x)) else erfc(x): = exp(−x^2)*gautschi(x);
```

[4] Sedimentation Velocity Analysis of Macromolecular Assemblies

By LENNY M. CARRUTHERS, VIRGIL R. SCHIRF, BORRIES DEMELER, and JEFFREY C. HANSEN

Introduction

The power of sedimentation velocity experiments performed in the analytical ultracentrifuge originates from the ability of this approach to directly and precisely measure sedimentation coefficients (s) of macromolecules under a wide variety of solution conditions. The use of this technique has been greatly facilitated in recent years by the availability of state-of-the-art analytical ultracentrifuges that detect the sample concentration in real time during sedimentation using absorbance or interference optical systems, and subsequently acquire the data in digital format. Historically sedimentation velocity experiments have not proven particularly useful for analysis of structurally complex macromolecular assemblies because, by definition, macromolecular assemblies inherently consist of multiple individual components. Furthermore, the function of a macromolecular assembly invariably is linked to either conformational changes of the intact assembly and/or association and dissociation of the individual components under

0076-6879/00 $30.00

specific solution conditions. Because both of these situations produce samples with significant sedimentation coefficient heterogeneity, the boundary spreading that occurs during a sedimentation velocity experiment due to diffusion (see below) often masks the presence of the multiple conformations and/or components, making analysis by traditional methods virtually impossible.[1,2]

In 1978, van Holde and Weischet[3] developed a data analysis method that removes the contributions of diffusion from sedimentation velocity boundaries, yielding an integral distribution of diffusion-corrected sedimentation coefficients, $G(s)$. This method originally was developed as a diagnostic for sample homogeneity/heterogeneity. However, it is now clear that $G(s)$ distributions not only permit the identification of sedimentation coefficient heterogeneity, but also allow one to exploit sample heterogeneity to help elucidate the structure–function relationships of complex multimolecular systems.[4–12] With this in mind, this chapter first provides a brief overview of the theory that underlies the $G(s)$ data analysis method. We then describe the steps required to (1) best perform a sedimentation velocity experiment that will be analyzed by this method, (2) correctly analyze raw absorbance and interference optical data using the public domain UltraScan II software program, and (3) generate and interpret $G(s)$ plots. This information will allow any researcher with access to one of the new generation analytical ultracentrifuges to characterize the solution properties and behavior of multimolecular systems whose properties otherwise would be too complex for productive analysis by sedimentation velocity.

Theory

In a sedimentation velocity experiment, application of sufficient centrifugal force leads to movement of solute molecules toward the bottom of

[1] J. C. Hansen, J. Lebowitz, and B. Demeler, *Biochemistry* **33,** 13155 (1994).

[2] J. C. Hansen, J. I. Kreider, B. Demeler, and T. M. Fletcher, *Methods* **12,** 62 (1997).

[3] K. E. van Holde and W. O. Weischet, *Biopolymers* **25,** 1981 (1978).

[4] J. A. Mendoza, B. Demeler, and P. M. Horowitz, *J. Biol. Chem.* **269,** 2447 (1994).

[5] R. H. Behal, M. S. DeBuysere, B. Demeler, J. C. Hansen, and M. S. Olson, *J. Biol. Chem.* **269,** 31372 (1994).

[6] J. C. Hansen and D. Lohr, *J. Biol. Chem.* **268,** 5840 (1993).

[7] A. Musatov and N. C. Robinson, *Biochemistry* **33,** 13005 (1994).

[8] H. T. H. Beernink and S. W. Morrical, *Biochemistry* **37,** 5673 (1998).

[9] P. M. Schwarz and J. C. Hansen, *J. Biol. Chem.* **269,** 16284 (1994).

[10] C. Tse and J. C. Hansen, *Biochemistry* **36,** 11381 (1998).

[11] C. Tse, T. Sera, A. P. Wolffe, and J. C. Hansen, *Mol. Cell. Biol.* **18,** 4629 (1998).

[12] L. M. Carruthers, J. Bednar, C. L. Woodcock, and J. C. Hansen, *Biochemistry* **37,** 14776 (1998).

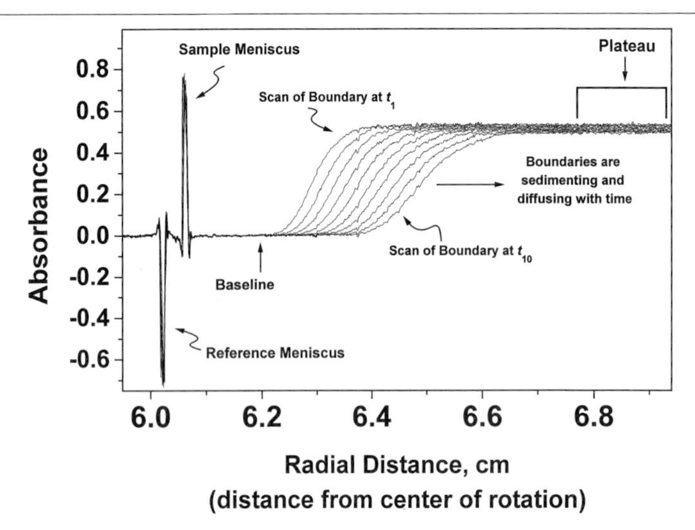

FIG. 1. Raw absorbance data acquired for a sedimentation velocity experiment. The formation and movement of the solute boundary is observed as absorbance plotted versus radial distance. Labeled are important features for subsequent data analysis. t_n represents time (i.e., t_1 is an earlier scan and t_{10} is a later scan).

the centrifuge cell. With time, movement of solute molecules away from the meniscus creates a solute concentration gradient, traditionally termed a *boundary* (Fig. 1). Hence, macromolecules in the presence of a centrifugal field not only will sediment, but also will diffuse. Mathematically, the combination of sedimentation and diffusion in the ultracentrifuge cell is described in terms of the flow, J:

$$J = s\omega^2 rc - D\frac{\partial c}{\partial r} \tag{1}$$

where s is the sedimentation coefficient, D is the diffusion coefficient, ω is the angular velocity ($\pi \cdot \text{rpm}/30$), r is the radial distance from the center of rotation, c is the solute concentration, and $\partial c/\partial r$ is the solute concentration gradient.[13] As alluded to in the Introduction, boundary spreading due to diffusion during a sedimentation velocity experiment has the potential to mask the presence of multiple components in a sample, and for such multicomponent samples it is highly advantageous to eliminate the effect of diffusion. As originally described by van Holde and Weischet,[3] the equation

[13] C. R. Cantor and P. R. Schimmel, "Biophysical Chemistry, Part II." Freeman, San Francisco, California, 1980.

that provides the basis for removing diffusion from sedimentation velocity boundaries is

$$s_{app} = s_{actual} - \frac{D^{1/2}}{r_0} \frac{2}{\omega^2} \Phi^{-1}(1 - 2w)t^{-1/2} \tag{2}$$

where t is time, D is the diffusion coefficient, r_0 is the meniscus position, ω is the angular velocity, Φ is the error function, and w represents $C(r)/C_p$, where $C(r)$ is the concentration at a selected radius and C_p is the plateau concentration. Importantly, for any given sedimentation velocity experiment, Eq. (2) can be rearranged and simplified into the following form:

$$s_{app} = s_{actual} - \frac{(D^{1/2} C)}{t^{1/2}} \tag{3}$$

where C represents an amalgam of the terms that are constant for that particular experiment. As can be seen from Eq. (3), at infinite time the term containing D will become equal to zero and s_{app} will become equal to s_{actual}. Furthermore, Eq. (3) is in the form of $y = mx + b$.

The general basis of the van Holde and Weischet analysis method is as follows. Initially one divides each boundary into 20–50 equally spaced fractions from top to bottom (i.e., along the concentration axis). Equation (2) is used to calculate apparent s (s_{app}) at the equivalent boundary fraction for each scan, and the s_{app} are plotted against $t^{-1/2}$ (see below, Fig. 4). A linear regression is used to globally fit the s_{app} at each equivalent boundary fraction, and the s_{actual} is obtained from the intersection of the linear regression with y intercept (i.e., the s_{app} at infinite time). Finally, the integral distribution of sedimentation coefficients, $G(s)$, is obtained by plotting the boundary fraction versus s_{actual} (see below, Fig. 5).

Figure 2 shows a general flowchart illustrating the steps required to carry out a sedimentation velocity experiment that will be analyzed by the method of van Holde and Weischet.[3] Details of sample and centrifuge cell preparation have been described elsewhere.[14,15] The following sections are designed to guide a user through (1) setting up the data acquisition program, (2) carrying out the sedimentation velocity experiment optimally, and (3) editing and analyzing the acquired data for the purposes of obtaining extrapolation plots and $G(s)$ distributions.

[14] J. C. Hansen and C. Turgeon, *in* "Methods in Molecular Biology: Chromatin Protocols" (P. Becker, ed.), p. 127. Humana Press, New Jersey, 1999.

[15] Beckman XL-A/XL-I Manual, Palo Alto, California.

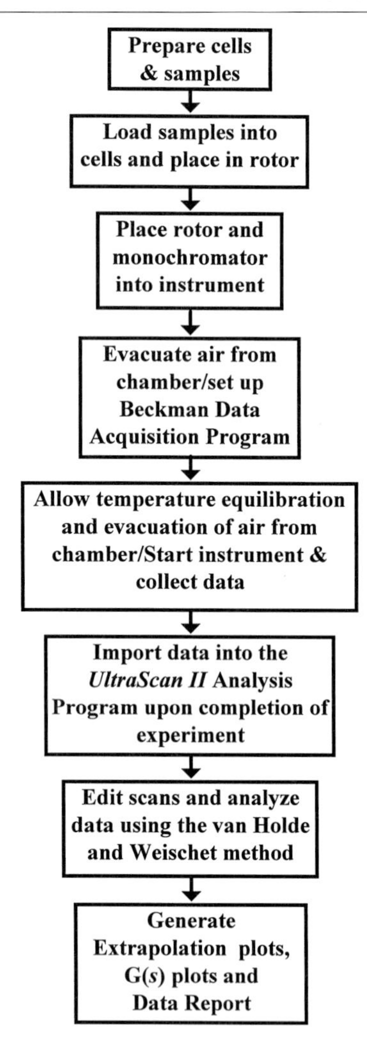

FIG. 2. Flowchart illustrating the steps for a sedimentation velocity experiment. Shown here is a simplified outline of the steps involved in preparing, executing, and analyzing a sedimentation velocity experiment.

Performing and Optimizing Sedimentation Velocity Experiment

First, we set up the experiment using the Beckman user interface.

1. Assemble the cells, fill the sample and reference sectors, insert the cells into the rotor, and place the rotor and monochromator in the analytical ultracentrifuge chamber as described.[14,15]

2. Start the Beckman Acquisition and Analysis program (note that the following steps are for version 3.01h). Click on FILE, and then ⟨New File⟩ from the pull-down menu. A parameter box appears showing a default file called Noname*.scn, where * is a sequential number. All subsequent run parameters (i.e., steps 3–8) are specified from this box.

3. At the top center of the parameter box are general run parameters. Select the type of rotor (four or eight hole), time, speed, and temperature for each run.

4. The parameters for each cell are located underneath the general run parameters. Click on Velocity and then Absorbance, Interference, or both depending on which optical system will be used to collect the data. The R_{max} default setting is 5.8 cm, and the R_{min} is 7.3 cm; change the latter to 7.2 cm (with proper radial calibration of the instrument, the inside radius is 5.85 cm and the outside radius is 7.15 cm). Select the wavelength to be used in conjunction with the absorbance optical system (the choice of wavelength will be influenced by the sample and buffer conditions being utilized). It is important to note that 1.2 cm is the path length of the cells used in an XL-A/XL-I. Finally, a sample and buffer description are entered in the Comment box.

5. Next, click on the ⟨Detail⟩ button to the right of the Comment box. If desired, second and third wavelengths may be chosen from the section on the left-hand side of the box under Absorbance. Note, the instrument may not reset the wavelength accurately between scans (i.e., it may be off by 1–2 nm). This is not a problem if the sample has a broad peak at the wavelength selected, but it could present a problem if the peak is narrow. In the latter case, the recommended procedure is to perform separate velocity runs at the two different wavelengths. Continuing with the Detail box settings, the recommended Radial Step Size is 0.001 cm and the number of Replicates is 1. The scan mode is set to Continuous. Standard two-channel centerpieces should be used. The directory where the data are to be stored is entered under Data. If only the absorbance optical system is being used, click OK to save and exit. To expedite copying the information from the cell 1 setup to the other cell setups, click on All Settings Identical To Cell 1, which will copy the same information to the other cell setups. Click on it again to allow modifications.

6. If Interference also was chosen on the main setup page for that particular cell, click on Laser Setup. Set the Laser Duration (generally 0.5° to 0.6°), the Brightness (0 or 1), and the Contrast (127 is recommended by Beckman). Make sure that the Simulate rpm is off; otherwise, you will see simulated fringes and not those of the sample. For most samples where no boundary is formed at low speeds, the instrument is set to spin at 3000 rpm and the Laser Delay is either adjusted manually or by using Auto Adjust

Laser Delay in order to optimize the fringes. If large molecules which require low speeds are run, use buffer to set the Laser Delay and then replace with the samples. There is approximately 90° between holes in the four-hole rotor and 45° between holes in the eight-hole rotor. Once the laser settings are completed, the rotor is stopped.

7. Now click the ⟨Options⟩ button to the left of the general run parameters and an option box should appear. For all absorbance runs, it is preferential to perform a radial calibration before the first scan and, in most runs, you select the option to STOP the XL-A/XL-I after the last scan. The Overlay last ____ scan(s) may vary from 1 to 20 and does not influence the data acquired; however, the higher number of scans overlaid may deplete computer RAM during the run. Click on OK.

8. Now click on the ⟨Methods⟩ box. Delayed start is for scanning only and is in minutes. The Time Between Scans is in hours : minutes : seconds. The Number of Scans is user/sample dependent. Each absorbance scan is acquired in approximately 3-min intervals, whereas each interference scan is acquired in approximately 10 sec. It is better to overestimate the number of scans because scanning can be stopped manually; however, if the number of scans is underestimated, no additional scans can be added.

9. If the Noname*.scn is to be used in subsequent runs, save the file with a descriptive name (e.g., chromatin.scn). The saved file can be recalled for future usage by clicking on FILE and ⟨Open⟩, then selecting the desired file that contains all the saved experimental parameters.

10. Once the instrument is set correctly, click on Start Method Scan button to the right of the general parameters in the main parameter box. Wait at least 1 hr to allow complete evacuation of the chamber and temperature equilibration of the samples. Press the Enter button on the XL-A/XL-I console and press Start.

11. The data are collected in a subdirectory with a specific path; e.g., /xlawin/(user)/(date)/(time run started). Absorbance scan files are named for the number of the scan with an extension that describes what type of scan (*.ra for radial absorbance or *.ip for interference) and the cell that was scanned. Thus, file 0001.ra1 would be the first radial absorbance scan file of cell 1.

12. For a detailed description of the optimal run parameters to be used for a sedimentation velocity experiment that is to be analyzed by the $G(s)$ method, refer to a previous report.[16]

[16] B. Demeler, H. Saber, and J. C. Hansen, *Biophys. J.* **72,** 397 (1997).

Editing and Analysis of Sedimentation Velocity Data Using UltraScan II for UNIX

After the experiment is completed, the next step is to edit and analyze the acquired sedimentation velocity data. These steps must be performed properly in order to avoid misinterpretation of the resulting $G(s)$ plots. As of this writing, the only software program available for editing and analysis of sedimentation velocity data by the method of van Holde and Weischet is UltraScan II for UNIX, which has been developed by Dr. Borries Demeler. This program is in the public domain and can be downloaded at http://ftp.biochem.uthscsa.edu/UltraScan/. A detailed description of UltraScan II and an on-line tutorial of the van Holde and Weischet analysis can be found at links within http://biochem.uthscsa.edu/auc/ultrascan/.

Prior to analysis of either absorbance or interference data, the UltraScan II program configuration settings must be set. This is accomplished by using the mouse to click on the FILE pull-down menu in the toolbar of the UltraScan II program and selecting ⟨Configuration⟩. The configuration setup allows one to enter the desired path and file names for the WWW browser, the tar archiver and gzip compression utilities, the data directory, the help directory, the UltraScan root directory, the archive directory, and the result directory. In addition, from within the ⟨Configuration⟩ box the user is able to set temperature tolerances, time corrections if necessary, and whether the scan file extension is in capital letters.

Absorbance Data

1. For both absorbance and interference data, the UltraScan II program has step-by-step instructions for the user that are located at the bottom right corner of the screen as well as a Help button that directs the user to a specific URL in a chosen Internet browser. Both can be used to assist and guide the user through the process of analyzing the sedimentation velocity data.
2. From the EDIT pull-down menu in the toolbar, first select ⟨Velocity⟩ then the Absorbance Data option. Choose the directory containing the data to be analyzed using either the Enter keystroke or by double clicking on the appropriate directory. The program subsequently analyzes the file architecture and loads the files.
3. The properties of the raw data are now shown on the screen as plotted values of temperature, time, rotor speed, and scan number. These data should be checked carefully to make sure they fall within acceptable limits of variation (e.g., $\pm 0.5°$) or have not been recorded incorrectly (e.g., sometimes a zero gets dropped from the last column of the scan times).

Enter a unique run identification number, click the Accept button, and proceed to the data edit screen.

4. Clicking on Start Editing will open the data files for cell 1 and plot them on the screen (see Fig. 1). The radial position of the sample meniscus (r_0) should be identified by using the mouse to click first to the left and then to the right of the meniscus peak as prompted by the program. The program automatically detects the maximal absorbance in this range and designates it as r_0. The next step is to select the range of data to be included in the analysis. This is done by using the mouse to first define the left limit (which should be just to the *right* of the sample meniscus) followed by clicking the mouse near the bottom of the cell to define the right limit.

5. If necessary, flaws in the scans (e.g., absorbance spikes) can be corrected or entire scans can be excluded from the analysis. Next, click on a region of the data that defines a true plateau for *all* the scans. It is important to avoid choosing a region where the plateau absorbencies of the latter scans are sloping upward; this is not a true plateau region and the subsequent $G(s)$ plot will be adversely affected.[16] Finally, use the mouse to select the region that best delineates the baseline absorbance of the scans. Note that if more than one data set is to be edited (i.e., more than one cell was included in the run), the program will prompt the user to repeat each of the steps described immediately above in steps 4 and 5 for each new set of scans. When all available data sets are edited, the program will indicate that editing is complete.

6. Return to the main screen, click on the VELOCITY command from the toolbar, and select ⟨van Holde–Weischet⟩ analysis from the pull-down menu.

7. After the new screen appears, click on the Load Data button and select the file that contains the desired edited data.

8. After the chosen data have been loaded, the program automatically displays in separate windows the edited data, the extrapolation plot (see Fig. 4 later), and the experimental parameters for the first edited data set; parameters can be adjusted accordingly as described in the following steps.

9. Enter the partial specific volume for the solute, as well as the density and viscosity of the buffer for that run, which may be calculated using the program's built in utility module. UltraScan II will calculate the partial specific volume of a protein from its amino acid sequence, which can either be entered as a file or directly downloaded by the program from the Expasy Swiss-Prot database.

10. Set the smoothing factor (1–50) *prior* to selecting the number of boundary divisions, percentage of the boundary to be analyzed, and scan(s) to be excluded from the analysis.

11. The $G(s)$ plot (also referred to as the integral distribution plot) and the data report (Fig. 3) containing all the information relating to the edited experiment subsequently are generated by clicking on the respective toolbar buttons.

12. A printout of the extrapolation plot is made by simply clicking on the print data button. To print a $G(s)$ plot choose UTILITIES from the pull-down menu in the toolbar and select ⟨Combine Dist. Plots⟩ and then click on the Print Data button. Finally, a printout of the data report (Fig. 3) is made by clicking on the data report button to open the data report and then pressing the Print Data button.

Interference Data

The ability of UltraScan II to analyze interference data by the $G(s)$ method is still in its early stages. As they become available, additional changes and improvements in the protocol will be incorporated into the latest version of the program, complete with corresponding on-line help.

1. Editing interference data is the same as described above in steps 1–3 for absorbance data except that after clicking on EDIT in the toolbar, ⟨Interference⟩ is selected. Once the appropriate directory is opened and the data assigned a unique file name, select Start Editing. A plot of the scans derived from the interference fringes is displayed for data set 1. Designate the sample meniscus, referring to the program's step-by-step instructions for details on zooming in on the sample meniscus to assist in its identification.

2. Click on the left and right boundaries of the air-to-air region to align the scans, then use the mouse to define the left- and right-most limits of the data range.

3. An optional step includes subtracting a baseline from the data set (if one is available).

4. The remainder of the editing and analysis procedure is the same as described above in steps 5–12 for editing and analyzing absorbance data.

Data Interpretation

A $G(s)$ plot is obtained from its corresponding extrapolation plot (Fig. 4) by graphing the boundary fraction versus s_{actual}. Any given point on the $G(s)$ plot represents the fraction of the sample (as indicated on the y axis) that has an s_{actual} equal to or less than the value indicated on the x axis. As such, the $G(s)$ plot provides both a rigorous diagnostic for the degree of homogeneity/heterogeneity of the sample, as well as a means to interpret

```
***************************************************
*       van Holde - Weischet Analysis          *
***************************************************

Data Report for Run "001v", Cell 1, Wavelength 1

Detailed Run Information:

Cell Description:         208-12 nuc. arrays (TE)
Raw Data Directory:       /user/lenny/us/data/
Rotor Speed:              20000 rpm
Average Temperature:      21.1425 °C
Temperature Variation:    Within Tolerance
Time Correction:          0 minutes 36 seconds
Run Duration:             1 hours 33 minutes
Wavelength:               260 nm
Baseline Absorbance:      -0.000263636 OD
Meniscus Position:        6.091 cm
Edited Data starts at:    6.108 cm
Edited Data stops at:     7.136 cm

Hydrodynamic Settings:

Viscosity correction:         1
Viscosity (absolute):         0.972677
Density correction:           0.9987 g/ccm
Density (absolute):           0.998455 g/ccm
Vbar:                         0.650486 ccm/g
Vbar corrected for 20 °C:     0.65 ccm/g
Buoyancy (Water, 20 °C):      0.351148
Buoyancy (absolute):          0.350519
Correction Factor:            0.972534

Data Analysis Settings:

Divisions:                50
Smoothing Frame:          1
Analyzed Boundary:        95 %
Boundary Position:        2.5 %
Early Scans skipped:      0 Scans

Scan Information:

Scan: Corrected Time:  Plateau Concentration:
  1:    9 min 48 sec            0.578682 OD
  2:   24 min 31 sec            0.554527 OD
  3:   29 min 10 sec            0.568736 OD
  4:   33 min 52 sec            0.563282 OD
  5:   38 min 40 sec            0.547282 OD
  6:   43 min 17 sec            0.545655 OD
  7:   48 min  0 sec            0.550591 OD
  8:   52 min 38 sec            0.538555 OD
  9:   57 min 19 sec            0.535209 OD
 10:   62 min 11 sec            0.540036 OD
```

FIG. 3. Data report. Shown is typical output for a data report obtained after editing and analyzing a sedimentation velocity experiment using UltraScan II software. This document contains a description of the run information details, hydrodynamic settings, data analysis settings, and the individual scan information.

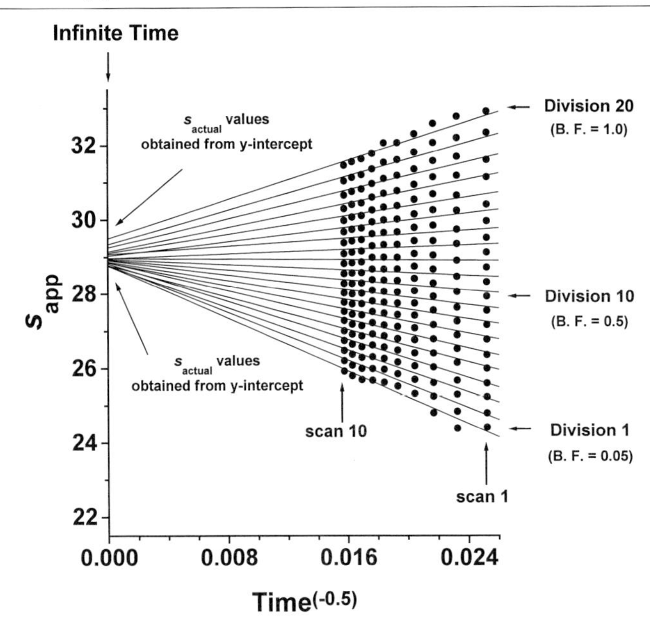

FIG. 4. Extrapolation plot. For each scan, the boundary is divided into at least 20 equal divisions (division 1 is at the bottom, 20 at the top). The apparent sedimentation coefficient (s_{app}), which is influenced by both sedimentation and diffusion [see Eq. (2)], is then calculated at each boundary division for each scan. The data are then plotted as the apparent s (s_{app}) at each division versus the inverse square root of time. Because sedimentation is proportional to the first power of time, whereas diffusion is only proportional to the square root of time, at infinite time the contribution of diffusion is factored out, thus the "true" or actual s is obtained at each boundary division from linear extrapolation to the y intercept. The y intercepts obtained from the extrapolation plot are subsequently graphed against the boundary fractions (B.F.) to obtain a $G(s)$ plot (Fig. 5).

the solution-state behavior for complex heterogeneous samples. The latter is what makes this method so useful for analysis of macromolecular assemblies.

Figure 5 shows $G(s)$ plots that are characteristic of a sample exhibiting nonideal behavior, i.e., concentration dependence of s (I), a single ideal component system (II), and a noninteracting sample containing multiple components (III). The origin of the effect of nonideality on the shape of the $G(s)$ plot has been described in detail.[16] The homogeneous sample yields a vertical $G(s)$ distribution since in this case the diffusion-corrected sedimentation coefficients at all points in the boundary are identical. In contrast, a heterogeneous sample will exhibit a broad distribution that has an increasing positive slope due to the presence of multiple sedimenting components. The shape of a heterogeneous $G(s)$ plot can take many forms

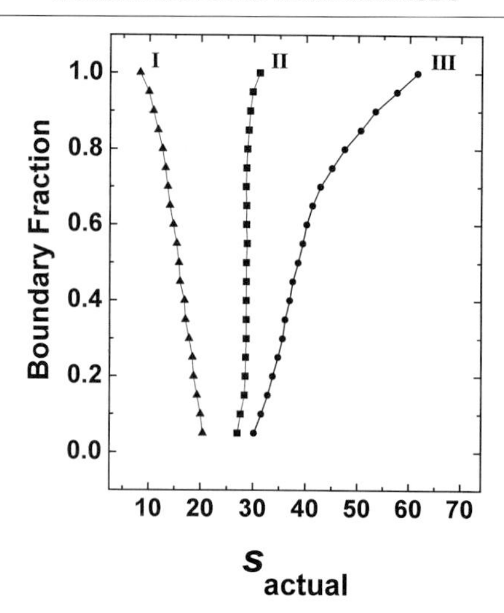

FIG. 5. $G(s)$ plot. The boundary fraction is graphed versus the data points taken from the extrapolation plot at infinite time. This plot is also referred to as an integral distribution plot. If the sample is nonideal, the point of convergence will be in front of the y axis, and the material at the bottom of the boundary will appear to sediment faster than the material at the top of the boundary (I). If the sample is homogeneous, the extrapolations from each boundary division will converge at the same point on the y axis and give rise to a vertical $G(s)$ plot (II). If the sample is heterogeneous, the extrapolations will intersect the y axis at multiple points, with the material at the top of the boundary sedimenting faster than the material at the bottom of the boundary (III).

depending on the composition of the sample (i.e., how many components are present, what their sedimentation and diffusion coefficient values are, and what fraction of the total sample consists of any given component).[2] Consequently, to allow the reader to best understand the utility of $G(s)$ plots for solving biological problems, several illustrative examples of how the van Holde and Weischet method has been used to characterize heterogeneous samples consisting of complex macromolecular systems are described briefly below.

Analysis of cpn60 Assembly, Disassembly, and Stability

In this study,[4] cooperativity of cpn60 chaperonin complex assembly was observed as the monomeric cpn60 species formed the active dodecameric complex. This active complex is made up of 14 identical subunits weighing 60 kDa each. In this case, the $G(s)$ plots showed that only discrete populations of monomers and 14-mers were present in the same sample during assembly. In contrast, urea-induced disassembly was noncooperative as judged by $G(s)$ plots indicating the presence of more than two components.

Analysis of Assembly Intermediates of the Pyruvate Dehydrogenase Multienzyme Complex

The assembly of the core subcomplex of pyruvate dehydrogenase was examined in this study.[5] This complex has a molecular mass of approximately 4.4 million daltons. The $G(s)$ plots demonstrated that several discrete assembly intermediates could be identified as well as showing that the assembly was noncooperative. On the basis of the $G(s)$ plots, these investigators were able to identify for the first time the previously unknown mechanism of assembly of this large multicomponent complex.

Examining Assembly and Structural Properties of Subsaturated Nucleosomal Arrays

Several studies have used sedimentation velocity experiments to probe the conformational changes of chromatin. In this case,[6] the assembly of chromatin reconstituted from purified defined length DNA and core histones was studied. The DNA template consists of 12 tandemly repeated 208-base-pair (bp) fragments and the core histone octamer is a protein complex containing two copies of four different proteins. The macromolecular assembly, referred to as a 208-12 nucleosomal array, consists of 96 proteins and ~2500-bp of DNA that have relative molecular mass of ~3.0 million daltons. By varying the input ratio of octamer, the $G(s)$ plots were able to confirm that the loading process of core histones onto a DNA lattice is not highly cooperative since the shift in s was uniform. More importantly regarding the study of heterogeneous samples, two of these samples that typically sediment homogeneously at 14 and 24, respectively, were mixed in a 1:1 ratio and a $G(s)$ plot was able to show that two distinct populations of samples exist (also refer to Ref. 16).

Analysis of Reversible Association and Aggregation of Detergent-Solubilized Cytochrome bc_1

In this study the monomer and dimer forms of cytochrome bc_1 were studied under varying conditions.[7] One cytochrome bc_1 monomer contains 11 different subunits, thus making this a complex macromolecular assembly weighing $\sim 2.35 \times 10^5$ Da. While the dimer form of this complex sediments homogeneously, when it was exposed to certain treatments, dissociation or aggregates of the dimer formed, causing broad, continuous heterogeneous $G(s)$ plots to be observed.

Analysis of uvsY Hexameric Assembly through Distinct Intermediates

The reversible self-association of uvsY hexamers was studied as a function of both salt and protein concentration.[8] In this study, stable homogeneous hexamers, sedimenting at 6S and weighing ~95 kDa were shown to self-associate as a function of increasing salt as evidenced by broad, $G(s)$ plots ranging from 6S to 19S. At sufficiently high salt, a discrete 19S oligomerized species was formed. The $G(s)$ plots also indicated that the hexamer multimers were undergoing rapid formation and dissociation.

Analysis of Salt-Dependent Folding of Nucleosomal and Chromatin Arrays

The method of van Holde and Weischet analysis has been integral in delineating the folding mechanisms of *in vitro* reconstituted nucleosomal and chromatin arrays.[6,9–12] $G(s)$ plots of nucleosomal arrays, which lack linker histones, were shown to have broad shapes indicating that many different conformations of the chromatin were being formed. This in turn indicated that folded nucleosomal arrays are unstable in the absence of linker histones. Since this study investigators have been systematically examining the structural and molecular determinants of nucleosomal array folding. Studying the contributions of the four core histone N termini to nucleosomal array folding was accomplished by assembling hybrid trypsinized nucleosomal arrays.[10] $G(s)$ plots subsequently allowed the identification of multiple functional roles of the core histone N termini in nucleosomal array folding. In another related study,[11] $G(s)$ plots showed differences in the extent of folding of nucleosomal arrays containing either acetylated or nonacetylated core histones, demonstrating that acetylation causes nucleosomal array decondensation. Finally, results from $G(s)$ plots played an integral role in delineating linker histone contributions to chromatin structure.[12] First, histone H5 was incorporated into 208-12 nucleosomal arrays to form chromatin. In both cases, the nucleosomal and chromatin arrays had to be purified, which was monitored using the $G(s)$ plots. Finally, the salt-dependent folding of nucleosomal and chromatin arrays could be directly compared by examining the $G(s)$ plots. The results showed that linker histones stabilize chromatin folding because at each equivalent fraction of the $G(s)$ plot chromatin exhibited a higher s value than the nucleosomal arrays. Only the method of van Holde and Weischet could elucidate the complexity of these samples since many different kinds of heterogeneous distributions were observed in these studies.

[5] Analysis of Weight Average Sedimentation Velocity Data

By JOHN J. CORREIA

Introduction

This chapter presents a detailed discussion of the analysis and fitting of weight average sedimentation coefficient data for reversibly interacting systems. It is well established that protein–protein interactions are important for the formation of functional oligomeric enzymes and structural proteins. The energetics of these interactions provide essential information required to understand the regulation of activity. The number of subunits, the presence of intermediates, the relationship to gain of function/activity, the role of ligands, the presence of cooperative structural and functional transitions, and the nature of the covalent/noncovalent forces involved are readily investigated by sedimentation techniques.[1–5]

The methods described here represent current applications of weight average analysis techniques that have been developed over four decades and are now appropriate for the new generation of Beckman analytical ultracentrifuges, the XLA/XLI. These techniques are highly appropriate for determination of the stoichiometry, mode of association, energetics, and thermodynamics of the interactions. A brief summary of previous applications with emphasis on important developments and the use of DCDT is presented first. Then rules to guide the investigator, practical issues, and complexities are discussed followed by a section on the details of developing a model and a fitter. Finally a brief consideration of active and future developments is presented.

Summary of Previous Applications

The majority of cases where weight average sedimentation coefficient data has been applied to macromolecular association come from the laboratories of Serge Timasheff and Jim Lee. Timasheff and co-workers used this technique to investigate self-association of tubulin induced by magnesium

[1] R. W. Oberfelder, T. C. Consler, and J. C. Lee, *Methods Enzymol.* **117,** 27 (1985).
[2] L. K. Hesterberg and J. C. Lee, *Methods Enzymol.* **117,** 96 (1985).
[3] G. C. Na and S. N. Timasheff, *Methods Enzymol.* **117,** 459 (1985).
[4] J. R. Cann, *Methods Enzymol.* **48,** 242 (1978).
[5] G. Kegeles and J. R. Cann, *Methods Enzymol.* **48,** 248 (1978).

0076-6879/00 $30.00

ions[6–8] and by vinca alkaloids.[9–13] Mg^{2+} induces tubulin to form rings in a nucleotide-dependent manner and the sedimenting boundary formed is bimodal at high concentrations consistent with a Gilbert type monomer–N-mer association scheme.[6–8] This system also includes appreciable concentrations of intermediates. (The two papers by Frigon and Timasheff[6,7] are remarkable for their complete hydrodynamic and thermodynamic characterization of the system while serving as a manual for how to collect and analyze sedimentation velocity data and as a condensed source of references to many of the fundamental papers in the macromolecular characterization field.)

Vinca alkaloids are antimitotic, antineoplastic drugs that induce tubulin to form indefinite spiral polymers and paracrystals.[14] Timasheff and co-workers performed the seminal studies that provided the mechanistic basis for understanding this clinically significant ligand-induced self-association process.[3,9–13] Lobert, Correia, and co-workers have extended this work to a wider family of drug congeners, solution conditions, and tubulin preparations utilizing an XLA and the methods discussed in this review.[14–20] Lee and co-workers have used methodologies similar to those of Timasheff to investigate a number of self-association processes linked to enzyme activity. These include pyruvate kinase,[21] phosphofructokinase,[22–27] DNA-dependent RNA polymerase,[28] and the formation of a transcription factor III

[6] R. P. Frigon and S. N. Timasheff, *Biochemistry* **25,** 8292 (1975).
[7] R. P. Frigon and S. N. Timasheff, *Biochemistry* **25,** 8292 (1975).
[8] W. D. Howard and S. N. Timasheff, *Biochemistry* **25,** 8292 (1986).
[9] J. C. Lee, D. Harrison, and S. N. Timasheff, *J. Biol. Chem.* **250,** 9276 (1975).
[10] G. C. Na and S. N. Timasheff, *Biochemistry* **19,** 1347 (1980).
[11] G. C. Na and S. N. Timasheff, *Biochemistry* **25,** 6222 (1986).
[12] V. Prakash and S. N. Timasheff, *Biochemistry* **24,** 5004 (1985).
[13] V. Prakash and S. N. Timasheff, *Biochemistry* **30,** 873 (1991).
[14] S. Lobert and J. J. Correia, *Methods Enzymol.* **323,** [4], in press.
[15] S. Lobert, A. Frankfurter, and J. J. Correia, *Biochemistry* **34,** 8050 (1995).
[16] S. Lobert, B. Vulevic, and J. J. Correia, *Biochemistry* **35,** 6806 (1996).
[17] S. Lobert, C. A. Boyd, and J. J. Correia, *Biophys. J.* **72,** 416 (1997).
[18] B. Vulevic, S. Lobert, and J. J. Correia, *Biochemistry* **42,** 12828 (1997).
[19] S. Lobert, A. Frankfurter, and J. J. Correia, *Cell Motil. Cytoskel.* **39,** 107 (1998).
[20] S. Lobert, J. W. Ingram, B. T. Hill, and J. J. Correia, *Mol. Pharm.* **53,** 908 (1998).
[21] R. W. Oberfelder, L. L.-Y. Lee, and J. C. Lee, *Biochemistry* **23,** 3813 (1984).
[22] L. K. Hesterberg and J. C. Lee, *Biochemistry* **19,** 2030 (1980).
[23] L. K. Hesterberg and J. C. Lee, *Biochemistry* **20,** 2974 (1981).
[24] L. K. Hesterberg and J. C. Lee, *Biochemistry* **21,** 216 (1982).
[25] L. K. Hesterberg, J. C. Lee, and H. P. Erickson, *J. Biol. Chem.* **256,** 9724 (1981).
[26] M. A. Luther, H. F. Gilbert, and J. C. Lee, *Biochemistry* **22,** 5494 (1983).
[27] M. A. Luther, G.-Z. Cai, and J. C. Lee, *Biochemistry* **25,** 7931 (1986).
[28] S. J. Harris, R. C. Williams, and J. C. Lee, *Biochemistry* **34,** 8752 (1995).

A-5S RNA complex.[29] A number of these papers also serve as excellent guides to conducting velocity sedimentation analysis of an interacting system and offer a general set of rules to follow throughout the process.[23,27]

The work on the RNA complex[29] is the first quantitative application of weight average sedimentation velocity techniques to a mixed associating system. As pointed out by Schachman 30 years earlier,[30] this is not a real weight average but an absorbance average since the two species have different extinction coefficients. (The experiments by Callaci *et al.*[29] were conducted with a scanner at 260 nm where the RNA contributes 94% of the OD in a 1:1 complex.) This method is to be compared with the work by Revzin and Von Hippel[31] where they used absorbance optics to measure weak protein/DNA binding curves by observing the absorbance in the supernatant after sedimentation. It was noted in the work of Callaci *et al.*[29] that more advanced techniques needed to be developed to understand heterologous systems and specifically indicated the need to monitor both the RNA and the protein species. Appropriately, Stafford has now developed methods for analyzing heteroassociating systems[32] that use a combination of simulating and fitting reactant concentrations, sedimentation coefficients of the complexes, and the shape of the reacting boundary. The absorbance option of this fitter (ABCDfitter) corrects for different extinction coefficients of reactants.

Use of DCDT and *g*(*s*) to Generate Weight Average Sedimentation Coefficients

A number of significant experimental differences exist between these previous studies and current methodologies. First, data scans are acquired directly by computer interface to the absorbance and/or interference optical system. Samples are spun at appropriate speeds to maximize the number of scans that can be used in the analysis. Typically at least 20 data scans are analyzed with absorbance data (and as many as 100 scans with interference data). To achieve this with the absorbance system, we collect velocity data at a radial spacing of 0.002 cm with 1 average in a continuous scan mode. This replaces time-consuming mechanical reading of the Schlieren patterns utilized by Lee and Timasheff. Second, rather than calculate the second moment of the reacting boundary, data are analyzed using DCDT,

[29] T. P. Callaci, G.-Z. Cai, J. C. Lee, T. J. Daly, and C. Wu, *Biochemistry* **29,** 4653 (1990).

[30] H. K. Schachman, "Ultracentrifugation in Biochemistry." Academic Press, New York, 1959.

[31] A. Revzin and P. H. Von Hippel, *Biochemistry,* **16,** 4769 (1977).

[32] W. F. Stafford III, *in* "Modern Analytical Ultracentrifugation" (T. M. Schuster and T. M. Laue, eds.), pp. 119–137. Birkhauser, Boston, 1994.

software that generates a distribution of sedimentation coefficients, $g(s^*)$.[32–35] These patterns are equivalent to Schlieren patterns being derived from slopes dc/dr converted to time derivatives dc/dt and finally sedimentation coefficient distributions dc/ds^* by the numerical equivalent of the chain rule. These distributions are then converted to a weight average sedimentation coefficient by integration in a spread sheet (e.g., Origin) or directly by DCDT.

$$\bar{s}_{20,w} = \int g(s^*)s^*ds / \int g(s^*)ds \tag{1}$$

Newer versions of DCDT (DCDT_30z or _60z) also output the uncertainty in $\bar{s}_{20,w}$, which can be appropriately used in the fitting. The advantage of using DCDT is the ease of analysis and the dramatically improved signal-to-noise ratio derived from averaging multiple pairs of scans. (Beckman-provided software will also calculate a second moment position but there has been no direct systematic comparison using this second moment program versus the weight average method described here. From our limited experience this second moment software will give less precise results with absorbance data while interference data analyzed with this method will require significant baseline corrections.[36])

Choosing the best scans for an analysis requires selecting scans pairwise from those that have cleared the meniscus up to scans that still retain a plateau. This is evident in the $g(s^*)$ because early scans will broaden the pattern and create an inversion near $s = 0$, whereas late scans will cut off the boundary before it comes back down to the baseline. It is essential that the $g(s^*)$ generated comes down to the baseline on both the centrifugal and the centripetal side of the boundary to ensure an accurate estimate of $\bar{s}_{20,w}$. We usually make a family of list files that select over various ranges of scans and then choose the best set by a graphical comparison of the *.gs0 or output files. Scans later in the run will give a narrower and taller $g(s^*)$ and thus an improvement in the resolution (Fig. 1). Due to radial dilution the area under the curve will decrease, but as long as the patterns come down to the baseline the $\bar{s}_{20,w}$ will be relatively unaffected within error (Fig. 1). The experimental weight average sedimentation coefficient corresponds directly to a molecular model derived from a distribution of discrete species.

[33] W. F. Stafford III, *Anal. Biochem.* **203,** 295 (1992).

[34] W. F. Stafford III, *in* "Analytical Ultracentrifugation in Biochemistry and Polymer Science" (S. E. Harding, A. J. Rowe, and J. C. Horton, eds.), pp. 359–353. Royal Society of Chemistry, Cambridge, UK, 1992.

[35] W. F. Stafford III, *Methods Enzymol.* **240,** 478 (1994).

[36] P. Schuck and B. Demeler, *Biophys. J.* **76,** 2288 (1999).

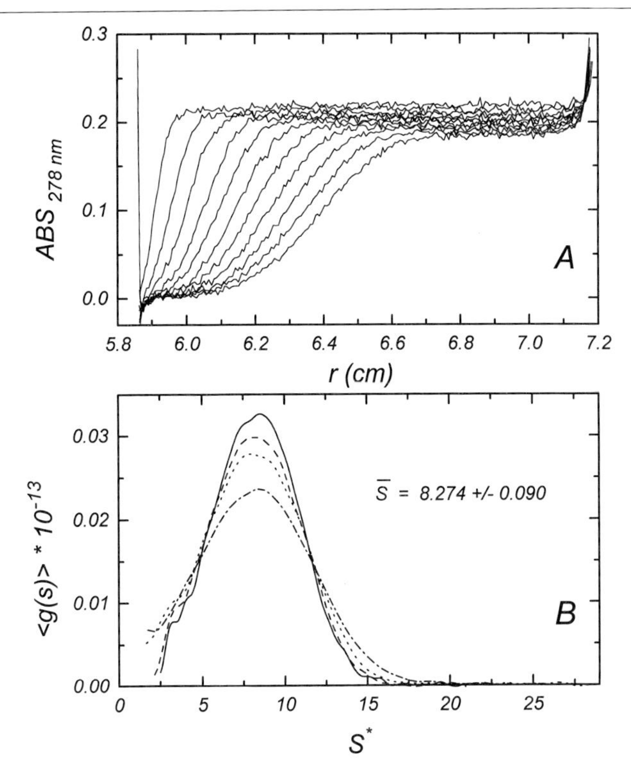

FIG. 1. Conversion of a sedimentation velocity data set to a $g(s^*)$. (A) Every third radial scan is presented from a sedimentation velocity run at 35,000, 24.7° with 2 μM tubulin in 10 mM PIPES, 100 mM NaCl, 1 mM $MgSO_4$, 50 μM GTP, pH 6.9, and 1 μM vinblastine. (B) A family of $g(s^*)$ curves generated from these data demonstrates that using scans from later in the run generates a narrower and taller distribution. This sharpening during the course of the run improves the resolving power and thus the information content of the analysis. Note the weight average s value is within error unaffected by this process ($\bar{s}$ = 8.274 ± 0.090).

$$\bar{s}_{20,w} = \Sigma c_i s_i / \Sigma c_i \tag{2}$$

Note that in Eqs. (1) and (2), s^* is a (pseudo) continuous distribution that relates to the discrete distribution s_i, $g(s^*)$ is the derivative of concentration with respect to s^* (dc/ds^*), and $\int g(s^*)\, ds$ and Σc_i correspond to the total macromolecule concentration in absorbance (or fringes) and weight concentration units, respectively. The second moment corresponds to the weight average polymer size and is thermodynamically linked through the energetics of the process to the experimental plateau concentration.[3,37] The second

[37] R. J. Goldman, *J. Phys. Chem.* **57,** 194 (1953).

moment techniques used by Timasheff and Lee and the weight average method described here give equivalent results.[3,37,38] Data are collected as a function of macromolecule and/or ligand concentration and fit to appropriate molecular models.

Rules to Follow, Practical Issues, and Complexities

The quantitative interpretation of sedimentation velocity data for an associating system requires a systematic approach. Although this has been suggested before,[23,27] a detailed reappraisal in light of recent advances is presented here. First, one must be certain of reversible association with the complete absence of irreversible aggregation or inactivation. Irreversible behavior is usually assessed by looking at the speed dependence of $g(s^*)$. Often an increase in $\bar{s}_{20,w}$ at lower protein concentrations will occur for systems that are prone to inactivation and irreversible aggregation.[39,40] One should also look for evidence of a change in the $g(s^*)$ pattern as a function of time or aging of the sample. Differences in sample history or purification methods[26,40] must also be considered. The advantage of sedimentation velocity is often the rapidity with which experiments can be conducted. Thus samples that are 80–90% active after 3–5 hr are still potential candidates for analysis. In addition inactivation of many proteins is often strongly coupled to large aggregate formation. Although these aggregates would dramatically interfere with light scattering studies, they often pellet out in sedimentation studies and allow the viewing and analysis of the active fraction. Nonetheless the presence of inactivation, either inactive monomers or aggregates, is expected to cause increased uncertainty in the fitted parameters or the choice of a model.[41]

The choice of an appropriate model often can be arrived at by a careful look at plots of $g(s^*)$ as a function of protein or ligand concentration. A plot of increasing weight average sedimentation coefficient ($\bar{s}_{20,w}$) versus loading concentration or ligand concentration is a definitive indication of a concentration-dependent association. The shape of the $g(s^*)$ pattern as a function of concentration is often diagnostic of certain well-described mechanisms, such as Gilbert type monomer–N-mer, indefinite or ligand mediated polymerization.[42] The shapes of the $g(s^*)$ patterns can be simu-

[38] W. F. Stafford III, *Methods Enzymol.* **323,** [13], in press.

[39] S. S. Rosenfeld, J. J. Correia, J. Xing, B. Rener, and H. C. Cheung, *J. Biol. Chem.* **271,** 30212 (1996).

[40] K. A. Foster, J. J. Correia, and S. P. Gilbert, *J. Biol. Chem.* **273,** 35307 (1998).

[41] D. A. Yphantis, J. J. Correia, M. L. Johnson, and G.-M. Wu, *in* "Physical Aspects of Protein Interactions" (N. Catsimpoolas, ed.), pp. 275. Elsevier/North Holland, New York, 1978.

[42] V. Prakash and S. N. Timasheff, *Methods Enzymol.* **130,** 3 (1986).

lated by finite element methods to strengthen (but not uniquely verify) the interpretation, and Stafford[32] has described software for simulating $g(s^*)$ patterns for both monomer–N-mer, heteroassociating (A + B → C; C + B → D) and pressure-dependent systems. (This is an active area of new developments and user-friendly software applicable to a wide range of interacting systems is becoming increasingly available.)

The summary of previous applications utilizing these methods presented above should serve as a guide to the types of behavior one might anticipate observing. Na and Timasheff[3] give an abbreviated list of the types of interacting systems that have been observed and/or simulated in the sedimentation literature. The shape of the reacting boundary is concentration dependent in all cases and thus a wide range of concentrations should always be explored to properly reveal asymmetry or bimodality in the $g(s^*)$ pattern. An essential component is the presence of cooperativity, e.g., ring closure, that readily gives rise to bimodality, versus ligand linkage, which tends to give rise to asymmetric boundaries.[3]

It is important to point out that the appropriate concentration range depends on the strength of the interaction, which may also dictate the optical system that is most appropriate for the study. Depending on the reference buffer, the extinction coefficient of the macromolecule, and the wavelength chosen, absorbance optics are useful from approximately 0.1 to 30 μM. (The XLA is essentially a spectrophotometer spinning at up to the speed of sound, and thus as in spectroscopic studies one should strive to work between 0.1 and 1.0 OD.) This concentration range corresponds, for a simple binding reaction, to a K_d of 1.0×10^7, a relatively tight interaction, to $3.3 \times 10^4\ M^{-1}$, a weak interaction.[3] Interference optics extend this range by approximately 1 order of magnitude in each direction. (The XLF being developed by Tom Laue and co-workers[43] currently extends this range down to at least 300 pM or $3.3 \times 10^{10}\ M^{-1}$).

Many definite polymerization schemes are difficult to describe uniquely because of the nature of an interacting boundary. In cases where the sedimentation coefficient of the largest species is not obtained, intermediates make significant contributions to the $g(s^*)$ pattern, which is always a superposition of the interacting species present. This is especially true of weak interactions. Stafford has demonstrated by simulation that to achieve the weight average sedimentation coefficient of the largest species requires that experiments be done 3 orders of magnitude above the K_d.[32] An especially instructive example is the work by Senear *et al.*[44] on the self-association of

[43] I. K. MacGregor and T. M. Laue, *Biophys. J.* **76,** A357 (1999).

[44] D. F. Senear, T. M. Laue, J. B. A. Ross, E. Waxman, S. Eaton, and E. Rusinova, *Biochemistry* **32,** 6179 (1993).

bacteriophage λ cl repressor protein. The average sedimentation velocity behavior in the micromolar range was initially interpreted as being consistent with tetramer formation, if assumptions are made about the shape of the oligomer. However, the sedimentation equilibrium data are best described by a dimer–tetramer–octamer scheme, with concerted octamer formation. This work has generated some disagreement[45] demonstrating that the analysis of these systems can be complicated and, where possible, sedimentation velocity and sedimentation equilibrium methods should be used in combination.[46] Nonetheless, it must be stressed that weight average data is an accurate reflection of the interacting species present in solution.

This is based on the original derivation by Goldman[37] where he showed that the second moment position rigorously reflects the weight average behavior of the boundary corresponding to the plateau concentration. For an interacting system the plateau concentration determines the degree of association. This can be seen in a direct comparison of sedimentation equilibrium and weight average sedimentation velocity data for kinesin motor domain constructs, K401.[47] Figure 2 presents a superposition of the experimental weight average velocity data on the curves generated from the K_2 and K_4 (and K_8) values derived from sedimentation equilibrium experiments. The agreement between the weight average velocity data and the values calculated from the equilibrium results are excellent, confirming the validity of the weight average approach. [Similar results were obtained for a histidine-tagged construct, K413,[39] but the agreement is only reasonable (data not shown) in part because K413 exhibited significant irreversible aggregation during sedimentation equilibrium and thus there is large uncertainty in the equilibrium constants.]

Furthermore, Stafford has shown that the weight average value for a 1:2:4:8 system accurately reflects the molecular distribution even in the absence of rapid reequilibration.[38] Thus the boundary may vary from four discrete peaks (reflecting extremely slow kinetic exchange between the monomer, dimer, tetramer, and octamer) to an asymmetric $g(s^*)$ pattern that resembles the inset in Fig. 2 (reflecting rapid kinetics between species) and yet, at equilibrium at the same total concentration, the $\bar{s}_{20,w}$ value is identical. This allows us to generalize and conclude that weight average sedimentation coefficients reflect the molecular distribution regardless of the nature of the interaction. As long as the system is at equilibrium at the start of the run then fitting to extract equilibrium constants will be

[45] D. S. Burz and G. K. Ackers, *Biochemistry* **35,** 3341 (1996).
[46] J. J. Correia, *Chem Tracts* **13,** 944 (1998).
[47] J. J. Correia, S. P. Gilbert, M. L. Moyer, and K. A. Johnson, *Biochemistry* **34,** 4898 (1995).

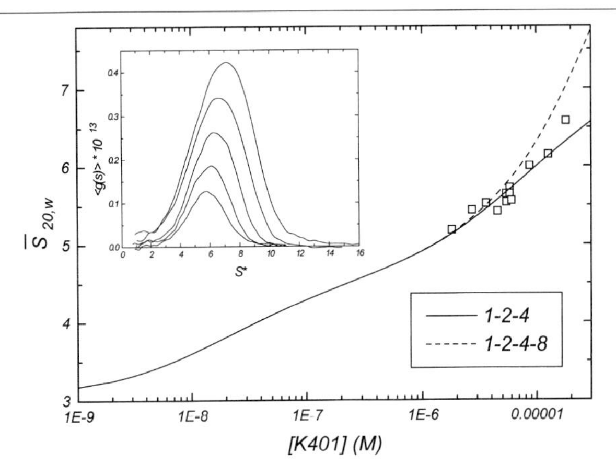

FIG. 2. Simulation of weight average sedimentation velocity data as a function of total K401 concentration for a 1-2-4 and a 1-2-4-8 mechanism. The equilibrium constants used correspond to the best fitted values derived from global fitting of multiple speed and concentration data.[47] To convert these data to $\bar{s}_{20,w}$, assumptions had to be made about the size of the discrete species (s_1 = 3.09S, s_2 = 4.9S, s_4 = 7.79S, s_8 = 12.36S), thus introducing uncertainty into the analysis. s_2 was directly determined from samples of inactive K401.[47] The data symbols (□) represent the experimental values derived from sedimentation velocity experiments partially shown in the inset. Note that the concentration dependence reveals a shift in the peak position and the centrifugal portion of the boundary to larger species, but the identity of the individual species is indiscernible in these patterns.

appropriate. Otherwise, the accuracy of the model will be constrained or compromised.

When to Use Sedimentation Velocity for Self-Associating Systems

The use of sedimentation velocity for the study of interacting systems is often criticized because sedimentation equilibrium is a more thermodynamically rigorous method. This statement is often misinterpreted to mean that sedimentation velocity is inappropriate for the study of self-association. The statement is actually a reference to the fact that the signal in sedimentation equilibrium is rigorously connected to the molecular weight of each species and thus in principle analysis is more robust when determining the best fitted model. This is in contrast to sedimentation velocity, a hydrodynamic technique, where S is a function of molecular weight and f (or D) and thus the fitting requires information or assumptions about (at least) two parameters per species thus making the analysis less well determined. Nonetheless there are cases where sedimentation velocity is actually preferred over sedimentation equilibrium studies.

As stated earlier, one of the best arguments for using sedimentation velocity is sample instability. The presence of irreversible aggregation is quickly and visually identified by sedimentation velocity and thus one can avoid the time of an equilibrium run and its analysis. (This does not preclude irreversible behavior occurring during the time of the equilibrium run, which is a minor problem for sedimentation velocity analysis.)

In addition, in the absence of additional information, it is often easier and quicker to propose a molecular model from $g(s^*)$ than from molecular weight as a function of concentration. For example, a cooperative Gilbert type system is readily apparent by the shapes of the $g(s^*)$ (see Ref. 28 for a discussion of the choice of models), but arriving at the correct cooperative monomer–N-mer model might take creativity and luck if analyzed by sedimentation equilibrium first.

Furthermore, there are cases where the degree of self-association is so large that sedimentation equilibrium studies are extremely difficult. For large complexes the time to equilibrium is extremely slow and the distribution of sizes is often so broad that many speeds must be combined to sample the full distribution. For example, the Mg^{2+}-induced ring formation of tubulin involves a 100-kDa protein self-associating into a 26-mer or a 2600-kDa complex.[6–8] Molecular weight measurements would be dominated by the largest species at low speed, especially if M_z values were estimated by Nonlin with short column data. Furthermore, the largest species would pellet at any speeds required to observe curvature due to the tubulin subunits and intermediate species. [This, of course, may also be true of sedimentation velocity. Large aggregates or assemblies up to 5000S can be analyzed by DCDT at low speeds (3000 rpm) if done in a single cell with the XLA or with multiple cells in an XLI. Then, on acceleration to appropriate speeds smaller species can be observed and studied. In these cases it is often necessary to make composite $g(s^*)$ curves where the graphs are presented sequentially or spliced together to make an apparent continuous distribution.]

Many viral or capsid assembly problems involve the formation of intermediate structures. These intermediates are readily seen in the $g(s^*)$ derived from a velocity run, but a sedimentation equilibrium experiment would reveal a continuous increase in molecular weight without revealing the presence of discrete intermediates. Once the presence of intermediates is determined, then proper analysis models can also be applied to the sedimentation equilibrium data.

Systems that undergo slow kinetic rearrangements can be analyzed by sedimentation velocity because during the time of the run little reequilibration occurs and thus the $g(s^*)$ is an accurate snapshot of the distribution.[5,38]

However, a system that relaxes over 12 to 24 hr would be in constant flux and take weeks to reach sedimentation equilibrium. An example of this is the assembly of tobacco mosaic virus (TMV) helical capsid, a kinetically controlled nucleated polymerization process, which can overshoot to extremely long spiral polymers that relax to an equilibrium distribution of polymers over the course of days and weeks.[48]

Finally, systems where conformational transitions occur in the absence of self-association are excellent candidates for sedimentation velocity studies. For example, the folding of nucleosomal arrays into highly condensed 30-nm fibers has been extensively studied by Hansen and co-workers by sedimentation velocity methods. By monitoring the integral distribution of s^* of the system, $G(s)$, derived from van Holde–Weischet analysis, chromatin compaction as a function of solution conditions and specific histone types and domains is readily investigated.[49] Sedimentation equilibrium is relatively insensitive to these conformational isomerizations since they involve only changes in f.

Note that extensive heterogeneity in both M and f often associated with studies of the structure and stability of complex macromolecular assemblies also makes sedimentation velocity experiments in combination with the van Holde–Wieschet analysis much more useful than sedimentation equilibrium approaches for these types of investigations.[50]

Sequential or cooperative folding models could be developed to further analyze the energetics of compaction by fitting the weight average of either the integral distribution $G(s)$ or the differential distribution $g(s^*)$. (There appears to be no direct application of weight average techniques to integral distributions of s^* but, in principle, it is feasible.) Thus in general, sedimentation velocity is preferred for unstable systems, complex assembly pathways involving intermediates, cooperative monomer–N-mer systems where N is large, systems with extremely slow kinetic transitions and systems that involve conformational changes in the absence of self-association. Often simply looking at the $g(s^*)$ as a function of concentration or time of incubation tells the qualitative story. More importantly, the quantitative analysis of $\bar{s}_{20,w}$ data will give results that are identical within error to sedimentation equilibrium analysis, if the system is well behaved and appropriate for both techniques (Fig. 2). Note that the simulation and analysis of reacting bound-

[48] S. S. Shire, J. J. Steckert, and T. M. Schuster, *J. Mol. Biol.* **127,** 487 (1979).

[49] L. C. Carruthers, J. Bednar, C. L. Woodcock, and J. C. Hansen, *Biochemistry* **37,** 14776 (1998).

[50] L. C. Carruthers, V. Schirf, B. Demeler, and J. C. Hansen, *Methods Enzymol.* **321,** [4], 2000 (this volume).

aries correctly assumes, in the absence of kinetic effects, thermodynamic equilibrium at each concentration, at each radial position, and in the sedimenting boundary.[4,5,32,38,51–53]

Fitters 101: How to Fit Weight Average Data to Extract the Energetics of an Interacting System

Na and Timasheff[3] give a detailed description of the mathematics required to develop a model, focusing primarily on ligand-mediated processes. Here we present sufficient details to reproduce and generalize the required steps for nonlinear least-squares fitting of weight average sedimentation coefficient data. We have developed a large number of fitting functions within the context of a commercial software package, FITALL. This is a Turbo Pascal-based fitter available directly from the programmer (MTR Software, Toronto, Canada). Functions developed include Gilbert type monomer–*N*-mer with or without appreciable intermediates, discrete association models like 1-2-4-8 where the degree of association can be varied (1-3-6-9, etc.), and a family of models appropriate to indefinite association schemes, including ligand-mediated processes.[14]

The general equation required to fit the $\bar{s}_{20,w}$ data is

$$\bar{s}_{20,w} = \sum c_i s_i^0 (1 - g_i C_t)/C_t \tag{3}$$

where s_i^0 is the sedimentation coefficient of each species, g_i is the concentration dependence of sedimentation due to hydrodynamic nonideality, C_t is the total concentration in mg/ml, and c_i is the concentration of I-mer in mg/ml.

The identity and the concentration of each I-mer is determined from the model, the equilibrium constants for the process and the total protein concentration. Since this is a fit to a weight average dependent parameter all concentrations must be expressed in weight/volume units or mg/ml. For a monomer–dimer–tetramer–octamer scheme like the data simulated in Fig. 2,

$$c_1 = K_1 c_1 \tag{4}$$
$$c_2 = K_2 (c_1)^2 \tag{5}$$
$$c_4 = K_4 (c_1)^4 \tag{6}$$
$$c_8 = K_8 (c_1)^8 \tag{7}$$

[51] L. M. Gilbert and G. A. Gilbert, *Methods Enzymol.* **48,** 195 (1978).
[52] J. R. Cann *Methods Enzymol.* **48,** 299 (1978).
[53] D. J. Cox, *Methods Enzymol.* **48,** 212 (1978).

where K_i is the overall association constant between I monomers and the I-mer in units of $(\text{ml/mg})^{I-1}$ and K_1 is by definition unity. Alternatively, K_i can be expressed as the product of each successive intrinsic association step between a single monomer and a polymer, k_i', such that

$$K_i = k_2' k_3' k_4' \ldots k_i' \tag{8}$$

This requires knowledge of the individual events or preference of a path for self-association. For example, it might be preferred to assume a more likely molecular mechanism and thus output information about monomers dimerizing, dimers associating to tetramers, and tetramers associating to octamers. For the purpose of numerically calculating the individual species and total protein concentration the choice is not critical. From the perspective of the best fit and the ability to resolve parameters and models it is very likely to be critical.

Many investigators prefer to report the equilibrium constants in weight concentration units,[6,7,22–28] but it is often more appropriate to convert them to molar units for the purpose of comparison with other techniques. In general, we use molar units as input and output parameters and mathematically convert them to weight concentration units only within the fitter.[15–20] (The use of molar units within the fitter occasionally creates overflow problems for large N.) For intrinsic association constants,

$$k_i^{\prime,\text{weight}} = \{I/[(I-1)\,\text{MW}]\} k_i^{\text{molar}} \tag{9}$$

where MW is the molecular weight of the monomer and I and $I - 1$ correct for the total mass in the I-mer and I − 1-mer, respectively. Thus,

$$C_t = \Sigma\, c_i = \Sigma\, K_i (c_1)^I = \Sigma\, k_2' k_3' k_4' \ldots k_i' (c_1)^I \tag{10}$$

where c_1 is in mg/ml. Inputting k_i in molar units [Eq. (9)] into this equation causes elimination of most of the terms so Eq. (10) reduces to

$$C_t = \Sigma\, c_i = \Sigma\, I (k_i/\text{MW})^{I-1} (c_1)^I \tag{11}$$

Thus the final equation for calculating $\bar{s}_{20,\text{w}}$ using the overall K_i in weight units is

$$\bar{s}_{20,\text{w}} = \Sigma\, s_i^0 (1 - g_i C_t)\, K_i\, (c_1)^I / \Sigma\, K_i (c_1)^I \tag{12}$$

or using the successive intrinsic k_i in molar units is

$$\bar{s}_{20,\text{w}} = \Sigma\, s_i^0 (1 - g_i C_t) I (k_i/\text{MW})^{I-1} (c_1)^I / \Sigma\, I (k_i/\text{MW})^{I-1} (c_1)^I \tag{13}$$

When developing and coding models the details are in the mathematics. Check units and also develop simulation routines to check the accuracy of the fitter. Data points are input as protein concentration, ligand concentration if appropriate, and $\bar{s}_{20,w}$ values with an uncertainty or weighting factor are available. Note, this is essentially a titration curve and for best results the data should span a reasonable enough concentration range to define the self-association process. FITALL can accommodate multiple independent parameters, for example, C_t and drug or ligand concentration. Initial equilibrium parameters are guessed and the first iteration is performed. For each data point the concentration of all species must be determined for those independent parameters and that estimate of K values, plugged into Eq. (11) and compared with the experimental values by least squares.

The method of converging on the correct species concentration for that iteration is a matter of mathematical expertise and preference. For models like 1-2-4-8 we prefer halves-ies routines. The monomer concentration, c_1, controls all other species concentrations. Guess its value at halfway between 0 and C_t (by conservation of mass it must be between these two values) and calculate the sum of all species estimated, c_{total}:

$$c_1 = (c\text{low} + c\text{high})/2 \tag{14}$$

where initially $c$low = 0 and $c$high = C_t. If this sum is too large, $>C_t$, then c_1 is too big. Guess its new value as halfway between 0 and $C_t/2$, the initial estimate, and repeat the comparison:

$$\text{If } c_{\text{total}} > c_t, \text{ then } c\text{high} = (c\text{low} + c\text{high})/2; \qquad \text{go to Eq. (14)} \tag{15}$$

Alternatively, if the sum is too small, $<C_t$, then c_1 is too small. Guess its new value as halfway between $C_t/2$, the initial estimate, and C_t, and repeat the comparison.

$$\text{If } c_{\text{total}} < C_t, \text{ then } c\text{low} = (c\text{low} + c\text{high})/2; \qquad \text{go to Eq. (14)} \tag{16}$$

Continue this way for at least 20–25 cycles and the value of c_1 will be accurately determined to within 1 part in 10^6 to 10^7. This nested cycle of halves-ies routines occurs for each data point within each iteration of the fitter. Halves-ies will converge for essentially all models.

Some models like monomer–dimer obviously have algebraic solutions for the determination of c_1 that will save computer time. Even more complex models like indefinite association with identical equilibrium constants can be reduced to closed-form solutions.[3] In this case the indefinite series converges to a simple binomial, which can be solved for c_1:

$$C_t = c_1/[1 - (K_2 c_1/\text{MW})]^2 \tag{17}$$

For indefinite polymerizations it is often a problem that one does not extend the sum over all species out far enough to account for all the mass. Convergence to C_t in these cases should also be performed as a fractional difference to 1 part in 10^6 or 10^7. For the example of vinca alkaloid-induced tubulin self-association, even at 2 μM tubulin, we found it necessary for some congeners to include up to 200 spiral polymers to recover the total mass, C_t.[14] Finally, convergence on the equilibrium constants occurs when the parameters and the rms of the fit do not change within a preset limit.

The selection of a model requires choosing s_i^0 and g_i values. The g_i values reflect hydrodynamic nonideality of the solutes, due either to excluded volume or charge, and can be experimentally determined by the concentration dependence of s_1 under nonpolymerizable conditions, if possible.[6] For globular proteins g_i is usually 0.01 to 0.03 ml/mg and thus a 1% change in s_i^0 will occur at 1 to 0.33 mg/ml. For a 100-kDa protein this corresponds to 10–3.3 μM, experimentally useful concentrations in the XLA/XLI. Although nonideality effects are often small we always include them in the fitter.

The choice of s_i^0 values is initially model dependent and the smallest and largest species can often be determined by extrapolation. For discrete, indefinite, or complex models this choice typically requires assumptions about the polymer shape. The simplest assumption is to treat all polymers as if they have the same axial ratio, often referred to as the spherical approximation. Mathematically this reduces to

$$s_i^0 = s_1^0(I)^{2/3} \tag{18}$$

This method assumes identical frictional ratios for all species. Alternatively, assumptions about the polymer shapes, oblate or prolate, can be explicitly introduced by including the frictional ratio that corresponds to the assumed axial ratio:

$$s_i^0 = s_1^0(I)^{2/3}\,(f/f_0) \tag{19}$$

Harris *et al.*[28] present a detailed example of this in their study of DNA-dependent RNA polymerase. This modification typically reduces the value of s_i^0 and thus increases the value of the equilibrium constants required to fit and describe the $\bar{s}_{20,\mathrm{w}}$.[15] Note, this analysis also assumes that s_i^0 is a constant as a function of ligand or solution conditions. Oberfelder *et al.*[21] describe an example where an inhibitor induces an allosteric change in the enzyme quaternary structure. This requires introducing ligand dependence into the equations for s_i^0 and $\bar{s}_{20,\mathrm{w}}$.

For the case where discrete oligomeric species are formed, it is useful to predict hydrodynamic parameters from the X-ray crystal structure, if known, to discriminate between different quaternary structures.[54,55] In com-

[54] G. De la Torre, J. S. Navarro, M. C. Lopez Martinez, F. G. Diaz, and J. J. Lopez Cascales, *Biophys. J.* **67,** 530 (1994).

[55] O. Bryon *Biophys. J.* **72,** 408 (1997).

bination with other approaches like neutron scattering[56] these predictions should improve the reliability of the weight average analysis. For indefinite associations and for cases where significant concentrations of intermediates exist, this approach requires the prediction of parameters that are essentially untestable experimentally. That is to say, the individual species cannot be isolated because a mixture of reversibly interacting species always exists in the reacting boundary. Depending on the assumptions and the range of structures utilized, there will be a range of equilibrium constants that are consistent with the data, similar to assuming cruder models like oblate and prolate structures. This situation is similar to the use of isodesmic indefinite models, where the K values are assumed to be equal for each step, versus isoenthalpic models, where the successive equilibrium constants are attenuated by the change in entropy of the polymer.[15,57] While theoretically preferred, an isoenthalpic analysis of this system requires knowledge of the shapes of polymers to calculate the entropy effect. Because the individual species cannot be isolated, the shapes must be assumed or modeled without resource to direct testing. While theoretically pleasing, this house of cards may or may not provide more accurate estimates of the energetics. Often it is sufficient to investigate systems of this type in a relative sense, for example, comparing the ability of different drugs to induced ligand-mediated indefinite association,[14,16,20] rather than attempt to determine absolute but intractable solutions to the problem.

The fitting of weight average sedimentation coefficient data requires data as a function of protein concentration and ligand concentration, if appropriate. For example, in the case of vinca alkaloid-induced tubulin self-association Timasheff and co-workers varied protein concentration at fixed ligand concentrations.[3,8–13] Lobert and Correia varied ligand concentration at fixed protein concentrations.[14–20] Note both variables are still required in the fitter. The Schlieren optical system is more amenable to a wide range of protein (macromolecule) concentrations, whereas in the XLA varying drug concentration at an optimal protein OD is preferable. A global analysis varying both parameters may provide the best results. Because there are two independent parameters, C_{protein} and C_{drug}, the fitting is to a 3-D surface, and 2-D presentations of the best fitted parameters necessarily are done at an average protein concentration. This often does not represent the best visual appearance of the data although the rms accurately reflects deviations from the surface. The interference system can easily accommodate either approach although, due to increased sensitivity, it is less forgiving

[56] R. J. C. Gilbert, J. Rossjohn, M. W. Parker, R. K. Toweten, P. J. Morgan, T. J. Mitchell, N. Errington, A. J. Rowe, P. W. Andrew, and O. Byron, *J. Mol. Biol.* **284,** 1223 (1999).

[57] R. C. Chatelier, *Biophys. Chem.* **28,** 121 (1987).

of poorly equilibrated samples. A major advantage of interference optics is the ability to work at essentially any buffer composition including millimolar nucleotide concentrations. With absorbance optics at 280 nm one is typically limited to <100 μM AXP or GXP.

Future Developments

Weight average is but one moment that can be used to analyze a reacting boundary. One could also transform $g(s^*)$ into a z and a $z + 1$ average sedimentation coefficient as defined here:

$$\bar{s}_{20,z} = \int g(s^*)(s^*)^2\, ds / \int g(s^*)s^*\, ds \tag{20}$$

$$\bar{s}_{20,z+1} = \int g(s^*)(s^*)^3\, ds / \int g(s^*)(s^*)^2\, ds \tag{21}$$

There has been one recent study that used these moments in the analysis of self-association of TMV coat protein,[58] although they only compared the values to confirm the presence of self-association ($\bar{s}_{20,w} < \bar{s}_{20,z} < \bar{s}_{20z+1}$), something that was already evident by visual inspection of the $g(s^*)$. Fitting all three moments simultaneously to a molecular model should in principle improve the robustness and information content of the analysis. This was elegantly demonstrated by Adams and co-workers years ago.[59,60] Combining moments of the sedimentation coefficient boundary with the equilibrium constants derived from sedimentation equilibrium allows robust determination of the sedimentation coefficient and hydrodynamic nonideality of the individual species. While this technique has been available for 25 years it has only been applied to simulated data and a single test case, β-lactoglobulin A.[60] The data in Fig. 2 could not be successfully analyzed in this way because it does not span enough of the concentration range to determine s_1 and s_8 uniquely. The use of additional moments still has potential utility for systems where equilibrium studies are inappropriate (see above) and thus cannot be performed under the same solution conditions as velocity experiments.

Most of the systems analyzed by weight average techniques have boundaries that are unimodal or bimodal with a limited number of distinct equilibrium constants. For these systems and for more complex associating systems the additional information provided by multiple moments would greatly enhance the robustness of the analysis as it does for sedimentation equilib-

[58] J. M. Toedt, E. H. Braswell, T. M. Schuster, D. A. Yphantis, Z. F. Taraporewala, and J. N. Culver, *Protein. Sci.* **8,** 261 (1999).

[59] C. A. Weirich, E. T. Adams, and G. H. Barlow, *Biophys. Chem.* **1,** 35 (1973).

[60] J. M. Beckerdite, C. A. Weirich, E. T. Adams, and G. H. Barlow, *Biophys. Chem.* **17,** 203 (1983).

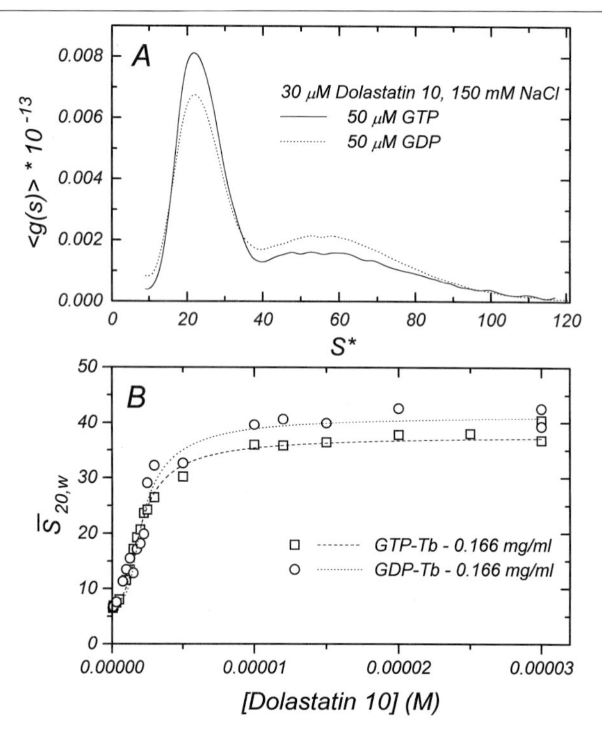

FIG. 3. (A) Sedimentation coefficient distribution, $g(s^*)$, plots with 2 μM PC-tubulin and 30 μM dolastatin 10 in 10 mM PIPES, 1 mM $MgSO_4$, 2 mM EGTA, 50 μM GXP, 0.06% dimethyl sulfoxide (DMSO), pH 6.9, plus 150 mM NaCl and 24.7°. Note the broad, biphasic character of the distribution as well as the nucleotide dependence of the self-association. The broadness of the second phase is consistent with an indefinite process. (B) Weight average $\bar{s}_{20,w}$ values versus drug concentration for GTP-tubulin (□) and GDP-tubulin (○). The lines are simulations of the best fits with a two-phase model (details not shown) at 1.66 μM tubulin. Note the cooperative transition at low drug concentration; the 20S species probably involves ring closure.

rium analysis.[61,62] This would be especially true for complex examples involving two phases of self-association where a polymer, a ring, or a rod forms and then associates in a different mode, lateral sheets or longitudinal stakes of rings or filaments. [Tubulin and Rev are systems that exhibit this behavior. For example, Fig. 3 presents the $g(s^*)$ for dolastatin 10-induced tubulin self-association. The fitting of the $\bar{s}_{20,w}$ versus drug concentration with a two-phase model is excellent, but the values of the parameters are poorly determined. Additional moments should improve the reliability of

[61] J. J. Correia, S. J. Shire, D. A. Yphantis, and T. M. Schuster, *Biochemistry* **24,** 3292 (1985).
[62] W. F. Stafford, *Biophys. J.* **29,** 149 (1980).

the analysis (Correia *et al.,* manuscript in preparation).] One can generalize that the more inflections in the boundary shape, and/or the more equilibrium constants in the model, the more moments required to solve the problem. The caveat is the higher moments, s_z and s_{z+1} are more likely to be imprecise, especially with absorbance optics, due to propagation of errors, but this has not been rigorously tested. [Note that versions of DCDT can be easily modified to output s_z and s_{z+1}, and a program under development by Philo (DCDT+) calculates the z and $z + 1$ moments and their uncertainty.]

Simulations and direct fitting of the shape of the reacting boundary are becoming state of the art in this field.[32,36,38,63–65] Fast computers, improvements in the simulation algorithms (primarily the moving hat grid approach[64] and the removal of systematic baseline noise[36]) and user-friendly graphical software[38,64,65] have made this approach exceedingly more tractable. This type of analysis is limited by many of the same issues discussed earlier. The presence of microheterogeneity and/or aggregation may prove to be exceedingly difficult to deal with in this approach due to significant propagation of error into the individual fitting parameters, especially D_i (unless M_i is constrained). Thus the quantitative analysis and fitting of sedimentation velocity reacting boundary shapes will have similar (if not more stringent) requirements of homogeneity and purity to that of sedimentation equilibrium.

In addition, the complexity of determining individual s_i and g_i parameters for complex cases may very well require similar approaches to that described by Adams and co-workers.[59,60] Multiple sample concentrations and species ratios for heteroassociation should be explored so that the concentration of macromolecules spans the K_d of the interaction. This increases the information content of the data set by populating all potential species in the system. For well-behaved systems, in the absence of irreversible behavior, direct fitting should prove to be a far more robust and accurate method than weight average techniques for the analysis of interacting systems. What is required initially is the simultaneous analysis of the same data sets by multiple techniques to establish the limitations. For example, how many species and equilibria (or rates if kinetic effects occur) can be reasonably resolved given the total number of parameters to be determined? Nonetheless, weight average analysis techniques, for many interacting systems, may in the future be relegated to providing a quick and more forgiving analysis, at the very least, the selection of potential

[63] G. P. Todd and R. H. Haschemeyer, *Proc. Natl. Acad. Sci. U.S.A.* **78,** 6739 (1981).

[64] B. Demeler and H. Saber, *Biophys. J.* **74,** 466 (1998).

[65] P Schuck, *Biophys. J.* **75,** 1503 (1998).

models prior to more rigorous investigations by sedimentation equilibrium and quantitative boundary analysis. However, as described earlier, there will always be interacting systems that are uniquely appropriate for weight average sedimentation coefficient analysis to extract valid energetic and thermodynamic information.

Acknowledgments

I wish to thank Sharon Lobert, Walter Stafford, John Philo, Peter Schuck, and Jeff Hansen for critical discussions during the writing of this manuscript. The author takes full responsibility for the content and the opinions expressed therein.

[6] Sedimentation Equilibrium Analysis of Mixed Associations Using Numerical Constraints to Impose Mass or Signal Conservation

By JOHN S. PHILO

Introduction

Sedimentation equilibrium can be a very useful and powerful tool for characterizing the stoichiometry and strength of many physiologically important binding interactions. Many such interactions involve the binding of two or more different molecules (protein–protein, protein–ligand, protein–DNA, etc), and are therefore commonly termed "mixed" or "heterogeneous" associations. Despite the intense interest in fields such as signal transduction, which are governed by networks of specific binding interactions, the application of sedimentation equilibrium to characterize mixed associations has been somewhat limited by difficulties in analyzing the data. This chapter demonstrates how to alleviate some of those difficulties through the use of numerical constraints during data analysis. Methods developed by others as well as those used in our laboratory are described in order to present a broader range of methodologies and discuss their relative advantages and disadvantages.

Analyzing Mixed Associations

One of the principal difficulties in analyzing mixed associations such as $A + B \leftrightarrow AB$ in the centrifuge is the fact that the absorbance and Rayleigh interference optical systems in the centrifuge usually only provide informa-

0076-6879/00 $30.00

tion about the *sum* of the contributions of all species, rather than the concentration distribution of the A and B entities individually. Particularly for associations more complex than A + B ↔ AB, this information about the summed contributions is unfortunately generally insufficient to provide a unique fit of the data and thus derive the association constants.

To see one reason why this is true, let us consider the case of an antibody (A) binding its antigen (B). Because the antibody has two binding sites, this is an A + B ↔ AB, AB + B ↔ AB_2 system. Figure 1 shows how the overall weight-average molecular weight, M_w, will vary as we titrate antibody into a fixed concentration of antigen, at a relatively high concentration of antigen ($20 \times K_d$) so binding is strongly favored. As the ratio of antibody to antigen increases, M_w rises until we reach the preferred 1 : 2 stoichiometry, and then falls. The key point is that because M_w is a *multivalued* function of the mixing ratio, in general, there are two mixing ratios that give rise to the same M_w, as illustrated by the dotted lines. In particular, this multivalued property applies for most mixing ratios that one would be likely to employ experimentally.

How does this relate to how data for such a system would be analyzed? In analyzing associating systems one first obtains an estimate for the free

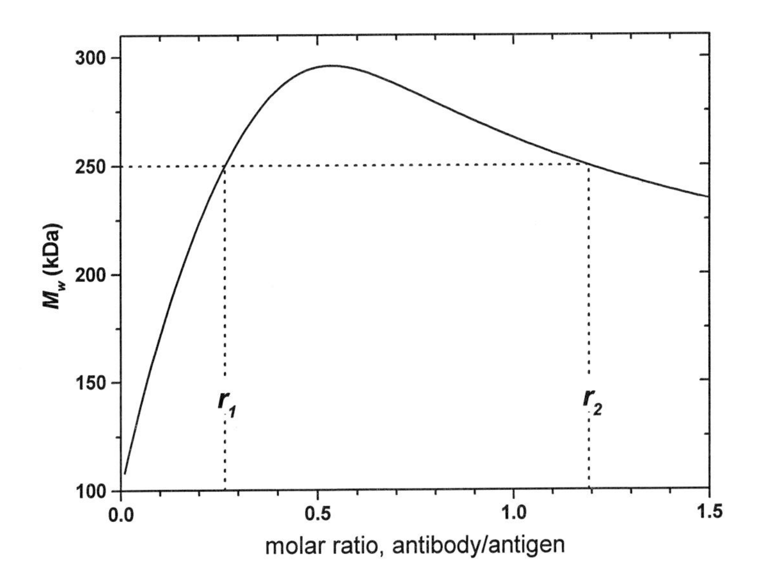

FIG. 1. Theoretical weight-average mass versus mixing ratio for an antigen–antibody system. Note that in general there will be two different mixing ratios that give the same M_w. The antigen and antibody masses are assumed to be 100 and 150 kDa, respectively. Each antibody is assumed to have two equivalent, independent binding sites. The antigen concentration was held fixed at 20 times the dissociation constant and the antibody concentration was varied.

monomer concentrations at some reference radius, then distributes each monomer across the cell according to its buoyant mass, and next calculates the concentrations of each associated state at each radius from the monomer concentrations, the association constant(s), and the law of mass action. In our work those estimates of the monomer concentration at the reference radius are obtained iteratively using least squares fitting of the raw data, using methods analogous to those in the well-known NONLIN program for analysis of self-associating systems.[1] Another noniterative "direct" approach was recently described that obtains the estimates of the monomer concentrations from a graphical extrapolation of plots of the function.[2] However, regardless of how the monomer concentrations are estimated, since the overall summed concentration distribution measured by the centrifuge is primarily governed by the overall M_w, what Fig. 1 implies is that in general for each sample there will be *two* mixing ratios, and two different sets of monomer concentrations that give rise to essentially identical experimental data and thus potentially represent reasonable fits of the data. In the context of least squares fitting procedures, the consequence is that if one attempts to do a global analysis of N data sets there may be $\sim 2^N$ local minima in the error surface, making finding the correct global minimum a formidable problem. Even for simpler association schemes such as A + B $\leftrightarrow$ AB, and under conditions when M_w is not multivalued, our experience and that of others[3] is that in attempting to directly fit total concentration scans using least squares methods often one has an ill-conditioned fitting problem with poorly defined minima in the variance surface and consequently poor convergence. Further, even when convergence is obtained, the parameters obtained are often physically unrealistic.

Another significant problem in data analysis for mixed associations is the fact that in many cases at least one of the monomer species contributes very little to the signals that are measured, and it is therefore difficult to get any direct estimate of that concentration. This circumstance arises, for example, at concentrations high enough to promote strong association. Weak signal contributions from one free monomer will also commonly arise when one type of monomer has a much lower molecular weight than the other. Despite the fact that the free monomer itself may contribute little to the signal, the overall distribution of species can be extremely

[1] M. L. Johnson, J. J. Correia, D. A. Yphantis, and H. R. Halvorson, *Biophys. J.* **36,** 575 (1981).

[2] P. R. Wills, M. P. Jacobsen, and D. J. Winzor, *Biopolymers* **38,** 119 (1996).

[3] S. P. Becerra, A. Kumar, M. S. Lewis, S. G. Widen, J. Abbotts, E. M. Karawya, S. H. Hughes, J. Shiloach, and S. H. Wilson, *Biochemistry* **30,** 11707 (1991).

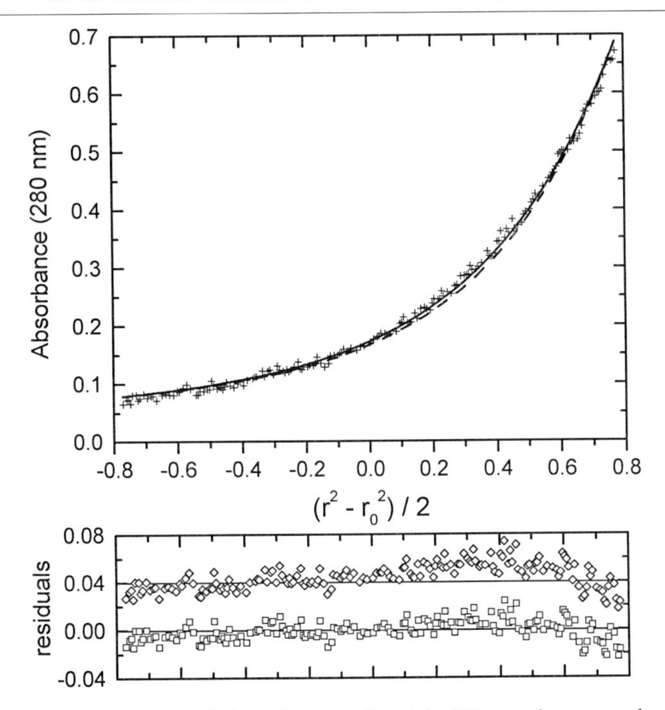

FIG. 2. Sedimentation equilibrium data and fits at 11,000 rpm for a sample containing 3 μM SCF and 3 μM sKit. Two different fits are shown, both using the same (correct) association constant, along with their residuals in the lower panel. The fit shown as a dashed line (above) and open diamonds (below) is an invalid solution that corresponds to a vastly different mixing ratio of SCF with sKit. The physically realistic fit is shown as a solid line (above) and open squares (below).

sensitive to the free monomer concentration, especially for systems with multivalent interactions.

These theoretical difficulties actually do arise for real experiments. Figure 2 shows some data for a receptor–hormone system. In this case the hormone is the dimeric protein hormone stem cell factor (SCF), and we are using the soluble extracellular ligand-binding domain of its receptor, *c-Kit*.[4] Because each SCF dimer molecule has two binding sites for receptor, this represents an A + B $\leftrightarrow$ AB, AB + B $\leftrightarrow$ AB_2 system, just like the antigen–antibody system discussed above, but here A corresponds to an SCF dimer and B is the soluble receptor, sKit. The data in Fig. 2 are for

[4] J. S. Philo, J. Wen, J. Wypych, M. G. Schwartz, E. A. Mendiaz, and K. E. Langley, *J. Biol. Chem.* **271,** 6895 (1996).

a mixture made with 3 μM SCF dimer plus 3 μM sKit. Depending on the initial guesses for the concentrations, a NONLIN-type fitting routine converges to two solutions that are reasonably good representations of the data (Fig. 2, upper panel), even when the association constant is held fixed at the known optimal value. One solution gives concentrations for SCF and sKit at the reference radius (the center of the cell in this case) of 588 and 10.3 nM, respectively, while the other gives concentrations of 0.12 and 838 nM. While the residuals (Fig. 2, lower panel) are somewhat worse for the latter pair of concentrations, without other information it would certainly be hard to say which of these solutions is right or wrong. For either solution, at least one of the monomers contributes less than 1% of the total absorbance even in the region near the meniscus.

Approaches to Resolving Fitting Problems

Because of the problems outlined above, to analyze these types of systems it has generally been considered to be necessary to have some means of directly measuring the concentration of at least one of the individual components. This may be done, for example, in a preparative microcentrifuge by postrun collection of individual fractions followed by separation and analysis of components using electrophoresis,[5] or by monitoring the concentration of a single component in those fractions labeled with a radioactive or fluorescent "tag."[6] In the analytical ultracentrifuge this may be done either by attaching a chromophore to one component and monitoring concentrations at a wavelength where only the chromophore is detected,[7] or by taking absorbance scans at many wavelengths and then exploiting differences in the intrinsic spectral properties of the species.[8]

An alternative to finding a means to determine concentrations of individual components is to directly attack these problems with analyzing the total concentration data by imposing mass conservation on the system. That is, returning again to Fig. 1, although in trying to fit the data one may find two sets of monomer concentrations that both fit the data reasonably well (corresponding to ratios r_1 and r_2), because the composition of the sample that was placed into the cell is known, only one of these sets of concentrations is physically realistic. The "false" solution will imply total amounts of antigen and antibody, integrated over the cell, that are quite different

[5] S. Darawshe, G. Rivas, and A. P. Minton, *Anal. Biochem.* **215,** 236 (1993).

[6] S. Darawshe, G. Rivas, and A. P. Minton, *Anal. Biochem.* **209,** 130 (1993).

[7] T. M. Laue, D. F. Senear, S. Eaton, and J. B. A. Ross, *Biochemistry* **32,** 2469 (1993).

[8] M. S. Lewis, R. I. Shrager, and S.-J. Kim, *in* "Modern Analytical Ultracentrifugation" (T. M. Schuster and T. M. Laue, eds.), p. 94. Birkhauser, Boston, Massachusetts, 1994.

from the initial makeup of the sample, and this false solution arises precisely because nothing in the equations giving simultaneous sedimentation and association equilibrium requires the conservation of mass.

Imposing Mass Conservation with Constraints

The previous discussion suggests that a good way to eliminate the multiple minima and "false" solution problems might be to put conservation of mass back into the equations, and indeed this approach has been used successfully by several investigators, although the implementations differ in detail and in the cell types for which they are designed.

One approach used for many years in the laboratory of Marc Lewis[3,9,10] has been to use mass conservation to provide an implicit constraint that eliminates one monomer concentration as a fitting parameter. This method is based on the fact that conservation of mass requires that the average molar concentration over the region from the meniscus at r_m to the cell base at r_b for each type of monomer in all its forms must equal the initial loading concentration. It is well known that at equilibrium a species with buoyant molecular weight M_b will have a concentration distribution given by

$$c(r) = c_m \exp[M_b\omega^2(r^2 - r_m^2)/2RT] \tag{1}$$

where c_m is the concentration at the meniscus (which will be used here as the reference radius), ω is the rotor angular velocity, R is the gas constant, and T is the absolute temperature. By good fortune, it turns out that for a sector-shaped cell, where the cell width grows linearly with radius, the integral of this concentration distribution has an exact analytical solution. Considering for the moment a single species A with buoyant weight M_A, then for a loading concentration $c_{0,A}$ mass conservation will be achieved if

$$\int_{r_m}^{r_b} c_A\, r\, dr = c_{0,A}(r_m^2 - r_b^2)/2 = \int_{r_m}^{r_b} c_{m,A} \exp[FM_A(r^2 - r_m^2)]\, r\, dr \tag{2}$$

where we have defined a factor $F \equiv \omega^2/2RT$. After evaluating the integral and rearranging to solve for the concentration at the meniscus, one obtains

$$c_{m,A} = \frac{c_{0,A}(r_b^2 - r_m^2)}{FM_A\{\exp[FM_A(r_b^2 - r_m^2)] - 1\}} \tag{3}$$

Thus mass conservation can be used to calculate the concentration at the reference position from the known loading concentration and the buoyant

[9] M. S. Lewis and R. J. Youle, *J. Biol. Chem.* **261,** 11571 (1986).
[10] M. R. Bubb, M. S. Lewis, and E. D. Korn, *J. Biol. Chem.* **266,** 3820 (1991).

mass (which could be the current estimate for that mass in an iterative fitting procedure).

This concept can be also be usefully employed even for mixed associations. Consider a simple A + B ↔ AB case with an association constant K_{AB}. At every point in the cell the concentration of the AB heterodimer can be calculated from K_{AB} times the product of the free monomer concentrations, and those monomer concentrations are given using Eq. (1) with the appropriate buoyant masses and reference concentrations. Thus we may write an equation for mass conservation for species A analogous to Eq. (2), accounting for both free monomer and heterodimer forms of A, obtaining

$$
\begin{aligned}
\int_{r_m}^{r_b} c_A r \, dr &= c_{0,A}(r_m^2 - r_b^2)/2 \\
&= \int_{r_m}^{r_b} c_{m,A}(1 + K_{AB}c_{B,free}) \exp[FM_A(r^2 - r_m^2)] r \, dr \qquad (4) \\
&= \int_{r_m}^{r_b} c_{m,A}(1 + K_{AB}c_{m,B} \exp[FM_B(r^2 - r_m^2)]) \exp[FM_A(r^2 - r_m^2)] r \, dr
\end{aligned}
$$

Assuming M_A and M_B are known (by running sedimentation equilibrium for each separately, from the sequence, or as the current value of fitting parameters) these integrals can again be evaluated and solved to obtain $c_{m,A}$ as a function of $c_{0,A}$, $c_{m,B}$, and K_{AB}. Still more complex association schemes can also be treated similarly to derive the reference concentration of one species from its loading concentration, the association constants, and the reference concentrations of the other type(s) of monomer(s).

At first glance this may seem a useless exercise, since the derived concentration for monomer A depends on the (presumably unknown) reference concentrations for other monomers and association constants. However, this is really quite useful, since it allows the elimination of $c_{m,A}$ as a fitting variable, through calculation of $c_{m,A}$ at each iteration from the loading concentration and the current values of other parameters. This reduction in the number of fitting parameters will by itself often convert an ill-conditioned fitting situation into one with a well-defined minimum and rapid convergence. Furthermore, by linking this monomer concentration directly to the true loading concentration and thus eliminating potential nonphysical values for that concentration, one usually eliminates nonphysical values for other parameters as well.

Problems with Implicit, "Hard" Mass Conservation Constraints

Although the above implicit constraint method is fairly easy to implement and has certainly been used successfully, it has a number of disadvantages. One drawback to the implicit constraints is that an analytic solution for the integrated concentration is only available for the dual-channel centerpiece with sector-shaped channels, whereas many investigators now employ the six-channel centerpieces (rectangular channels) for equilibrium studies to obtain more samples (and incorporate more data sets into global analyses). Fortunately, because the sample column height is small compared to the radius, the error in monomer concentration resulting from treating a six-channel centerpiece as a sector-shaped one should only be 2–3%.

A more significant problem with this implicit constraint method results from the fact that in practice mass is usually not truly conserved in the centrifuge. Most proteins will to some extent irreversibly stick to the windows and centerpieces, and many are unstable and exhibit a slow loss over the course of the experiment. Moreover, it is not actually possible to test the degree of mass conservation because the optical systems in the centrifuge do not allow the accurate observation of all the mass that was placed into the cell. Optical distortions near the meniscus and base of the cell require the exclusion of data points in those regions from the fit, and the fractional amount of "nonobservable" sample at the cell base can be quite large when the concentration gradient is steep, particularly for absorbance data.

These implicit constraints are also "hard" constraints that assume there is no uncertainty or error in the value for the loading concentration. This is obviously never true, and indeed since the values required are absolute molar concentration, often there will be a significant uncertainty in how to convert between the optical signals (absorbance or fringes) and molar concentrations.

"Conservation of Signal" Method

Allen Minton has developed a related technique which he calls "conservation of signal" that addresses some of the drawbacks of a strict conservation of mass.[11] In this technique an experimental integrated signal is computed, where the region of integration excludes the meniscus and base regions and covers exactly the region over which data are being fitted. This experimental integral is then equated to the integral of the signals from each species, to again implicitly solve for the monomer concentration at the reference radius using the analytical integrals available for sector-shaped cells. This procedure is thus also able to eliminate a concentration term as a fitting parameter, but does not require true mass conservation or the ability to observe the entire solution column.

[11] A. P. Minton, *in* "Modern Analytical Ultracentrifugation" (T. M. Schuster and T. M. Laue, eds.), p. 81. Birkhauser, Boston, Massachusetts, 1994.

The use of the conservation of signal approach in strategies to analyze mixed associations has been discussed elsewhere,[12] and this strategy can be incorporated into both multiwavelength analysis strategies or situations where one can directly monitor the distribution of one type of monomer species. This method was successfully employed to analyze dissociation of the tryptophan synthase heterotetramer.[13]

When only the total signal can be observed for mixed associations, one potential problem with applying the conservation of signal approach is that the unobservable regions near the meniscus and base will generally contain a ratio of species that is different than the mixing ratio loaded into the cell. Since one is "losing" a different amount of each type of reactant, the amount of integrated signal to assign to each reactant species may not be accurately known. A second potential problem from a statistical point of view is that, in effect, this approach assumes that the correct fit will have the property that the integral of the theoretical signals calculated from the fitting parameters will equal the integral of the experimental data. This assumption is equivalent to requiring that the integral of the residuals be zero, which certainly is often true for an excellent fit. However, there is no guarantee that this requirement necessarily gives the true best fit in the least squares sense, and indeed we find for many global least squares fits of centrifuge data at least some of the data sets will have nonzero integrated residuals.

"Soft" Mass or Signal Conservation Constraints

An alternative approach we have employed is to use general numerical constraint methods to guide an iterative least squares fitting procedure toward sets of fitting parameters that are approximately consistent with mass or signal conservation. To implement this generalized constraint, as described by Johnson and Frasier,[14] the constraint is incorporated as an additional data point that stores the desired value for the physical quantity that is being constrained (a quantity related to some combination of parameters or value known from independent experiments). During each iteration of a least squares procedure, when this data point is encountered the program uses the current parameters to calculate the value of the constrained quantity, and compares this to the "experimental" point in the usual way. Thus if the current parameters are *not* consistent with the con-

[12] A. P. Minton, *Progr. Colloid Polym. Sci.* **107,** 11 (1997).

[13] S. Darawshe, D. B. Millar, S. A. Ahmed, E. W. Miles, and A. P. Minton, *Biophys. Chem.* **69,** 53 (1997).

[14] M. L. Johnson and S. G. Frasier, *Methods Enzymol.* **117,** 301 (1985).

straint, the constraint data point will have a large residual, and the least squares minimization procedure will naturally drive the system toward parameters that are more consistent with the constraint.

A significant advantage of this form of constraint is that the constraint can have an uncertainty associated with it just like any other data point. These constraints are "soft" in the sense that the optimal fit does not necessarily have to *exactly* fulfill the constraint condition (just as the fitted curves will not exactly pass through all the "normal" data points). Another advantage is that this approach does not rely on the assumption of a sector-shaped cell. Our implementation is designed for the rectangular six-channel centerpieces, but in principle any cell shape could be readily accommodated. A significant drawback of this approach relative to "hard" mass or signal conservation is that the monomer concentrations at the reference radius must still be fitted, i.e., these constraints do not reduce the number of fitting parameters.

The actual implementation that we use is to numerically integrate the total molar concentration of each type of monomer (free as well as in all its associated forms) across the cell, based on the current values of the fitting parameters, in order to calculate the mean concentration over the radial range that is being included in the analysis. That is, the molar concentration of each species as a function of radius, $c_i(r)[i = 1, \ldots, N]$, is calculated using the current values for the concentrations of free monomers at the reference radius, the association constants, and the buoyant masses. Designating the number of monomers of type j in the ith species as n_{ij}, next $c_{\mathrm{av},j}$, the mean concentration of species j, is computed over the fitting region from radius r_1 to r_2 by

$$c_{\mathrm{av},j} = \frac{\sum_{i=1}^{N} \int_{r_1}^{r_2} c_i(r) n_{i,j} \, \mathrm{d}r}{\int_{r_1}^{r_2} \mathrm{d}r} \tag{5}$$

This theoretical value of $c_{\mathrm{av},j}$ is compared with $c_{0,j}$, the concentration that was initially loaded into the cell, which is input by the user as part of defining the data to be included in the fit. Thus for systems containing two types of monomers there are two extra data points added per scan, corresponding to the constraint on each type of monomer.

It is worth noting some details of how this algorithm has been incorporated into a Gauss–Newton fitting algorithm. The extra data point(s) representing the constraint(s) on each data set are added as the *last* point(s) of that set. This means that by the time the Gauss–Newton algorithm needs to evaluate the theoretical function for the constraint point, the fitting function will already have been evaluated for all the normal data points.

This ordering thereby permits the integral in the numerator of Eq. (5) to be calculated step by step as the normal fitting function is being calculated at each radius, thus reducing the computational time. Another potential problem arises because the Gauss–Newton method requires derivatives of the fitting function. The derivatives of the fitting function with respect to the n fitting parameters, p_i, for the normal data points are derived numerically by temporarily adding a small increment to each parameter, recalculating the fitting function, and evaluating

$$\frac{\partial f}{\partial p_i} = \frac{f(p_1, \ldots, p_i + \delta p, \ldots, p_n) - f(p_1, \ldots, p_i, \ldots, p_n)}{\delta p} \tag{6}$$

During these function evaluations related to derivative calculation, a flag is set that causes the sums needed for evaluating the numerator of Eq. (5) with concentrations corresponding to the altered parameter $p_i + \delta p_i$ to be stored in separate array elements indexed to the parameter number. Thus if n parameters are being fitted there are actually $n + 1$ integrals like Eq. (5) being accumulated. Hence when it is time to evaluate the derivatives of the constraint data point(s), the integrals corresponding to all the temporarily incremented parameters have already been evaluated, and the partial derivatives of the fitting function at the constraint points can then also be easily calculated numerically using Eq. (6).

One other somewhat difficult problem in implementing this type of constraint is that it has totally different units and a different numerical range than the normal experimental points. Further, since for each scan this constraint data point is usually only one among hundreds, even a large violation of the constraint could potentially have little impact on the overall variance. Therefore another related issue is how to weight the constraint points relative to the others (which effectively determines whether the constraint is "tight" or "loose").

In our implementation for each constraint the user also inputs a fractional tolerance or uncertainty value, δc_j (usually 10%), that is used to control the overall weighting of the constraint point. (We do not actually apply weights to the experimental points.) For simplicity we will hereafter assume we are working with absorbance scans, but the same approach can be applied to interference data or even postrun analysis methods. To put the constraint value in a numerical range comparable to the scan data, $c_{0,j}$ is also multiplied by the value of the total absorbance signal at the highest radius included in the fit, $y(r_2)$. That is, the actual value stored as the experimental value for the constraint point is $c_{0,j}y(r_2)/(\delta c_j c_{0,j})$ and the function that is computed for the constraint points is $c_{\mathrm{av},j}y(r_2)/(\delta c_j c_{0,j})$. The result is that if the computed average concentration is in error by the fractional tolerance c_j, then when the constraint is evaluated it will produce an increase in variance equivalent to that from a 100% error in a single point at the base of the cell (or a 10% error in 100 data points at the base).

This increase in variance when the constraint is not met is significant enough to drive the fit toward satisfying the constraint. In particular, it is usually sufficient to eliminate the false minima such as the one shown in Fig. 2. On the other hand, such a constraint is loose enough to allow for errors in determining the loading concentrations or violations of mass conservation. In fact, one of the nicest features of this form of constraint is that the user can go from strict mass conservation and trying to include data all the way from meniscus to base, to an implementation analogous to the conservation of signal method, simply by changing the fitting region and choosing to input values of $c_{0,j}$ that are consistent with the total integrated signal over that fitting region.

In applying these constraints we generally take the latter "conservation of signal" approach, setting the constraint concentrations so they approximately match (within 3–5%) the integrated total absorbance (allowing for the molar extinction coefficients for each type of monomer at the measurement wavelength). In setting these concentrations for signal conservation, we hold the *ratio* of concentrations at the value intially loaded into the cell. As noted previously, the true ratio may actually differ from the loading ratio due to the unobservable regions near the meniscus and base and/or solutes preferentially lost to cell surfaces. However, since this is only a "soft" constraint, possible errors in setting these constraint values should have less influence on the results.

Examples of Applications

The generalized constraint procedures described above worked successfully at Amgen in the analysis of at least 22 different pairs of interacting proteins, as well as for some protein–peptide and protein–ligand pairs. These studies provided important new insights about receptor–ligand interactions for a number of protein hormones that are currently in clinical use or under development,[4,15–17] as well as providing data useful for developing improved versions of these drugs. For most of these systems at least one component exhibits multivalent binding, and in every case we were able to complete the analysis using only total absorbance data, without labeling

[15] T. P. Horan, J. Wen, L. O. Narhi, V. Parker, A. Garcia, T. Arakawa, and J. S. Philo, *Biochemistry* **35,** 4886 (1996).

[16] J. S. Philo, J. Talvenheimo, J. Wen, R. Rosenfeld, A. A. Welcher, and T. Arakawa, *J. Biol. Chem.* **269,** 27840 (1994).

[17] J. S. Philo, K. H. Aoki, T. Arakawa, L. O. Narhi, and J. Wen, *Biochemistry* **35,** 1681 (1996).

or exploiting spectral differences between components. It is important to note, however, that when only total concentration data are available, extraction of the association constants will generally require inclusion of significantly more data sets, covering a wider range of concentration and mixing ratios, into the global analysis than would be needed when individual components can be measured.

One example that will hopefully illustrate how we have approached analyzing these systems and applied the generalized constraint methodology comes from studies of the interactions of the hormone erythropoietin (EPO) with a soluble extracellular binding domain of its cognate receptor, a protein we call sEPOR. One of the things that makes this system interesting is that although EPO is a monomeric protein (in contrast to the dimeric SCF discussed above) it is still able to bind to two molecules of receptor, and on cells this dimerization of receptor brings together the intracellular domains and thus initiates a chain of signaling events that ultimately leads to alterations in gene expression.

To analyze such a system, what data do we need before we start? For any mixed association it is always important to study the behavior of each component by itself to determine its self-association behavior. In this case both EPO and sEPOR sediment as simple ideal monomers, at least for concentrations up to ~1 mg/ml. Particularly when the only available data are for total concentrations, we also believe it is quite important to know the buoyant mass of each component. Those buoyant masses can then be held fixed during the analysis of data from mixtures, thus limiting the number of free parameters and improving discrimination of the association constants. In some cases it may be sufficient to calculate the buoyant mass based on a sequence mass and a calculated partial specific volume. However, it is always best to have an accurate experimental value, particularly for glycoproteins where there is usually a large uncertainty about both the correct mass and partial specific volume.

It is also extremely helpful to know in advance the stoichiometry of the interactions. Obviously this will help tremendously in choosing a correct model for analyzing the data, but in addition, the stoichiometry dictates what range of mixing ratios should be prepared (a point for further discussion below) and what rotor speeds will be appropriate. In this case we had already demonstrated the formation of 2:1 receptor–ligand complexes using on-line laser light scattering in conjunction with size-exclusion chromatography,[18] and indeed for most mixed associations we have investigated we found this technique to be a very helpful and complementary adjunct to analytical ultracentrifugation. One of the surprises from the light scattering

[18] J. Wen, T. Arakawa, and J. S. Philo, *Anal. Biochem.* **240,** 155 (1996).

studies, however, was that these 2 : 1 complexes were only seen if the mixture was injected at quite a high protein concentration. It was not entirely clear whether this was a dynamic phenomenon resulting from the physical separation occurring on the column, or truly reflected weak binding, and thus we were anxious to probe these interactions with a true equilibrium method and to determine the association constants.

Armed with this knowledge, we prepared and ran 9 samples made over a range of receptor : EPO ratios and total protein concentrations, collecting data at two rotor speeds, and all 18 resulting data sets were then analyzed globally, using the soft conservation of signal restraints. Since we knew that each EPO molecule has two binding sites for receptor, and there is no known internal symmetry in EPO, we first tried a simple model where these two sites have potentially different association constants, but are independent of each other (no binding cooperativity). This model turned out to provide quite a good fit of the data, gave consistent results for independent experiments, and provided a framework for interpreting other experiments. Because these results have been described in detail elsewhere,[17] here we merely summarize the model and the dissociation constants derived from the fitting, and present a few of the data sets to illustrate some aspects of analyzing a system of this type.

The somewhat surprising finding is that the two binding sites on EPO have very different affinities, and that one is fairly weak. As shown in Fig. 3, the binding site we call site 1 has a dissociation constant of $\sim$1 nM, whereas site 2 is some three orders of magnitude weaker. The much weaker binding at site 2 explains why the 2 : 1 complex was only seen by chromatography at high concentrations, and why other research groups had failed to detect it at all. The site 2 affinity is well determined, but the site 1 affinity is too high to be determined accurately, especially because we used relatively high protein concentrations in order to promote binding at site 2. All that can really be said with 95% confidence about site 1 from these data is that its K_d is $<$3.4 nM, and this result is quite consistent with values of $\sim$1 nM reported by others. (Better discrimination of the site 1 affinity would be possible by including data at lower concentrations, but in this case we were more interested in characterizing site 2.)

Before discussing the use of the concentration constraints, it is worthwhile to examine some other aspects of these data. Three of the data sets and their fitted curves are shown in Fig. 4, along with the contributions of different species to the signals. As would be expected, although theoretically a 1 : 1 complex with receptor on site 2 is possible (the species at the lower left in Fig. 3), the amount of this species is always negligible. In addition, the high site 1 affinity means that there is very little free EPO present, and in all three experiments its contribution to the total absorbance is also

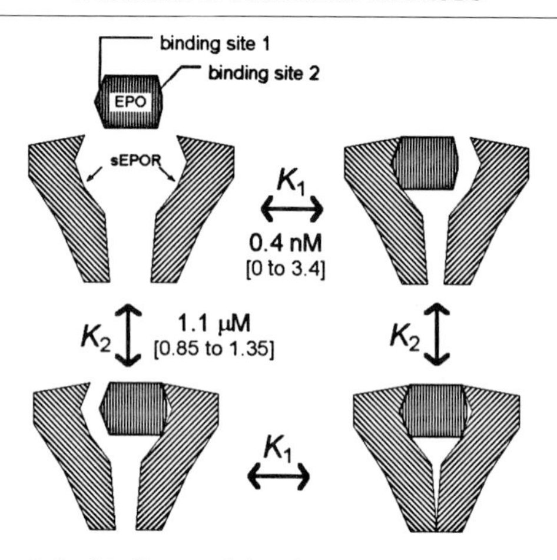

FIG. 3. Summary of the binding model and fitting results for the interactions of EPO (made in *Escherichia coli*) with deglycosylated sEPOR.[17] The dissociation constants were obtained from a global analysis of 18 data sets; their 95% confidence limits are shown in square brackets.

negligible. Figure 4A shows a high concentration sample made at a 2:1 receptor:EPO ratio. Most of the signal for this sample comes from 2:1 complexes, with lesser but significant amounts of a 1:1 complex in which the receptor is occupying site 1. Figure 4B shows a sample with the same amount of receptor but twice as much EPO (a 1:1 mixing ratio). What is very striking is that the extra EPO leads to a strong decline in the amount of 2:1 complex, and the sample is virtually entirely 1:1 complex. To understand why this happens, consider what will happen if we titrate more EPO into the 2:1 sample from panel A. As we add more EPO we are adding more and more of the strong site 1 sites, and it is thermodynamically highly favored for the receptors occupying the weak site 2 sites in 2:1 complexes to leave those sites in order to occupy site 1 on the new EPO molecules, leading to a net loss of most of the 2:1 complexes. Figure 4C again shows a 2:1 mixture, but at a fourfold lower concentration. The drop in concentration leads to a significant loss of 2:1 complexes compared to Fig. 4A.

Thus it is obvious from comparing Figs. 4A and C why the global analysis can determine K_2. What is far less obvious is how we can obtain *any* information about a nanomolar K_1 from data at micromolar protein concentrations. The interesting point is that we get this information *only* from the samples made with more EPO than needed for the 2:1 stoichiometry. Figure 5 shows from theory how the weight-average mass varies with EPO:receptor mixing ratio at a receptor concentration of 5 μM, assuming

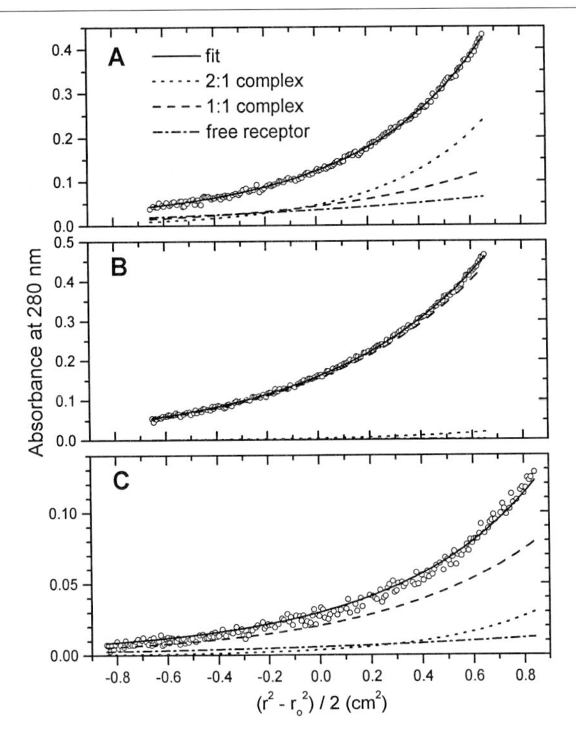

FIG. 4. Selected data sets at 18,000 rpm from the EPO–sEPOR studies. The fitted curves from the global analysis are shown as solid lines. The amount of absorbance arising from 2:1 receptor EPO complexes is shown as a dotted line, that from 1:1 complexes (site 1 on EPO occupied) is shown as a dashed line, and that from free receptor is shown as a dashed-dot line. The contributions from free EPO and 1:1 complexes with site 2 occupied are negligible. (A) Sample made with 4 μM sEPOR + 2 μM EPO; (B) 4 μM sEPOR + 4 μM EPO; (C) 1 μM sEPOR + 0.5 μM EPO.

$K_2 = 1$ μM and using K_1 values of either 1, 3, or 10 nM. Clearly for EPO:sEPOR ratios above 0.5 the average mass *is* sensitive to K_1 even when the concentration exceeds K_1 by over 3 orders of magnitude. For this system the asymmetry of the binding constants causes M_w to fall off much more rapidly on the right-hand side of the curve than it does for a symmetric case (Fig. 1), and the extent to which this happens depends on the ratio of the two binding constants.

Figures 1 and 5 also illustrate an important point about strategies for studying these mixed associations. One common strategy is to purify and isolate a stoichiometric complex (for example, using size-exclusion chromatography) and then to study the dissociation of this complex in the centrifuge. By thus obtaining samples which are exactly at the stoichiometric

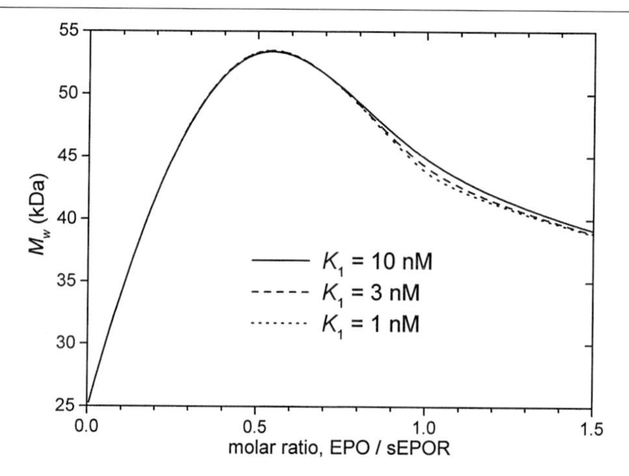

FIG. 5. Theoretical weight-average masses for mixtures of EPO (18.4 kDa) with sEPOR (24.8 kDa) versus mixing ratio. The sEPOR concentration was held at 5 μM, and K_2 at 1 μM. The solid line shows the results when $K_1 = 1$ nM, the dashed line is for $K_1 = 3$ nM, and the dotted line is for $K_1 = 10$ nM. These curves illustrate that the data are actually sensitive to K_1 at protein concentrations >1000-fold higher than K_1, but only for samples where EPO is in excess over the one EPO per two sEPOR stoichiometry.

mixing ratio, and staying on the maximum of the M_w versus mixing ratio curve, one can avoid the problem of multiple solutions, and fitting is simplified. However, in our experience much of the critical information needed for extracting affinities and for discriminating among models comes from samples that are far from the preferred stoichiometry, and this will be true particularly for more complex models and when the only information available is for total concentration. Another good example of this arose during the global analysis of data for SCF–sKit interactions.[4] In this case we wanted to know whether the two binding interactions of the SCF dimer were truly equivalent, i.e., is there any binding cooperativity? We found that by incorporating into the analysis more data for samples made with greater excess SCF we were able to significantly improve discrimination of whether the first and second receptors bind equivalently, and we were thereby able to say with 95% confidence that any difference in affinity is less than 15%. In general, the judicious use of simulations is probably the best way to truly optimize the choice of mixing ratios and protein concentrations to get the most information and best discrimination among models, and Hsu and Minton have described an elegant numerical method for doing this.[19]

[19] C. S. Hsu and A. P. Minton, *J. Mol. Recog.* **4,** 93 (1991)

Assessing Impact of Signal Conservation Constraints on this Type of Analysis

This important role for off-stoichiometry samples then leads us directly back to the use of conservation constraints, because it is for such samples where the multiple solutions problem arises. For this EPO–sEPOR data we used the soft generalized conservation of signal approach discussed above, specifying total concentrations to match the ratio loaded into the cell, and a concentration tolerance of 10%. When using this approach we try to keep the maximum rotor speed fairly low to limit the steepness of the gradient (Fig. 4) and thus limit the amount of material in the unobservable region near the base. Nonetheless, the total signal in the fitted regions is often 5–15% less than what was originally put into the cell for shallow concentration gradients, and 25–35% less for gradients like those in Fig. 4. Thus in general we find it is impractical to use conservation of mass constraints, and hence we fall back on conservation of signal.

How well does the global fit actually fulfill the signal conservation constraints, and do the constraints contribute significantly to the overall variance? In this case 4 of the 36 constraints are not satisfied within the nominal 10% tolerance. The worst case is actually for the sample shown in Fig. 4C, where the returned total EPO concentration is 26% higher than was specified. However, because this is a low concentration sample, and also because the specified EPO concentration represented only 22% of that small total absorbance, even this level of deviation accounts for only 0.02% of the total variance of the fit, i.e., less than the average variance for the other 3280 data points. (The insensitivity of the variance to meeting this constraint is, of course, exactly why this is a situation where the concentration deviation can be larger without producing a bad fit.) One can, of course, choose to set a tighter tolerance for those situations where the constraint is perhaps too loose (one of the nice features of this type of constraint) and thus force a tighter match to the concentrations loaded into the cell. Together, all 36 constraints in this fit contribute 4.9% of the total variance. Reviewing our results for other mixed associating systems, typically 15–20% of constraints fall outside the nominal tolerance, and these are typically the ones where the corresponding species contributes little to the total signal.

Another interesting question is how much do the constraints actually influence the binding affinities that are returned? Another nice feature of this approach is that once we have fitted the data using the constraints, and thus have hopefully avoided local minima and nonphysical solutions, the constraints can be relaxed or eliminated entirely and fitting then continued in order to examine this issue. For these EPO data the complete removal of constraints leads to an approximately twofold increase in K_1 and a similar decrease in K_2. Significantly, for some samples the unconstrained fit implies mixing ratios up to approximately twofold different than the initial composition of the cell. Thus even though the unconstrained fit may fit the scans more closely, such fits may stray significantly from physical reality, and for that reason we always report results for the constrained fits. For other mixed associating systems we have studied, the changes on removing the constraints are typically smaller than for these EPO–sEPOR data, with unconstrained association parameters usually falling within the 95% confidence limits for the constrained fit.

Removal of the constraints typically also causes a significant increase in the cross-correlation between parameters and slower convergence. Because there are so many parameters being fitted (36 monomer concentrations and 2 association constants in this case), this reduction of parameter cross-correlation by the constraints can be a major help in converging to a unique solution. For strongly associating systems such as this, where nearly all the signal is due to associated species, it also can be quite difficult to obtain good starting guesses for the monomer concentrations (it is not unusual for the guess to predict signal levels that are off by several orders of magnitude). A poor initial guess can easily make convergence extremely slow or impossible. When using the concentration constraints the initial guess is fortunately far less critical, because the constraint will quickly guide the fit toward a monomer concentration that gives signals in a reasonable range.

Dealing with Baseline Offsets

For fitting complex mixed associations we strongly recommend against fitting a baseline offset as a free parameter, and instead advocate directly determining an experimental value for the offset, for two reasons. First, baseline offsets are somewhat problematic for all approaches to imposing mass or signal conservation. In effect, an offset produces a concentration error that will be integrated across the entire cell. Thus even the small offsets of 0.01–0.02 AU, which seem to be typical for the XL-A, can cause a significant error in the computed total concentration, especially when the total concentration is low and/or the meniscus is highly depleted. This could be particularly troublesome for conservation of signal approaches when the baseline offset is allowed to vary as a fitting parameter, since the varying baseline directly affects the magnitude of the total "real" signal to be conserved. While this is certainly not an insoluble problem, it can easily be avoided by using fixed, experimental values for the baseline offsets.

Probably a more important reason for not fitting baseline offsets is to reduce the number of fitting parameters and to put the maximum amount of real experimental data into the analysis to enhance the extraction of meaningful values from the fit. For the EPO–sEPOR data discussed previously, fitting the baseline offsets would increase the number of parameters from 38 to 56, which can be done but is highly undesirable. In our work we always determine the offset experimentally by taking the rotor to high speed at the end of the run to completely deplete the solutes over a substantial fraction of the cell. The corresponding scan data are then used in a 1-parameter fit to obtain an average experimental value for the offset, and this value is used and held fixed during the global analysis to obtain association constants. Another advantage of this overspeed approach is that it also accounts for any absorbance by the buffer or other small molecules. An alternative approach to obtaining experimental offsets is to take a scan at a wavelength where the solutes do not absorb. However this latter approach requires an assumption that the offset is independent of wavelength, and our experience indicates this is not true.

When experimental offsets are lacking and the offsets are going to be fitted, then we have found a different type of fitting constraint to be useful. When data are available at two or more rotor speeds for the same cell, the offset applied to those data sets can be constrained to be the same. Thus only a single offset is fitted that is common to a particular cell and channel at all rotor speeds. This approach is particularly helpful at obtaining more reasonable estimates of the offsets for data sets with shallow concentration gradients. It is easily implemented by setting the offsets for the constrained data sets to match that of the fitted one after each fitting iteration, and adjusting the calculation of the partial derivative for the fitted offset parameter so it also applies to the constrained data sets.

How important are the baseline offsets in practice? Since they are usually small, why not ignore them altogether? Considering first fits for the simpler SCF–sKit system,[4] where there is only a single type of interaction, setting all the offsets to zero reduces the returned affinity by 2.3-fold (about 10 times the standard error). Allowing the offsets to "float" as fitting parameters gives an affinity about 25% lower than with the experimental offsets, and also increases the uncertainty in that affinity by about 30%. For the EPO data, with the offsets fixed at zero the value for K_1 is essentially indeterminate and the fit will not converage. If K_1 is held fixed at the value obtained with experimental offsets and the offsets are then set to zero, there is a twofold change in the optimal value for K_2. If the offsets are fitted this produces only a ~10% change in the optimal value for K_2, but the value for K_1 drops by ~1.5 orders of magnitude, and its uncertainty increases by ~1 order of magnitude. Thus these two examples show that baseline offsets, and how they are treated, can have a significant impact on the accuracy of the analysis and on the ability to resolve binding parameters for complex assembly models.

Other Potential Approaches and Applications of Concentration Constraints

Rather than directly constraining the total concentrations of each type of reactant, an alternative strategy would be to use a generalized constraint on the *ratio* of the reactants, forcing it to match within some tolerance the ratio placed into the cell. As noted above, it is already our usual practice to set the constraints on total signals so they match the initial mixing ratio, and thus a ratio constraint may be a superior approach, but one we have not yet explored.

An interesting potential use of this constraint methodology would be in analyzing systems where one reactant does not contribute at all to the signals. This might arise, for example, for protein–carbohydrate or protein–small molecule interactions. The use of the constraint would ensure a physically meaningful total concentration of the "invisible" component even in situations where it is nearly all in associated forms. Certain protein–protein systems may come close to this type of situation when one component is much more massive and also has a higher extinction coefficient, as was true for example for granulocyte colony-stimulating factor and its receptor.[15]

Conclusion

Both "hard" and "soft" mass or signal conservation constraints have proven to be very useful in the analysis of mixed associations by sedimentation equilibrium. With such approaches it is possible to analyze even quite complex associations using only total concentration data, i.e., without requiring labeling or multiwavelength approaches to obtain concentrations of individual species. Each of these methodologies has certain strengths and weaknesses, and each deserves a place among every scientist's tool kit for analytical ultracentrifugation data analysis.

Acknowledgment

All of the experiments, and much of the development of the algorithms described here, took place while the author was in the Protein Chemistry Department at Amgen, Inc., and would not have been possible without the skills and support of many scientists there.

[7] Optimal Data Analysis Using Transmitted Light Intensities in Analytical Ultracentrifuge

By EMILIOS K. DIMITRIADIS and MARC S. LEWIS

Introduction

Analytical ultracentrifugation machines, such as the Beckman XL-A, may be equipped with absorbance optics that is essentially a dual-beam spectrophotometer with a xenon lamp and a monochromator. The lamp illuminates through solution-containing, sector-shaped channels in the cylindrical cells, which are sealed on opposite sides with quartz windows. The intensity of the transmitted light through pairs of channels in every cell is measured by a photomultiplier. Most commonly, double-sector cells are used, but other configurations exist. One sector contains a buffer solution to serve as reference and is usually very transparent to the wavelengths of the light used; the other contains the same buffer plus one or more species of solutes, such as proteins, peptides, or oligonucleotides, whose behavior is being investigated. To reduce instrument noise, the sectors are scanned a number of times at each of an array of closely spaced (typically 0.002-mm step size) radial positions. The intensity data are then converted to absorbance by taking the base-10 logarithm of the ratio between the intensity transmitted through the reference and the solution sectors, and the computed absorbance data at every position are averaged. The acquired absorbance data (the dependent variable) correspond to the profile of solute concentrations as a function of radial position (the independent variable) in the sector.

To investigate the state of the solution, solute behavior models are assumed and corresponding mathematical models for the expected concentration profiles are written. The suitability of the assumed models is then tested by attempting to curve fit the mathematical models to the collected absorbance data. Usually, quantities such as molecular masses and association constants or sedimentation and diffusion coefficients are used as the fitting parameters for modeling sedimentation equilibrium or velocity data, respectively. This study focuses exclusively on optimizing the data analysis of sedimentation equilibrium experiments. The final decision for the most probable behavior of the solute in the particular solution is based on the quality of the fit as described by various statistical measures. The method used most often for curve fitting is that of least

0076-6879/00 $30.00

squares,[1] which seeks estimates of the unknown model parameters so that the sum of squares of the deviations of the actual data from the model curve is minimized. In addition, it is known that optimal least squares estimation is achieved if the data are weighted by the inverse of the variance. Hence, weighted least squares has been the method of choice. This, however, requires an iterative scheme because the optimal weights are not known at the outset. It is known from estimation theory[2,3] that, although the procedure will invariably minimize the variance of the fit, the least squares method results in unbiased estimation of the unknown parameters under a number of specific conditions. Assuming that we are using the correct model, that the number of data points is adequate, and that uncertainty exists only in the dependent variable, we must still examine the remaining assumptions which are (1) the data are independent of each other, (2) there is no other systematic error, and (3) the noise probability distribution is Gaussian. The first two assumptions are generally satisfied except for the fact that often the enclosing windows may have not been properly cleaned or the solution itself may contain particles that will present an image to the measuring system. It would be desirable if the mathematical analysis scheme "cleaned" the data of such systematic noise. The last assumption is the main issue in this chapter.

Fortunately, the design of the Beckman XL-A analytical ultracentrifuge (AUC) used here permits the acquisition of raw intensity data, and if enough samples are collected it should be possible to examine the probability distribution of the noise in some detail. We have performed such measurements and found that while the intensity is contaminated by Gaussian noise, the absorbance noise is not normally distributed. The fundamental statistical principles explaining this observation are presented and the non-Gaussian distribution for the absorbance data is derived. The conclusion is that the least squares method estimates are statistically biased. Unbiased estimations can be achieved either by fitting the intensity data directly or, since the absorbance noise characteristics are known, by using the maximum likelihood method. The latter is equivalent to least squares only when the noise is Gaussian but otherwise it results in better, more reliable estimates of the model parameters than least squares would give in this case.

[1] C. R. Rao, "Linear Statistical Inference and Its Applications." John Wiley and Sons, New York, 1965.

[2] P. Eykhoff, "System Identification: Parameter and State Estimation." John Wiley and Sons, New York, 1974.

[3] E. B. Manukian, "Modern Concepts and Theorems of Mathematical Statistics," Springer-Verlag, Berlin, 1986.

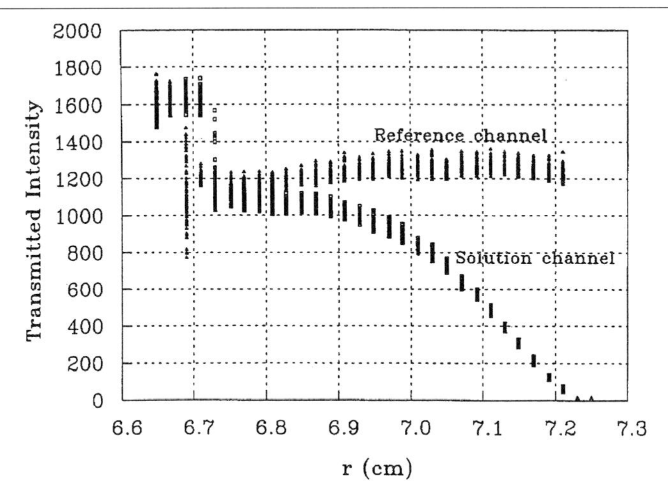

FIG. 1. Raw intensity data collected using the Method Scan portion of software on the XL-A AUC. Ninety-nine data points collected at each radial position. From E. K. Dimitriadis and M. S. Lewis, *Progr. Colloid Polym. Sci.* **107,** 20 (1997) © Springer-Verlag, Berlin (previously Steinkopff Verlag, Darmstadt).

Methods

All the measurements presented here are performed with a Beckman XL-A AUC equipped with absorbance optics. The Method Scan procedure in the XL-A data acquisition software is invoked to collect a large number of intensity data at each of a number of radial locations for the purpose of establishing the noise characteristics of the instrument. Four-hole rotors are used where, in each of the three cells, one of the sectors is loaded with 185 μl phosphate-buffered saline (PBS) buffer solution and the other with 180 μl of one of three concentrations of equine myoglobin in PBS. The myoglobin is not expected to self-associate. A total of 99 samples of transmitted intensity are collected from each sector and at about 30 radial positions between the meniscus and the sector bottom. The collected data are shown in Fig. 1 where the scatter is seen to be significant.

The data are stored and processed by a 486-based data acquisition personal computer. Histograms of the intensity data, transmitted intensity versus frequency of occurrence, are constructed and attempts are made to fit the data at different radial locations and along the whole solution column with Gaussian distributions. The fit is very good in all cases and an example of the data along with the fitted Gaussian is shown in Fig. 2. It may be argued that this should be expected by virtue of the central limit theorem,[4]

[4] A. Papoulis, "Probability, Random Variables, and Stochastic Processes," 3rd ed. McGraw-Hill, New York, 1991.

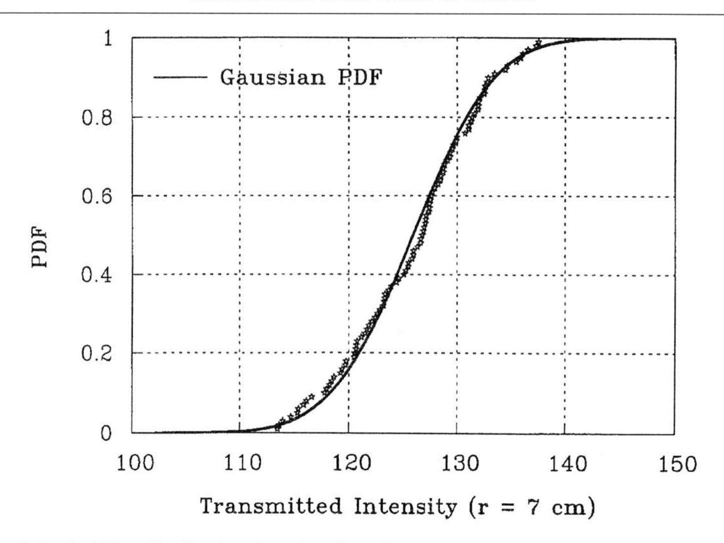

FIG. 2. Probability distribution function (PDF) of intensity data at a single radial position. From E. K. Dimitriadis and M. S. Lewis, *Progr. Colloid Polym. Sci.* **107,** 20 (1997) © Springer-Verlag, Berlin (previously Steinkopff Verlag, Darmstadt).

there are a fair number of noise contributors like electronic noise, suspended particles in the two channels, temperature fluctuations, and inadvertent deposits on windows to name a few.

Note also that the Gaussian noise distributions vary along the solution column. In fact, the variances of these distributions appear to be inversely proportional to the transmitted intensity itself. The noise variance is plotted as a function of intensity and then fitted by a low-order polynomial to quantitatively establish their relationship.

Figure 3 shows the experimentally observed variances versus signal strength and the fitted curve, which is linear. The availability of such information is useful for the generation of realistic data for computer simulations of any number of associating or nonassociating systems in the ultracentrifuge. The generated intensity data that correspond to the two sectors are then combined to compute absorbances at a large number of radial positions (200–250 points). Absorbance distribution histograms, however, departed from Gaussian distributions sufficiently to motivate a more detailed investigation into the issues involved from a theoretical, statistical point of view.

Theory

An important condition under which the method of least squares is known to perform optimal data noise filtering,[1] or otherwise is known as an unbiased estimator, is that the statistical characteristics of the experimental

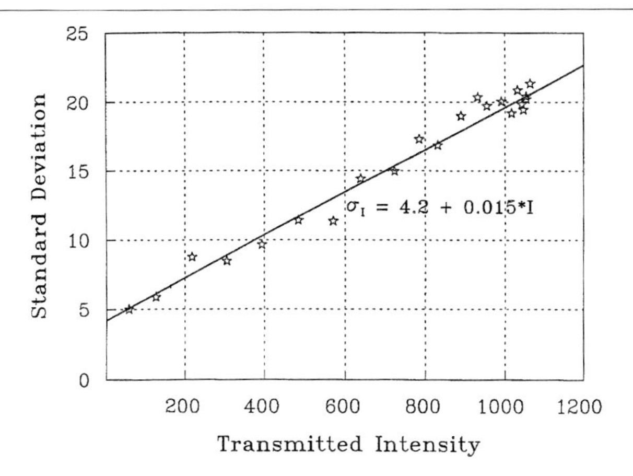

FIG. 3. Noise standard deviation as a function of transmitted intensity signal strength: raw data and least squares linear fit. From E. K. Dimitriadis and M. S. Lewis, *Progr. Colloid Polym. Sci.* **107,** 20 (1997) © Springer-Verlag, Berlin (previously Steinkopff Verlag, Darmstadt).

noise must be Gaussian. If this condition is not met and the noise can be described statistically by a different distribution, unbiased parameter estimation can still be performed by directly using the more general maximum likelihood estimation (MLE) method.[3] It is easy to show that when the signals are contaminated by Gaussian noise, the least squares method is a special case of the general MLE method.[1] It is therefore said that the least squares is a maximum likelihood estimator for Gaussian signals. Both the general MLE method, which needs *a priori* knowledge of the noise probability distribution, and the least squares method applied to signals with Gaussian noise have been shown to give parameter estimations whose error covariance asymptotically tends to the theoretical lowest limit achievable, known as the Cramer–Rao lower bound,[3] as the number of data points is increased. Application of the least squares method, however, to signals contaminated by anything but Gaussian noise would give biased estimations with lowest possible bound of the estimated parameter variance higher than the Cramer–Rao bound.

Here, the above-described measurements have ascertained that the noise in the transmitted intensity data from both sectors is indeed Gaussian in distribution. Any linear combination of two Gaussian random variables, x and y, results in a new random variable whose probability distribution is the two-dimensional Gaussian distribution

$$P(x, y) = \frac{U(x)\,U(y)}{2\pi\sigma_1\sigma_2\sqrt{1-r^2}} \exp\left\{\frac{1}{2(1-r^2)}\left[\frac{(x-\bar{x})^2}{\sigma_1^2} - \frac{2rxy}{\sigma_1\sigma_2} + \frac{(y-\bar{y})^2}{\sigma_2^2}\right]\right\} \tag{1}$$

where σ_1 and σ_2 are the standard deviations of x and y respectively, $\bar{x}$ and $\bar{y}$ are their respective mean values, and r is their correlation coefficient which, for the intensity data, would be zero since the noise in the two sectors is presumed independent. The unit step functions $U(x)$ and $U(y)$ signify the fact that no negative intensity measurements are possible. The intensity data, however, are not combined linearly when absorbances are computed. In fact, taking the ratio of two zero-mean, Gaussian signals results in a new signal whose noise is distributed according to what is known as the Cauchy distribution, which, in this case, can be written as,[4]

$$P_z(Z) = \frac{\sqrt{1-r^2}\,\sigma_1\sigma_2/\pi}{\sigma_2^2\left(z - \frac{r\sigma_1}{\sigma_2}\right)^2 + \sigma_1^2(1-r^2)} \tag{2}$$

where $z = x/y$ and the other symbols are as defined above. Yet another variable transformation defined by $w = \log_{10}(z)$ will result in a signal whose noise characteristics should be similar to those of absorbance. The probability distribution of the new variable is in general given by[4]:

$$P_w(w) = \sum_{i=1}^{n} \frac{P_z[z(w)]}{(\partial w/\partial z)_{z=z_i(w)}} \tag{3}$$

where $z = \exp(w)$ is the inverse transformation and z_j are the n real roots of the transformation. Fortunately, in this case there is only one such root ($n = 1$). Using the new transformation and Eq. (2) into Eq. (3) leads to the probability distribution of w,

$$P_w(w) = \frac{2\sqrt{1-r^2}\sigma_1}{\pi\sigma_2} \frac{e^w}{e^{2w} - 2r\frac{\sigma_1}{\sigma_2}e^w + \left(\frac{\sigma_1}{\sigma_2}\right)^2} \tag{4}$$

with a mean value

$$\overline{w} = \frac{2\cos^{-1}(r)}{\pi}\ln\left(\frac{\sigma_1}{\sigma_2}\right) \tag{5}$$

which, interestingly enough, is not zero unless the variances of the two parameters are identical or the two variables are totally dependent ($r = 1$). Here, $r = 0$, and this simplifies Eq. (4) to

$$P_w(w) = \frac{1}{\pi\cosh(w - \overline{w})} \tag{6}$$

with a distribution mean $\overline{w} = \ln(\sigma_1/\sigma_2)$. The shift in the mean of the transformed variable is a clear indication that the use of the least squares approach on the absorbance data will introduce a systematic bias in the estimation process.

The above result has to be generalized for intensity signals that are not zero mean. The procedure is slightly modified from the above and is developed using the two signals defined in Eq. (1) except that they are assumed to be independent ($r = 0$). Two new random variables, X and Y, are defined as the 10-base logarithms of x and y, respectively. Based on Eq. (3), the probability density function of X will be given by

$$f_x(X) = \ln(10)\, 10^X f_x(10^X) \tag{7}$$

with a similar expression for the pair (y, Y), where $f_x(x)$ and $f_y(y)$ are the Gaussian distributions of x and y, respectively. If x and y represent the transmitted intensity signals for the reference and solution sectors, the absorbance signal will have characteristics of a new random variable defined by $z = \log_{10}(x) - \log_{10}(y) = X - Y$ whose probability density is given by the correlation of the two distributions,[4]

$$p_z(z) = \int_{-\infty}^{\infty} f_x(Y+z) f_y(Y)\, \mathrm{d}Y \tag{8}$$

When the equation for the Gaussian distribution and Eq. (7) are introduced into Eq. (8), the indicated integration results in[5]

$$\begin{aligned} p_z(z) = {} & \frac{\ln(10)10^z}{2\beta(z)\pi\sigma_1\sigma_2} \exp(-\delta) \\ & \left\{1 + \gamma(z)\sqrt{\frac{\pi}{\beta(z)}} \exp\left[\frac{\gamma(z)^2}{4\beta(z)}\right] \right. \\ & \left. \left[1 + \phi\left(\frac{\gamma(z)}{2\sqrt{\beta(z)}}\right)\right]\right\} \end{aligned} \tag{9}$$

where

$$\beta(z) = \frac{10^{2z}\sigma_2^2 + \sigma_1^2}{2\sigma_1^2\sigma_2^2}, \quad \gamma(z) = \frac{10^z \overline{x}\sigma_2^2 + \overline{y}\sigma_1^2}{\sigma_1^2\sigma_2^2}, \quad \delta = \frac{\overline{x}^2\sigma_2^2 + \overline{y}^2\sigma_1^2}{2\sigma_1^2\sigma_2^2} \tag{10}$$

[5] I. S. Gradshteyn and I. M. Ryszhik, "Tables of Integrals, Series and Products," 4th ed. Academic Press, New York, 1980.

and the special function $\phi(x)$ is the well-known probability integral or error function[5] defined by

$$\phi(x) = \frac{2}{\sqrt{\pi}} \int_0^x \exp(-t^2)\, dt \tag{11}$$

When the original variables, x and y, have zero-mean values, β and γ are zero, and, as expected, Eq. (10) becomes identical to Eq. (6). Unfortunately, no closed-form expression for the distribution mean is available but numerical computations using Eqs. (10) and (11) indicate that the shift of the mean is approximately still a function of the ratio of the two variances, its value is effectively independent of the mean values of x and y and the nonzero-mean values of x and y lead to a sharpening of the distribution of z compared to the distribution in Eq. (6). As discussed above, the variances in both sectors are functions of the signal strengths and, from the fitted equation in Fig. 2, it is seen that for typical data the standard deviation can exhibit up to a fourfold variation, which implies a nonnegligible shift of the distribution mean. This shift would normally be positive because the noise variance in the reference channel is generally larger than that in the sample channel. Moreover, as the sample channel meniscus is being depleted, the transmitted intensity through it approaches that of the reference channel, and the mean shift is minimum. The mean shift on the other hand is maximum at the cell bottom where the noise variance ratio is maximum. The least squares method fits a curve through the mean values of the data point distribution at each radial position. The variable transformation, therefore, used to obtain absorbance from transmitted intensities introduces a systematic bias to the least squares estimator that is monotonically increasing along the channel. This translates to steeper gradients along the column, which means that the bias in the least squares method should result in overestimation of the molecular masses and association constants.

The cost function for the general MLE problem is the conditional joint probability distribution of the measurements at all the radial locations given the values of the parameters. For random noise, the measurements at the different radial locations are independent and the cost function is simply the product of the probability distributions at each measurement point. This cost function has to be maximized. If the noise probability distribution at a measurement point is Gaussian it is easy to show that maximization of the cost function leads to the same algorithm as does least squares. When the noise is not Gaussian, one needs to go directly to the MLE method using the known noise distribution. Here the distribution is given by Eq.

(10) and the MLE cost function to be maximized becomes

$$P[A(\mathbf{r})/f_{\mathrm{A}}(\mathbf{r};\mathbf{a})] = \prod_{i=1}^{n} p_z[A(r_i); f_{\mathrm{A}}(r_i;\mathbf{a})] \tag{12}$$

where $\mathbf{r}$ is the vector of radial locations, $A(r_i)$ is the absorbance measured at location r_i, f_{A} is the function to be fitted to the absorbance data, and $\mathbf{a}$ is the vector of the unknown parameters in the model. It is convenient to maximize the logarithm of this function, which is equivalent to maximizing the function itself. The scheme results in a set of equations for the parameters (a_i, $i = 1, m$) in the form of

$$\frac{\partial}{\partial a_i} \ln P[A(\mathbf{r})/f_{\mathrm{A}}(\mathbf{r};\mathbf{a})] = 0, \qquad \text{for } i = 1, m \tag{13}$$

The scheme requires explicit expression of the probability density function in terms of its mean value, which for the expression in Eq. (10) is not possible in closed form. Instead, the method could be developed numerically. For fitting absorbance data this approach should give better results than the weighted least squares method. However, the complexity of the probability distribution function would most likely be a discouraging factor especially given the fact that the direct analysis of transmitted intensities with the weighted least squares method presents no significant additional complexity and gives maximum likelihood estimated parameters.

Thus, in view of the noise characteristics in the ultracentrifuge and the mathematical manipulations associated with parameter estimation, the user is given two choices. The first choice is to implement the MLE method to fit the usual absorbance data and the second choice is to acquire intensity data and to fit these directly using the least squares approach. This second approach is complicated by the fact that there are no mathematical/physical models for the intensity data in the two sectors. Instead, there is a model that relates the two intensity data sets, $I_0(r)$ and $I_s(r)$, with the usual mathematical function used to model absorbance data

$$I_s(\mathbf{r}) = I_0(\mathbf{r}) 10^{f_{\mathrm{A}}(\mathbf{r};\mathbf{a})} \tag{14}$$

where subscripts s and 0 denote the solution and reference sectors, respectively, and the exponent represents the mathematical model for the solute concentration distribution on an absorbance scale. Two methods were developed here to resolve the problem of the lack of independent models for the two intensities. These are described next.

Method 1: Spline Smoothing of Reference Sector Intensity

Cubic splines are used to construct a smooth curve that optimally fits the reference intensity data. The smoothed intensity is constructed as the sum of scaled cubic spline functions.[6,7] Cubic splines are compactly supported functions in the form of cubic polynomials which satisfy the second derivative continuity condition. These functions are constructed piecewise; they are bell shaped and symmetrical about their maximum. The reference channel can be fit by a sequence of N, spatially overlapping, cubic spline functions centered at equally spaced point x_n so that $x_{n+1} - x_n = 2p$. Such a function is written

$$F(x) = \sum_{n=1}^{N} a_n f^{(3)}(x - x_n) \tag{15}$$

A cubic spline centered at the radial position x having half-width equal to $2p$ may be written

$$\begin{aligned}
&f^{(3)}(x) = q_{\mathrm{A}}(x) = [(x + 2p)/p]^{3/4} && p > x \geq 0 \\
&f^{(3)}(x) = q_{\mathrm{B}}(x) = 1 - \frac{3}{2}(x/p)^2 - \frac{3}{4}(x/p)^3 && 2p > x \geq p \\
&f^{(3)}(x) = q_{\mathrm{B}}(-x) && -p > x \geq -2p \\
&f^{(3)}(x) = q_{\mathrm{A}}(-x) && 0 > x \geq -p \\
&f^{(3)}(x) = 0 && -2p > x > 2p
\end{aligned} \tag{16}$$

where p is an arbitrary length scale that can be optimized for smoothing outlier points caused by, for example, small inadvertent deposits on the window over the reference channel. The scaling coefficients a_n are estimated by fitting the function $F^{(3)}(x)$ using the weighted least squares method. This computation has to be performed once for each data set at the beginning of the analysis and the scaling coefficients stored for use when fitting the data in the solution sector based on the model in Eq. (14).

A problem with the spline smoothing of the reference data is that the condition demanding independence among successive data points is violated. The introduction of a correlation distance of order p in the reference data would cause bias in the estimated parameter values but, in our experience, this bias is balanced out by the gain of smoothing the systematic noise that may be the result of, for example, undesired deposits on the window. Generally, the sensor photocathode transfer function should be a rather smooth curve without abrupt changes. Thus, without introducing significant bias one can estimate optimal spline amplitudes and correlation distance, p, by minimizing a weighted combination of the sum of the squares of the errors (as in least squares) and the total curvature in $F^{(3)}(x)$.[6,7] Even if one does not go to such lengths for choosing the best parameters for the

[6] M. Unser, A. Aldroubi, and M. Eden, *IEEE Trans. Signal Proc.* **41**(2), 821 (1993).
[7] M. Unser, A. Aldroubi, and M. Eden, *IEEE Trans. Signal Proc.* **41**(2), 834 (1993).

reference channel smoothing data, the length scale p should be chosen conservatively so as not to introduce smoothing that is too severe.

Method 2: Simultaneous Fitting of Data from Both Channels

A new, four-dimensional function is constructed from Eq. (10) as follows:

$$G_{\mathrm{I}}(i_{\mathrm{s}}, i_0, r) = i_{\mathrm{s}} - i_0\, 10^{f_{\mathrm{A}}(r;\mathbf{a})} \quad (17)$$

where the new variables, i_{s} and i_0 represent the sample and reference intensities, respectively, and $f_{\mathrm{A}}(r; \mathbf{a})$ is the function representing the concentration distribution along the channel in absorbance scale. Given the intensity data and the vector of radial locations, the vector of unknown parameters **a** can be estimated so that the deviation of the function G_{I} from zero is minimized in the least squares sense. This method has the theoretical advantage that the data in the reference sector are treated as uncorrelated as opposed to the previous method, which introduces correlations as discussed in connection with method 1. On the other hand, the ability of the spline method to smooth systematic noise caused by the inadvertent deposits on the window over the reference channel is lost. Such smoothing is also not possible when absorbance data are fit directly.

Based on the above methods and considerations, a series of simulations are performed to demonstrate and compare the behavior of the various approaches to the problem of estimating equilibrium parameters for a single nonassociating species and for a variety of interacting systems in ultracentrifuge experiments. Obviously, analysis of actual data from the centrifuge would not be useful in the context of this effort but the expected error variance minimization is reflected clearly in an example of an actual experiment discussed later.

Results

In all the simulations, reference channel intensity data were constructed by first fitting actual data from the reference sector using cubic splines as discussed in the Methods section. The scaling coefficients were stored and used as benchmarks to generate reference sector noisy data for all subsequent simulations. In all simulations, experiments with two cells were assumed to have a 2:1 concentration ratio in the solution sectors. In the following, the method of spline smoothing of the reference sector data is designated SSLS for spline smoothing least squares, the multidimensional data fitting method MDLS for multidimensional least squares, and the usual least squares directly on the absorbance data AWLS for absorbance weighted least squares. The computer package MLAB, by Civilized Software, Inc. (Bethesda, MD), was used in all cases. The least squares algorithm uses a Levenberg–Marquardt technique for solving the nonlinear equations for the parameters sought.

Monomer Solution

A solution of myoglobin was assumed for the first simulation. It is known that myoglobin does not self-associate and has a compositional molecular mass of 16,952 Da. The molecular mass M, the cell bottom concentration c_b, and a baseline error offset e_b were the fitting parameters in the model given by

$$f_A(r; M, c_b, e_b) = c_b \exp[AM(r^2 - r_b^2)] + e_b \tag{18}$$

where $f_A(r)$ is total solute concentration, expressed as absorbance at 280 nm, as a function of radial position r_b; $A = (1 - \overline{U}\rho)\ \omega^2/2RT$ where $\overline{U}$ is the compositional partial specific volume,[8] ρ is the solvent density, ω is the rotational speed in rad/sec, R is the gas constant, and T is the absolute temperature. Intensity data were generated using the nominal molecular mass and with zero-mean Gaussian noise, with variance according to Fig. 2 added. Ten samples were generated and then averaged for each radial position. The resulting intensity values were used to compute absorbance profiles for the two channels.

Curve fitting was performed using the three methods described above and starting with the same initial guesses for all methods. The results for the molecular mass were as follows: 16,971 Da for SSLS, 16,963 Da for MDLS, and 17,758 Da for AWLS. The quality of fit was excellent in all cases. It is clear that the first two methods perform better since their error is of the order of 0.1% while the error for ALS is of the order of 5%, which is not negligible. Equally important is the measurement of the buoyancy term, $1 - \overline{U}\rho$, for a monomeric species for which the molecular mass is reliably known. This is often done to correct compositional values of $1 - \overline{U}\rho$ for each species before analyzing data from an association experiment.

Monomer–Dimer Association

A macromolecule with the same molecular mass as myoglobin was assumed to self-associate forming a dimer. The fitting parameters for this case were the cell bottom monomer concentrations, the baseline errors, and the natural logarithm of the equilibrium association constant $\ln(k_{12})$, which was given on the absorbance scale. The appropriate model for this association is

$$f_A[r; \ln(k_{12}), c_b, e_b] = c_b \exp[g(r)] + c_b^2 \exp[\ln(k_{12}) + 2g(r)] + e_b \tag{19}$$

[8] H. Fujita, "Foundations of Ultracentrifugal Analysis." John Wiley and Sons, New York, 1975.

where $g(r) = AM_1(r^2 - r_0^2)$ and the remaining symbols were defined earlier. For data generation, three values of the association constant were used: $\ln(k_{12}) = 2, 4,$ and 8 covering a wide range of association strengths. Excellent quality fits were achieved in all cases. Using the first two methods the error in estimating $\ln(k_{12})$ was consistently less than 1%. Fitting the absorbance data, the error was of the order of 3% for the weakest association, 1.5% for the intermediate, and 1% for the strongest. Note, however, that this is the error in the estimated value of $\ln(k_{12})$ and that the actual error in the association constant k_{12} itself is roughly proportional to the product of k_{12} with the magnitude of the estimation error for its logarithm. Therefore, the approximate percentage error for k_{12} for the weakest association $[\ln(k_{12}) = 2]$ is 6% (2×0.03); for the intermediate strength of association the error is also 6% (4×1.5), and it is 8% (8×0.01) for the strongest association. The possible estimation errors are greater here than in the single species analysis since, with effectively the same amount of information (same number of data points), estimation is performed in a higher dimensional parameter space.

For this reason and because it has been observed that quite often the optimization surface gradients are very small over an extensive area around the optimal point, reaching true convergence is more critical. Here, every effort was made to achieve true convergence by making the convergence criteria increasingly more stringent. It is certain that the above computed errors are systematic, and due to the biased nature of the least squares estimator as it is applied to absorbance data.

Two Interacting Proteins Forming a Heterodimer

Two macromolecules, α and β, with molecular mass ratio of 3 : 1 in a solution with 1 : 1 molar concentration ratio were assumed to undergo one-to-one association forming a heterodimer. The natural logarithm of the equilibrium association constant $\ln(k_{\alpha\beta})$, the concentrations at the cell bottom of the two monomers $c_{b,\alpha}$ and $c_{b,\beta}$, and the baseline offset $e_{\alpha\beta}$ were used as fitting parameters. The association model was written as

$$f_A[r; \ln(k_{\alpha\beta}), c_{b,a}, c_{b,\beta}, e_{\alpha\beta}] = c_{b,\alpha} \exp[g_\alpha(r)] \\ + c_{b,\beta} \exp[g_\beta(r)] + c_{b,\alpha}c_{b,\beta} \exp[\ln(k_{\alpha\beta}) + [g_\alpha(r) + g_\beta(r)] + e_{\alpha\beta}, \quad (20)$$

where $g_\alpha(r) = A_\alpha M_\alpha(r^2 - r_b^2)$ and $g_\beta(r) = A_\beta M_\beta(r^2 - r_b^2)$. Again, three values of the natural logarithm of the equilibrium association constant $\ln(K_{\alpha\beta}) =$ 2, 4, and 8 were used for data generation. Equilibrium association simulations were again performed and the estimated parameters were compared to the data generation parameter values. It was systematically observed that the SSLS method gave the best results. For the weakest association the greatest error in the value of $\ln(k_{\alpha\beta})$ resulted when using the AWLS and it was about 5%. For the intermediate association strength the corresponding estimation error was of the order of 3% and for the strongest of about 2%.

A final computation of interest is shown in Fig. 4 for the PDF of the residuals obtained after fitting the simulated absorbance data for the interacting system and for the cases of strongest and weakest interaction. The residual distributions were then fitted by both Gaussian functions and by the log-ratio PDF given in Eq. (9). Notice that the log-ratio PDF describes the noise much more closely as the association strength decreases which is consistent with the error magnitudes in the simulation results.

Discussion

It has been shown that the noise probability distribution in the usual absorbance data collected in an XL-A ultracentrifuge is not Gaussian. The noise in the transmitted intensity, however, is Gaussian. The nonlinear transformation between the two forms of the data allows for the non-Gaussian distribution to be analytically derived in closed form. The form of the new distribution is not as convenient as one might desire for easy implementation of the MLE method, which is known to perform unbiased estimation for any noise distribution function, provided that the information

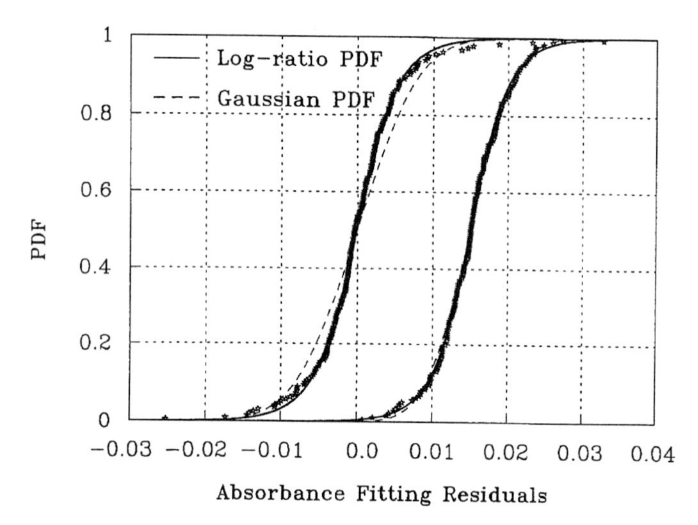

FIG. 4. Absorbance data fitting residuals' PDF for two association strengths: raw data, fitted Gaussians and log-ratio PDFs. The data for the strongest association have been shifted to the right to facilitate visualization. From E. K. Dimitriadis and M. S. Lewis, *Progr. Colloid Polym. Sci.* **107,** 20 (1997) © Springer-Verlag, Berlin (previously Steinkopff Verlag, Darmstadt).

is available. It has also been observed that the variance of the Gaussian noise in the AUC intensity data is proportional to the signal strength itself. Here, this variation was measured for the purpose of generating realistic centrifuge data for simulations. In general, of course, clear understanding of the noise characteristics and of the noise sources is the necessary prerequisite for the long-term implementation of more powerful and robust estimation methods.

The performed simulations show that a small but clear advantage exists in using intensity data to fit for the desired parameters. The fact that the estimation error variance is expected to be smaller when intensities are analyzed instead of absorbance has been manifested in actual experiments where data ovcr a range of temperatures were collected to thermodynamically characterize interactions. Figure 5 represents one such example. Here, the self-association of human polymerase-β was studied over a range of temperatures, and the association constant for each of the seven temperature settings was estimated and then converted into the Gibbs free energy of association, $\Delta G^0(T) = -RT\ln(K_{12})$, where R is the gas constant, T is the absolute temperature, and K_{12} is the monomer–dimer association constant on molar scale. The resulting seven points from each of the three analysis methods discussed above are shown in Fig. 5 and each set is fitted

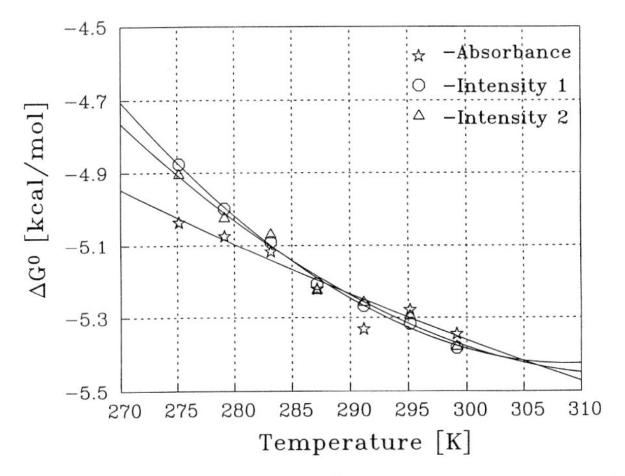

FIG. 5. Gibbs free energy of association (ΔG^0) for human polymerase-β self-association. The three sets of data were obtained by fitting absorbance data and by using the two methods of fitting intensities described in the "Theory" section. From E. K. Dimitriadis, R. Prasad, M.-K. Vaske, L. Chen, A. E. Tomkinson, M. S. Lewis and S. H. Wilson, *J. Biol. Chem.* **273**(32), 20540 (1998) © American Society for Biochemistry and Molecular Biology, Bethesda, Maryland.

by the function relating ΔG^0 to the standard enthalpic, entropic, and heat capacity changes taking place on association. Notice that the scatter of points about the fitted curves is significantly smaller for the intensity fits. Also, the values of the thermodynamic parameter changes estimated by the different fits differ significantly between intensity and absorbance, to the point that the thermodynamic characterization based on absorbances is much less definitive.

There is an additional advantage when using AUC intensity data that has to do with the truncation of absorbance data required to avoid cell bottom effects and nonlinear photocathode response at very high absorbance values. Additionally, the bias introduced in the analysis by the application of least squares estimation on the absorbance data is maximized near the cell bottom where the variance ratio in the two sectors is largest, thus introducing a maximal distribution mean shift away from zero. There is minimal need for truncation at the cell bottom when using intensity data, since the data from the solution sector are the least noisy in that region and aberrant points are easily identified. In contrast, however, absorbance data at the cell bottom are the most noisy because of the transformation from intensities to absorbance. This condition leads to uncertainty as to where truncation of absorbance data should be implemented and that uncertainty is reduced with the use of intensity.

The improvement in accuracy is always welcome but it would be even more so in cases that will certainly be borne with time as more and more experimental systems are studied by fitting both forms of data. For example, it might prove critical in cases where it is difficult to choose between two competing models either by the quality of the fit or by the thermodynamic analysis. This issue becomes even more important when parameter estimation errors are compounded with deviations between compositional and actual parameters of the macromolecules such as their specific volumes.

[8] Ultracentrifugal Analysis of Molecular Mass of Glycoproteins of Unknown or Ill-Defined Carbohydrate Composition

By MARC S. LEWIS and RICHARD P. JUNGHANS

Introduction

The presence of carbohydrate on protein is widely characteristic of cell surface proteins, receptors, and secreted proteins in eukaryotic systems.

0076-6879/00 $30.00

The carbohydrate moiety is thought to provide solubility, conformational stability, and resistance to degradative protease activities *in vivo,* and to participate in biological recognition mechanisms.[1] With modern techniques of molecular biology, it has become a simple matter to determine the precise amino acid sequence and thus the molecular mass of many proteins. However, comparable rules for predicting carbohydrate substitutions have not evolved, and the existing techniques for determining composition and structure are complex and the results are subject to many uncertainties. The requirement to have precise values for the relative amounts of protein and carbohydrate adds other sources of error, as protein mass determinations by the Lowry method, ultraviolet absorbance, and other procedures may be in error by as much as 10–50%.

Alternatively, one could determine the molecular mass of the whole glycoprotein from which the carbohydrate contribution might be deduced by subtraction of the protein molar mass. Yet, there are also major uncertainties in the values of the whole molecule molecular mass of glycoproteins as determined by such quantitative techniques as gel electrophoresis and exclusion chromatography. The biochemistry of protein binding of sodium dodecyl sulfate (SDS), a negatively charged detergent, forms the basis for size estimation in SDS–polyacrylamide gel electrophoresis (SDS–PAGE) in which a proportionate increase in negative charge and total mass is conferred with each molecule of SDS bound. Carbohydrate does not bind SDS and the presence of carbohydrate on a protein increases the mass and resistance to electrophoretic mobility without a proportionate increase in negative charge. As a result, glycoprotein molecular masses are regularly overestimated by this technique.[2] An additional drawback of SDS–PAGE is that the mass of noncovalent protein complexes cannot be assessed in the presence of SDS due to its strong dissociating properties. Gel exclusion chromatography is less subject to such errors, but standardization of the contribution of carbohydrate mass on such columns is not well established. Moreover, shape and charge of the glycoprotein imposes new uncertainties because these may lead to differences in penetration of and interaction with the solid media of the column.[3] Mass spectroscopy offers the possibility of a very accurate result, which is rapidly and relatively inexpensively available from commercial services.

However, if the solution properties of the glycoprotein are of impor-

[1] J. C. Paulson, *in* "Proteins: Form and Function" (R. A. Bradshaw and M. Purton, eds.), pp. 209–218. Elsevier, Cambridge, 1990.

[2] B. D. Hames, *in* "Gel Electrophoresis of Proteins" (B. D. Hames and D. Rickwood, eds.), pp. 17–18, 72. IRL Press, Oxford and Washington, D.C., 1982.

[3] H. G. Barth, *J. Chromat. Sci.* **18,** 409 (1980).

tance, and particularly if the study of molecular interactions is of importance, analytical ultracentrifugation also offers the possibility of providing accurate measurements of the molecular mass because equilibrium sedimentation is completely independent of shape and charge under the usual conditions of buffer pH and ionic strength and because the absence of detergent permits assessment of the possibility of reversible aggregation in the native state.[4] Yet even this method may be operationally hampered by a paucity of information regarding the carbohydrate composition and its contribution to the total compositional partial specific volume of many glycoproteins. While it is possible to calculate the apparent compositional partial specific volume for the protein portion of a glycoprotein by conventional techniques and thus determine the value of that apparent molecular mass with considerable confidence, these techniques are not applicable if the carbohydrate portion is ill defined, thus creating a significant level of uncertainty with regard to the molecular mass of the whole molecule.

We have developed an ultracentrifugal procedure for determining the molecular mass of a glycoprotein within acceptable limits when the amino acid sequence of the protein is known. The method is demonstrated for estimation of the molecular mass of Tac, the alpha (α) chain of the interleukin 2 (IL-2) receptor, a protein of central importance to immunobiology. Tac exists as a transmembrane protein of 251 amino acids (Tac^{251}) with a calculated protein molecular mass of 29,209 g mol^{-1}. Using SDS–PAGE, it appears as a band with a value for apparent molecular weight of 55,000, from which it has been designated the "p55" subunit. The intracellular maturation of Tac reveals a complex pattern of carbohydrate substitution including both O- and N-linked glycosylations and sialic acid residues. No mass estimate currently exists either for the complete glycoprotein or for the carbohydrate contribution.

For the purposes of our study, we have employed a soluble genetic construct of Tac (Tac^{224}) that is 27 amino acids shorter than the transmembrane molecule and 32 amino acids longer than the natural shed soluble form. This molecule preserves the entire extracellular domain, including all glycosylation sites, plus four amino acids of the transmembrane domain. Extensive studies of this product have shown it to preserve ligand binding characteristics and antibody recognition sites and to have a apparent molecular weight on SDS–PAGE compatible with the protein mass reduction of the engineered molecule. The soluble IL-2 receptor protein (soluble Tac, Tac^{224}) used here was the kind gift of Dr. John Hakimi (Hoffmann-LaRoche, Nutley, NJ). It is a genetically engineered protein of 224 amino

[4] M. S. Lewis, R. I. Shrager, and S.-J. Kim, *in* "Modern Analytical Ultracentrifugation" (T. M. Schuster and T. M. Laue, eds.), pp. 94–118. Birkhäuser, Boston, Massachusetts, 1994.

acids with a calculated molecular mass of 25,253 g mol^{-1} that includes a C-terminal Gly → Pro substitution. The protein was expressed in mammalian cell transfectants and purified from cell culture supernatants. It preserves the ligand binding and epitope character of the naturally occurring molecule.[5] This product was extensively characterized with confirmation of the intactness of N- and C-terminal residues.[6,7]

We have demonstrated the superiority of this biophysical methodology in a comparison with other procedures that yielded misleading results for the molecular mass of the IL-2 receptor.[8] A method quite similar to ours, but lacking the detailed consideration of the carbohydrate portion of the molecule described here was applied to the study of FcR–Fc interactions by Keown, Ghirlando, and co-workers[9] and by Ghirlando *et al.*[10] that included one of us (MSL) as a co-author. The present article represents the derivations of the method used in the aforementioned applications, describes its modification for the range of known glycoprotein carbohydrate contents, and provides a model sensitivity and error analysis. This treatment has key features not represented in the other citations applying this method and should be most useful to investigators who are considering methods to derive glycoprotein molecular masses.

Two further applications have been published that approach the problem of mass definition for glycoproteins.[11,12] Our contribution can be distinguished from these as follows. The Shire description[11] proposes an iterative method for obtaining the solution of the molecular mass, which is intrinsically more cumbersome to apply, whereas our solution is in an exact, closed algebraic form that simplifies the method's application. Further, no allowance is made for the reduction of partial specific volume that accompanies the loss of water in polymerization of the sugars in the glycosylation process, to which the total mass estimates are sensitive. The second article, by Philo *et al.*,[12] addresses the issues of glycoprotein mass estimation

[5] J. Hakimi, C. Seals, L. E. Anderson, F. J. Podlaski, P. Lin, W. Danho, J. C. Jenson, A. Perkins, P. E. Donadio, P. C. Familletti, Y. E. Pan, W. H. Truen, R. A. Chizzonite, L. Casabo, D. L. Nelson, and B. R. Cullen, *J. Biol. Chem.* **262,** 17336 (1987).

[6] M. C. Miedel, J. D. Hulmes, D. V. Weber, P. Bailon, and Y. E. Pan, *Biochem. Biophys. Res. Com.* **154,** 372 (1988).

[7] M. C. Miedel, J. D. Hulmes, and Y. E. Pan, *J. Biol. Chem.* **264,** 21097 (1989).

[8] R. P. Junghans, A. Stone, and M. S. Lewis, *J. Biol Chem.* **271,** 10453 (1996).

[9] M. B. Keown, R. Ghirlando, R. J. Young, A. J. Beavil, R. J. Owens, S. J. Perkins, B. J. Sutton, and H. J. Gould, *Proc. Natl. Acad. Sci. U.S.A.* **92,** 1841 (1995).

[10] R. Ghirlando, M. B. Keown, G. A. Mackay, M. S. Lewis, J. C. Unkeless, and H. J. Gould, *Biochemistry* **34,** 13320 (1995).

[11] S. J. Shire, Technical Note DS-837, Beckman Instruments, Palo Alto, Califonia, 1992.

[12] J. Philo, J. Talvenheimo, J. Wen, R. Rosenfeld, A. Welcher, and T. Arakawa, *J. Biol. Chem.* **269,** 27840 (1994).

only passingly in the derivation of their own mass estimates, and does not attempt to derive partial specific volume values for the carbohydrate fraction in any systematic fashion. It does not appear to be their purpose to serve as a primary reference of special utility to the biochemist who contemplates application of this method.

In contrast, the present treatment addresses the problem of glycoprotein mass estimation in a general way, with the intention of providing the basis for an approach that will be comprehensible to the general biochemist, while providing important details on the carbohydrate $\bar{v}$ estimation. Further, we go beyond the other studies to examine the effect on partial specific volume of carbohydrate condensation, not considered in the other reports, and we provide a paradigm for refined estimates of glycoprotein mass when partial information is available on the carbohydrate composition. An error analysis is presented that reflects uncertainties on the molecular mass estimates as a function of the sedimentation analysis and, more significantly, as a function of the error range on the partial specific volume of the carbohydrate component. In sum, it is the aim of this article to be both more usable and more accessible to scientific investigators while guiding the method's application and expected errors, and to provide more accurate estimates due to improved derived constants for complex carbohydrate partial specific volume values. The new analytical strategy that we demonstrate here is general and should be applicable for any soluble glycoprotein of known amino acid composition.

Methods

Analytical ultracentrifugation is carried out using a Beckman Model E analytical ultracentrifuge with an absorption optical system and a digital data acquisition system. A four-hole rotor is used with a counterbalance and three cells with carbon-filled Epon double-sector centerpieces. The protein is in phosphate-buffered saline (PBS), pH 7.4, which had the NaCl concentration raised by an additional 1.00 *M* to 1.15 *M*, and the initial protein concentrations are 0.12, 0.20, and 0.28 absorbance units at 280 nm. Solution column lengths are approximately 5 mm. The ultracentrifuge is run at 20,000 rpm and 20.0° for 71 hr, by which time the concentration distributions in all three cells has been invariate within the limits of experimental measurement for 23 hr, indicating that centrifugal equilibrium has been obtained. All computations, including data editing and global data analysis by means of weighted nonlinear least squares curve fitting, are performed using MLAB (Civilized Software, Silver Spring, MD), a mathematical modeling program.[13] Weights are obtained by initially globally fitting the three data sets with appropriate mathematical models and then examining the distribution of the residuals to obtain the standard error as a function of the concentration expressed as absorbance at 280 nm. For this study, this is given by

$$SE_c = 0.003 + 0.005c \tag{1}$$

Because the variance is the square of the standard error and the appropriate weight for a datum is the reciprocal of its variance, a weighting vector of n reciprocal variances is obtained. These weights are then normalized so that the sum of the weights in a given weight vector is equal to the number of weights in the vector, since, for an unweighted fit, a weight of one for each datum is implicit. Normalization is necessary if the root-mean-square (rms) errors are to be used as criteria for the quality of fits or if the rms errors of weighted and unweighted fits are to be compared.

For a homogeneous and thermodynamically ideal solute, the concentration expressed as a function of radial position in the cell at ultracentrifugal equilibrium is given by

$$c_{\mathrm{r}} = c_{\mathrm{b}} \exp[AM(r^2 - r_{\mathrm{b}}^2)] + \varepsilon \tag{2}$$

where c_{b} is the concentration at the radial position of the cell bottom, r_{b}, which is taken as the reference radius; M is the solute molecular mass; ε is a small baseline error term; $A = (1 - \bar{v}\rho)\, \omega^2/2RT$, where $\bar{v}$ is the solute partial specific volume, ρ is the solvent density, ω is the angular velocity of the rotor, R is the gas constant, and T is the absolute temperature. Ordinarily, the value of the molecular mass is best obtained by nonlinear least squares curve fitting, globally fitting the data of concentration as a function of radial position of three data sets obtained for different initial loading concentrations to three mathematical models defined by Eq. (2) and using M, and three values of c_{b} and ε as fitting parameters. This assumes that the value of the partial specific volume is known.

When the partial specific volume is not known, as is the case for glycoproteins of unknown or incompletely defined carbohydrate composition, ultracentrifugal homogeneity can still be demonstrated. Equation (2) can be rewritten in the form

$$c_{\mathrm{r}} = c_{\mathrm{b}} \exp[A'M'(r^2 - r_{\mathrm{b}}^2)] + \varepsilon \tag{3}$$

where $A' = \omega^2/2RT$ and where the reduced mass $M' = M(1 - \bar{v}\rho)$. The value of the reduced mass is readily obtained by the same curve-fitting

[13] G. D. Knott, *Comput. Programs Biomed.* **10,** 271 (1979).

procedure, now using Eq. (3) as the mathematical model. Homogeneity can be assessed on the basis of a low value for the rms error and, of greater importance, a nominally random distribution of the residuals about the fitting line with no apparent systematic deviations. If this model is applied when there is dimer or a higher aggregate present, the rms error will be significantly increased and significant systematic deviation of the residuals will be observed, particularly in the region near the cell bottom.

The molecular mass of the glycoprotein $M_{GP} = M_G + M_P$, where M_G is the molecular mass of the carbohydrate portion of the molecule and M_P is the molecular mass of the protein portion. The latter can be readily calculated from the amino acid sequence data. We now make the assumption that the partial specific volume of the glycoprotein is a weight-average partial specific volume defined in the relationship

$$M_{GP}\bar{v}_{GP} = M_G\bar{v}_G + M_P\bar{v}_P \tag{4}$$

and using the definition $M' = M(1 - \bar{v}\rho)$, we obtain

$$\begin{aligned} M' = M_{GP} - M_{GP}\bar{v}_{GP}\rho = M_{GP} - M_G\bar{v}_G\rho - M_P\bar{v}_P\rho \\ = M_{GP} - \bar{v}_G\rho(M_{GP} - M_P) - M_P\bar{v}_P\rho \end{aligned} \tag{5}$$

With appropriate rearrangement we now obtain

$$M_{GP} = [M' + M_P(\bar{v}_P - \bar{v}_G)\rho]/(1 - \bar{v}_G\rho) \tag{6}$$

Equation (6) can now be used to calculate the molecular mass of the glycoprotein as a function of the possible values of the partial specific volume of the carbohydrate using the experimentally obtained value of the reduced mass and the molecular mass and apparent partial specific volume of the protein portion of the glycoprotein calculated from the amino acid composition.

The simultaneous curve fit gives a global value of 9620 ± 20 for the reduced mass M'. Because the mathematical model used for fitting is nonlinear with respect to the parameters, the standard error returned by the Levenberg–Marquardt algorithm is a linear approximation. It has been shown, using Monte Carlo analyses, that when this error is a very small fraction of the value of the molecular mass, as is the case here, this error value is a very good approximation of the standard error.[4] Both the individual and the combined fits had rms errors of approximately 0.004 absorbance units. These fits are of excellent quality with apparently random distributions of the residuals. The quality of these fits is such that it appears appropriate to conclude that ultracentrifugal homogeneity consistent with the presence of a single monomeric species exhibiting apparent thermodynamic ideality of the solute has been demonstrated.

Discussion

Characteristics of Carbohydrates on Proteins

The analytical method that we propose is most powerful when applied in the context of what is known about carbohydrates on proteins. With carbohydrate compositional information, the range of applicable partial specific volumes is restricted and the estimate of the total molecular mass of the glycoprotein is correspondingly more precise. Using Eq. (6), where the other terms in this equation are either calculatable or determined experimentally, the range of values of M_{GP} is dependent only on the range of values of $\bar{v}_G$ and not on the compositional fraction of carbohydrate.

Carbohydrate substitution is either N linked or O linked. Substitution at N-addition sites [Asn–X–Ser(Thr)] occurs on the nascent peptide chain in the rough endoplasmic reticulum and begins with an *en bloc* transfer of the core oligosaccharide unit $(Glc)_3(Man)_3(GlcNAc)_2$ to Asn. Terminal Glc residues are removed and the molecule is transferred to the Golgi subsite where the core oligosaccharide is modified to a high-mannose type of chain with little modification or to a more complex branched structure that can be extensively modified. Monesin blocks transfer out of the rough endoplasmic reticulum and traps the complex in its N-substituted state without terminal modifications or O-linked substitutions. Tunicamycin blocks N addition completely but permits subsequent O-linked processing. O-linked oligosaccharides are directly substituted at Ser or Thr residues with GalNAc in the Golgi, followed by further substitution with stepwise addition of monosaccharides. In general, the major difference between N- and O-linked carbohydrates lies in the core structures. The remainder of the molecule is typically comparable with a predominance of mannose and/or galactose with varying proportions of *N*-acetyl sugars, until the terminal substitutions are added. Sialic acid and fucose are found only as terminating residues. Partial specific volumes for these substituent sugars are listed in Table I as are the partial specific volumes of some other pertinent molecules.[14]

Several conclusions can be formed from examination of Table I. The first of these is that the value of the partial specific volume for neutral sugar residues is approximately 0.610 $cm^3\ g^{-1}$, whereas those for the deoxy and *N*-acetyl forms lies between 0.671 and 0.684 $cm^3\ g^{-1}$. Sialic acid (*N*-acetylneuraminic acid) is the principal charged residue and has the lowest partial specific volume, 0.584 $cm^3\ g^{-1}$. Finally, the condensation of monomeric sugars into dimeric, trimeric, and linear polymeric forms reduces the

[14] H. Durchschlag, *in* "Thermodynamic Data for Biochemistry and Biotechnology" (H.-J. Hinz, ed.), pp. 45–128. Springer-Verlag, Berlin, 1986.

TABLE I
PARTIAL SPECIFIC VOLUMES OF CARBOHYDRATES, PROTEINS, AND GLYCOPROTEINS[a]

Compound	Partial specific volume[b]
Carbohydrates	
Glucose	0.622
Galactose[c]	0.622
Fructose[c]	0.614
Mannose[c]	0.607
Fucose[c]	0.671
N-Acetylglucose[c]	0.684
N-Acetylgalactose[c]	0.684
Sialic acid[c]	0.584
Sucrose (Glu-Fru)	0.613
Lactose (Glu-Gal)	0.606
Raffinose (Fru-Glu-Gal)	0.608
Amylose (Glu_{12})	0.610
Inulin (poly-Fru)	0.600
Starch (poly-Glu)	0.600
Proteins: extreme examples	
Pyruvate kinase	0.754
Hemoglobin (unliganded)	0.749
Aldolase (muscle)	0.742
Fibrinogen	0.706
Trypsin inhibitor	0.698
Ribonuclease A	0.696
Glycoproteins	
Bovine submaxillary glycoprotein (44% carbohydrate)	0.654
Human α_1-seromucoid (41% carbohydrate)	0.672
Protein	0.762
Carbohydrate	0.621
Human α_1-glycoprotein (30% carbohydrate)	0.646
Human corticosteroid binding globulin (26% carbohydrate)	0.708
Tac^{224} (this work)	0.688
Protein	0.716
Carbohydrate	0.613

[a] Data from the tables of Durchschlag.[14]
[b] Partial specific volumes expressed in units of $cm^3\ g^{-1}$.
[c] Sugars normally found in glycoproteins.

partial specific volume by approximately 0.011 $cm^3\ g^{-1}$ on average; this is deduced by comparing the values for fructose, galactose, and glucose with those for lactose and sucrose and with those for raffinose and amylose. The values for inulin and starch suggest that branched forms appear to have a further reduction of partial specific volume of approximately 0.009 $cm^3\ g^{-1}$.

The listed partial specific volumes for proteins reflect some extreme values at the high and low ends. The values for the overwhelming majority of proteins lie between 0.700 and 0.750 $cm^3 g^{-1}$, from which a value of 0.725 $cm^3 g^{-1}$, is commonly assumed in the absence of other information. The amino acid composition of a protein is the primary determinant of the value of its partial specific volume.[15] An above-average abundance of such amino acids as valine, leucine, isoleucine, phenylalanine, lysine, and proline and/or a paucity of glycine, serine, threonine, cysteine, aspartic acid, glutamic acid, asparagine, and glutamine will give a relatively high value for the partial specific volume; the reverse will give a relatively low value. The glycoproteins listed illustrate some typical values and demonstrate the dependence of the value of the partial specific volume on the fraction of carbohydrate present.

Predicting Compositional Partial Specific Volume of the Carbohydrate

The compositional partial specific volume of a protein is calculated as a weight-average partial specific volume using the contributions of the constituent amino acid residues with the equation

$$\bar{v} = \sum \bar{v}_i n_i M_i / \sum n_i M_i \tag{7}$$

where $\bar{v}_i$, n_i, and M_i denote the partial specific volume, number, and molecular mass, respectively, of the ith amino acid residue. This same equation can be used for calculating the contributions of the various carbohydrates. The major problem is apportioning the contributions of the component sugars to the composite carbohydrate partial specific volume. We can apply the principles of carbohydrate substitution to obtain more refined expectations.

We consider first the extremes of the values listed in Table I where sialic acid has the lowest value (0.584) and *N*-acetylglucose and *N*-acetylgalactose have the highest values (0.684) for partial specific volume. Because sialic acid is normally a terminal residue, there cannot be a greater number of sialic acid residues than the number of uncharged unsubstituted residues; generally sialic acids are much fewer. The lowest net partial specific volume involving the coupling of sialic acid to a single neutral sugar would be expected to be $(0.584 + 0.607)/2 - 0.011 = 0.586$ $cm^3 g^{-1}$. The value of the mean partial specific volume has been reduced by 0.011 in order to reflect the probable volume change associated with sugar polymerization. With a one-to-one ratio of sialic to neutral sugars, this would be the absolute

[15] A. A. Zamyatnin, *in* "Annual Review of Biophysics and Bioengineering" (D. M. Engelman, C. R. Cantor, and T. D. Pollard, eds.), Vol. 13, pp. 145–166. Annual Reviews, Inc., Palo Alto, California, 1984.

lower limit for the value of the partial specific volume contribution of the carbohydrate to a glycoprotein. However, no such constructs exist, as all include at least one *N*-acetyl sugar with a partial specific volume of 0.684 $cm^3\ g^{-1}$ to begin the chain. The lowest practically encountered limit on partial specific volume not including sialic acid is probably represented by the high mannose ($Man_9NAcGlc_2$) core as found in the A peptide of bovine thyroglobulin[16] and predicts a partial specific volume of $(9 \times 0.607 + 2 \times 0.684)/11 - 0.011 = 0.610\ cm^3\ g^{-1}$. If this, in turn, were fully sialic acid substituted (3 residues), the minimum net partial specific volume would be 0.602 $cm^3\ g^{-1}$. At the high end of values, a survey of a large number of defined structures shows that the sum of deoxy and *N*-acetyl sugars rarely exceeds the number of neutral unsubstituted residues.[14] In the absence of sialic acid substitution, these molecules would be expected to have a maximum partial specific volume of approximately $(0.622 + 0.684)/2 - 0.011 = 0.642\ cm^3\ g^{-1}$, and would be lower with any sialic acid addition. Thus, the entire range of "normally" substituted glycoproteins would be expected to have carbohydrate partial specific volumes in the range of 0.602–0.642 $cm^3\ g^{-1}$. We recommend use of this range of values of $\bar{v}_G$ in Eq. (7) for all applications of this method. Knowledge of sialic acid content would permit further refinement of these limits, as shown later.

Data on Tac Protein

Tac contains two N-linked carbohydrate addition sites,[17] and two discrete intermediate forms are noted in [^{35}S]methionine pulse-chase experiments that are compatible with use of both sites.[18] Subsequently, the molecule undergoes a large retardation shift on gel electrophoresis with little change in p*I*, and finally undergoes a large p*I* shift from 6.2–6.5 to 4.2–4.7 with only a small change in mobility; this is compatible with the addition of approximately 10 sialic acid residues.[19,20] On the other hand, tunicamycin treatment, which blocks N substitution, induces only a small change in gel mobility and no change in p*I*, indicating that the two N sites are not sialic substituted and that most of the carbohydrate is actually O linked.[21]

[16] R. Kornfeld and S. Kornfeld, *in* "The Biochemistry of Glycoproteins and Proteoglycans" (W. J. Lennarz, ed.), pp. 1–34. Plenum, New York, 1980.

[17] W. J. Leonard, J. M. Deppner, M. Kanehisa, M. Kronke, N. J. Peffer, P. B. Svetlik, M. Sullivan, and W. C. Greene, *Science* **230,** 633 (1985).

[18] W. J. Leonard, J. M. Depper, R. J. Robb, T. A. Waldmann, and W. C. Greene, *Proc. Natl. Acad. Sci. U.S.A.* **80,** 6957 (1983).

[19] Y. Wano, T. Uchiyama, K. Fukui, M. Maeda, H. Uchino, and J. Yodoi, *J. Immunol.* **132,** 3005 (1984).

[20] Y. Wano, T. Uchiyama, J. Yodoi, and H. Uchino, *Microbiol. Immunol.* **29,** 451 (1985).

[21] G. Lambert, E. A. Stura, and I. A. Wilson, *J. Biol. Chem.* **264,** 451 (1989).

However, this information is not useful to our calculations, because there is relatively little difference in the values of the partial specific volumes of the O-linked and N-linked sugars, unless we know *a priori* (which we do not) that the N-linked sugar is a high mannose type. The apparent absence of sialic acid on the N-linked residues suggests, however, that these are the high mannose simple type of substitution since the complex forms are more likely to be sialic acid substituted.

Estimation of Molecular Mass of Tac[224]

Using standard handbook data for the molecular masses of the amino acids and the data of Perkins[22] for their partial specific volumes, we calculate a value of 25253 g mol^{-1} for the compositional molecular mass, M_p, and a value of 0.718 cm^3 g^{-1} at 20° for $\bar{v}_p$, the compositional apparent partial specific volume of the protein portion of the molecule. Experimentally, we obtain a value for M' of 9620 ± 20 in PBS augmented to 1.15 *M* NaCl (ρ = 1.047 g cm^{-3} at 20°). We use these values for calculation of the molecular mass of Tac[224] making various assumptions concerning the carbohydrate composition, and then we estimate the most probable range of values and the effect of the uncertainty of M' and the uncertainty of $\bar{v}_p$ on these values.

The extreme limits may be obtained by using the smallest and the largest values of the carbohydrate partial specific volumes listed in Table I. Under these conditions (assumption set I, Table II), the lowest listed value is 0.584 cm^3 g^{-1} for pure sialic acid, giving a value for M_{GP} of 33,700 g mol^{-1}. The highest listed value is 0.684 cm^3 g^{-1} for pure *N*-acetylglucose (or galactose) giving a value for M_{GP} of 36,900 g mol^{-1}. These calculations establish the lower and upper limits of the molecular mass of the glycoprotein but are unrealistic because of the assumptions made with regard to the composition of the carbohydrate and, hence, the values of the partial specific volumes.

If we now use the range of values for compositional partial specific volume of 0.602–0.642 cm^3 g^{-1}, calculated for the carbohydrate of "normally substituted" glycoproteins, then the value of M_{GP} ranges more narrowly from 34,200 to 35,300 g mol^{-1} (assumption set II, Table II). This range of values may be further refined if we now assume that there are 10 sialic acid residues[19,20] attached to Tac[224] (assumption set III, Table II). This "partial glycoprotein" has a compositional molecular mass of 28346 g mol^{-1} with an apparent compositional partial specific volume of 0.702 cm^3 g^{-1}; as expected, this is slightly lower than the $\bar{v}$ = 0.718 stated above for the unsubstituted protein. For the remainder of the carbohydrate, we use the

[22] S. J. Perkins, *Eur. J. Biochem.* **157,** 169 (1986).

TABLE II
TAC MOLECULAR MASS AS FUNCTION OF ESTIMATED CARBOHYDRATE PARTIAL SPECIFIC VOLUME[a]

Assumption set	Assumption $\bar{v}_G$	Derived[b] $\bar{v}_{GP}$	Derived M_{GP}
I. Extreme values of $\bar{v}_G$ (unrealistic)			
$M_P = 25{,}253$ g mol^{-1}, $\bar{v}_P = 0.716$ cm^3 g^{-1}			
All sialic acid	0.584	0.682	33700
All *N*-acetylglucose (or *N*-acetylgalactose)	0.684	0.706	36900
Means ± limits			35300 ± 1600
II. Extreme values of $\bar{v}_G$ (realistic)			
$M_P = 25{,}253$ g mol^{-1}, $\bar{v}_P = 0.716$ cm^3 g^{-1}			
Normal carbohydrate (low)	0.602	0.686	34200
Normal carbohydrate (high)	0.642	0.695	35300
Means ± limits			34800 ± 600
III. Refined values of $\bar{v}$: 10 sialic acid residues per Tac[224]			
M_{P+SA} 28346 g mol^{-1}, $\bar{v}_{P-SA} = 0.702$ cm^3 g^{-1}			
Normal carbohydrate (low)	0.610	0.688	34200
Normal carbohydrate (high)	0.642	0.691	34800
Mean ± limits			34500 ± 300

[a] Three different sets of assumptions of the composition of carbohydrate lead to different estimates of partial specific volumes and to different estimates of the total molecular mass of the Tac[224] glycoprotein.

[b] The partial specific volume of the glycoprotein is calculated by using the equation $\bar{v}_{GP} = (1 - M'/M_{GP})/\rho$, which is obtained by rearranging $M' = M_{GP}(1 - \bar{v}_{GP}\rho)$.

same range of values for partial specific volume as above but without sialic acid (0.610–0.642), and the resulting, "refined" value of M_{GP} ranges from 34,200 to 34,800 g mol^{-1}. Thus, we obtain a mean value of 34,500 g mol^{-1} for the molecular mass of Tac[224] with maximum uncertainty of ±300 g mol^{-1} reflecting the possible range of values of the compositional partial specific volumes of the carbohydrates other than sialic acid.

Using a standard statistical procedure, we estimate a standard error of ±400 for M_{GP} calculated from the estimated standard error of M' obtained during fitting and from Perkins's estimate of ±0.005 cm^3 g^{-1} for the standard error of $\bar{v}_p$ calculated using his consensus values.[22] Although it is rather improbable that the remainder of the carbohydrate would be represented by only one type of sugar, we still cannot treat the uncertainties representing the extremes of carbohydrate partial specific volume values as representing statistically validatable confidence limits. If we use standard procedures for the estimation of the propagation of errors, then with the aforementioned fitting errors and uncertainties in partial specific volume values, the apparent uncertainty of the value of the molecular mass of Tac^{224} is estimated to be ±500, which is 1.45% of the molecular mass. Thus, it can be seen that this procedure restricts the probable mean molecular mass of the glycoprotein to a very narrow range. From this the percentage of carbohydrate comprising the molecule may be readily calculated as 26.8%, or as 36.6% with respect to the protein.

From this study we conclude that Tac^{224} is monomeric with a probable molecular mass of 34,500 ± 500 g mol^{-1}. Bearing in mind caveats about cellular differences in protein glycosylation patterns, we infer a nominal total molecular mass for native transmembrane Tac of 38,100 g mol^{-1} with a fractional carbohydrate value of 24.3%, or 32.0% with respect to protein. The role of carbohydrate on Tac is unknown. It is not essential for IL-2 binding or for recognition by monoclonal antibodies, but it may provide stability or control hydration of the molecule.

[9] Irregularity and Asynchrony in Biologic Network Signals

By STEVEN M. PINCUS

Introduction

Series of sequential data arise throughout biology, in multifaceted contexts. Examples include (1) hormonal secretory dynamics based on frequent, fixed-increment samples from serum, (2) heart rate rhythms, (3) electroencephalograms (EEGs), and (4) DNA sequences. Enhanced capabilities to quantify differences among such series would be extremely valuable, since in their respective contexts, these series reflect essential biological information. Although practitioners and researchers typically quantify mean levels, and oftentimes the extent of variability, it is recognized that in many instances, the persistence of certain patterns, or shifts in an "apparent ensemble amount of randomness," provides the fundamental insight of subject status. Despite this recognition, formulas and algorithms to quantify an "extent of randomness" have not been developed and/or utilized in the above contexts, primarily since even within mathematics itself, such quantification technology was lacking until very recently. Thus except for the settings in which egregious (changes in) serial features presented themselves, which specialists are trained to detect visually, subtler changes in patterns would largely remain undetected, unquantified, and/or not acted on.

0076-6879/00 $30.00

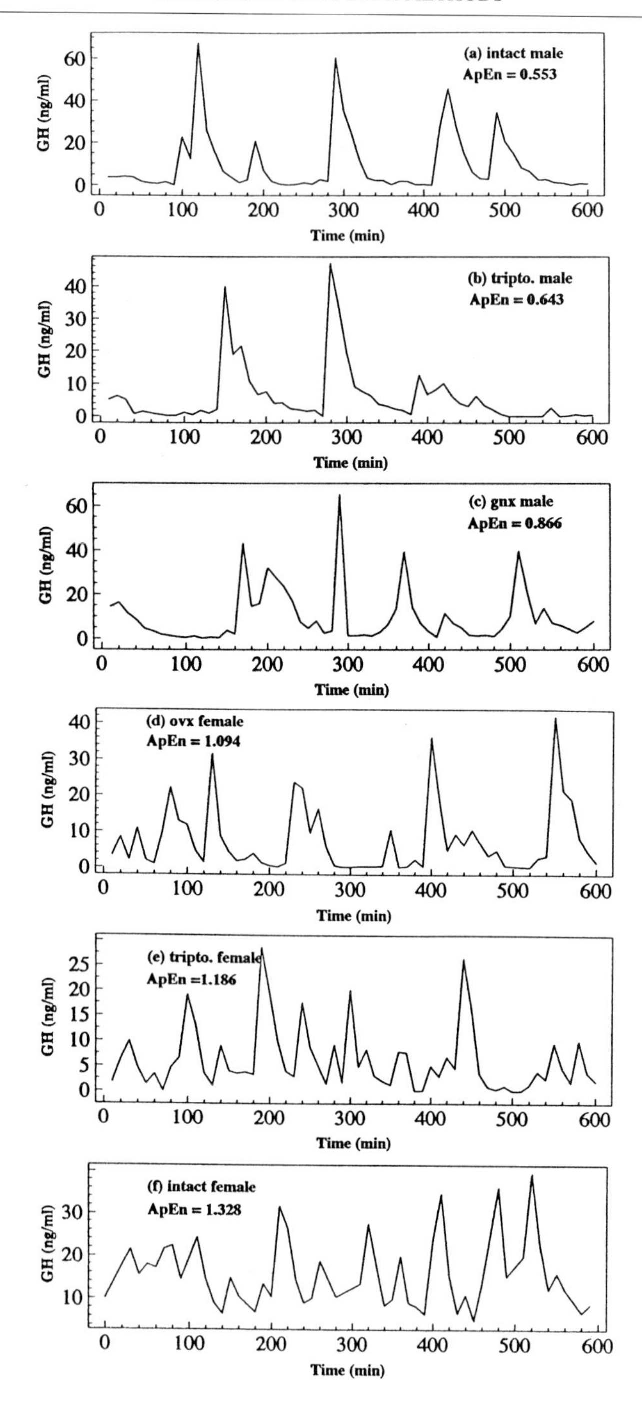
(a) intact male
ApEn = 0.553
GH (ng/ml)
Time (min)
(b) tripto. male
ApEn = 0.643
GH (ng/ml)
Time (min)
(c) gnx male
ApEn = 0.866
GH (ng/ml)
Time (min)
(d) ovx female
ApEn = 1.094
GH (ng/ml)
Time (min)
(e) tripto. female
ApEn =1.186
GH (ng/ml)
Time (min)
(f) intact female
ApEn = 1.328
GH (ng/ml)
Time (min)

Recently, a new mathematical approach and formula, approximate entropy (ApEn), has been introduced as a quantification of *regularity* of data, motivated by both the above application needs[1] and by fundamental questions within mathematics.[2,3] This approach calibrates an ensemble extent of sequential interrelationships, quantifying a continuum that ranges from totally ordered to completely random. The central focus of this review is to discuss ApEn, and subsequently cross-ApEn,[2,4] a measure of two-variable asynchrony that is thematically similar to ApEn.

Before presenting a detailed discussion of regularity, we consider two sets of time series (Figs. 1 and 2) to illustrate what we are trying to measure. In Fig. 1a–f, the data represent a time series of growth hormone levels from rats in six distinct physiologic settings, each taken at 10-min samples during a 10-hr lights-off ("dark") state.[5] The end points (a) and (f) depict, respectively, intact male and intact female serum dynamics; (b) and (c) depict two types of neutered male rats; and (d) and (e) depict two classes of neutered female rats. It appears that the time series are becoming increasingly irregular as we proceed from (a) to (f), although specific feature differences among the sets are not easily pinpointed. In Fig. 2, the data represent the beat-to-beat heart rate, in beats per minute, at equally spaced time intervals. Figure 2A is from an infant who had an aborted SIDS (sudden infant death syndrome) episode 1 week prior to the recording, and Fig. 2B is from a healthy infant.[6] The standard deviations (SD) of these two tracings are approximately equal, and while the aborted SIDS infant has a somewhat higher mean heart rate, both are well within the normal range. Yet tracing (A) appears to be more regular than tracing (B). In both of these instances, we ask these questions: (1) How do we quantify the apparent differences in regularity? (2) Do the regularity values significantly

[1] S. M. Pincus, *Proc. Natl. Acad. Sci. U.S.A.* **88,** 2297 (1991).

[2] S. Pincus and B. H. Singer, *Proc. Natl. Acad. Sci. U.S.A.* **93,** 2083 (1996).

[3] S. Pincus and R. E. Kalman, *Proc. Natl. Acad. Sci. U.S.A.* **94,** 3513 (1997).

[4] S. M. Pincus, T. Mulligan, A. Iranmanesh, S. Gheorghiu, M. Godschalk, and J. D. Veldhuis, *Proc. Natl. Acad. Sci. U.S.A.* **93,** 14100 (1996).

[5] E. Gevers, S. M. Pincus, I. C. A. F. Robinson, and J. D. Veldhuis. *Am. J. Physiol.* **274** (*Regul. Integrat.* **43**), R437 (1998).

[6] S. M. Pincus, T. R. Cummins, and G. G. Haddad, *Am. J. Physiol.* **264** (*Regul. Integrat.* **33**), R638 (1993).

FIG. 1. Representative serum growth hormone (GH) concentration profiles, in ng/ml, measured at 10-min intervals for 10 hr in the dark. From, in ascending order of ApEn values, and hence increasing irregularity or disorderliness: (a) intact male, (b) triptorelin-treated male, (c) gonadectomized male, (d) ovariectomized female, (e) triptorelin-treated female, and (f) intact female rats.

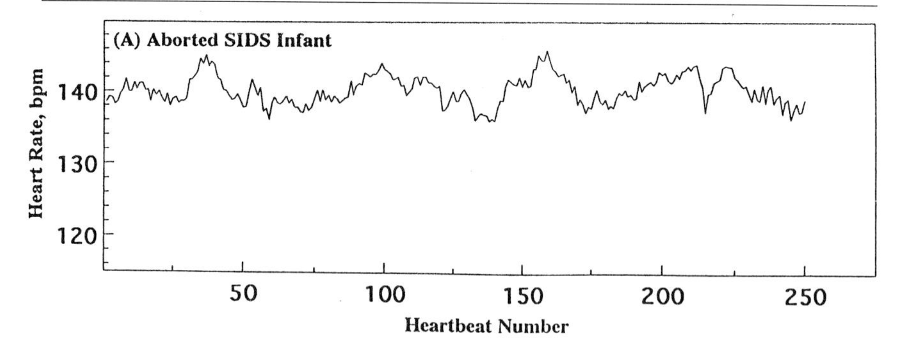

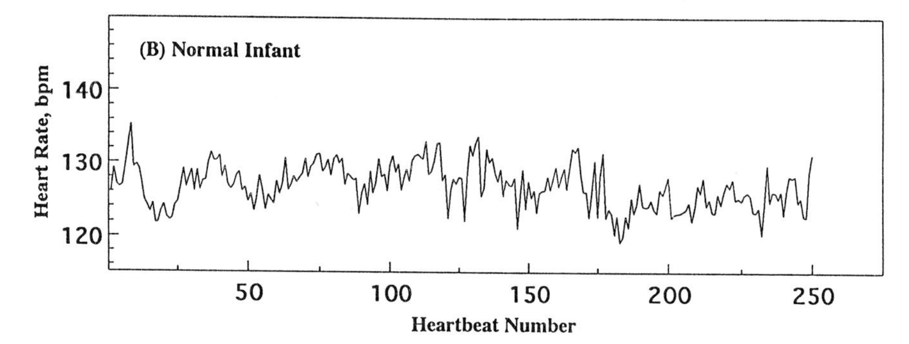

FIG. 2. Two infant quiet sleep heart rate tracings with similar variability, SD (A) aborted SIDS infant, SD = 2.49 beats per minute (bpm), ApEn(2, 0.15 SD, 1000) = 0.826; (B) normal infant, SD = 2.61 bpm, ApEn(2, 0.15 SD, 1000) = 1.463.

distinguish the data sets? (3) How do inherent limitations posed by moderate length time series, with noise and measurement inaccuracies present as shown in Figs. 1 and 2, affect statistical analyses? (4) Is there some general mechanistic hypothesis, applicable to diverse contexts, that might explain such regularity differences?

The development of ApEn evolved as follows. To quantify time series regularity (and randomness), we initially applied the Kolmogorov–Sinai (K–S) entropy[7] to clinically derived data sets. The application of a formula for K–S entropy[8,9] yielded intuitively incorrect results. Closer inspection of the formula showed that the low magnitude noise present in the data

[7] A. N. Kolmogorov, *Dokl. Akad. Nauk. S.S.S.R.* **119,** 861 (1958).
[8] P. Grassberger and I. Procaccia, *Phys. Rev. A* **28,** 2591 (1983).
[9] F. Takens "*Atas do 13.Col. Brasiliero de Matematicas.*" Rio de Janerio, Brazil, 1983.

greatly affected the calculation results. It also became apparent that to attempt to achieve convergence of this entropy measure, extremely long time series would be required (often 1,000,000 or more points), which even if available, would then place extraordinary time demands on the computational resources. The challenge was to determine a suitable formula to quantify the concept of regularity in moderate-length, somewhat noisy data sets, in a manner thematically similar to the approach given by the K–S entropy.

Historical context further frames this effort. The K–S entropy was developed for and is properly employed on truly chaotic processes (time series). Chaos refers to output from deterministic dynamic systems, where the output is bounded and aperiodic, thus appearing partially "random." Recently, there have been myriad claims of chaos based on analysis of experimental time series data, in which correlation between successive measurements has been observed. Because chaotic systems represent only one of many paradigms that can produce serial correlation, it is generally inappropriate to infer chaos from the correlation alone. The mislabeling of correlated data as "chaotic" is a relatively benign offense. Of greater significance, complexity statistics that were developed for application to chaotic systems and are relatively limited in scope have been commonly misapplied to finite, noisy, and/or stochastically derived time series, frequently with confounding and nonreplicable results. This caveat is particularly germane to biologic signals, especially those taken *in vivo,* because such signals usually represent the output of a complicated network with both stochastic and deterministic components. We elaborate on these points in the later section titled Statistics Related to Chaos. With the development of ApEn, we can now successfully handle the noise, data length, and stochastic/composite model constraints in statistical applications.

As stated, we also discuss cross-ApEn,[2,4] a quantification of asynchrony or conditional irregularity between two signals. Cross-ApEn is thematically and algorithmically quite similar to ApEn, yet with a critical difference in focus: it is applied to two time series, rather than a single series, and thus affords a distinct tool from which changes in the extent of synchrony in interconnected systems or networks can be directly determined. This quantification strategy is thus especially germane to many biological feedback and/or control systems and models for which cross-correlation and cross-spectral methods fail to fully highlight markedly changing features of the data sets under consideration.

Importantly, we observe a fundamental difference between regularity (and asynchrony) statistics, such as ApEn, and variability measures: most short- and long-term variability measures take raw data, preprocess the data, and then apply a calculation of SD (or a similar, nonparametric variation) to the processed data.[10] The means of preprocessing the raw data varies substantially with the different variability algorithms, giving rise to many distinct versions. However, once preprocessing of the raw data is completed, the processed data are input to an algorithm for which the order of the data is immaterial. For ApEn, the order of the data is the essential factor; discerning changes in order from apparently random to very regular is the primary focus of this statistic.

Finally, an absolutely paramount concern in any practical time series analysis is the presence of either artifacts or nonstationarities, particularly clear trending. If a time series is nonstationary or is riddled with artifacts, little can be inferred from moment, ApEn, or power spectral calculations, because these effects tend to dominate all other features. In practice, data with trends suggest a collection of heterogeneous epochs, as opposed to a single homogeneous state. From the statistical perspective, it is imperative that artifacts and trends first be removed before meaningful interpretation can be made from further statistical calculations.

Quantification of Regularity

Definition of ApEn

Approximate entropy (ApEn), was introduced as a quantification of regularity in time series data, motivated by applications to relatively short, noisy data sets.[1] Mathematically, ApEn is part of a general development of approximating Markov chains to a process[11]; it is furthermore employed to refine the formulations of i.i.d. (independent, identically distributed) random variables, and normal numbers in number theory, via rates of convergence of a deficit from maximal irregularity.[2,3,12] Analytical properties for ApEn can be found in Refs. 1, 2, 13, and 14; in addition, it provides a finite sequence formulation of randomness, via proximity to maximal irregularity.[2,3] Statistical evaluation is given in Refs. 13 and 14.

ApEn assigns a nonnegative number to a sequence or time series, with larger values corresponding to greater apparent process randomness (serial irregularity) and smaller values to more instances of recognizable patterns or features in data. ApEn measures the logarithmic likelihood that runs of

[10] W. J. Parer, J. T. Parer, R. H. Holbrook, and B. S. B. Block, *Am. J. Obstet. Gynecol.* **153,** 402 (1985).

[11] S. M. Pincus, *Proc. Natl. Acad. Sci. U.S.A.* **89,** 4432 (1992).

[12] S. Pincus and B. H. Singer, *Proc. Natl. Acad. Sci. U.S.A.* **95,** 10367 (1998).

[13] S. M. Pincus and W. M. Huang, *Commun. Statist. Theory Methods* **21,** 3061 (1992).

[14] S. M. Pincus and A. L. Goldberger, *Am. J. Physiol.* **266** (*Heart Circ. Physiol.* **35**), H1643 (1994).

patterns that are close for m observations remain close on next incremental comparisons. From a statistician's perspective, ApEn can often be regarded as an ensemble parameter of process autocorrelation: smaller ApEn values correspond to greater positive autocorrelation; larger ApEn values indicate greater independence. The opposing extremes are perfectly regular sequences (e.g., sinusoidal behavior, very low ApEn) and independent sequential processes (very large ApEn).

Formally, given N data points $u(1), u(2), \ldots, u(N)$, two input parameters, m and r, must be fixed to compute ApEn [denoted precisely by ApEn(m, r, N)]. The parameter m is the "length" of compared runs, and r is effectively a filter. Next, form vector sequences $x(1)$ through $x(N - m + 1)$ from the $\{u(i)\}$, defined by $x(i) = [u(i), \ldots, u(i + m - 1)]$. These vectors represent m consecutive u values, commencing with the ith point. Define the distance $d[x(i), x(j)]$ between vectors $x(i)$ and $x(j)$ as the maximum difference in their respective scalar components. Use the sequence $x(1), x(2), \ldots, x(N - m + 1)$ to construct, for each $i \leq N - m + 1$,

$$C_i^m(r) = [\text{number of } x(j) \text{ such that } d[x(i), x(j)] \leq r]/(N - m + 1) \quad (1)$$

The $C_i^m(r)$ values measure within a tolerance r the regularity, or frequency, of patterns similar to a given pattern of window length, m. Next, define

$$\Phi^m(r) = (N - m + 1)^{-1} \sum_{i=1}^{N-m+1} \ln C_i^m(r) \quad (2)$$

where ln is the natural logarithm. We define approximate entropy by

$$\text{ApEn}(m, r, N) = \Phi^m(r) - \Phi^{m+1}(r) \quad (3)$$

Via some simple arithmetic manipulation, we deduce the important observation that

$$\begin{aligned} -\text{ApEn} &= \Phi^{m+1}(r) - \Phi^m(r) \\ &= \text{average over } i \text{ of } \ln[\text{conditional probability} \\ &\quad \text{that } |u(j + m) - u(i + m)| \leq r \end{aligned} \quad (4)$$

given that $|u(j + k) - u(i + k)| \leq r$ for $k = 0, 1, \ldots, m - 1$].

When $m = 2$, as is often employed, we interpret ApEn as a measure of the difference between the probability that runs of value of length 2 will recur within tolerance r, and the probability that runs of length 3 will recur to the same tolerance. A high degree of regularity in the data would imply that a given run of length 2 would often continue with nearly the same third value, producing a low value of ApEn.

ApEn evaluates both dominant and subordinate patterns in data; notably, it will detect changes in underlying episodic behavior that do not reflect in peak occurrences or amplitudes,[15] a point that is particularly germane to numerous diverse applications. Additionally, ApEn provides a direct barometer of feedback system change in many coupled systems.[15,16]

ApEn is a relative measure of process regularity, and can show significant variation in its absolute numerical value with changing background noise characteristics. Because ApEn generally increases with increasing process noise, it is appropriate to compare data sets with similar noise characteristics, i.e., from a common experimental protocol.

ApEn is typically calculated via a short computer code; a FORTRAN listing for such a code can be found in Pincus *et al.*[17] ApEn is nearly unaffected by noise of magnitude below r, the *de facto* filter level; and it is robust or insensitive to artifacts or outliers: extremely large and small artifacts have little effect on the ApEn calculation, if they occur infrequently.

Finally, to develop a more intuitive, physiological understanding of the ApEn definition, a multistep description of its typical algorithmic implementation, with figures, is presented in Ref. 14.

Implementation and Interpretation

Choice of m, r, and N

The value of N, the number of input data points for ApEn computations, is typically between 50 and 5000. This constraint is usually imposed by experimental considerations, not algorithmic limitations, to ensure a single homogeneous epoch. Based on calculations that included both theoretical analysis[1,13,15] and numerous clinical applications,[6,17–20] we have concluded that for both $m = 1$ and $m = 2$, and $50 \leq N \leq 5000$, values of r between 0.1 and 0.25 SD of the $u(i)$ data produce good statistical validity of ApEn(m, r, N). For such r values, we demonstrated[1,13,15] the theoretical utility of ApEn(1, r) and ApEn(2, r) to distinguish data on the basis of regularity for both deterministic and random processes, and the clinical utility in the aforementioned applications. These choices of m and r are made to ensure that the conditional frequencies defined in Eq. (4) are reasonably estimated from the N input data points. For smaller r values than those indicated,

[15] S. M. Pincus and D. L. Keefe, *Am. J. Physiol.* **262** (*Endocrinol. Metab.* **25**), E741 (1992).
[16] S. M. Pincus, *Math. Biosci.* **122,** 161 (1994).
[17] S. M. Pincus, I. M. Gladstone, and R. A. Ehrenkranz, *J. Clin. Monit.* **7,** 335 (1991).
[18] L. A. Fleisher, S. M. Pincus, and S. H. Rosenbaum, *Anesthesiology* **78,** 683 (1993).
[19] D. T. Kaplan, M. I. Furman, S. M. Pincus, S. M. Ryan, L. A. Lipsitz, and A. L. Goldberger, *Biophys. J.* **59,** 945 (1991).
[20] S. M. Pincus and R. R. Viscarello, *Obstet. Gynecol.* **79,** 249 (1992).

one usually achieves poor conditional probability estimates as well, while for larger r values, too much detailed system information is lost.

To ensure appropriate comparisons between data sets, it is strongly preferred that N be the same for each data set. This is because ApEn is a *biased* statistic; the expected value of ApEn(m, r, N) generally increases asymptotically with N to a well-defined, limit parameter denoted ApEn(m, r). Restated, if we had 3000 data points, and chose $m = 2$, $r = 0.2$ SD, we would expect that ApEn applied to the first 1000 points would be smaller than ApEn applied to the entire 3000-point time series. Biased statistics are quite commonly employed, with no loss of validity. As an aside, it can be shown that ApEn is asymptotically unbiased, an important theoretical property, but that is not so germane to "everyday" usage. This bias is discussed elsewhere,[13,14] and techniques to reduce bias via a family of ε estimators are provided. However, note that ultimately, attempts to achieve bias reduction are model specific; thus, as stated earlier, it is cleanest to impose a (nearly) fixed data length mandate on all ApEn calculations.

Family of Statistics

Most importantly, despite algorithmic similarities, ApEn(m, r, N) is not intended as an approximate value of K–S entropy.[1,13,17] It is imperative to consider ApEn(m, r, N) as a *family* of statistics; for a given application, system comparisons are intended with fixed m and r. For a given system, there usually is significant variation in ApEn(m, r, N) over the range of m and r.[6,13,17]

For fixed m and r, the conditional probabilities given by Eq. (4) are precisely defined probabilitistic quantities, marginal probabilities on a coarse partition, and contain a great deal of system information. Furthermore, these terms are finite, and thus allow process discrimination for many classes of processes that have infinite K–S entropy (see below). ApEn aggregates these probabilities, thus requiring relatively modest data input.

Normalized Regularity

ApEn decrease frequently correlates with SD decrease. This is not a "problem," as statistics often correlate with one another, but typically we desire an index of regularity decorrelated from SD. We can realize such an index, by specifying r in ApEn(m, r, N) as a fixed percentage of the sample SD of the *individual* subject data set (time series). We call this *normalized regularity*. Normalizing r in this manner gives ApEn a translation and scale invariance to absolute levels[6] in that it remains unchanged under uniform process magnification, reduction, or constant shift higher or lower. Choosing r via this procedure allows sensible regularity comparisons of processes with substantially different SDs. In most clinical applications, it is the normalized version of ApEn that has been employed, generally with $m = 1$ or $m = 2$ and $r = 20\%$ of the SD of the time series.

Relative Consistency

Earlier we commented that ApEn values for a given system can vary significantly with different m and r values. Indeed, it can be shown that for many processes, ApEn(m, r, N) grows with decreasing r like log($2r$), thus exhibiting infinite variation with r.[13] We have also claimed that the utility of ApEn is as a relative measure; for *fixed m* and *r*, ApEn can provide useful information. We typically observe that for a given time series, ApEn(2, 0.1) is quite different from ApEn(4, 0.01), so the question arises as to which parameter choices (m and r) to use. The guidelines above address this, but the most important requirement is consistency. For noiseless, theoretically defined deterministic dynamic systems, we have found that when K–S entropy(A) $\leq$ K–S entropy(B), then ApEn(m, r)(A) $\leq$ ApEn(m, r)(B) and conversely, for a wide range of m and r. Furthermore, for both theoretically described systems and those described by experimental data, we have found that when ApEn(m_1, r_1)(A) $\leq$ ApEn(m_1, r_1)(B), then ApEn(m_2, r_2)(A) $\leq$ ApEn(m_2, r_2)(B), and conversely. This latter property also generally holds for parameterized systems of stochastic (random) processes, in which K–S entropy is infinite. We call this ability of ApEn to preserve order a relative property. It is the key to the general and clinical utility of ApEn. We see no sensible comparisons of ApEn(m, r)(A) and ApEn(n, s)(B) for systems A and B unless $m = n$ and $r = s$.

From a more theoretical mathematical perspective, the interplay between meshes [(m, r) pair specifications] need not be nice, in general, in ascertaining which of (two) processes is "more" random. In general, we might like to ask this question: Given no noise and an infinite amount of data, can we say that process A is more regular than process B? The *flip-flop pair* of processes[13] implies that the answer to this question is "not necessarily": in general, comparison of relative process randomness at a prescribed level is the best one can do. That is, processes may appear more random than processes on many choices of partitions, but not necessarily on all partitions of suitably small diameter (r). The *flip-flop pair* is two i.i.d. processes A and B with the property that for any integer m and any positive r, there exists $s < r$ such that ApEn(m, s)(A) $<$ ApEn(m, s)(B), and there exists $t < s$ such that ApEn(m, t)(B) $<$ ApEn(m, t)(A). At alternatingly small levels of refinement given by r, process B appears more random and less regular than process A followed by appearing *less* random and more regular than process A on a still smaller mesh (smaller r). In this construction, r can be made arbitrarily small, thus establishing the point that process regularity is a relative [to mesh, or (m, r) choice] notion.

Fortunately, for many processes A and B, we can assert more than relative regularity, even though both A and B will typically have infinite K–S entropy. For such pairs of processes, which have been denoted as a *completely consistent pair,*[13] whenever ApEn(m, r)(A) $<$ ApEn(m, r)(B) for any specific choice of m and r, then it follows that ApEn(n, s)(A) $<$ ApEn(n, s)(B) for *all* choices of n and s. Any two elements of $\{\text{MIX}(p)\}$ (defined below), for example, appear to be completely consistent. The importance of completely consistent pairs is that we can then assert that process B is more irregular (or random) than process A, without needing to indicate m and r. Visually, process B appears more random than process A at any level of view. We indicate elsewhere[14] a conjecture that should be relatively straightforward to prove, that would provide a sufficient condition to ensure that A and B are a completely consistent pair, and would indicate the relationship to the autocorrelation function.

Model Independence

The physiologic modeling of many complex biological systems is often very difficult; one would expect accurate models of such systems to be complicated composites, with both deterministic and stochastic components, and interconnecting network features. The advantage of a *model-independent* statistic is that it can distinguish classes of systems for a wide variety of models. The mean, variability, and ApEn are all model-independent statistics in that they can distinguish many classes of systems, and all can be meaningfully applied to $N > 50$ data points. In applying ApEn, therefore, we are not testing for a particular model form, such as deterministic chaos; we are attempting to distinguish data sets on the basis of regularity. Such evolving regularity can be seen in both deterministic and stochastic models.[1,13,15,16]

Statistical Validity: Error Bars for General Processes

Ultimately, the utility of any statistic is based on its replicability. Specifically, if a fixed physical process generates serial data, we would expect statistics of the time series to be relatively constant over time; otherwise, we would have difficulty ensuring that two very different statistical values implied two different systems (distinction). Here, we thus want to ascertain ApEn variation for typical processes (models), so we can distinguish data sets with high probability when ApEn values are sufficiently far apart. This is mathematically addressed by SD calculations of ApEn, calculated for a variety of representative models; such calculations provide "error bars" to quantify probability of true distinction. Via extensive Monte Carlo calculations we established the SD of ApEn(2, 0.2 SD, 1000) < 0.055 for a large class of candidate models.[13,15] It is this small SD of ApEn, applied to 1000 points from various models, that provides its utility to practical data analysis of moderate-length time series. For instance, applying this analysis, we deduce that ApEn values that are 0.15 apart represent nearly 3 ApEn SDs, indicating true distinction with error probability nearly $p = 0.001$. Similarly, the SD of ApEn(1, 0.2 SD, 100) < 0.06 for many diverse models,[13,15] thus providing good replicability of ApEn with $m = 1$ for the shorter data length applications.

Analytic Expressions

For many processes, we can provide analytic expressions for ApEn(m, r). Two such expressions are given by Theorems 1 and 2 (Ref. 1):

Theorem 1. Assume a stationary process $u(i)$ with continuous state space. Let $\mu(x, y)$ be the joint stationary probability measure on $\mathbf{R}^2$ for this process, and $\pi(x)$ be the equilibrium probability of x.

Then

$$\mathrm{ApEn}(1, r) = -\int \mu(x, y) \log \left[\int_{z=y-r}^{y+r} \int_{w=x-r}^{x+r} \mu(w, z)\, \mathrm{d}w\, \mathrm{d}z \Big/ \int_{w=x-r}^{x+r} \pi(w)\, \mathrm{d}w \right] \mathrm{d}x\, \mathrm{d}y$$

Theorem 2. For an i.i.d. process with density function $\pi(x)$ for any $m \geq 1$,

$$\mathrm{ApEn}(m, r) = -\int \pi(y) \log \left[\int_{z=y-r}^{y+r} \pi(z)\, \mathrm{d}z \right] \mathrm{d}y$$

Theorem 1 can be extended in straightforward fashion to derive an expression for ApEn(m, r) in terms of the joint [($m + 1$)-fold] probability distributions. Hence we can calculate ApEn(m, r) for Gaussian processes, since we know the joint probability distribution in terms of the covariance matrix. This important class of processes (for which finite sums of discretely sampled variables have multivariate normal distributions) describes many stochastic models, including solutions to ARMA (autoregressive-moving average) models and to linear stochastic differential equations driven by white noise.

Moreover, from a different theoretical setting, ApEn is related to a parameter in information theory, *conditional entropy*.[21] Assume a finite

[21] R. E. Blahut, "Principles and Practice of Information Theory," pp. 55–64. Addison-Wesley, Reading, Massachusetts, 1987.

state space, where the entropy of a random variable X, $\text{Prob}(X = a_j) = p_j$, is $H(X) := -\Sigma\, p_j \log p_j$, and the entropy of a block of random variables $X_1, \ldots, X_n = H(X_1, \ldots, X_n) := -\Sigma\, \Sigma \ldots \Sigma\, p^n\, (a_{j1}, \ldots, a_{jn}) \log p^n(a_{j1}, \ldots, a_{jn})$. For two variables, the conditional entropy $H(Y \| X) = H(X, Y) - H(X)$; this extends naturally to n variables. Closely mimicking the proof of Theorem 3 of Ref. 1, the following theorem is immediate: for $r < \min_{j \neq k} |a_j - a_k|$, $\text{ApEn}(m, r) = H(X_{m+1} \| X_1, \ldots, X_m)$; thus in this setting, ApEn is a conditional entropy. Observe that we do not assume that the process is mth-order Markov, i.e., that we fully describe the process; we aggregate the mth-order marginal probabilities. The rate of entropy $= \lim_{n \to \infty} H(X_n \| X_1, \ldots, X_{n-1})$ is the discrete state analog of the K–S entropy. However, we cannot go from discrete to continuous state naturally as a limit; most calculations give ∞. As for differential entropy, there is no fundamental physical interpretation of conditional entropy (and no invariance; see Ref. 21, p. 243) in continuous state.

Representative Biological Applications

ApEn has recently been applied to numerous settings both within and outside biology. In heart rate studies, ApEn has shown highly significant differences in settings in which moment (mean, SD) statistics did not show clear group distinctions,[6,17–20] including analysis of aborted SIDS infants and of fetal distress. Within neuromuscular control, e.g., ApEn showed that there was greater control in the upper arm and hand than in the forearm and fingers.[22] In applications to endocrine hormone secretion time series data based on as few as $N = 60$ points, ApEn has shown vivid distinctions ($P < 10^{-10}$; nearly 100% sensitivity and specificity in each study) between normal and tumor-bearing subjects for growth hormone, GH,[23] adrenocorticotropin (ACTH) and cortisol,[24] and aldosterone,[25] with the tumorals markedly more irregular, a pronounced and consistent gender difference in GH irregularity in both human and rat,[26] highly significant differences between follicle stimulating hormone (FSH) and luteinizing

[22] S. Morrison and K. M. Newell, *Exp. Brain Res.* **110,** 455 (1996).

[23] M. L. Hartman, S. M. Pincus, M. L. Johnson, D. H. Mathews, L. M. Faunt, M. L. Vance, M. O. Thorner, and J. D. Veldhuis, *J. Clin. Invest.* **94,** 1277 (1994).

[24] G. van den Berg, S. M. Pincus, J. D. Veldhuis, M. Frolich, and F. Roelfsema, *Eur. J. Endocr.* **136,** 394 (1997).

[25] H. M. Siragy, W. V. R. Vieweg, S. M. Pincus, and J. D. Veldhuis, *J. Clin. Endocr. Metab.* **80,** 28 (1995).

[26] S. M. Pincus, E. Gevers, I. C. A. F. Robinson, G. van den Berg, F. Roelfsema, M. L. Hartman, and J. D. Veldhuis. *Am. J. Physiol.* **270** (*Endocrinol. Metab.* **33**), E107 (1996).

hormone (LH) in both sheep[27] and in women and men,[28] and a positive correlation between advancing age and each of greater irregularity of (1) GH[29] and of (2) LH and testosterone.[4] We next discuss briefly the gender difference findings in GH, to further develop intuition for ApEn in an application context.

Sample Application: Gender Differences in GH Serum Dynamics

In two distinct human subject studies (employing, respectively, immunoradiometric assays and immunofluorimetric assays), females exhibited significantly greater irregularity than their male counterparts, $P < 0.001$ in each setting, with almost complete gender segmentation via ApEn in each context.[26] ApEn likewise vividly discriminates male and female GH profiles in the adult intact rat,[5,26] $P < 10^{-6}$, with nearly 100% sensitivity and specificity (Fig. 3A). More remarkably, in rats that had been castrated prior to puberty, the ApEn of GH profiles in later adulthood is able to separate genetically male and female animals.[5] Among intact animals and rats treated prepubertally either with a long-acting GnRH agonist or surgical castration, the following rank order of ApEn of GH release emerged, listed from maximally irregular to maximally regular: intact female, GnRH-agonist-treated female, ovariectomized female, orchidectomized male, GnRH-agonist-treated male, and intact male,[5] illustrated in Fig. 1. ApEn was highly significantly different between the pooled groups of neutered females and neutered males, $P < 10^{-4}$, confirmed visually in Fig. 3B.

More broadly, this application to the rat studies indicates the clinical utility of ApEn. ApEn agrees with intuition, confirming differences that are visually "obviously distinct," as in the comparisons in Figs. 1a and f, intact males versus females. Importantly, ApEn can also uncover and establish graded and oftentimes subtle distinctions, as in comparisons of Figs. 1b–e, the neutered subject time series. Furthermore, these analyses accommodated both a point-length restriction of $N = 60$ samples (10-hr dark period, 10-min sampling protocol) and a typically noisy environment (due to assay inaccuracies and related factors), representative of the types of constraints that are usually present in clinical and laboratory settings.

[27] S. M. Pincus, V. Padmanabhan, W. Lemon, J. Randolph, and A. R. Midgley, *J. Clin. Invest.* **101,** 1318 (1998).

[28] S. M. Pincus, J. D. Veldhuis, T. Mulligan. A. Iranmanesh, and W. S. Evans, *Am. J. Physiol.* **273** (*Endocrinol. Metab.* **36**), E989 (1997).

[29] J. D. Veldhuis, A. Y. Liem, S. South, A. Weltman, J. Weltman, D. A. Clemmons, R. Abbott, T. Mulligan, M. L. Johnson, S. Pincus, M. Straume, and A. Iranmanesh, *J. Clin. Endocrinol. Metab.* **80,** 3209 (1995).

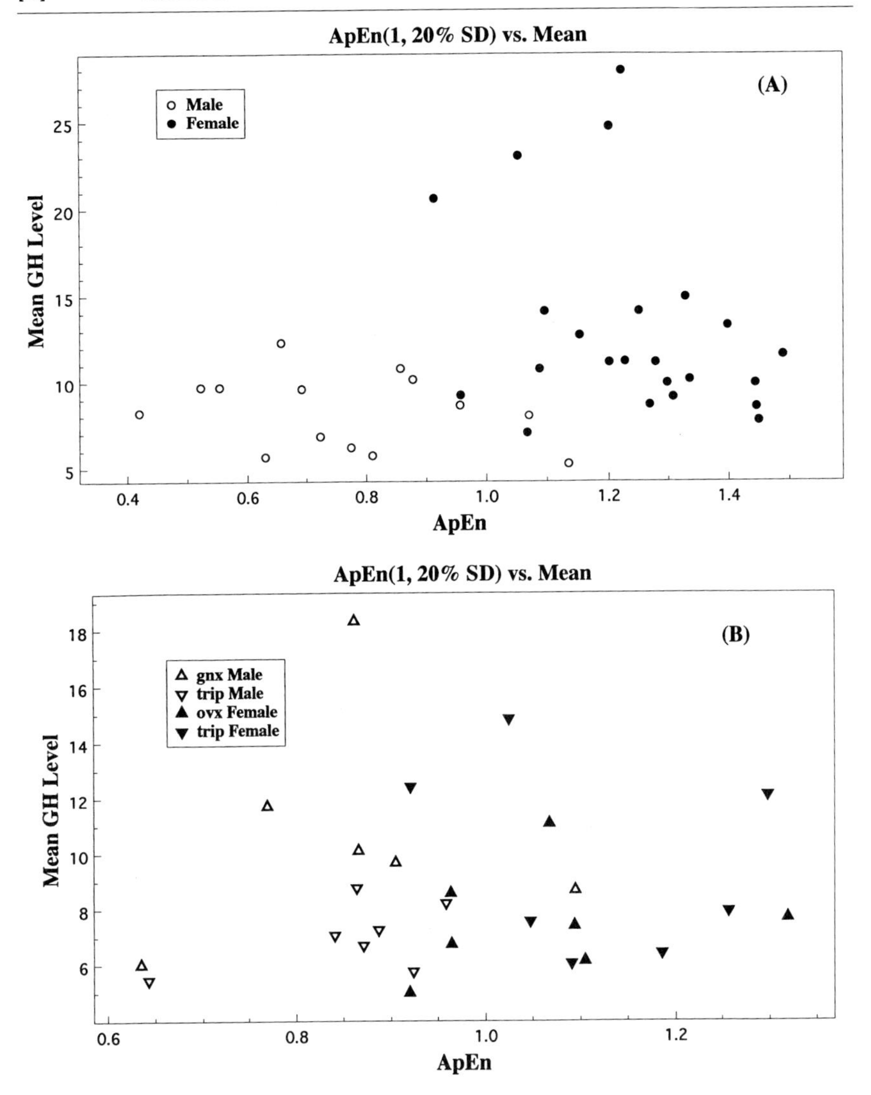

FIG. 3. Scatterplots of mean serum GH concentrations (ng/ml) versus ApEn(1, 20% SD) values in (A) individual intact male and female rats; (B) in surgically gonadectomized (gnx) or pharmocologically neutered (GnRH agonist triptorelin treatment) male and female rats.

Relationship to Other Approaches

Feature Recognition Algorithms

The orientation of ApEn is to quantify the amount of regularity in time series data as a single parameter (number). This approach involves a different and complementary philosophy than do algorithms that search for particular pattern features in data. Representative of these latter algorithms are the pulse detection algorithms central to endocrine hormone analysis, which identify the number of peaks in pulsatile data and their locations.[30] When applied to clearly pulsatile data, such pulse detection algorithms have provided significant capability in the detection of abnormal hormone secretory patterns. However, such algorithms often ignore "secondary" features whose evolution may provide substantial information. For instance, ApEn will identify changes in underlying episodic behavior that do not reflect changes in peak occurrences or amplitudes, whereas the aforementioned pulse identification algorithms generally ignore such information. Also, ApEn can be applied to those signals in which the notion of a particular feature, such as a pulse is not at all clear, e.g., an EEG time series. We recommend applying feature recognition algorithms in conjunction with ApEn when there is some physical basis to anticipate repetitive presence of the feature.

Statistics Related to Chaos

The historical development of mathematics to quantify regularity has centered around various types of entropy measures. Entropy is a concept addressing system randomness and predictability, with greater entropy often associated with more randomness and less system order. Unfortunately, there are numerous entropy formulations, and many entropy definitions cannot be related to one other.[1] K–S entropy, developed by Kolmogorov and expanded on by Sinai, allows one to classify *deterministic* dynamic systems by rates of information generation.[7] It is this form of entropy that algorithms such as those given by Grassberger and Procaccia[8] and by Eckmann and Ruelle[31] estimate. There has been keen interest in the development of these and related algorithms[9] in the last few years, since entropy has been shown to be a parameter that characterizes chaotic behavior.[32]

[30] R. J. Urban, W. S. Evans, A. D. Rogol, D. L. Kaiser, M. L. Johnson, and J. D. Veldhuis, *Endocr. Rev.* **9,** 3 (1988).

[31] J. P. Eckmann and D. Ruelle, *Rev. Mod. Phys.* **57,** 617 (1985).

[32] R. Shaw, *Z. Naturforsch. A* **36,** 80 (1981).

However, K–S entropy was not developed for statistical applications, and it has major debits in this regard. The original and primary motivation for K–S entropy was to handle a highly theoretical mathematics problem: determining when two Bernoulli shifts are isomorphic. In its proper context, this form of entropy is primarily applied by ergodic theorists to well-defined theoretical transformations, for which no noise and an infinite amount of "data" are standard mathematical assumptions. Attempts to utilize K–S entropy for practical data analysis represent out-of-context application, which often generates serious difficulties, as it does here. K–S entropy is badly compromised by steady, (even very) small amounts of noise, generally requires a vast amount of input data to achieve convergence,[33,34] and is usually infinite for stochastic (random) processes. Hence a "blind" application of the K–S entropy to practical time series will only evaluate system noise, not underlying system properties. All of these debits are key in the present context, since most biological time series likely are comprised of both stochastic and deterministic components.

ApEn was constructed along lines thematically similar to those of K–S entropy, though with a different focus: to provide a widely applicable, statistically valid formula that will distinguish data sets by a measure of regularity.[1,17] The technical observation motivating ApEn is that if joint probability measures for reconstructed dynamics that describe each of two systems are different, then their marginal probability distributions on a fixed partition, given by conditional probabilities as in Eq. (4), are likely different. We typically need orders of magnitude fewer points to accurately estimate these marginal probabilities than to accurately reconstruct the "attractor" measure defining the process. ApEn has several technical advantages in comparison to K–S entropy for statistical usage. ApEn is nearly unaffected by noise of magnitude below r, the filter level; gives meaningful information with a reasonable number of data points; and is finite for both stochastic and deterministic processes. This last point gives ApEn the ability to distinguish versions of composite and stochastic processes from each other, whereas K–S entropy would be unable to do so.

Extensive literature exists about understanding (chaotic) deterministic dynamic systems through reconstructed dynamics. Parameters such as correlation dimension,[35] K–S entropy, and the Lyapunov spectrum have been much studied, as have techniques to utilize related algorithms in the presence of noise and limited data.[36–38] Even more recently, prediction (forecast-

[33] A. Wolf, J. B. Swift, H. L. Swinney, and J. A. Vastano, *Physica D* **16,** 285 (1985).
[34] D. S. Ornstein and B. Weiss, *Ann. Prob.* **18,** 905 (1990).
[35] P. Grassberger and I. Procaccia, *Physica D* **9,** 189 (1983).
[36] D. S. Broomhead and G. P. King, *Physica D* **20,** 217 (1986).

ing) techniques have been developed for chaotic systems.[39–41] Most of these methods successfully employ embedding dimensions larger than $m = 2$, as is typically employed with ApEn. Thus in the purely *deterministic dynamic system* setting, for which these methods were developed, they reconstruct the probability structure of the space with greater detail than does ApEn. However, in the general (stochastic, especially correlated stochastic process) setting, the statistical accuracy of the aforementioned parameters and methods is typically poor; see Refs. 1, 2, and 42 for further elucidation of this operationally central point. Furthermore, the prediction techniques are no longer sensibly defined in the general context. Complex, correlated stochastic and composite processes are typically not evaluated, because they are not truly chaotic systems. The relevant point here is that because the dynamic mechanisms of most biological signals remain undefined, a suitable statistic of regularity for these signals must be more "cautious," to accommodate general classes of processes and their much more diffuse reconstructed dynamics.

Generally, changes in ApEn agree with changes in dimension and entropy algorithms for low-dimensional, deterministic systems. The essential points here, ensuring broad utility, are that (1) ApEn can potentially distinguish a wide variety of systems: low-dimensional deterministic systems, periodic and multiply periodic systems, high-dimensional chaotic systems, stochastic and mixed (stochastic and deterministic) systems[1,15]; and (2) ApEn is applicable to noisy, medium-sized data sets, such as those typically encountered in biological time series analysis. Thus ApEn can be applied to settings for which the K–S entropy and correlation dimension are either undefined or infinite, with good replicability properties as discussed below. Evident, yet of paramount importance, is that the data length constraint is key to note; e.g., hormone secretion time series lengths are quite limited by physical (maximal blood drawing) constraints, typically <300 points.

Power Spectra, Phase Space Plots

Generally, smaller ApEn and greater regularity correspond in the spectral domain to more total power concentrated in a narrow frequency range, in contrast to greater irregularity, which typically produces broader banded

[37] A. M. Fraser and H. L. Swinney, *Phys. Rev. A* **33,** 1134 (1986).

[38] G. Mayer-Kress, F. E. Yates, L. Benton, M. Keidel, W. Tirsch, S. J. Poppl, and K. Geist, *Math. Biosci.* **90,** 155 (1988).

[39] M. Casdagli, *Physica D* **35,** 335 (1989).

[40] J. D. Farmer and J. J. Sidorowich, *Phys. Rev. Lett.* **59,** 845 (1987).

[41] G. Sugihara and R. M. May, *Nature* **344,** 734 (1990).

[42] S. M. Pincus, *Chaos* **5,** 110 (1995).

spectra with more power spread over a greater frequency range. The two opposing extremes are (1) periodic and linear deterministic models, which produce very peaked, narrow-banded spectra, with low ApEn values; and (2) sequences of independent random variables, for which time series yield intuitively highly erratic behavior, and for which spectra are very broad banded, with high ApEn values. Intermediate to these extremes are autocorrelated processes, which can exhibit complicated spectral behavior. These autocorrelated aperiodic processes can be either stochastic or deterministic chaotic. In some instances, evaluation of the spectral domain will be insightful, when pronounced differences occur in a *particular* frequency band. In other instances, there is oftentimes more of an ensemble difference between the time series, both viewed in the time domain and in the frequency domain, and the need remains to encapsulate the ensemble information into a single value to distinguish the data sets.

Also, greater regularity (lower ApEn) generally corresponds to greater ensemble correlation in phase space diagrams. Such diagrams typically display plots of some system variable $x(t)$ versus $x(t - T)$, for a fixed "time lag" T. These plots are quite in vogue, in that they are often associated with claims that correlation, in conjunction with aperiodicity, implies chaos. A cautionary note is strongly indicated here. The labeling of bounded, aperiodic, yet correlated output as *deterministic* chaos has become a false cognate. This is incorrect; application of Theorem 6 in Ref. 11 proves that any n-dimensional steady-state measure arising from a deterministic dynamic system model can be approximated to arbitrary accuracy by that from a *stochastic* Markov chain. This then implies that any given phase space plot could have been generated by a (possibly correlated) stochastic model. The correlation seen in such diagrams is typically real, as is geometric change that reflects a shift in ensemble process autocorrelation in some comparisons. However, these observations are entirely distinct from any claims regarding underlying model form (chaos versus stochastic process). Similarly, in power spectra, decreasing power with increasing frequency (oftentimes labeled $1/f$ decay) is also a property of process correlation, rather than underlying determinism or chaos.[16]

Mechanistic Hypothesis for Altered Regularity

It seems important to determine a unifying theme suggesting greater signal regularity in a diverse range of complicated neuroendocrine systems. We would hardly expect a single mathematical model, or even a single family of models, to govern a wide range of systems; furthermore, we would expect that *in vivo,* each physiologic signal would usually represent the output of a complex, multinodal network with both stochastic and deterministic components. Our mechanistic hypothesis is that in a variety of systems, greater regularity (lower ApEn) corresponds to greater component and subsystem autonomy. This hypothesis has been mathematically established via analysis of several very different, representational (stochastic and deterministic) mathematical model forms, conferring a robustness to model form of the hypothesis.[15,16] Restated, ApEn typically increases with greater system coupling and feedback, and greater external influences, thus providing an explicit barometer of autonomy in many coupled, complicated systems.

Many endocrine hormone findings, including those indicated above, suggest that hormone secretion pathology usually corresponds to greater signal *irregularity*. Accordingly, a possible mechanistic understanding of such pathology, given this hypothesis, is that healthy, normal endocrine systems function best as relatively closed, autonomous systems (marked by regularity and low ApEn values), and that accelerated feedback and too many external influences (marked by irregularity and high ApEn values) corrupt proper endocrine system function.

It would be very interesting to attempt to experimentally verify this hypothesis in settings where some of the crucial network nodes and connections are known, via appropriate interventions to normal neuroendocrine (more generally, biological network) flow, coupled with signal analysis at one or more output sites.

Cross-ApEn

Cross-ApEn is a measure of asynchrony between two time series.[2,4] As for ApEn, it is a two-parameter family of statistics, with m and r taking the same meaning as in the ApEn setting, herein fixed for application to the paired time series $\{u(i)\}$, $\{v(i)\}$. Cross-ApEn measures, within tolerance r, the (conditional) regularity or frequency of v patterns similar to a given u pattern of window length m. It is typically applied to standardized u and v time series. Greater asynchrony indicates fewer instances of (sub)pattern matches, quantified by larger cross-ApEn values. Figure 4, taken from a recent study of paired ACTH–cortisol dynamics in Cushing's disease,[43] illustrates the cross-ApEn quantification, with greater ACTH–cortisol secretory asynchrony in the diseased subject, compared to the control.

Cross-ApEn is generally applied to compare sequences from two distinct yet intertwined variables in a network. Thus we can directly assess network, and not just nodal, evolution under different settings; e.g., to evaluate uncoupling and/or changes in feedback and control. Hence, cross-ApEn facilitates analyses of output from myriad complicated networks, avoiding

[43] F. Roelfsema, S. M. Pincus, and J. D. Veldhuis, *J. Clin. Endocr. Metab.* **83,** 688 (1988).

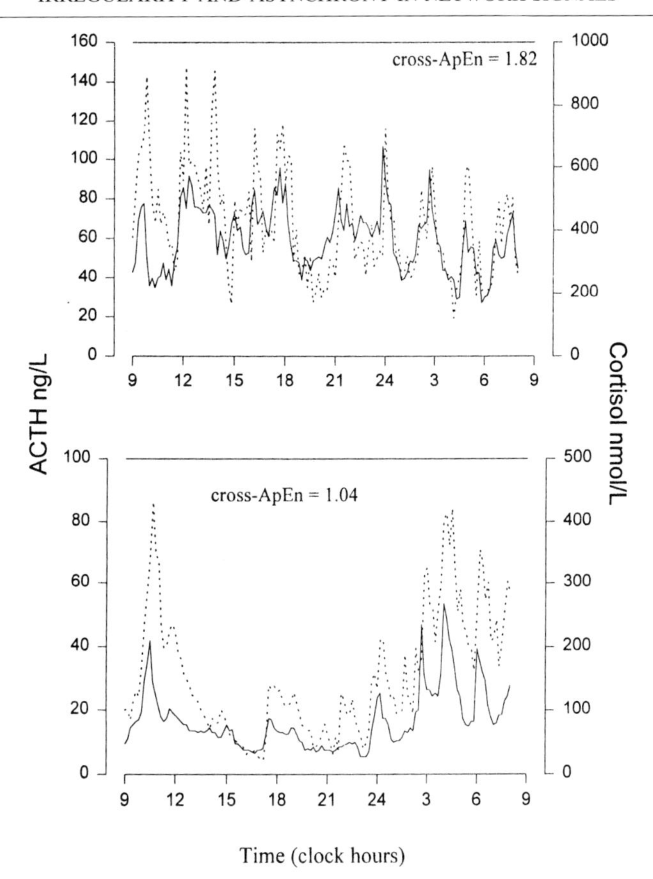

FIG. 4. Plasma concentrations of ACTH (dotted line) and cortisol (continuous line) in a female patient with Cushing's disease (upper panel) and a control subject (lower panel), each sampled at 10-min intervals for 24 hr.

the requirement to model the underlying system. This is especially important, since accurate modeling of (biological) networks is often nearly impossible—even a full description of all system nodes and pathways is typically unknown in most biologic systems, to say nothing of subsequent good mathematical approximations of the resultant internetwork dynamics. The key point, similarly for ApEn, is that full model specification is not required to realize an effective discrimination strategy. Furthermore, of course, there is a paucity of general *multi*variate time series statistical tools, as discussed further below.

In addition to the evident means to potentially discriminate network aspects of systems, cross-ApEn allows us to now address the following critical, yet generic, network issue: Are system changes primarily nodal (one-variable) or, rather, pathway or central control alterations (multivariate)? An answer to this question is not only essential to basic system understanding, but also a prime determinant in choosing, e.g., therapy/intervention strategies to attempt to restore pathobiologic milieus to more normative settings. Also, given multiple-node networks, we can successively probe pairwise, via cross-ApEn, to determine the weakest or altered (paired) links in the system. Furthermore, in cross-ApEn applications, paired (u, v) signal inputs can be within known subnetworks, e.g., the FSH–LH hormone secretory system; or alternatively, from less obviously related, broader networks, e.g., the EEG–LH system. Analysis of this latter setting allows us to address obliquely central control changes, in instances in which direct evaluation of the same would be effectively impossible. Concomitantly, to generate the u, v paired time series, one can utilize (quite) distinct sampling frequencies for each series. The technical point is that because we are interested in discrimination, rather than full model specification of the joint measure, we only require a fixed (common) protocol applied throughout to all data sets in a study.

The precise definition, introduced in Ref. 2, Definition 5, given next, is thematically similar to that for ApEn.

Definition of Cross-ApEn

Let $u = [u(1), u(2), \ldots, u(N)]$ and $v = [v(1), v(2), \ldots, v(N)]$ be two length $- N$ sequences. Fix input parameters m and r. Form vector sequences $x(i) = [u(i), u(i + 1), \ldots, u(i + m - 1)]$ and $y(j) = [v(j), v(j + 1), \ldots, v(j + m - 1)]$ from u and v, respectively. For each $i \leq N - m + 1$, set $C_i^m(r)(v\|u) =$ (number of $j \leq N - m + 1$ such that $d[x(i), y(j)] \leq r$)/$(N - m + 1)$, where $d[x(i), y(j)] = \max_{k=1,2,\ldots,m} [|u(i + k - 1) - v(j + k - 1)|]$, i.e., the maximum difference in their respective scalar components. The $C_i^m(r)$ values measure within a tolerance r the regularity, or frequency, of ($v-$) patterns similar to a given ($u-$) pattern of window length m.

Then define $\Phi^m(r)(v\|u)$ as the average value of $\ln C_i^m(r)(v\|u)$, and, finally, define cross-ApEn$(m, r, N)(v\|u) = \Phi^m(r)(v\|u) - \Phi^{m+1}(r)(v\|u)$.

Typically, we apply cross-ApEn with $m = 1$ and $r = 0.2$ to *standarized* u and v time series data, i.e., for each subject, we apply cross-ApEn(1, 0.2) to the $\{u^*(i), v^*(i)\}$ series, where $u^*(i) = [u(i) - \text{mean } u]/\text{SD } u$ and $v^*(i) = [v(i) - \text{mean } v]/\text{SD } v$. This standardization, in conjunction with the choice of m and r, ensures good replicability properties for cross-ApEn for the data lengths to be studied. To establish a theoretical statistical validity of cross-ApEn as so employed, we studied a range of two-variable vector AR(2) processes, and several types of coupled two-variable analogs of the "variable lag" process described below, for each of which we applied cross-ApEn(1, 0.2) to standardized time series (x, y pair) outputs, 50 replicates of $N = 150$-point data lengths per process. For each process studied, SD (cross-ApEn) was ≤ 0.06, the SD calculated from the cross-ApEn values from the 50 replicates; this imparts reasonable replicability properties similar to that for ApEn.[4,13,15] This degree of reproducibility is not unexpected, since qualitatively, cross-ApEn is a parameter that aggregates low-order, two-variable joint distributions at a moderately coarse resolution (determined by r).

As a representative example of application of cross-ApEn to biological data, we now consider the following study.

LH-T Study, Males

In recent years, many studies have been concerned with LH and testosterone (T) serum concentration time series in both younger and older males, both to better understand the physiology of reproductive capacity and, clinically, to assess, e.g., a loss of libido, or decreased reproductive performance. Furthermore, there is considerable interest in determining whether a hypothesized male climacteric (or so-called andropause) at least partially analogous to menopause in the female exists and, if so, in what precise sense. While considerable insight has already been gained from many studies, nontrivial controversies remain concerning several classes of findings, including primary determinations of whether overall mean levels of LH and T decrease with increasing age.

A study was performed to determine possible secretory irregularity shifts with aging within the LH–T axis.[4] Serum concentrations were derived for LH and T in 14 young (21–34 yr) and 11 older (62–74 yr) healthy men. For each subject, blood samples were obtained at frequent (2.5-min) intervals during a sleep period, with an average sampling duration of 7 hr. Although mean (and SD) of LH and T concentrations were indistinguishable in the two age groups, for each of LH and T, older males have consistently and highly significantly more irregular serum reproductive-hormone concentrations than younger males: for LH, aged subjects had greater ApEn values (1.525 +/− 0.221) than younger individuals (1.207 +/− 0.252), $P < 0.003$, while for testosterone, aged subjects had greater ApEn values (1.622 +/− 0.120) than younger counterparts (1.384 +/− 0.228), $P < 0.004$.

Probably a yet mechanistically more important finding in this study[4] was seen via cross-ApEn analysis. Cross-ApEn was applied to the paired LH–T time series; statistically, even more vividly than for the irregularity (ApEn) analyses, older subjects exhibited greater cross-ApEn values (1.961 +/− 0.121) compared to younger subjects (1.574 +/− 0.249), $P <$

10^{-4}, with nearly 100% sensitivity and specificity, indicating greater LH–T asynchrony in the older group (Fig. 5). Moreover and notably, no significant LH–T linear correlation (Pearson R) differences were found between the younger and older cohorts, $P > 0.62$ (Fig. 5). Several possibilities for the source of the erosion of LH–testosterone synchrony are discussed,[4] although a clear determination of this source awaits future study. Mechanistically, the results implicate (LH–T) network uncoupling as marking male reproductive aging, which we now have several quantifiable means to assess.

As another endocrinologic example of cross-ApEn utility, in a study of 20 Cushing's disease patients versus 29 controls,[43] cross-ApEn of ACTH–cortisol was greater in patients (1.686 +/− 0.051) than in controls (1.077 +/− 0.039), $P < 10^{-15}$, with nearly 100% sensitivity and specificity, suggesting compromise of hormonal pathways and feedback control in diseased subjects, atop that previously seen for more localized nodal secretory dynamics of each hormone individually.[24] Figure 4 displays representative serum profiles from this study. Additionally, healthy men and women showed progressive erosion of bihormonal ACTH–cortisol synchrony with increased aging[43] via cross-ApEn, similar to the LH–T erosion of synchrony in men noted above, suggesting that increased cross-ApEn (greater asynchrony) of paired secretory dynamics is an ubiquitous phenomenon with advancing age.

Complementarity of ApEn and Cross-ApEn to Correlation and Spectral Analyses

Mathematically, the need for ApEn, and particularly for cross-ApEn, is clarified by considering alternative parameters that might address similar

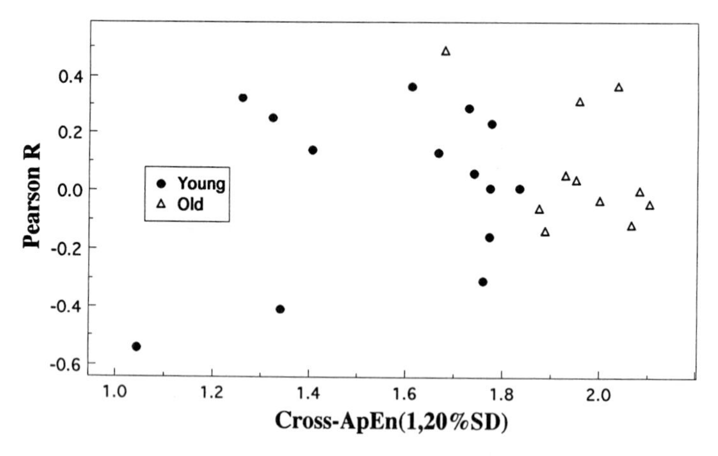

FIG. 5. Individual subject cross-ApEn values versus cross-correlation (Pearson R), applied to the joint LH–testosterone time series in healthy young (●) versus older men (△).

concepts. In comparing two distinct signals or variables (e.g., to assess a degree of synchrony), primary parameters that one might employ include the cross-correlation function (including Pearson R) and the cross-spectrum,[44] with single-variable counterparts, the autocorrelation function and the power spectrum. Evaluation of these parameters often is insightful, but with relatively short length data sets, statistical estimation issues are nontrivial and, moreover, interpretation of the sample cross-correlation function is highly problematic, unless one employs a model-based prefiltering procedure (Ref. 44, p. 139). Furthermore, "standard" spectral estimation methods such as the FFT (fast Fourier transform) can be shown to be inconsistent and/or so badly biased that findings may be qualitatively incorrect, especially in the presence of outliers and nonstationarities. This is vividly demonstrated by Thomson,[45] who recently developed a superior multiple-data-window technique with major advantages compared to other spectral estimation techniques.[45,46] These difficulties are mirrored in the cross-spectrum, in addition to an often serious bias in estimation of coherency in short series.

Most importantly, the autocorrelation function and power spectrum, and their bivariate counterparts, are most illuminating in linear systems, e.g., SARIMA (seasonal autoregressive integrated moving average) models, for which a rich theoretical development exists.[47] For many other classes of processes, these parameters are often relatively ineffective at highlighting certain model characteristics, even apart from statistical considerations. To illustrate this point, consider the following simple model, which we denote as a "variable lag" process: this consists of a series of quiescent periods, of variable length duration, interspersed with identical positive pulses of a fixed amplitude and frequency. Formally, we recursively define an integer time-valued process denoted VarLag whose ith epoch consists of (a quiescent period of) values $= 0$ at times $t_{i-1} + 1, t_{i-1} + 2, \ldots, t_{i-1} + \text{lag}_i$, immediately followed by the successive values $\sin(\pi/6)$, $\sin(2\pi/6)$, $\sin(3\pi/6)$, $\sin(4\pi/6)$, $\sin(5\pi/6)$, $\sin(6\pi/6) = 0$ at the next 6 time units, where lag_i is a random variable uniformly distributed on (randomly chosen between) the integers between 0 and 60, and t_{i-1} denotes the last time value of the $(i - 1)$st sine-pulse. Figure 6A displays representative output from this process, with Fig. 6B giving a closer view of this output near time $t = 400$. The power spectrum and autocorrelation function calculations shown in Figs. 6C and E were calculated from a realization of length $N = 100{,}000$

[44] C. Chatfield, "The Analysis of Time Series: An Introduction," 4th Ed. Chapman and Hall, London, 1989.

[45] D. J. Thomson, *Phil. Trans. R. Soc. Lond. A* **330,** 601 (1990).

[46] C. Kuo, C. Lindberg, and D. J. Thomson, *Nature* **343,** 709 (1990).

[47] G. E. P Box and G. M. Jenkins, "Time Series Analysis, Forecasting and Control." Holden-Day, San Francisco, 1976.

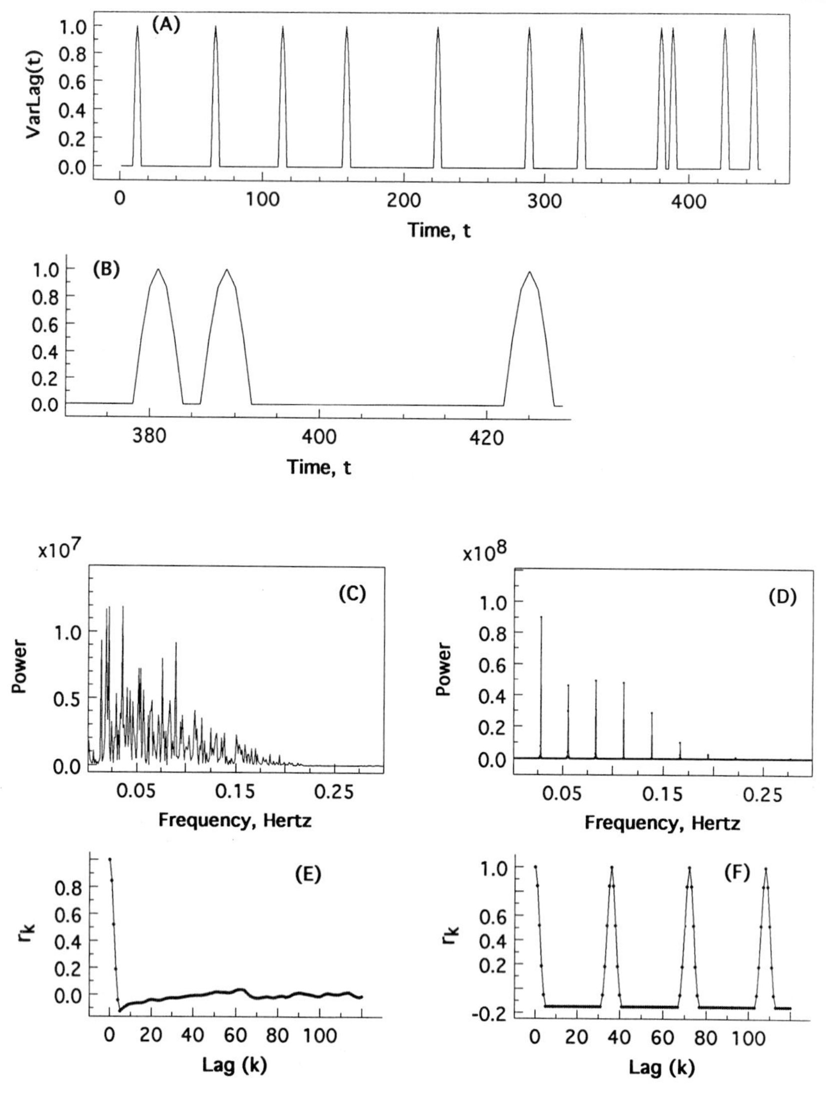

FIG. 6. (A) Representative time series for a "variable lag" sine-wave process denoted VarLag (see text for formal definition); (B) close-up view of (A), near time $t = 400$; (C) power spectrum for VarLag; (D) power spectrum for a constant (fixed) lag analog of VarLag; (E) autocorrelogram corresponding to (C); (F) autocorrelogram corresponding to (D). Parts (C)–(F) are all derived from time series of length $N = 100{,}000$ points.

points. (The somewhat coarse sampling of the pulse in the above process definition was chosen to approximate typical sampling resolution in actual clinical studies.)

Processes consisting of alternatingly quiescent and active periods would seem reasonable for biologists to consider, as they appear to model a wide variety of phenomena. However, within mathematics, such processes with a variable quiescent period are not commonly studied. To the biologist, output from the above model would be considered smoothly pulsatile, especially with the identical pulses; the variable lag process would be most readily distinguished from its constant lag counterpart (for which $\text{lag}_i = 30$ time units for all i) via a decidedly positive SD for the interpulse duration time series, in the variable lag setting, as opposed to SD = 0 (constant interpulse duration) in the constant lag setting. The essential point here, however, is that for VarLag, the power spectrum and autocorrelation function somewhat confound, as seen in Figs. 6C and E. Based on these figures alone, the pulsatile nature of the time series realizations is hardly evident, and for all $k \geq 6$, the autocorrelation coefficient r_k at lag k is insignificantly different from 0. In contrast, the power spectrum and autocorrelation function confirm the periodicity of the constant lag analog, shown in Figs. 6D and F, as expected. Significantly, the issues here are in the parameters, rather than statistical inadequacies based on an insufficiently long output, or on artifacts (outliers), since Figs. 6C–F were derived from calculations based on 100,000 points from a purely theoretical model.

Similar limitations of the spectra and autocorrelation function are inherent to wide classes of processes. From a general mathematical framework, we can construct large classes of variable lag processes simply by considering point processes,[48] in which we replace the "point" occurrence by a pulse occurrence, the pulse itself of either a fixed or variable form. The associated counting process could be of any character, and need not be so special as Poisson or renewal (as in the above example). Also, variable lags between events to be compared are the normative case in nonlinear (deterministic and stochastic) differential equations, in Poisson clumping models,[49] and in output variables in typical (adaptive) control theory models and queueing network models. Notably, for many two-dimensional analogs of variable lag processes, and indeed for many two-dimensional systems in which no small set of dominant frequencies encapsulates most of the total power, the cross-spectrum and the cross-correlation function often will similarly

[48] J. P. Bremaud, "Point Processes and Queues: Martingale Dynamics." Springer-Verlag, New York, 1981.

[49] D. Aldous, "Probability Approximations via the Poisson Clumping Heuristic." Springer-Verlag, Berlin, 1989.

fail to highlight episodicities in the underlying model and data, and thus fail to highlight concomitant changes to such episodic components.

In contrast to the autocorrelation function and spectral differences between the above variable lag and constant lag processes, the respective ApEn(1, 20% SD) values for the two processes are in close agreement: mean ApEn = 0.195 for the variable lag process, while ApEn = 0.199 for the constant lag setting. This agreement in ApEn values manifests the primary requirement of matching (sub)patterns within data, while relaxing the requirement of a dominant set of frequencies at which these subpatterns occur. The two-variable analog of ApEn, given by cross-ApEn, similarly enables one to assess synchrony in many classes of models. It thus should not be surprising that in many studies, e.g., the LH-T study,[4] cross-correlation (Pearson R) does not show significant group differences, whereas cross-ApEn does (as in Fig. 5).

It should be emphasized, nonetheless, that Figs. 6C–F neither invalidate spectral power and (lagged) autocorrelation calculations, nor do they violate a properly oriented intuition. The broad-banded spectrum in Fig. 6C, and the negligible lagged autocorrelation in Fig. 6E for lag $\geq$ 6 time units, primarily reflect the independent, identically distributed, relatively broad distribution of the variable lag_i. Visually this conforms to viewing Fig. 6A from afar, in effect (nearly) ignoring the nature of each pulse, instead *de facto* primarily focusing on the "random" timing of the peaks as the process of interest. The viewpoint taken by ApEn is thus complementary to the spectrum and correlogram, more *de facto* focusing on (close-up) similarities between active pulses, e.g., from the perspective given in Fig. 6B, while in effect nearly ignoring the nature of the quiescent epoch aspect of the process. The utility of ApEn and cross-ApEn to biologists is based on the recognition that in many settings, changes in the episodic character of the *active* periods within time series appear to mark physiologic and pathophysiologic changes—thus there is a concomitant need for quantitative methods that primarily address this perspective, e.g., ApEn and cross-ApEn.

Toward More Faithful Network Modeling

In modeling general classes of real biologic networks, we anticipate that any single homogeneous model form, such as deterministic dynamic systems, ARMA models, stochastic differential equations, or Markov chains or processes, is inadequate. At the least, we would expect faithful models in many settings to incorporate queueing network and (adaptive) control theory considerations, as well as the likelihood that different components in a network are defined by polymorphous and distinct mathematical model forms. Queueing models arise naturally in multinode network analysis with interconnections; control models arise from considering the brain (or some focal component) as an intelligent central processor, possibly altering system characteristics based on, e.g., a threshold response. Queueing theory has developed largely within communication (traffic) theory[50] and computer network analysis,[51,52] while control theory has developed toward optimizing performance in engineering systems.[53] Notably, analytical developments from these fields may not be directly suitable to physiologic network modeling, not surprisingly since these fields were not driven by biological context. Two physiologically motivated problems within these fields that seem worthy of significant effort are to describe the behavior of (1) queueing networks in which some nodes are coupled oscillators; (2) (adaptive) control systems in which there is a balking probability p with which the control strategy is not implemented. Problem (2) could model some diseases, in which messages may not reach the controller or the controller may be too overwhelmed to respond as indicated.

Several "decision theoretic" modeling features that fall under the umbrella of queueing theory seem especially appropriate (and timely) to impose on many biological networks, to achieve faithful characterizations of true network protocols, both qualitatively as well as quantitatively. These aspects of traffic theory include (1) "broadcast" signaling (of a central controller); (2) priority service; (3) alternative routing hierarchies; (4) "finite waiting areas" for delayed messages, incorporating the possibility (and consequences) of "dropped" or lost messages; and (5) half-duplex transmission, in which on a given pathway between two sources, only one source at a time may use the transmission pathway. In addition, we must always clarify the "network topology" or routing configuration among nodes in a network, i.e., determine which pairs of nodes have (direct) pathways to one another, before beginning to address quantitative specifications of signal transmission along the putative pathways. All of these features can be described quantitatively, via decision-theoretic point processes and, typically, resultant network performance is then evaluated by large-scale numerical programs that "simulate" the stochastic environment.[54] General versions of such programs require considerable expertise and time to write,

[50] D. Gross and C. M. Harris, "Fundamentals of Queueing Theory," 2nd Ed. John Wiley and Sons, New York, 1985.

[51] A. O. Allen, "Probability, Statistics, and Queueing Theory: With Computer Science Applications," 2nd Ed. Academic Press, San Diego, 1990.

[52] J. R. Jackson, *Operations Res.* **5**, 518 (1957).

[53] W. H. Fleming and R. Rishel, "Deterministic and Stochastic Optimal Control." Springer-Verlag, Berlin, 1975.

[54] A. M. Law and W. D. Kelton, "Simulation Modeling and Analysis," 2nd Ed. McGraw-Hill, New York, 1991.

and are commercially available from a few sources, though regrettably, these are quite expensive to procure, and are usually targeted to specialists. Furthermore, only in very specialized settings do the mathematical descriptions of the decision-theoretic constraints allow for purely analytic (as opposed to simulation, i.e., so-called "numerical Monte Carlo" methods) solutions. This is quite possibly the reason why this essential yet specialized branch of applied probability theory is relatively unknown.

The above perspective strongly motivates the requirement that for effective and broadest utility, statistics developed for general network analysis be model independent, or at least provide robust qualitative inferences across a wide variety of network configurations. The observation that both ApEn and cross-ApEn are model independent, i.e., functionals of the presented sequences (time series), and are not linked to a prescribed model form, fits squarely with this perspective.

Spatial (Vector) ApEn

A spatial (vector) version of approximate entropy (ApEn) was recently developed to quantify and grade the degrees of irregularity of planar (and higher dimensional) arrangements.[55] Spatial ApEn appears to have considerable potential, both theoretically and empirically, to discern and quantify the extent of changing patterns, and the emergence and dissolution of traveling waves, throughout multiple contexts within both biology and chemistry. This is particularly germane to the detection of subtle or "insidious" structural differences among arrays, even where clear features or symmetries are far from evident.

One initial application of spatial ApEn will facilitate an understanding of both its potential utility and, simultaneously, of precisely what the quantification is doing. In Ref. 55, we clarified and corrected a fundamental ambiguity (flaw) in R. A. Fisher's specification of experimental design.[56,57] Fisher implicitly assumed throughout his developments that all Latin squares (n row $\times$ n column arrangements of n distinct symbols where each symbol occurs once in each row and once in each column) were equally and maximally spatially random, and subsets of such Latin squares provided the underpinnings of experimental design. In the example below, even in the small sized 4×4 Latin square case, we already see that spatial ApEn quantifies differences among the candidate squares. (We then proposed an

[55] B. H. Singer and S. Pincus, *Proc. Natl. Acad. Sci. U.S.A.* **95,** 1363 (1998).

[56] R. A. Fisher, "Statistical Methods for Research Workers." Oliver & Boyd, Edinburgh, UK, 1925.

[57] R. A. Fisher, "The Design of Experiments." Oliver & Boyd, Edinburgh, UK, 1935.

experimental design procedure based on *maximally irregular* Latin squares, eliminating the flaw.[55])

The precise definition of spatial ApEn is provided as Definition 1 in Ref. 55. Thematically, again, it is similar to that for ApEn, both in the form of comparisons (determining the persistence of subpatterns to matching subpatterns), and in the input specification of window length m and *de facto* tolerance width r. The critical epistemologic novelty is that in the planar and spatial case, given a multidimensional array A, and a function on A (spatial time series) u, we specify a vector direction $\mathbf{v}$, and consider irregularity in A along the vector direction $\mathbf{v}$. We denote this as vector-ApEn$_{\mathbf{v}}$ $(m, r)(u)$; in instances in which the array values are discrete, e.g., integers, as in the Latin square example below, we often set r to 0, thus monitoring precise subpattern matches, and suppress r in the vector-ApEn notation, with the resultant quantity denoted vector-ApEn$_{\mathbf{v}}(m)(u)$. Descriptively, vector-ApEn$_{\mathbf{v}}$ $(m)(u)$ compares the logarithmic frequency of matches of blocks of length m (for $m \geq 1$) with the same quantity for blocks of length $m + 1$. Small values of vector-ApEn imply strong regularity, or persistence, of patterns in u in the vector direction $\mathbf{v}$, with the converse interpretation for large values. The vector direction $\mathbf{v}$ designates arrangements of points on which the irregularity of u is specified, *a priori,* to be of particular importance. For example, if $\mathbf{v} = (0, 1)$, then vector-ApEn measures irregularity along the rows of A, and disregards possible patterns, or the lack thereof, in other directions; $\mathbf{v} = (1, 0)$ focuses on column irregularity; and $\mathbf{v} = (1, 2)$ or $(2, 1)$ or $(-1, 2)$ emphasizes knight's move (as in chess) patterns. In typical applications, it is necessary to guarantee irregularity in two or more directions simultaneously. This requires evaluation of vector-ApEn for a set $\mathbf{V}$ of designated vectors. For example, simultaneous row, column, and diagonal irregularity assessment entails calculation of vector-ApEn for all elements $\mathbf{v}$ in $\mathbf{V} = \{(1, 0); (0, 1); (1, -1); (1, 1)\}$.

Example 1 illustrates vector-ApEn for four Latin squares, and as noted above, also clarifies the remarks concerning Fisher's ambiguity in the specification of experimental design.

Example 1

Consider the following four 4×4 Latin squares.

A	B	C	D
1 2 3 4	1 2 3 4	1 2 3 4	1 2 3 4
2 3 4 1	3 4 1 2	4 3 2 1	3 1 4 2
3 4 1 2	4 3 2 1	3 1 4 2	2 4 1 3
4 1 2 3	2 1 4 3	2 4 1 3	4 3 2 1

For A, vector-ApEn$_{(1,0)}$(1) = vector-ApEn$_{(0,1)}$(1) = 0; for B, vector-ApEn$_{(1,0)}$(1) = vector-ApEn$_{(0,1)}$(1) = 0.637; for C, vector-ApEn$_{(1,0)}$(1) = 0.637, and vector-ApEn$_{(0,1)}$(1) = 1.099; and for D, vector-ApEn$_{(1,0)}$ (1) = vector-ApEn$_{(0,1)}$(1) = 1.099. These calculations manifest differing extent of feature replicability in the (1, 0) and (0, 1) directions, with A quite regular in both directions, B intermediately irregular in both directions, C maximally irregular in rows, yet intermediate in columns, and D maximally irregular in both rows and columns. Alternatively, in A, e.g., in rows, there are 3 occurrences each of 4 pairs [(1, 2), (2, 3), (3, 4), and (4, 1)], and no occurrences of the other 8 possible pairs. In B, in rows, 4 pairs occur twice [(1, 2), (2, 1), (3, 4), and (4, 3)], while 4 pairs occur once [(1, 4), (2, 3), (3, 2), and (4, 1)]. In D, in rows, each of the 12 pairs (i, j), $1 \leq i, j \leq 4$, $i \neq j$, occur precisely once. (Similar interpretation follows readily for columns.)

Several broad application areas illustrate the proposed utility of vector-ApEn to frequently considered settings within biology and chemistry. First, we anticipate that vector-ApEn will bear critically on image and pattern recognition determinations,[58] to assess the degree of repeatability of prescribed features. Sets of base atoms would be shapes of features of essential interest; moreover, these can be redefined (as indicated in Ref. 55) either on the same scale as the original atoms, or on a much larger scale, thus providing a more macroscopic assessment of spatial irregularity.

Also, many models within physics and physical chemistry are lattice-based systems, e.g., the nearest neighbor Ising model and the classical Heisenberg model, which have been employed to model a magnet (via spin), a lattice gas, alloy structure, and elementary particle interactions.[59] Determining relationships between changes in vector-ApEn in these models and physical correlates would seem highly worthwhile, either theoretically or experimentally. Also, within solid-state physical chemistry, we speculate that grading the extent of array disorder may prove useful in assessing or predicting (1) crystal and alloy strength and/or stability under stresses; (2) phase transitions, either liquid-to-gas, solid-to-liquid, or frigid-to-superconductive; and (3) performance characteristics of semiconductors.

Lastly, the analysis of traveling waves oftentimes requires a quantification of subtle changes, particularly as to the extent of formation and, conversely, the extent of dissolution or dissipation of wave fronts, above and beyond an identification of primary wave "pulses" and resultant statistical analyses. Although considerable signal-to-noise analysis methodology has been developed for and applied to this setting, to clarify wave fronts, in

[58] U. Grenander, "General Pattern Theory." Oxford University Press, UK, 1993.

[59] R. Israel, "Convexity in The Theory of Lattice Gases." Princeton University Press, New Jersey, 1979.

the ubiquitous instances where the extent of insidious or subordinate activity is the primary feature of interest, a critical and further assessment of the wave patterns is required, to which vector-ApEn should readily apply, both in two- and three-dimensional settings. This recognition may be particularly critical near the genesis of an upcoming event of presumed consequence. One representative, quite important application of this perspective is to (atrial) fibrillation and arrhythmia detection within cardiac physiology.

Summary and Conclusion

The principal focus of this chapter has been the description of both ApEn, a quantification of serial irregularity, and of cross-ApEn, a thematically similar measure of two-variable asynchrony (conditional irregularity). Several properties of ApEn facilitate its utility for biological time series analysis: (1) ApEn is nearly unaffected by noise of magnitude below a *de facto* specified filter level; (2) ApEn is robust to outliers; (3) ApEn can be applied to time series of 50 or more points, with good reproducibility; (4) ApEn is finite for stochastic, noisy deterministic, and composite (mixed) processes, these last of which are likely models for complicated biological systems; (5) increasing ApEn corresponds to intuitively increasing process complexity in the settings of (4); and (6) changes in ApEn have been shown mathematically to correspond to mechanistic inferences concerning subsystem autonomy, feedback, and coupling, in diverse model settings. The applicability to medium-sized data sets and general stochastic processes is in marked contrast to capabilities of "chaos" algorithms such as the correlation dimension, which are properly applied to low-dimensional iterated deterministic dynamical systems. The potential uses of ApEn to provide new insights in biological settings are thus myriad, from a complementary perspective to that given by classical statistical methods.

ApEn is typically calculated by a computer program, with a FORTRAN listing for a "basic" code referenced above. It is imperative to view ApEn as a family of statistics, each of which is a relative measure of process regularity. For proper implementation, the two input parameters m (window length) and r (tolerance width, *de facto* filter) must remain fixed in all calculations, as must N, the data length, to ensure meaningful comparisons. Guidelines for m and r selection are indicated above. We have found normalized regularity to be especially useful, as in the growth hormone studies discussed above; "r" is chosen as a fixed percentage (often 20%) of the subject's SD. This version of ApEn has the property that it is decorrelated from process SD—it remains unchanged under uniform process magnification, reduction, and translation (shift by a constant).

Cross-ApEn is generally applied to compare sequences from two distinct yet interwined variables in a network. Thus we can directly assess network, and not just nodal, evolution, under different settings—e.g., to directly evaluate uncoupling and/or changes in feedback and control. Hence, cross-ApEn facilitates analyses of output from myriad complicated networks, avoiding the requirement to fully model the underlying system. This is especially important, since accurate modeling of (biological) networks is often nearly impossible. Algorithmically and insofar as implementation and reproducibility properties are concerned, cross-ApEn is thematically similar to ApEn.

Furthermore, cross-ApEn is shown to be complementary to the two most prominent statistical means of assessing multivariate series, correlation and power spectral methodologies. In particular, we highlight, both theoretically and by case study examples, the many physiological feedback and/or control systems and models for which cross-ApEn can detect significant changes in bivariate asynchrony, yet for which cross-correlation and cross-spectral methods fail to clearly highlight markedly changing features of the data sets under consideration.

Finally, we introduce spatial ApEn, which appears to have considerable potential, both theoretically and empirically, in evaluating multidimensional lattice structures, to discern and quantify the extent of changing patterns, and for the emergence and dissolution of traveling waves, throughout multiple contexts within biology and chemistry.

[10] Distribution Methods and Analysis of Nonlinear Longitudinal Data

By MICHELLE LAMPL and MICHAEL L. JOHNSON

Introduction

This chapter considers the application of frequency distribution analysis to longitudinal data. The statistics of distribution functions have previously been employed in endocrinological studies aimed at identifying the frequency and variability of hormonal concentrations present in serial endocrine data. More recently, this approach has been applied to longitudinal data as a pattern identification method. The information content of such analyses is considered here with examples from infant body length data.

0076-6879/00 $30.00

Frequency Distribution Analysis of Pulsatile Data

Previous studies employing frequency distribution methods include those aimed at characterizing the nature and frequency of pulses contained in endocrine profiles. Some of these studies have as an objective the investigation of differences between clinical samples in underlying hormonal concentration patterns, whereas other studies focus on identifying the presence of pulse amplitude heterogeneity, with the goal of inferring mechanisms regarding pulse control and target signaling. An example of the first type of study includes the comparison of percentiles and first derivatives from the logit linear probits of cumulative probabilities.[1] All observations of hormonal concentrations or pulses identified by pulse detection programs are sorted into predetermined bins, irrespective of their temporal attributes. With the sequential relationships thus removed, a frequency distribution of concentrations results, providing an estimation of the total time duration of different concentrations. The cumulative frequency distribution of these data is calculated and the discrete probabilities, linear probits and their first derivatives, and the probability estimates of the peak (95%), intermediate (50%), and trough (5%) concentrations are employed to investigate differences in growth hormone characteristics between groups by age and height.[1,2]

An example of the second type of study includes a frequency distribution analysis of pulses that are identified by a pulse algorithm to investigate the presence of pulse subpopulations in studies by Lopez and colleagues concerning secretory patterns of prolactin.[3-5] They conclude that two subpopulations of pulses are present, a small and a large pulse, and it is suggested that these might reflect two different control systems.

These analyses focus on the frequency distribution of pulse amplitudes in serial data in order to quantify characteristic patterns within and between samples. Investigations into the time distribution of amplitudes is interesting in terms of the apparent simplicity of identifying subpopulations of events by inspection of modality, skewness, and cumulative probability densities. It has been suggested that frequency distribution methods might be helpful in the analysis of longitudinal data in general.[2] This chapter explores the

[1] P. C. Hindmarsh, D. R. Matthews, I. Stratton, P. J. Pringle, and C. G. Brook, *Clin. Endocr.* **36,** 165 (1992).

[2] D. R. Matthews, P. C. Hindmarsh, P. J. Pringle, and C. G. Brook, *Clin. Endocr.* **35,** 245 (1991).

[3] F. J. Lopez, J. R. Domingo, J. E. Sanchez-Criado, and A. Negro-Vilar, *Endocrinology* **124,** 536 (1989a).

[4] F. J. Lopez, J. R. Domingo, J. E. Sanchez-Criado, and A. Negro-Vilar, *Endocrinology* **124,** 527 (1989b).

[5] F. J. Lopez, J. E. Sanchez-Criado, and A. Negro-Vilar, *Endocrinology* **129,** 1471 (1991).

usefulness of frequency distributions in longitudinal data analysis. The data employed in the investigations are longitudinal infant body length data. The focus of investigation includes the application of frequency distribution analysis for identifying patterns of growth in both individuals and pooled sample data.

Frequency Distribution Analysis of Longitudinal Growth Data

One of the first applications of the distribution method to longitudinal changes was directed toward the identification of growth patterns. The approach involved analyzing data from a longitudinal study of growth and calculating a time distribution of growth rates. The proposition was that such an approach permits the identification of the temporal pattern by which the child is growing. This method was specifically used to distinguish between two competing theoretical models of how growth occurs: A model of growth as a continuous and linear process in which growth is accrued by approximately equal increments each day is compared to a pulsatile, saltatory process in which growth occurs by intermittently occurring growth saltation events that punctuate stasis intervals during which no growth occurs.[6] In theory, if observational errors of measurement are Gaussian and growth occurs at a constant rate, a frequency distribution of growth rates will provide a unimodal normal curve. This proposition was used for the analysis of both human and rabbit growth[7–9] and was based on two assumptions: (1) if incremental data are found to provide a frequency distribution that is unimodal and normal, the sequence from which they were drawn in necessarily a linear and continuous one; and (2) saltation and stasis data are necessarily bimodal, with one distribution reflecting growth saltations and the second distribution characterizing stasis events. The second proposition derives from the two-phase description of saltation and stasis growth.

In subsequent investigations of saltation and stasis growth data, this type of diagnostic use of frequency distribution methods was shown to be inaccurate.[10–12] One of the shortcomings of distribution function analysis

[6] M. Lampl, J. D. Veldhuis, M. L. Johnson, *Science* **258,** 801 (1992).

[7] K. O. Klein, P. J. Munson, J. D. Bacher, G. B. Cutler, Jr., and J. Baron, *Endocrinology* **134,** 1317 (1994).

[8] C. Heinrichs, P. J. Munson, D. R. Counts, G. B. Cutler, Jr., and J. Baron, *Science* **268,** 442 (1995).

[9] M. Hermanussen, and K. Geiger-Benoit, *Ann. Hum. Biol.* **22,** 341 (1995).

[10] M. Lampl, N. Cameron, J. D. Veldhuis, and M. L. Johnson, *Science* **268,** 445 (1995).

[11] M. Lampl, M. L. Johnson, and E. A. Frongillo, Jr., *Ann. Hum. Biol.* **24,** 65 (1997).

[12] M. L. Johnson, J. D. Veldhuis, and M. Lampl, *Endocrinology* **137,** 5197 (1996).

is that the temporal sequence of events is removed. Thus, the specific pattern followed by the events is nonrecoverable from the probability distribution of events. Many different time functions can result in similar probability distributions of events and this is an imprecise method for reconstructing the unique sequences in the original data.[10,13] In the growth pattern identification work, it was shown that bimodality of increments in serial growth data is a compatible, but not necessary condition of saltation and stasis growth.[12] This is because growth by saltation and stasis occurs in individual subjects by variable amplitude growth pulses at episodic, nonperiodic intervals. Such data can result in a number of distribution functions of increments and no unique distribution function can be expected. In general, distribution functions of increments from saltatory growth data are unimodal with an attenuated right shoulder. However, as the attenuation reflects the number and variable amplitude of growth saltations, both a unimodal normal and unimodal skewed function are compatible with saltatory growth.

The growth data are an illustrative model of the limits of distribution functions as a method of analysis for longitudinal data, particularly when the data are nonlinear and episodic. The present analyses further investigate this. The data employed are daily length measurements from three infants aged between 4 and 13 months. These data were collected under an approved human subjects protocol from the University of Pennsylvania as previously described.[14]

A useful initial question is this: What information does a distribution function analysis of growth increments provide? Statistical assessment of the means, medians, and cumulative probability distributions for these subjects' growth increments documents significant individual variability ($p < 0.01$, Kruskal–Wallis). Individual distribution functions are statistically significantly different from one another ($p < 0.001$, Kolmogorov–Smirnoff, Lilliefors). In this example, not only are each of the individual's distribution functions statistically significantly different from one another, but they are also statistically significantly different from the composite distribution function that characterizes the three as a sample. This analysis documents the uniqueness of individual growth patterns and statistically documents the inappropriateness of combining these individual subjects into a composite sample distribution function for pooled analysis of their growth pattern.

Figure 1 presents an overlay of three subjects' saltation and stasis growth distribution functions. All three are characterized by an apparent unimodal distribution with an attenuated right shoulder. For closer examination of the

[13] M. L. Johnson and M. Lampl, *Methods Neurosci.* **28,** 364 (1995).
[14] M. Lampl, *Am. J. Hum. Biol.* **5,** 641 (1993).

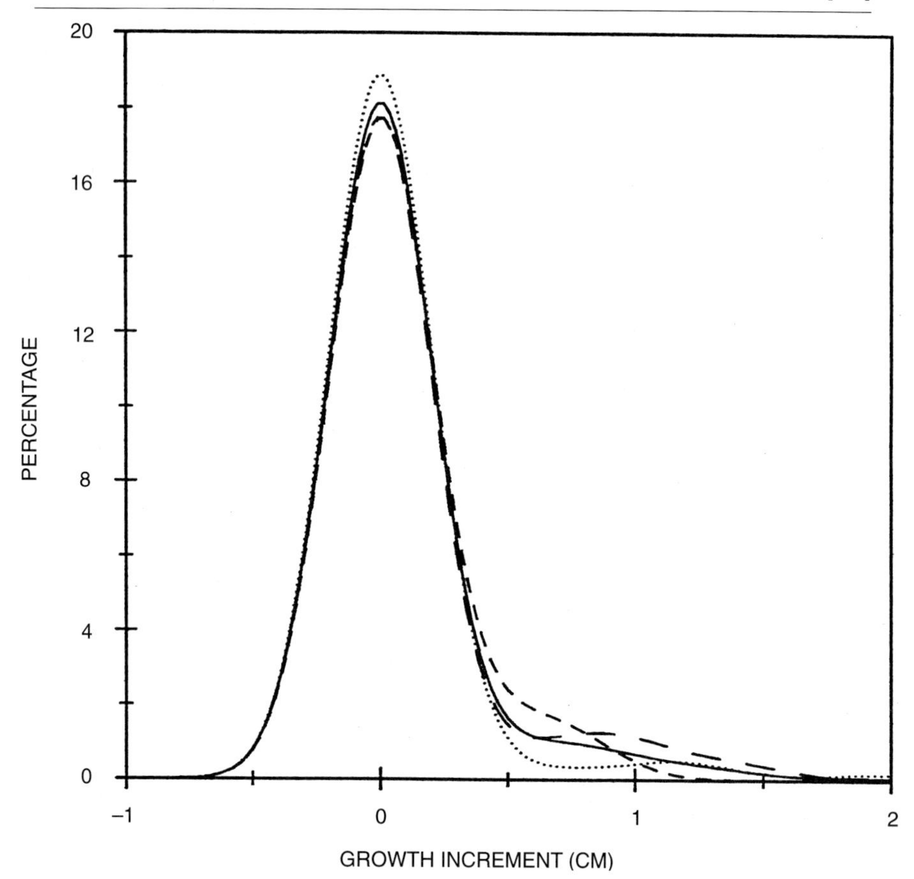

FIG. 1. Frequency distribution functions for daily length growth according to the saltation and stasis model for three infants. The short dashed line represents data from a male infant measured from 90 to 218 days of age, the longer dashed line represents data from a male infant followed from 142 to 281 days of age, and the dotted line represents data from a female infant followed daily between 312 and 433 days of age. The solid line represents their combined data. All are plotted at 0.2-cm measurement error, in line with the originally calculated observational uncertainties.

distinctive nature of each individual's saltation amplitudes, the overlapping right shoulders are expanded in Fig. 2. The significant interindividual variability in the amplitude ($p < 0.01$, Kruskal–Wallis) and timing of growth saltations produces frequency distributions that are unique to each subject's growth pattern. This violates the statistical assumptions permitting the use of a composite distribution function, and the visualization provided in Fig. 2 points to the loss of information and misrepresentation that can result

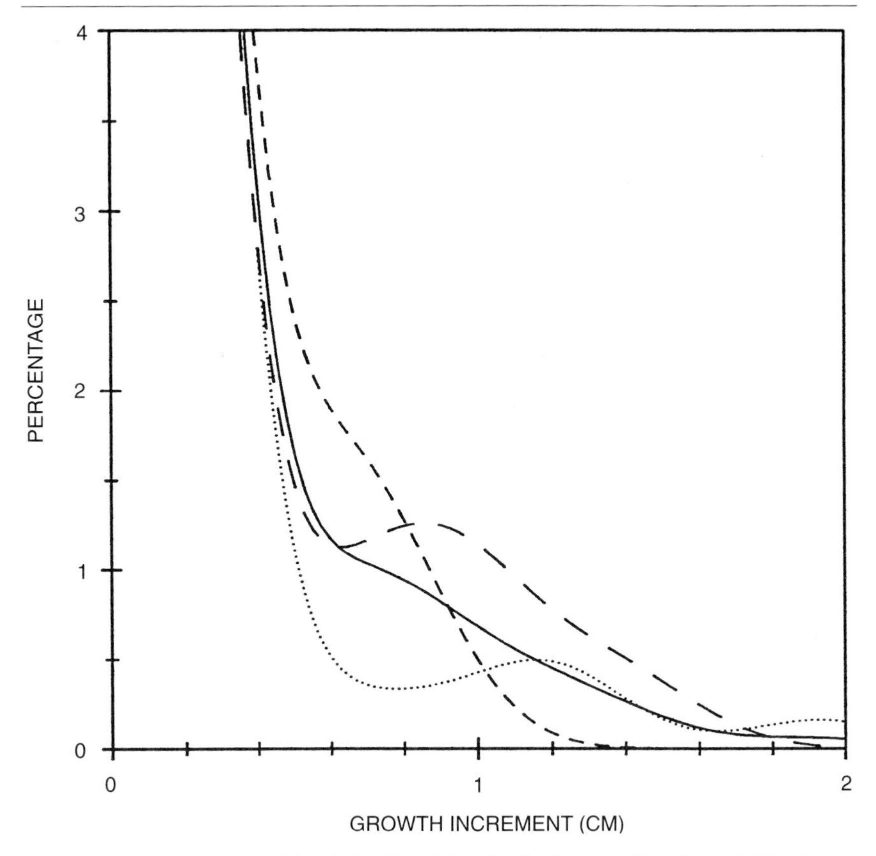

FIG. 2. A close-up of the right side of the distribution function shown in Fig. 1.

from pooling these individual data into a sample. A phase resolution effect has obscured the nature of the original data.

Thus, frequency distribution analysis is a good method for investigating characteristics of growth rates within and between individuals and for establishing the appropriateness of pooling. The uniqueness of individual patterns of growth, while often assumed and long held as a truism,[15] has rarely been specifically investigated. It is well documented by distribution analysis. The importance of this observation is relative to the research question under investigation. When the object of study is the nature of the biological process of growth, these analyses become critically important. For studies

[15] W. R. Dearborn and J. W. M. Rothney, "Predicting the Child's Development." Sci-Art Publishers, Cambridge, 1941.

aiming to describe the specific aspects of a nonlinear episodic process that proceeds with unique patterns, pooled analyses would be inappropriate: the unique patterns would be obscured.

These observations provoke consideration of just what factors contribute to unique patterns of individual growth rates: genetics, environmental input, and developmental age no doubt all contribute to the documented variability in the amount and timing of unique growth pulses. As a nonlinear pulsatile system, sampling frequency relative to underlying pulse patterns must significantly contribute to the observed variability. To further investigate the effects that sampling frequency would have on the distribution functions for these subjects, one data set was further explored.

Sampling Frequency

Figure 3 illustrates the results that variable sampling frequencies have on the daily data illustrated previously. Growth increments for 4-, 7-, and 10-day intervals are calculated from the original daily data set for one subject and distribution functions are derived from these calculated increments. Remarkably, each of these sampling schemas results in entirely

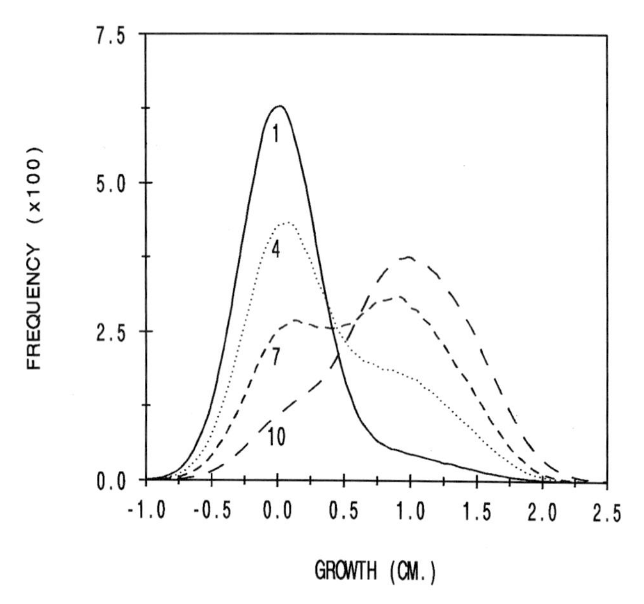

FIG. 3. Simulated frequency distributions for sampling frequencies of 4, 7, and 10 days by comparison with the original daily length data (solid line here). The overall observational uncertainty used here is 0.2 cm, in line with the experimental observations. Reproduced with permission from *Endocrinology* **137,** 5197 (1996).

different frequency distributions. For observation intervals of 1, 4, and 10 days, a unimodal distribution results, by comparison with an apparent bimodality in the 7-day derived increments. The distribution function is skewed to the right for observation intervals of 1 and 4 days, but skewed to the left for 7- and 10-day intervals. While all were derived from the same daily data sequence, each of these temporally distinctive data sets are characterized by different means, medians, and modes. Thus, sampling frequency to biological signal is a powerful determinant of the shape and statistics of a distribution function. For investigations aimed at quantifying the percentage of time a pulsatile process exhibits specific amplitude characteristics, the sampling frequency relative to biological pulsatility is a significant variable to consider. Because the latter is often unknown prior to investigation, this is a caveat to researchers in terms of distribution function interpretations.

The above results lead to a consideration of the nature of longitudinal data in general. As repeated measurements, analysis of longitudinal data raises specific statistical considerations. Moreover, when longitudinal data are differenced and analyzed as incremental data, any analysis method based on the assessment of raw increments is problematic. Sequential increments suffer as a data set from shared error components: any error at each data point effects two neighboring increments. For example, an overestimation at one time point will result in an inflated increment for the preceding time interval and an underestimation for the subsequent time interval. Thus, a series of erroneous increments results. This dependent negative correlation inherent in sequential increments necessitates attention to the magnitude of observational uncertainty in any data analysis scheme. A simple frequency distribution does not directly accommodate consideration of errors of measurement. In some studies, observational uncertainty can be measured and a confidence interval calculated for levels of probability within which to identify significant increments. To investigate the effects this factor may have on frequency distributions, the longitudinal length data are employed.

Observational Uncertainty

The effects of measurement error on the distribution function of the saltatory growth increments from the three subjects are presented in Fig. 4. In this analysis, Monte Carlo simulations are employed.[12] The results from the saltation and stasis analysis[6] were taken as a baseline (a known noise-free saltatory pattern). Gaussian-distributed pseudorandom noise of various magnitudes was added to the total length measurements. The pseudorandom noise was generated as the sum deviates equal to the number

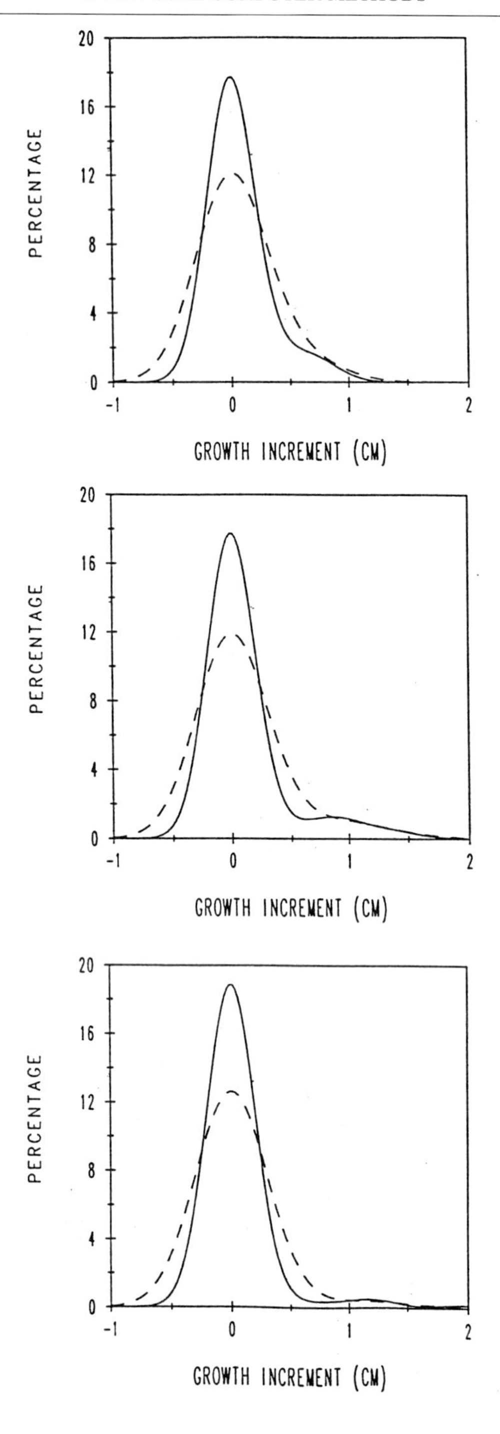

PERCENTAGE
GROWTH INCREMENT (CM)
PERCENTAGE
GROWTH INCREMENT (CM)
PERCENTAGE
GROWTH INCREMENT (CM)

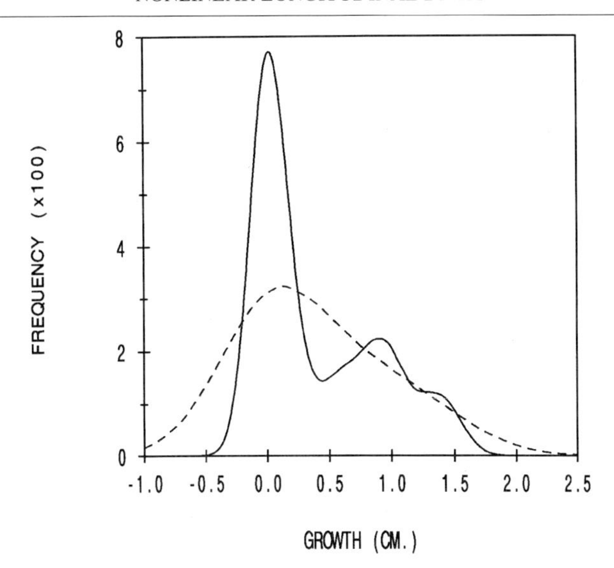

FIG. 5. The consequences of observational uncertainty for the growth data from a male infant with daily growth data from 90 to 218 days of age shown with a 4-day sampling interval and uncertainties of 1 mm (solid line) and 3 mm (dashed line). The data are Monte Carlo simulations from the saltatory growth data of a male infant followed from 90 to 218 days of age, as described in the text. Reproduced with permission from *Endocrinology* **137,** 5197 (1996).

of saltations produced by RAN3[16] routine and scaled appropriately. Mathematically, this is equivalent to a convolution integral of the noise-free pattern and a Gaussian distribution with a standard deviation given by the magnitude of the measurement error. This analysis is applied to the individual data and the increments of the individuals are combined for a sample.

Each Monte Carlo simulation used approximately 10^6 simulated growth measurements, or increments. Bid widths of 0.05 cm sorted the data and medians and modes are calculated from the binned data with the Interna-

[16] W. H. Press, B. P. Flanner, S. A. Teukolsky, and W. T. Vetterling, "Numerical Recipes, The Art of Scientific Computing," p. 199. MIT Press, Cambridge, Massachusetts, 1986.

FIG. 4. Frequency distribution functions for the daily length growth according to the saltation and stasis model for the subjects shown in Fig. 1, by individual. *Top:* Daily growth increments from a male infant followed from 90 to 218 days of age. *Middle:* Daily growth data from a male infant followed between 142 and 281 days of age. *Bottom:* Daily growth increments for a female infant for 312 to 433 days of age.

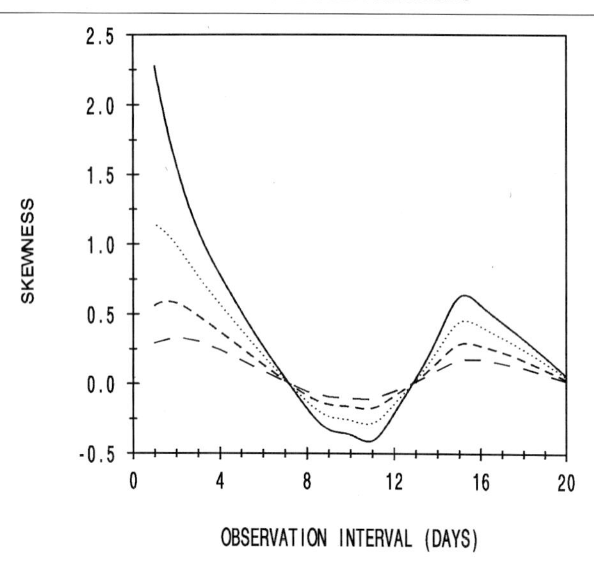

FIG. 6. The expected skewness of the distribution function as a function of experimental error and observation interval. Measurement errors of 1 mm (solid line), 2 mm (dotted line), 3 mm (short dashed line), and 4 mm (long dashed line) are illustrated. The data are Monte Carlo simulations from the saltatory growth data of a male infant followed from 90 to 218 days of age, as described in the text. Reproduced with permission from *Endocrinology* **137,** 5197 (1996).

tional Mathematics and Statistics Library (IMSL)[17] routine to compute basic statistics from grouped data. GRPES ref. Skewness and kurtosis are evaluated with the IMSL routine to compute basic univariate statistics (UVSTA).

For each individual, two different levels of observational uncertainty are shown in Fig. 4. For all subjects the increments assessed with 2-mm error result in a distribution function that appears to have a skewed right shoulder, while those at 3-mm error are ambiguous to visual inspection. The loss of detail in distribution function resolution contributed by observational uncertainty is further illustrated in Fig. 5, where the differences between a 1- and 3-mm experimental error component are compared for one of the subjects. Notably, the details of the distribution function are attenuated and entirely eliminated as the uncertainty level increases. Thus, observational uncertainty significantly affects any distribution function such that as noise levels increase, skewness and kurtosis decrease.

[17] Users' Manual, Version 1.1. International Mathematics and Statistics Library, Houston, Texas, 1989.

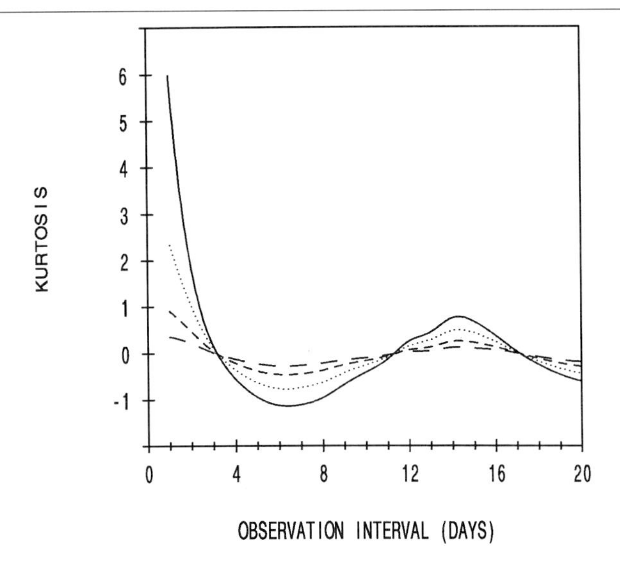

FIG. 7. The expected kurtosis of the distribution function as a function of experimental error and observation interval. Measurement errors of 1 mm (solid line), 2 mm (dotted line), 3 mm (short dashed line), and 4 mm (long dashed line) are illustrated. The data are Monte Carlo simulations from the saltatory growth data of a male infant followed from 90 to 218 days of age, as described in the text. Reproduced with permission from *Endocrinology* **137,** 5197 (1996).

These examples illustrate the dependency of these moment statistics and distribution function shapes on the noise level and measurement interval relative to the timing of a pulsatile process. Taken together, sampling frequency and observational uncertainty significantly alter the specific nature of the skewness and kurtosis characteristics of a distribution function. This is further illustrated in Figs. 6 and 7.

Study Duration

The simulations illustrated in Figs. 1–7 raise the issue of the importance of total number of observations. The results presented in Figs. 1–7 are based on the values for 4 months of daily data collection, with Figs. 5–7 illustrating the Monte Carlo simulations that provide an evaluation for approximately 10^6 growth increments. By contrast, real data are much shorter. Thus, further evaluation of what is to be expected from smaller sample sizes is also undertaken, by methods previously described.[12] What would happen if the study were shorter, and data were only collected for a month or two? Figure 8 presents a simulation for daily observations of

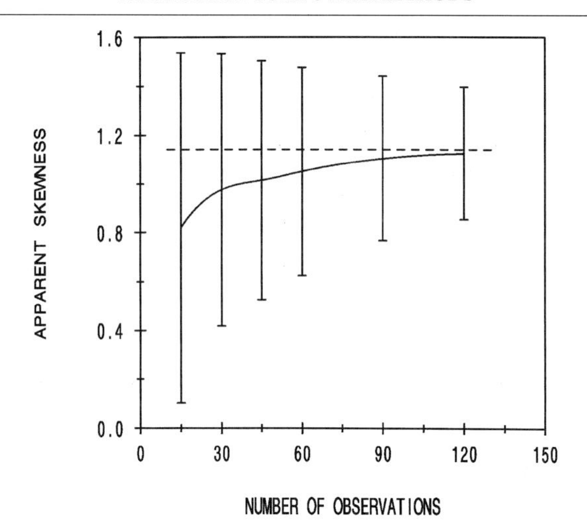

FIG. 8. The effect of study duration or total number of observations. The solid line represents the observed skewness as a function of the total number of observations. The error bars correspond to ± 1 SD. The dashed line is the expected value for an infinite number of observations. The simulation here corresponds to daily observations from the saltatory growth data of a male infant followed from 90 to 218 days of age, with 2-mm measurement error. Reproduced with permission from *Endocrinology* **137,** 5197 (1996).

length with an experimental uncertainty level of 2 mm. Total duration of a study, or total number of observations, significantly determines the skewness of a distribution function.

Conclusion

The investigations presented here provide perspective on how much information can be gleaned from a distribution function analysis of longitudinal data with particular consideration of nonlinear episodic data. Distribution function analyses are very useful investigative tools regarding statistical homogeneity of longitudinal data sets and provide an objective assessment regarding the appropriateness of pooling individual data into samples for analysis. This is an important contribution to investigations whose aim is the understanding of pulsatile processes as they unfold in individuals. It is unlikely that analyses aiming to identify the nature of individual patterns will successfully discriminate this from pooled data.

The drawbacks to the use of distribution functions and their associated statistics as a method for the analysis of longitudinal data include sensitivity to sampling interval, observational uncertainty, and total number of observations. Certainly, for studies in which those confounders are the object of study, the distribution function is a very useful tool.

In summary, consideration of the usefulness of a distribution function analysis for longitudinal data emphasizes the importance of careful consideration of the research question. The analyses presented here raise the question of the usefulness of distribution analyses for the clarification of pulsatile biological processes when the object of analysis is identifying the specific nature of the pulsatile patterning. This not the use for which the descriptive statistics were intended, and the omission of temporal data from their analytic strategy is at the basis of this limitation. The apparent visual characteristics of a histogram and the accompanying statistics describing modality, skewness, and cumulative probability densities provide no information about either the time series pattern in the original data or the number of distinctive populations contained in the data. The modality of a histogram may not be a reliable guide as to the true number of components or modes in a data set. When the object of investigation is the identification of mixtures, it is useful to note that modality and mixtures are not the same. A more appropriate method for the investigation of subpopulations in a data set is the direct investigation of mixtures, such as the mixed distribution function, a maximum likelihood method.[18] Previous studies have illustrated the better specificity and sensitivity of such an approach to mixed distribution, or subpopulation identification.[11,19,20] Overall, nonlinear longitudinal data have unique characteristics that are statistically challenging. Research questions, data collection techniques and variability of the underlying biological process provoke consideration when employing distribution functions as analytical methods.

Acknowledgments

The authors acknowledge the support of the National Science Foundation Science and Technology Center for Biological Timing at the University of Virginia (NSF DIR-890162), the General Clinical Research Center at the University of Virginia (NIH RR-00847), the University of Maryland at Baltimore Center for Fluorescence Spectroscopy (NIH RR-08119), and the Wenner-Gren Foundation for Anthropological Research.

[18] D. M. Titterington, A. F. M. Smith, and U. E. Makov, "Statistical Analysis of Finite Mixture Distributions." John Wiley & Sons, Chichester, 1985.

[19] L. D. Meyers, J.-P. Habicht, C. L. Johnson, and C. Brownie, *Am. J. Pub. Health* **73,** 1042 (1983).

[20] D. A. Tufts, J. D. Haas, J. L. Beard, and H. Spielvogel, *Am. J. Clin. Nut.* **42,** 1 (1985).

[11] Distinguishing Models of Growth with Approximate Entropy

By Michael L. Johnson, Michelle Lampl, and Martin Straume

Introduction

Researchers are commonly faced with differentiating two, or more, hypotheses (i.e., theories, models, etc.) based on how well they describe actual experimental data. Frequently this is done by translating the mechanistic theory, or hypothesis, into a mathematical model and then "fitting" the model to the experimental data. This "fitting" process is commonly done by a least squares parameter estimation procedure.[1–5]

When the quality of the fits of the experimental data is significantly different between the models (i.e., theories) they can usually be distinguished by goodness-of-fit criteria,[6–13] such as a runs test. However, if these tests are not conclusive how can the researcher distinguish the theories?

This chapter presents a unique application of approximate entropy,[13–17]

[1] M. L. Johnson and S. G. Frasier, *Methods Enzymol.* **117,** 301 (1985).
[2] M. L. Johnson and L. M. Faunt, *Methods Enzymol.* **210,** 1 (1992).
[3] M. L. Johnson, *Methods in Enzymol.* **240,** 1 (1994).
[4] J. A. Nelder and R. Mead, *Comput. J.* **7,** 308 (1965).
[5] D. M. Bates and D. G. Watts, "Nonlinear Regression Analysis and Its Applications." John Wiley and Sons, New York, 1988.
[6] M. Straume and M. L. Johnson, *Methods Enzymol.* **210,** 87 (1992).
[7] P. Armitage, "Statistical Methods in Medical Research," 4th Ed., p. 391. Blackwell, Oxford, 1977.
[8] W. W. Daniel, "Biostatistics: A Foundation for Analysis in the Health Sciences," 2nd Ed. John Wiley and Sons, New York, 1978.
[9] P. R. Bevington, "Data Reduction and Error Analysis in the Physical Sciences," p. 187. McGraw-Hill, New York, 1969.
[10] Y. Bard, "Nonlinear Parameter Estimation," p. 201. Academic Press, New York, 1974.
[11] N. R. Draper and R. Smith, "Applied Regression Analysis," 2nd Ed., p. 153. John Wiley and Sons, New York, 1981.
[12] G. E. P. Box and G. M. Jenkins, "Time Series Analysis Forecasting and Control," p. 33. Holden-Day, Oakland, California, 1976.
[13] M. L. Johnson and M. Straume, *Methods Enzymol.* **321,** [12], 2000, (this volume).
[14] S. M. Pincus, *Proc. Natl. Acad. Sci. U.S.A.* **88,** 2297 (1991).
[15] S. M. Pincus, *Proc. Natl. Acad. Sci. U.S.A.* **89,** 4432 (1992).
[16] S. M. Pincus, *Methods Enzymol.* **240,** 68 (1994).
[17] S. M. Pincus *Methods in Enzymol.* **321,** [9], 2000 (this volume).

0076-6879/00 $30.00

ApEn, to distinguish models of growth in children.[18–30] While this specific example is only applicable to some types of experimental data, it does illustrate the broad applicability of ApEn.

Historically, growth[18–30] has been considered to be a smooth continuous process that varies little from day to day. In this model, growth rates change gradually on a timescale of months or years, not hours or days. However, when Lampl *et al.* measured the lengths and heights of infants and adolescents at daily intervals it was observed that, instead, large changes in growth rates occur between some days and no growth at all occurs between other days.[18–30] These daily observations led to the development of the *saltation and stasis hypothesis* and mathematical model of growth.[18–30]

The saltation and stasis hypothesis states growth (i.e., saltation) occurs over a very short time and then the organism enters a refractory period of little or no growth (i.e., stasis). In this context "a very short time" means less than the interval between the measurements and "little or no growth" means less than could be measured.

Previously, we utilized classical goodness-of-fit criteria,[6–13] such as autocorrelation, to demonstrate that the saltation and stasis model provided a better description of the experimental observations that is obtainable with the more classical growth models. However, for some data sets, the goodness-of-fit tests do not provide a clear distinction between the models and hypotheses. With these data sets in mind, we developed a new method to distinguish models of growth that is based on a modified version of the approximate entropy, ApEn, metric of the experimental data.

The basic procedure for the use of ApEn to distinguish these models is to calculate the ApEn value for the original data sequence and the expected ApEn values, with standard errors, for each of the growth models.

[18] M. Lampl, J. D. Veldhuis, and M. L. Johnson, *Science* **258,** 801 (1992).
[19] M. Lampl and M. L. Johnson, *Ann. Hum. Biol.* **20,** 595 (1993).
[20] M. L. Johnson, *Am. J. Hum. Biol.* **5,** 633 (1993).
[21] M. L. Johnson and M. Lampl, *Methods Enzymol.* **240,** 51 (1994).
[22] M. Lampl, N. Cameron, J. D. Veldhuis, and M. L. Johnson, *Science* **268,** 445 (1995).
[23] M. L. Johnson and M. Lampl, *Methods Neurosci.* **28,** 364 (1995).
[24] M. L. Johnson, J. D. Veldhuis, and M. Lampl, *Endocrinology* **137,** 5197 (1996).
[25] M. Lampl, M. L. Johnson, and E. A. Frongillo, Jr., *Ann. Hum. Biol.* **24,** 65 (1997).
[26] M. Lampl and M. L. Johnson, *Am. J. Hum. Biol.* **9,** 343 (1997).
[27] M. Lampl and M. L. Johnson, *Ann. Hum. Biol.* **25,** 187 (1998).
[28] M. Lampl, K. Ashizawa, M. Kawabata, and M. L. Johnson, *Ann. Hum. Biol.* **25,** 203 (1998).
[29] M. Lampl and M. L. Johnson, *in* "Applications of Nonlinear Dynamics to Developmental Process Modeling" (K. M. Newell and P. C. Molenaar, eds.), pp. 15–38. Lawrence Erlbaum Associates, New York, 1998.
[30] M. L. Johnson, "Methods for the Analysis of Saltation and Stasis in Human Growth Data," *in* "Saltation and Stasis in Human Growth and Development," pp. 101–120. Smith-Gordon, London, 1999.

The observed ApEn value is then compared with the distributions of expected ApEn values for each of the growth models being tested. ApEn quantifies the regular versus irregular nature of a time series. This test of the adequacy of the growth models is based on the quantifiable degree of regularity of the experimental observations.

Definition and Calculation of Approximate Entropy

Approximate entropy (ApEn) was formulated by Pincus to discriminate statistically time series by quantifying time series regularity.[14–17] Of particular significance regarding the method is the ability of ApEn to quantify reliably the regularity of finite length time series, even in the presence of noise and measurement inaccuracy. This is a property unique to ApEn and not shared by other methods common to nonlinear dynamic systems theory.[16]

Specifically, ApEn measures the logarithmic likelihood that runs of patterns in a time series that are close for m consecutive observations remain close when considered as $m + 1$ consecutive observations. A higher probability of remaining close (i.e., greater regularity) yields smaller ApEn values whereas greater independence among sequential values of a time series yields larger ApEn values.

Calculation of ApEn requires prior definition of the two parameters m and r. The parameter m is the length of run to be compared (as alluded to above) and r is a filter (the magnitude that will discern "close" and "not close," as described below). ApEn values can only be validly compared when computed for the same m, r, and N values,[16] where N is the number of data points in the time series being considered. Thus, ApEn is specified as ApEn(m,r,N). For optimum statistical validity, ApEn is typically implemented using m values of 1 or 2 and r values of approximately 0.2 standard deviations of the series being considered.[16]

ApEn is calculated according to the following[16]: Given N data points in a time series, $u(1), u(2), \ldots, u(N)$, the set of $N - m + 1$ possible vectors, $x(i)$, are formed with m consecutive u values such that $x(i) = [u(i), \ldots, u(i + m - 1)]^T$, $i = 1, 2, \ldots, N - m + 1$. The distance between vectors $x(i)$, and $x(j)$, $d[x(i), x(j)]$, is defined as the maximum absolute difference between corresponding elements of the respective vectors. For each of the $N - m + 1$ vectors $x(i)$, a value for $C_i^m(r)$ is computed by comparing all $N - m + 1$ vectors $x(j)$ to vector $x(i)$ such that

$$C_i^m(r) = \frac{\text{number of } x(j) \text{ for which } d[x(i), x(j)] \leq r}{N - m + 1} \tag{1}$$

These $N - m + 1$ $C_i^m(r)$ values measure the frequency with which patterns were encountered that are similar to the pattern given by $x(i)$ of length m within tolerance r. Note that for all i, $x(i)$ is always compared relative to $x(i)$ (i.e., to itself), so that all values of $C_i^m(r)$ are positive. Now, define

$$\Phi^m(r) = \frac{\sum_{i=1}^{N-m+1} \ln C_i^m(r)}{N - m + 1} \tag{2}$$

from which the approximate entropy ApEn(m, r, N) is given by

$$\text{ApEn}(m, r, N) = \Phi^m(r) - \Phi^{m+1}(r) \tag{3}$$

ApEn defined in this way can be interpreted (with $m = 1$, for example) as a measure of the difference between (1) the probability that runs of length 1 will recur within tolerance r and (2) the probability that runs of length 2 will recur within the same tolerance.[16]

Modifications of Approximate Entropy Calculation for this Application

Two modifications were made to the standard methods for the calculation of the approximate entropy for the present use. First, the ApEn calculation is normally performed on a stationary time series (i.e., a series of data where the mean of the data is not a function of time). Clearly, measures of growth such as height generally increase with time and consequently the ApEn values were calculated on these nonstationary time series. As described above, ApEn(m, r, N) is a function of the run length size m; r the magnitude that will discern "close" and "not close"; and N the length of the time series. Normally, r is expressed in terms of the standard deviation of stationary time series; r is expressed in terms of the experimental variability, or uncertainty, of the time series. When applied to a nonstationary time series (e.g., growth) it is more logical to express r in terms of the actual measurement uncertainties. Thus, the second modification of the ApEn calculation was to express r in terms of the known measurement error levels.

Growth Models

Numerous mathematical models are available for describing the growth process. The saltation and stasis model is unique in that it is based on a mechanistic hypothesis about how growth proceeds. This mechanistic hypothesis was translated into a mathematical form where growth is described as a series of distinct instantaneous events of positive growth (i.e., saltations) separated by stasis periods of no growth:

$$\text{Height} = \sum_{k=1}^{i} G_i \qquad (4)$$

where the summation is over each of the observations, and G_i is zero during a stasis interval and a positive value for the measurement intervals where a saltation occurred.

This mathematical model does not require that the saltation events be instantaneous. It simply requires that the growth events occur in less time than the interval between observations. Under these conditions, the experimental observations do not contain any information about the actual shape of the saltation event and thus can simply be approximated as a step function.

Virtually all other growth models[31–33] assume that growth is a smooth continuous process that varies little from day to day. These models assume that small changes in growth rates occur on a timescale of months or years, not hours or days. Furthermore, virtually all of these models are not based on a hypothesis or theory. These models are simply empirical descriptions of growth. They are generally formulated to describe only a few observations per year. Thus, these models do not—and cannot—describe growth patterns that vary on a daily time frame.

It is impossible to test a set of experimental data against every possible slowly varying smooth continuous mathematical form. Consequently, we decided to use an exponential rise

$$\text{Height} = A_0 - A_1 e^{-k\text{Time}}$$

and/or polynomials of order 1 to 6

$$\text{Height} = \sum_{i=1}^{6} A_i \,\text{Time}^i$$

as surrogates for the infinite number of possible slowly varying empirical mathematical forms that might be utilized to describe growth.

Expected Model-Dependent Distribution of ApEn

The expected distributions of ApEn values for any growth model (e.g., saltation and stasis, polynomial, exponential) are evaluated by a Monte Carlo procedure. This involves simulating a large number of growth patterns and observing the distribution of resulting ApEn values.

[31] T. Gasser, A. Kneip, P. Ziegler, R. Largo, and A. Prader, *Ann. Hum. Biol.* **13,** 129 (1990).
[32] J. Karlberg, *Stat Med.* **6,** 185 (1987).
[33] M. A. Preece and M. J. Baines, *Ann. Hum. Biol.* **5,** 1 (1978).

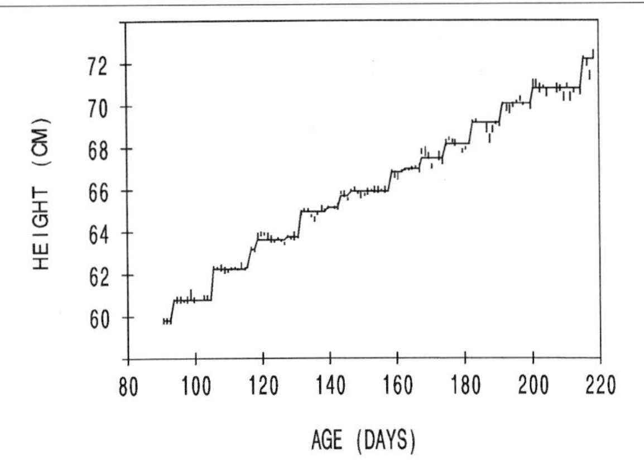

FIG. 1. A typical set of daily experimental observations of the height of an infant. The solid line corresponds to the saltation and stasis analysis of these data. The saltation and stasis model analysis indicated that there were 13 statistically significant ($P < 0.05$) saltations at 94, 106, 117, 119, 132, 144, 159, 168, 175, 183, 192, 201, and 216 days.

The first step in this Monte Carlo procedure is to least squares fit a particular growth model to a particular set of experimental observations. This provides an optimal model-dependent description of the growth pattern (i.e., the best calculated curve for the particular model) and a set of residuals. The residuals are the differences between the experimental observations and optimal calculated curve.

The next step is to simulate a large number (e.g., 1000) of calculated growth patterns for the particular growth model. These calculated growth patterns are the best calculated curves for the particular model with "pseudorandom noise" added. The expected model-dependent distribution of ApEn values is then calculated from the ApEn values for each of the large number of simulated model dependent growth patterns.

The "pseudorandom noise" can be generated by three possible methods. First, the noise can be calculated by generating Gaussian distributed pseudorandom numbers with a variance equal to the variance of the residuals and a mean of zero. Second, the actual residuals can simply be shuffled (i.e., selected in a random order) and added back to the best calculated curve in a different order. Third, the actual residuals can be shuffled with replacement as is done in bootstrap procedures.[34] The "shuffled with replacement" choice means that for each of the simulated growth patterns ~37% of the residuals are randomly selected and not used while the same

[34] B. Efron and R. J. Tibshirani, "An Introduction to the Bootstrap." Chapman and Hall, New York, 1993.

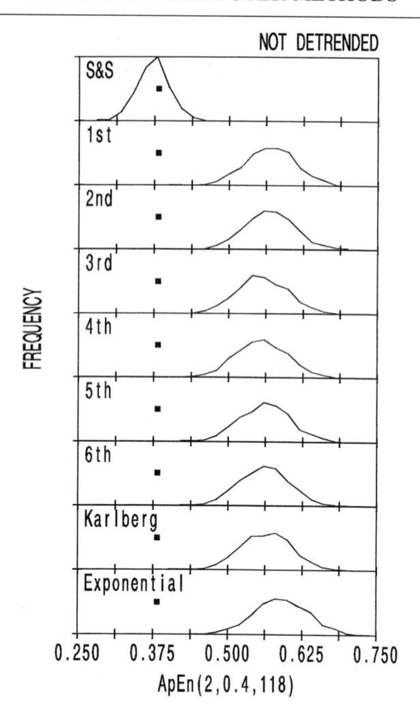

FIG. 2. The various model-dependent distributions of ApEn(2, 0.4, 118) for the data presented in Fig. 1. The panels from top to bottom correspond to the expected distribution for the saltation and stasis model; the expected distribution for first- through sixth-degree polynomials; the Karlberg infant model; and the exponential rise model. The square in each of the panels is the actual observed ApEn(2, 0.4, 118) value for the original data. These analyses are the results of 1000 Monte Carlo cycles with the noise being generated by shuffling the residuals. The nonstationary height series was not detrended.

number of the remaining residuals are used twice to obtain the requisite number of residuals. A different 37% of the residuals are selected for each of the simulated growth patterns. The order of the selected residuals is randomized (i.e., shuffled) before they are added back to the calculated growth pattern.

The expected distribution of ApEn values as determined for each of the growth models is compared with the observed ApEn value from the experimental observations. The observed ApEn value will be within either one, or more, distributions or none. The conclusion of this test is that if the observed ApEn value is within a distribution expected for a particular mathematical model, then the data are, by this test, consistent with that model. Conversely, if the observed ApEn value is not within a distribution expected for a particular mathematical model, then the data are, by this test, inconsistent with that model. If the observed ApEn is within the

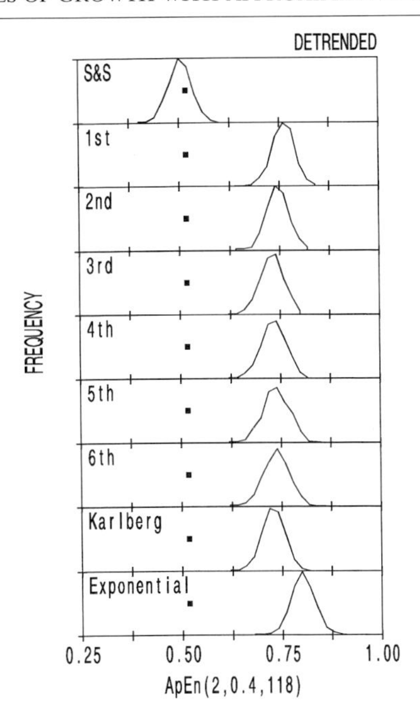

FIG. 3. The same analysis as presented in Fig. 2 except that the nonstationary time series was detrended by subtracting the best least squares straight line before the analyses were performed.

expected distributions for more than one model, then those models cannot be distinguished by this test.

Example of this Use of ApEn

Figure 1 presents a typical set of daily experimental observations of the height of an infant. The solid line in Fig. 1 corresponds to the saltation and stasis analysis of these data. The intervals of saltation and stasis are visually obvious. The analysis of these data by the saltation and stasis model indicated that there were 13 statistically significant ($P < 0.05$) saltations at 94, 106, 117, 119, 132, 144, 159, 168, 175, 183, 192, 201, and 216 days. These saltations are not at regular intervals; they are episodic but not periodic. The average saltation amplitude in this infant was 0.9 cm. The standard deviation of the differences (i.e., residuals) between the calculated optimal saltation and stasis model and the experimental observations is 0.3 cm.

The irregular nature of the distributions in Fig. 2, and those to follow, is due to using only 1000 Monte Carlo cycles for their generation. These distributions become increasingly smooth as the number of Monte Carlo cycles increases. However, 1000 cycles are usually sufficient to characterize the distributions and only require a few seconds on a 450-MHz Pentium II PC.

The various model-dependent distributions of ApEn(2, 0.4, 118) for the data shown in Fig. 1 are presented in Fig. 2. The panels from top to bottom correspond to the expected distribution for the saltation and stasis model; the expected distributions for first- through sixth-degree polynomials; the Karlberg infant model[32]; and the exponential rise model. These distributions are the results of 1000 Monte Carlo cycles with the noise being generated by shuffling the residuals. The nonstationary height series was not detrended. The value of r was set to 0.4 cm, not a fraction of the standard deviation of the data points. The square in each of the panels in Fig. 2 is the actual observed ApEn(2, 0.4, 118) value for the original data. It is clear that this value is consistent with the expected distribution for the saltation and stasis model and inconsistent with the others. Although not shown, the ApEn(1, 0.4, 118) values provide analogous results.

Figure 3 presents the same analysis as Fig. 2 except that the nonstation-

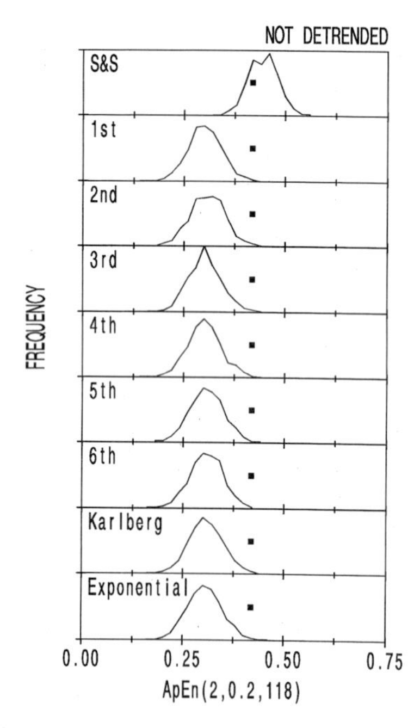

FIG. 4. The same analysis as presented in Fig. 2 except that the value of r was decreased to 0.2 cm. The nonstationary height series was not detrended.

ary time series was detrended by subtracting the best least squares straight line before the analyses were performed. Clearly while the results are numerically slightly different, the conclusions remain the same. Again, the ApEn(1, 0.4, 118) values provide analogous results (not shown).

When the Fig. 2 analysis was repeated with the pseudorandom noise being created by either shuffling with replacement (i.e., a bootstrap) or by a Gaussian distribution, the results were virtually identical to those shown in Fig. 2. Consequently, the plot of these is not repeated.

Figure 4 presents the same analysis as is presented in Fig. 2 except that the r value was set to 0.2 cm. Note that while the figure appears somewhat different, the conclusions remain the same. When r is increased to 0.6 cm the results are virtually identical to those shown in Fig. 2. It is interesting to note that the ApEn(1, 0.2, 118) distributions (not shown) do not exhibit the reversal of magnitude that the ApEn(2, 0.2, 118) distributions show.

From a comparison of Figs. 2 and 4 it appears that at some intermediate value of r (e.g., 0.27) the ApEn(2, 0.27, 118) distributions coincide and the models cannot be distinguished by that metric (not shown). However, as shown in Fig. 5, the ApEn(1, 0.27, 118) distributions are clearly distinguishable.

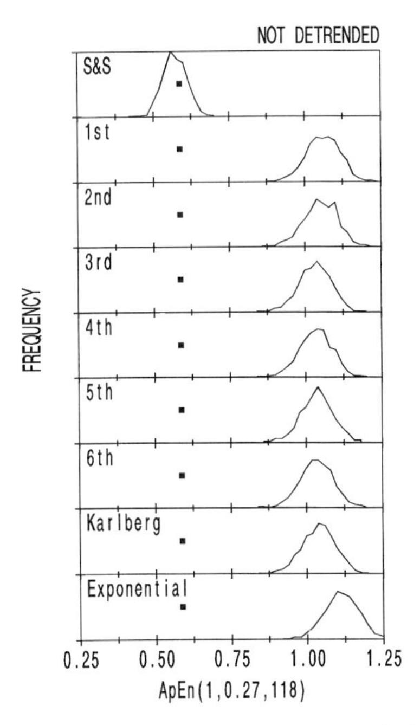

FIG. 5. The ApEn(1, 0.27, 118) distributions for the data in Fig. 1. The nonstationary height series was not detrended.

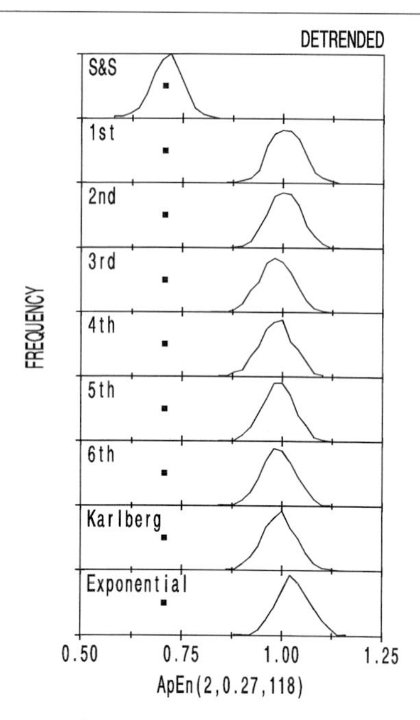

FIG. 6. The ApEn(2, 0.27, 118) distributions for the data in Fig. 1. The nonstationary time series was detrended by subtracting the best least squares straight line before the analyses were performed.

The ApEn(2, 0.27, 118) distributions are clearly distinguishable if the time series is detrended. This is shown in Fig. 6.

Conclusion

This chapter presents examples of how approximate entropy, ApEn, can be used to distinguish between mathematical models, and the underlying mechanistic hypotheses that purport to describe the same experimental observations. If the expected distributions of ApEn for the different models do not overlap, then it is expected that ApEn can be utilized to distinguish these models and hypotheses. However, if the distributions overlap significantly, then no conclusion can be drawn.

From the various figures it is obvious that ApEn distributions are fairly robust to variations in the value of m and r. The ApEn distributions do not appear to be sensitive, in these examples, to the method of generating the random noise (i.e., pseudorandom number generator, shuffling, or bootstrapping) for the Monte Carlo process.

Furthermore, the ApEn distributions appear to be relatively insensitive to the stationarity of the data when the value of r is expressed as an absolute quantity that is dependent on the known experimental measurement errors. However, as shown in Fig. 6, there are cases where the results are better if the data are detrended. Using a straight line as a detrending function is sufficient and preferable.

Acknowledgments

The authors acknowledge the support of the National Science Foundation Science and Technology Center for Biological Timing at the University of Virginia (NSF DIR-8920162), the General Clinical Research Center at the University of Virginia (NIH RR-00847), and the University of Maryland at Baltimore Center for Fluorescence Spectroscopy (NIH RR-08119).

[12] Approximate Entropy as Indication of Goodness-of-Fit

By MICHAEL L. JOHNSON and MARTIN STRAUME

Introduction

When biomedical researchers "fit" a mathematical equation to experimental data there are several components of the analysis. Some of these components are as follows:

The fitting process should provide the parameters of the mathematical model that will "best" describe the experimental data within the constraints of the mathematical model.

The biomedical researcher also desires realistic estimates of the precision of the "best" parameters.

We should also consider the goodness-of-fit of the mathematical model to the experimental data, that is: How well did the mathematical model describe the experimental data?

It is the last of these that is the main consideration of this chapter.

The "fitting" process is typically a least squares procedure.[1–5] Least

[1] M. L. Johnson and S. G. Frasier, *Methods Enzymol.* **117,** 301 (1985).
[2] M. L. Johnson and L. M. Faunt, *Methods Enzymol.* **210,** 1 (1992).
[3] M. L. Johnson, *Methods Enzymol.* **240,** 1 (1994).
[4] J. A. Nelder and R. Mead, *Comput. J.* **7,** 308 (1965).
[5] D. M. Bates and D. G. Watts, "Nonlinear Regression Analysis and Its Applications." John Wiley and Sons, New York, 1988.

squares procedures will adjust a set of parameters of the mathematical function such that the weighted sum of squares of the differences between the experimental data and the calculated function is minimized. The validity of the least squares procedure is dependent on a number of assumptions about the experimental data, and primarily about the nature of the experimental uncertainties contained within the data. One of these assumptions is that the experimental uncertainties of the data follow a Gaussian, or bell-shaped, distribution. For most applications these assumptions are reasonable and the least squares procedure will provide parameter values that have the highest probability (i.e., maximum likelihood) of being correct.

There are, however, several non–least squares parameter estimation procedures that make other assumptions about the nature of the experimental uncertainties of the data. Examples of such procedures are the Padé–Laplace algorithm,[6] the method of moments,[7] maximum entropy methods,[8] and robust methods for parameter estimation.[9–11]

One of the advantages of the least squares methods is that there are well-defined methods for the evaluation of the precision of the estimated parameters.[1–5,12–15] There are fewer methods for the evaluation of the precision of the estimated parameters for the non–least squares procedures. However, bootstrap methods[12] and Monte Carlo methods[13] can easily be applied to non–least squares parameter estimation procedures.

Clearly, if the mathematical model cannot provide a good description of the experimental data, then either the mathematical model or the experimental data are incorrect. If the mathematical model is incorrect, then it is likely that the mechanistic theory and hypotheses on which the model is based are also incorrect. Thus, it is the goodness-of-fit tests that can be used to test the validity or applicability of the theory and hypotheses.

Another advantage of the least squares methods is that there are numer-

[6] H. R. Halvorson, *Methods Enzymol.* **210,** 54 (1992).

[7] E. W. Small, *Methods Enzymol.* **210,** 237 (1992).

[8] J.-C. Brochon, *Methods Enzymol.* **240,** 262 (1994).

[9] P. J. Huber, "Robust Statistics." John Wiley and Sons, New York, 1981.

[10] D. C. Hoaglin, F. Mosteller, and J. W. Tukey, "Understanding Robust and Exploratory Data Analysis." John Wiley and Sons, New York, 1983.

[11] M. L. Johnson, *Methods Enzymol.* **321,** [23], 2000 (this volume)

[12] B. Efron and R. J. Tibshirani, "An Introduction to the Bootstrap." Chapman and Hall, New York, 1993.

[13] M. Straume and M. L. Johnson, *Methods Enzymol.* **210,** 117 (1992).

[14] D. M. Bates and D. G. Watts, "Nonlinear Regression Analysis and Its Applications." John Wiley and Sons, New York, 1988.

[15] D. G. Watts, *Methods Enzymol.* **240,** 23 (1994).

ous methods to test the goodness-of-fit.[16] One of the assumptions required for the least squares parameter estimation procedures is that the experimental uncertainties (e.g., measurement errors) follow a Gaussian (i.e., a normal or bell-shaped) distribution. Thus, if the assumptions are valid for the least squares parameter estimation procedure, then the resulting residuals should also follow a Gaussian distribution. The residuals are the standard deviation weighted differences between the fitted curve and the data points. The available goodness-of-fit tests are actually tests of whether the residuals follow a Gaussian distribution. If the residuals do not, then it can be assumed that the least squares procedure did not provide a good description of the experimental data. Typical examples of these goodness-of-fit tests are tests of χ^2 statistics,[17–19] most applications of the Kolmogorov–Smirnov test,[17] the runs test,[20] the Durbin–Watson test,[21] and autocorrelation analysis.[22] Unfortunately, only an uncommonly used form of the Kolmogorov–Smirnov test[17] can be applied with non–least squares parameter estimation procedures. Furthermore, this application still requires that the form of the distribution of experimental uncertainties be specified.

The purpose of this chapter is to introduce approximate entropy,[23–27] ApEn, as a goodness-of-fit criterion. ApEn is a measure of the regularity, or orderliness, of a sequence of numbers. It calibrates sequential relationships between the numbers in terms of the randomness of the numbers. ApEn does not require the assumption that the residuals follow a Gaussian distribution and it does not require knowledge of the actual distribution. Thus, ApEn can be applied as a goodness-of-fit criterion for non–least squares applications.

[16] M. Straume and M. L. Johnson, *Methods Enzymol.* **210,** 87 (1992).

[17] P. Armitage, "Statistical Methods in Medical Research," 4th Ed., p. 391. Blackwell, Oxford, 1977.

[18] W. W. Daniel, "Biostatistics: A Foundation for Analysis in the Health Sciences," 2nd Ed. John Wiley and Sons, New York, 1978.

[19] P. R. Bevington, "Data Reduction and Error Analysis in the Physical Sciences," p. 187. McGraw-Hill, New York, 1969.

[20] Y. Bard, "Nonlinear Parameter Estimation," p. 201. Academic Press, New York, 1974.

[21] N. R. Draper and R. Smith, "Applied Regression Analysis," 2nd Ed., p. 153. John Wiley and Sons New York, 1981.

[22] G. E. P. Box and G. M. Jenkins, "Time Series Analysis Forecasting and Control," p. 33. Holden-Day, Oakland, Californain, 1976.

[23] S. M. Pincus, *Proc. Natl. Acad. Sci. U.S.A.* **88,** 2297 (1991).

[24] S. M. Pincus, *Proc. Natl. Acad. Sci. U.S.A.* **89,** 4432 (1992).

[25] S. M. Pincus, *Methods Enzymol.* **240,** 68 (1994).

[26] S. M. Pincus, *Methods Enzymol.* **321,** [9], 2000 (this volume).

[27] M. L. Johnson, M. Lampl, and M. Straume, *Methods Enzymol.* **321,** [10], 2000 (this volume).

Definition and Calculation of Approximate Entropy

Approximate entropy (ApEn) was developed and formulated by Pincus to specifically address the issue of how to statistically discriminate time series by quantifying time series regularity.[23–26] Of particular significance is the property of ApEn to reliably statistically quantify the regularity of finite-length time series even in the presence of noise and measurement inaccuracy, a property unique to ApEn and not shared by other methods common to nonlinear dynamical systems theory.[25]

More specifically, ApEn measures the logarithmic likelihood that runs of patterns in a time series that are close for m consecutive observations remain close when considered as $m + 1$ consecutive observations. Greater regularity (i.e., higher probability of remaining close) yields smaller ApEn values, whereas greater independence among sequential values of a time series yields larger ApEn values.

Calculation of ApEn requires prior definition of the two parameters m and r. The parameter m is the length of run to be compared (as alluded to above) and r is a filter (the magnitude that will discern "close" and "not close," as described below). ApEn is thus formally defined as ApEn (m,r,N), where N is the number of data points in the time series being considered. ApEn values can only be validly compared when computed for the same m, r, and N values.[25] For optimum statistical validity, ApEn is typically implemented using m values of 1 or 2 and r values of approximately 0.2 SDs of the series being considered.[25]

ApEn is calculated according to the following[25]: Given N data points in a time series, $u(1), u(2), \ldots, u(N)$, the set of $N - m + 1$ possible vectors, $x(i)$, is formed with m consecutive u values such that $x(i) = [u(i), \ldots, u(i + m - 1)]^T$, $i = 1, 2, \ldots, N - m + 1$. The distance between vectors $x(i)$ and $x(j)$, $d[x(i), x(j)]$, is defined as the maximum absolute difference between corresponding elements of the respective vectors. For each of the $N - m + 1$ vectors $x(i)$, a value for $C_i^m(i)$ is computed by comparing all $N - m + 1$ vectors $x(j)$ to vector $x(i)$ such that

$$C_i^m(r) = \frac{\text{number of } x(j) \text{ for which } d[x(i), x(j)] \leq r}{N - m + 1} \tag{1}$$

These $N - m + 1$ $C_i^m(r)$ values measure the frequency with which patterns were encountered that are similar to the pattern given by $x(i)$ of length m within tolerance r. Note that for all i, $x(i)$ is always compared relative to $x(i)$ (i.e., to itself), so that all values $C_i^m(r)$ are positive. Now, define

$$\Phi^m(r) = \frac{\sum_{i=1}^{N-m+1} \ln C_i^m(r)}{N - m + 1} \tag{2}$$

from which the approximate entropy ApEn(m, r, N) is given by

$$\text{ApEn}(m, r, N) = \Phi^m(r) - \Phi^{m+1}(r) \tag{3}$$

ApEn defined in this way can be interpreted (with $m = 1$, for example) as a measure of the difference between (1) the probability that runs of length 1 will recur within tolerance r and (2) the probability that runs of length 2 will recur within the same tolerance.[25]

Approximate Entropy for a Randomized Sequence

ApEn provides a measure of the sequential relationships between the numbers in a sequence. The theoretical lower limit for ApEn values is zero, which corresponds to a highly ordered sequence. The upper limit of the ApEn value, corresponding to a totally random sequence, is variable and is a function of m, r, and N. For any sequence of numbers this upper limit, corresponding to a random sequence of numbers, can be approximated by simply shuffling the numbers and then recalculating ApEn. The mean, variance, standard deviation, and maximum of the shuffled (i.e., randomized) ApEn value can be evaluated by repeatedly calculating the ApEn values after repeatedly shuffling the series of numbers.

Runs Test: Quantifying Trends in Residuals

The existence of trends in residuals with respect to either the independent (i.e., experimental) or dependent (i.e., the experimental observable) variables suggests that some systematic behavior is present in the data that is not accounted for by the analytical model. Trends in residuals will often manifest themselves as causing too few runs (consecutive residual values of the same sign) or, in cases where negative serial correlation occurs, causing too many runs. A convenient way to assess quantitatively this quality of a distribution of residuals is to perform a runs test.[16,20] The method involves calculating the expected number of runs given the total number of residuals as well as an estimate of variance in this expected number of runs. A run is simply one, or more, residuals in a row of the same sign.

The expected number of runs, R, may be calculated from the total number of positive and negative valued residuals, n_p, and n_n, and as

$$R = [2n_\mathrm{p}n_\mathrm{n}/(n_\mathrm{p} + n_\mathrm{n})] + 1 \tag{4}$$

The variance in the expected number of runs, σ_R^2, is then calculated as

$$\sigma_R^2 = \frac{2n_\mathrm{p}n_\mathrm{n}(2n_\mathrm{p}n_\mathrm{n} - n_\mathrm{p} - n_\mathrm{n})}{(n_\mathrm{p} + n_\mathrm{n})^2 (n_\mathrm{p} + n_\mathrm{n} - 1)} \tag{5}$$

A quantitative comparison is then made between the expected number of runs, R, and the observed number of runs, n_R, by calculating an estimate for the standard normal deviate as

$$Z = \left| \frac{n_R - R \pm 0.5}{\sigma_R} \right| \tag{6}$$

When n_p and n_n are both greater than 10, Z will be distributed approximately as a standard normal deviate. In other words, the calculated value of Z is the number of standard deviations that the observed number of runs is from the expected number of runs for an independently distributed set of residuals of the number being considered. The value of 0.5 is a continuity correction to account for biases introduced by approximating a discrete distribution with a continuous one. This correction is +0.5 when testing for too few runs and −0.5 when testing for too many runs. The test is therefore estimating the probability that the number of runs observed is different from that expected from independently distributed residuals. The greater the value of Z, the greater the likelihood that there exists some form of correlation in the residuals relative to the particular variable being considered.

Approximate Entropy as Measure of Goodness-of-Fit

A Z score for goodness-of-fit can be calculated from the observed ApEn and the distribution of randomized ApEn as:

$$Z_{\text{random}} = \frac{\text{Mean}_{\text{random}} - \text{ApEn}}{\text{SD}_{\text{random}}} \tag{7}$$

where $\text{Mean}_{\text{random}}$ is the mean of the set of randomized ApEn values and $\text{SD}_{\text{random}}$ is the standard deviation of the set of randomized ApEn values. This Z score can then be translated into a probability of randomness (i.e., goodness-of-fit) with the standard statistical tables. Note that this translation should be as a one-sided test because all negative values indicate "random" residuals.

An alternate Z score for goodness-of-fit can be calculated from the observed ApEn and the maximum ApEn obtained during the shuffling process, $\text{Max}_{\text{random}}$, as:

$$Z_{\max} = \frac{\text{Max}_{\text{random}} - \text{ApEn}}{\text{SD}_{\text{random}}} \tag{8}$$

If the Z scores, as calculated in Eqs. (7) and (8), are behaving as expected, then approximately 5% of the time it is expected that a Z score will be greater than 1.645 when ApEn is applied to random sequences. Two simulations were performed to test this behavior. First, 1000 sequences of 100 Gaussian distributed (mean = 0 and SD = 1) random numbers were generated. The ApEn, $\text{Mean}_{\text{random}}$, $\text{SD}_{\text{random}}$, and Z were calculated for each of these 1000 sequences. $\text{Mean}_{\text{random}}$ and $\text{SD}_{\text{random}}$ were evaluated by shuffling each of the particular sequences 1000 times. The value of Z_{random} for ApEn(1, 0.2 SD,100) exceeded the 1.645 limit 4.9% of the time and the Z_{random} for ApEn(2, 0.2 SD,100) exceeded the 1.645 limit 4.7% of the time. The value of $Z_{\max}$ for ApEn(1, 0.2 SD,100) exceeded the 1.645 limit 93.2% of the time and the $Z_{\max}$ for ApEn(2, 0.2 SD,100) exceeded the 1.645 limit 94.4% of the time.

When the simulations were repeated with evenly distributed random numbers between −0.5 and 0.5, similar results were obtained. For the simulated rectangular noise it was observed that the value of Z_{random} for ApEn(1, 0.2 SD,100) exceeded the 1.645 limit 5.5% of the time while the Z_{random} for ApEn(2, 0.2 SD,100) exceeded the 1.645 limit 5.0% of the time. Again, the values of $Z_{\max}$ were much larger than expected. The $Z_{\max}$ for ApEn(1, 0.2 SD,100) exceeded the 1.645 limit 93.5% of the time while the $Z_{\max}$ for ApEn(2, 0.2 SD,100) exceeded the 1.645 limit 94.7% of the time. The randomized ApEn is behaving as expected while the maximum ApEn is not. Thus, the value of Z_{random} will be used for the calculation of probabilities for the remainder of this chapter. When used as a goodness-of-fit criterion, ApEn was not obviously sensitive to the form of the noise distribution of the residuals. This is not surprising because ApEn does not require Gaussian distributed sequences of numbers in its application.

An example of the use of ApEn as a goodness-of-fit criterion is shown in Fig. 1. The lower panel presents a series of 17 data points that were least-squares fit to a single exponential decay, i.e., Eq. (9):

$$Y = \text{Amplitude}\ \mathbf{e}^{-\text{decay}X} \tag{9}$$

For the present parameter estimation the apparent amplitude is 1.982 (±1 SD confidence interval is 1.972–1.993) and the apparent decay rate is 2.082 (±1 SD confidence interval is 2.058–2.096). The data, however, were simulated as the sum of two exponentials with equal amplitudes of 1.0, decay rates of 1.5 and 3.0, and Gaussian noise with an SD of 0.01. It is expected that this parameter estimation will not pass the goodness-of-fit tests since it utilized an incorrect fitting equation.

The upper panel of Fig. 1 presents the weighted residuals for this fit. It appears to have some trend in these residuals. According to a runs test[20] the expected number of runs is 9.24 ± 1.93 and the observed number of runs is 7. This corresponds to a probability of 0.82 that there are too few runs. Clearly not a significant result. It is important to note that

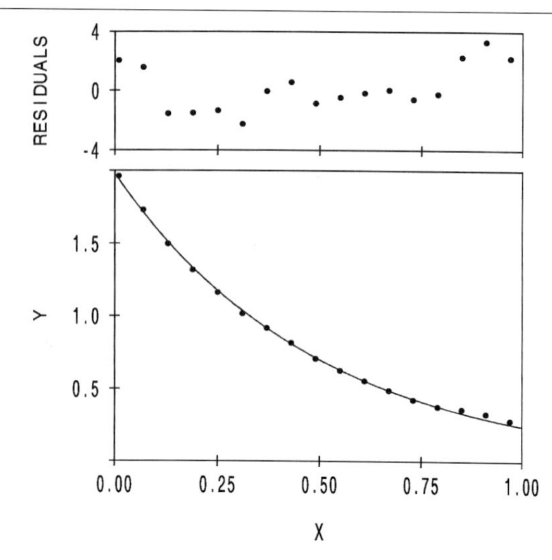

FIG. 1. Least Squares analysis of a set of data. The data are shown as ● in the lower panel. The curve in the lower panel is the best single exponential fit of these data. The upper panel displays the standard deviation weighted residuals for the parameter estimation. The residuals are the differences between the data points and the calculated optimal curve.

we cannot conclude from a 0.82 probability level that the fit is a good description of the data. All that has been accomplished is that we have failed to demonstrate that the fit is a bad description of the data. ApEn(1, 0.2 SD,17) = 0.55589 for this set of residuals while the randomized ApEn(1, 0.2 SD,17) = 0.725 ± 0.087, which corresponds to a single-sided probability of 0.975 based on the Z_{random} value. Clearly, for this example, ApEn provides a substantially more sensitive test for nonrandom residuals as compared to the runs test.

Figure 2 presents two analyses of a set of data containing evenly distributed experimental uncertainties instead of the more commonly assumed Gaussian distribution. Consequently, for these analyzes a robust, least-abs, fitting procedure was utilized. These data are shown in the lower panel. The solid line in the lower panel is the two-site Adair binding equation [see Eq. (10)] fit of the data by the least-abs procedure; $K_1 = 4.9$ and $K_2 = 1.1$. The dashed line in the lower panel corresponds to the least-abs fit of the data to a single exponential rise [see Eq. (11)]; baseline = 0.81, amplitude = −0.50, and $k = 0.46$. The upper panel in Fig. 2 presents the corresponding weighted residuals: open circles for the exponential fit and closed circles for the two-site Adair binding equation fit. By inspection it appears that the Adair equation provides a better description of the data as compared to the exponential rise.

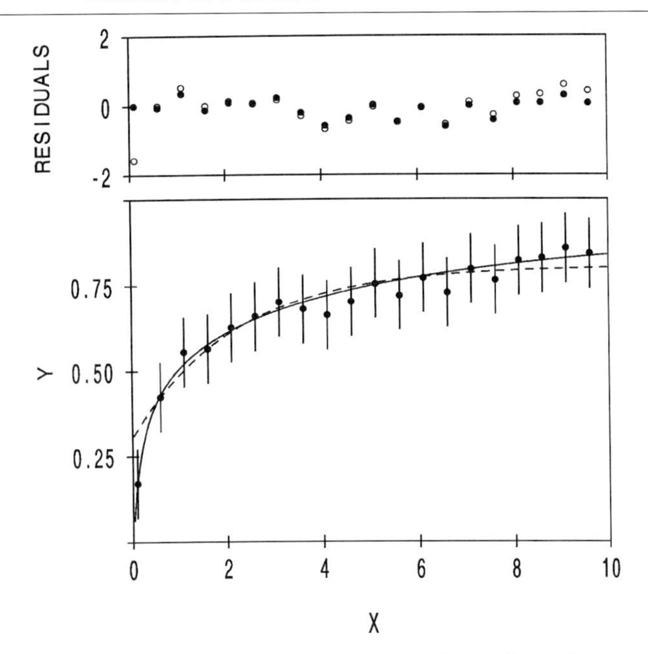

FIG. 2. Analysis of a set of data with evenly distributed experimental uncertainties. The data are shown as ● in the lower panel. The solid curve in the lower panel is the least-abs fit of a two-site Adair binding equation [Eq. (10)] to the data. The dashed curve in the lower panel is the best single exponential [Eq. (11)] least-abs fit of the data. The upper panel displays the standard deviation weighted residuals for these parameter estimations: ● for the two-site Adair binding equation and ○ for the exponential rise.

$$Y = \frac{1}{2}\frac{K_1X + 2K_2X^2}{1 + K_1X + K_2X^2} \tag{10}$$

$$Y = \text{Baseline} + \text{Amplitude}\, e^{-kX} \tag{11}$$

The ApEn analysis of the goodness-of-fit for the two curves shown in Fig. 2 indicates that the exponential is not a good description of the data. The Z_{random} based probability corresponding to ApEn(1, 0.2 SD,20) is 0.984 for the exponential analysis. This indicates that the exponential rise equation is not consistent with the data. The corresponding probability is 0.431 for the two-site Adair binding analysis. Thus, it cannot be concluded that the two-site Adair binding model is inconsistent with the data. Consequently, it appears that the two-site Adair equation provides a good description of these data.

Conclusion

This chapter describes the use of ApEn as a measure of the goodness-of-fit for parameter estimation procedures. Many possible goodness-of-fit tests can be applied to data where the residuals are expected to follow a Gaussian distribution. For these, ApEn is simply another tool that may have some applicability.

However, for non–least squares parameter estimation procedures, such as robust least-abs methods, there are very few available methods for the analysis of goodness-of-fit. ApEn is not restricted to considering Gaussian distributions of residuals. Thus, it is expected that ApEn will provide a useful goodness-of-fit criterion for use with robust parameter estimation procedures, such as least-abs.

Acknowledgments

The authors acknowledge the support of the National Science Foundation Science and Technology Center for Biological Timing at the University of Virginia (NSF DIR-8920162), the General Clinical Research Center at the University of Virginia (NIH RR-00847), and the University of Maryland at Baltimore Center for Fluorescence Spectroscopy (NIH RR-08119).

[13] Kinetic Models and Data Analysis Methods for Fluorescence Anisotropy Decay

By EDWARD L. RACHOFSKY and WILLIAM R. LAWS

Introduction

Time-resolved fluorescence anisotropy is a powerful technique for investigating macromolecular dynamics.[1] In a time-resolved anisotropy measurement, a sample is excited by linearly polarized light, and the anisotropy (or polarization) decay of the resulting emission is evaluated by observing the fluorescence decay at polarizations parallel and perpendicular to the excitation. The polarized excitation selectively excites a population of fluorophores that have excitation dipoles approximately parallel to itself. Depolarizing processes may rotate the emission dipole of the fluorophore while it is in the excited state, and consequently light is emitted with a different polarization.

The emission can be depolarized by a variety of dynamic and photophys-

[1] M. Badea and L. Brand, *Methods Enzymol.* **61,** 378 (1979).

0076-6879/00 $30.00

ical processes. The former include rotational diffusion of the macromolecule, segmental motion of a macromolecular domain, and local rotation of the fluorophore about a covalent bond or within a noncovalent binding site. Photophysical mechanisms of depolarization include the vibrational relaxation of the excited-state fluorophore as well as more unusual events such as resonance energy transfer.[2] Each of these depolarizing processes occurs with a characteristic correlation time, and the influence of a given process on the anisotropy decay depends on the relative values of that correlation time and the lifetime of the excited state, as determined by a fluorescence intensity decay experiment. Motions or other events that are much faster than the intensity decay will decrease the initial observed anisotropy without altering the apparent anisotropy decay kinetics, because depolarization by that process will be complete before the emission is observed. For example, the vibrational relaxation of a fluorophore from the initial Franck–Condon excited state to the emitting state occurs on the sub-picosecond timescale, which is much shorter than a typical fluorescence lifetime. Depolarizing processes that are much slower than the intensity decay also will not influence the anisotropy decay kinetics, because the excited state will be completely depopulated before any depolarization takes place. Various studies have shown that a depolarizing event can be detected in the anisotropy decay kinetics only if the associated rotational correlation time is within approximately an order of magnitude of the intensity decay lifetime.[3] Given the broad range of intensity decay lifetimes for various fluorophores, it is therefore possible to observe dynamic processes with correlation times as long as milliseconds and as short as picoseconds. This range includes many biochemically important motions, such as local dynamics, conformational changes, and macromolecular rotational diffusion.

Fluorescence anisotropy decay is subject to several technical difficulties that can hamper its application to biological macromolecules. The principal impediments involve data analysis and interpretation: formulating a theoretical model to describe the anisotropy decay kinetics; determining the kinetic parameters from data analysis with a reasonable degree of certainty; and discriminating between different models that all adequately fit the observed decays. While sophisticated approaches have been applied to each of these problems, the difficulties inherent in formulating kinetic models and analyzing data continue to hinder the recovery of accurate information about macromolecular dynamics. It is the objective of this review to summarize the limitations of time-resolved anisotropy, and to present the results

[2] Th. Förster, *Ann. Phys.* (*Leipzig*) **2,** 55 (1948).
[3] Ph. Wahl, *Biophys. Chem.* **10,** 91 (1979).

of recent investigations aimed at overcoming these limitations. The material is presented in two sections: (1) development and application of new kinetic models for fluorescence anisotropy decay and (2) presentation of methods to improve parameter recovery and model discrimination during data analysis.

Kinetic Models for Fluorescence Anisotropy Decay

Anisotropy Decay of Single Emitting State

A number of models have been formulated to describe the anisotropy decay kinetics of various macromolecular systems. These include the cases of (1) a fluorophore bound rigidly to a "free" asymmetric rotor, such as a protein, and therefore subject only to rotational diffusion of the macromolecule[4]; (2) a fluorophore attached to a macromolecule by a flexible linker and, therefore, subject both to rotational diffusion and to local angular motion[5]; and (3) a fluorophore undergoing "hindered rotation," such as when buried in a phospholipid membrane.[6] The objective of each of these models has been to describe the anisotropy decay kinetics that arise from a particular type of macromolecular dynamics. Details of these models are not specified here, but certain common features are reviewed for later reference.

In general, the anisotropy decay $r(t)$ of a fluorophore with single exponential intensity decay kinetics is given by

$$r(t) = \sum_{j=1}^{m} \beta_j e^{-t/\varphi_j} \tag{1}$$

where there are m rotational correlation times φ_j, and the preexponential factors β_j are functions of the excitation and emission transition dipoles.[4] The sum of the β_j factors is equal to the limiting anisotropy r_0. This quantity, which is determined by the angle between the excitation and emission transition dipoles, is theoretically limited to the range $-0.2 \leq r_0 \leq 0.4$, where the lower and upper limits represent perpendicular and parallel dipoles, respectively.

Experimentally, $r(t)$ is commonly obtained by exciting with vertically

[4] T. J. Chuang and K. B. Eisenthal, *J. Chem. Phys.* **57**, 5094 (1972).

[5] This model has been developed by a number of investigators. For a review and discussion of the literature, see Ref. 24.

[6] R. E. Dale, *in* "Time-Resolved Fluorescence Spectroscopy in Biochemistry and Biology" (R. B. Cundall and R. E. Dale, eds.), p. 555. Plenum, New York, 1983.

polarized light, and measuring fluorescence decays at three emission polarizations: vertical (V), horizontal (H), and the magic angle (M).[7] For a fluorophore with single exponential intensity decay, the fluorescence decay laws at these three polarizations are given by

$$M(t) = \frac{\alpha}{3} e^{-t/\tau} \tag{2}$$

$$V(t) = M(t)\{1 + 2r(t)\} \tag{3}$$

$$H(t) = M(t)\{1 - r(t)\} \tag{4}$$

where τ is the intensity decay lifetime and α is a normalization factor. Each of the aforementioned anisotropy decay models describing specific types of depolarizing motions can be derived from Eq. (1) with small modifications.

The common limitation of these models is that they each assume that a macromolecule contains only a single fluorophore, and that this fluorophore has a monoexponential intensity decay. In fact, the intensity decays of intrinsic or extrinsic fluorophores associated with proteins, nucleic acids, and membranes often are not monoexponential. This complexity can arise from diverse mechanisms. For instance, multiple fluorophores within a given macromolecule can decay with different lifetimes. Alternately, when a macromolecule is in equilibrium between multiple conformations or tautomeric states, a single fluorophore may exist in multiple environments that give rise to multiple lifetimes. Finally, there are a variety of excited-state processes, such as chemical reactions, dielectric relaxation, and resonance energy transfer, that can create complex intensity decay kinetics. Regardless of whether the intensity decay kinetics derive from ground-state or excited-state processes, the anisotropy decay law should account for the time-dependent heterogeneity of emitting states. However, the $r(t)$ models that are currently in common use are rigorously correct only for a system with a single exponential intensity decay.

Ground-state and excited-state processes require different modifications to the current theory of anisotropy decay. If there is ground-state heterogeneity—multiple fluorophores, conformers, or tautomers—then the kinetic model must acknowledge that each fluorophore may undergo a distinct set of depolarizing motions. For example, a protein might contain two fluorophores bound in different domains. Both fluorophores would be depolarized by the rotational diffusion of the protein. However, different segmental motions experienced by the two domains would cause the two fluorophores to have different anisotropy decays. The anisotropy decay law

[7] T. Azumi and S. P. McGlynn, *J. Chem. Phys.* **37,** 2413 (1962).

for the case of such ground-state heterogeneity has recently been derived[8] and is detailed below.

When complex intensity decay kinetics arise from excited-state processes, the anisotropy decay law must account for the possibility that each emitting state may undergo distinct depolarizing events. However, the analysis of these systems is complicated by two additional factors. The first consideration is that the apparent decay lifetimes cannot be associated with specific emitting states when there are excited-state reactions. The observed lifetimes are kinetically derived functions that depend on the mechanism of the excited-state process(es) and the rates of the various steps. For instance, in a two-state excited-state reaction the apparent lifetimes are functions of the forward and reverse reaction rates, as well as the radiative and nonradiative decay rates of both excited-state species.[9] Consequently, it becomes more difficult to account for differences in the depolarizing processes experienced by each emitting state. The second complication is that there may be an additional depolarization due to the excited-state process itself. For example, resonance energy transfer may depopulate an excited state in an orientation-dependent manner, consequently changing the angular distribution of emitters.[2] Because of these complications, formally correct kinetic models describing the anisotropy decay in the presence of any of the common excited-state processes have never been developed.

Anisotropy Decay Law for Case of Ground-State Heterogeneity

When there are multiple fluorophores (or multiple ground-state forms of a single fluorophore) the anisotropy decay of the ith fluorophore will be given by[8]

$$r_i(t) = \sum_{j=1}^{m} \beta_{ij} e^{-t/\varphi_j} \tag{5}$$

Because the preexponential β factors depend on transition dipoles, the values of these terms in general will differ for each fluorophore; therefore, they are doubly subscripted. A β_{ij} factor is the degree to which the ith fluorophore is depolarized by the process(es) giving rise to the jth correlation time. The sum of the β_{ij} terms for a given fluorophore over all correlation times ($j = 1$ to m) is the limiting anisotropy r_{0i} for that fluorophore, and is subject to the theoretical range of $-0.2 \leq r_{0i} \leq 0.4$, as described above for the case of a single fluorophore.

[8] C. N. Bialik, B. Wolf, E. L. Rachofsky, J. B. A. Ross, and W. R. Laws, *Biophys. J.* **75,** 2564 (1998).

[9] W. R. Laws and L. Brand, *J. Phys. Chem.* **83,** 795 (1979).

It is not necessary to specify the identity of the fluorophore when labeling the correlation times φ_j, because the correlation times are not functions of transition dipoles or other properties of the fluorophore. A common set of correlation times can describe the anisotropy decay of all fluorophores, with the extent of depolarization of each fluorophore by each process specified by the value of the appropriate β_{ij}. If a given β_{ij} is equal to zero, then the ith fluorophore is not depolarized by the process(es) giving rise to the jth correlation time. In this formulation, the β_{ij} can be regarded as "mapping" factors between intensity decay lifetimes and rotational correlation times. However, the meaning of these mapping factors differs from that of a previous formulation in which the association between the ith lifetime and the jth correlation time was specified by a binary switch L_{ij}.[10] In the current scheme, the mapping factor β_{ij} is a continuously variable function of the transition dipoles and dynamics of the fluorophore.

Based on Eq. (5), it is trivial to write the M, V, and H decay laws for the case of ground-state heterogeneity

$$M(t) = \frac{1}{3}\sum_{i=1}^{n} \alpha_i e^{-t/\tau_i} \tag{6}$$

$$V(t) = \frac{1}{3}\sum_{i=1}^{n} \alpha_i e^{-t/\tau_i}\left\{1 + 2\sum_{j=1}^{m} \beta_{ij} e^{-t/\varphi_j}\right\} \tag{7}$$

$$H(t) = \frac{1}{3}\sum_{i=1}^{n} \alpha_i e^{-t/\tau_i}\left\{1 - \sum_{j=1}^{m} \beta_{ij} e^{-t/\varphi_j}\right\} \tag{8}$$

Here, it must be emphasized that each ground state is assumed to give rise to a single excited state that decays with a single intensity decay lifetime τ_i. There can be no interactions between the various excited states, or else the complications described above for excited-state reactions will ensue.

The ground-state heterogeneity model differs in two respects from the "homogeneous" anisotropy decay model that is currently in common use. First, the homogeneous model assumes that each lifetime is associated with each correlation time; i.e., that all fluorophores are subject to all depolarizing processes. Second, the homogeneous model assumes that all fluorophores have identical β_j factors; i.e., that all fluorophores have the same limiting anisotropy. In contrast, the ground-state heterogeneity model acknowledges that each fluorophore may have a different limiting anisotropy and may undergo a distinct set of depolarizing events. The benefit of

[10] L. Brand, J. R. Knutson, L. Davenport, J. M. Beechem, R. E. Dale, D. G. Walbridge, and A. A. Kowalczyk, *in* "Spectroscopy and the Dynamics of Molecular Biological Systems" (P. M. Bayley and R. E. Dale, eds.), p. 259. Academic Press, London, 1985.

explicitly considering these complications is the assignment of depolarizing motions (or other processes) to specific fluorophores and, hence, to particular macromolecular domains. This localization of dynamics to particular structural elements cannot be accomplished using the homogeneous model.

With n lifetimes and m correlation times, a total of $(2^n - 1)^m$ "association models" can be defined based on the above kinetic scheme. These models describe all of the possible ways in which the lifetimes and correlation times can be associated. Obviously, the number of incorrect models that must be rejected during data analysis increases rapidly with n and m. Consequently, the problems of parameter recovery and model discrimination become significantly more difficult when ground-state heterogeneity is considered explicitly. The general problem of statistical uncertainty in anisotropy decay data analysis is discussed further in a later section. First, however, results are reviewed demonstrating the application of the ground-state heterogeneity model to both synthetic and experimental anisotropy decay data.

Simulated Anisotropy Decays with Ground-State Heterogeneity

To assess the feasibility of discriminating among a group of closely related τ–φ association models during data analysis, Bialik *et al.*[8] performed a series of simulation studies. These investigators synthesized data by each of the nine models for the case $n = 2, m = 2$ that are shown in Fig. 1, and then attempted to assign the proper association model to each data set (consisting of M, V, and H decays) based on the results of a blinded analysis. The decision to accept or to reject a particular model during data analysis was based both on statistical criteria and the requirement that the values of the recovered parameters fall within physically realistic limits.

Analysis of single data sets using the ground-state heterogeneity scheme was ineffective: the assignment of an association model was acceptable for only 3 of 54 simulated decays, and the 3 assignments that were made were all subsequently found to be incorrect. For all other data sets, there were at least two acceptable association models, so that unequivocal assignment of a model was impossible. Therefore, the authors concluded that τ–φ association models for these data could not be resolved from individual analyses of data sets.

Global analysis[11,12] improved the discrimination between models. Global analysis was performed on families of six data sets that had been generated by the same association model but with different α_1/α_2 ratios, a

[11] J. M. Beechem, J. R. Knutson, J. B. A. Ross, B. W. Turner, and L. Brand, *Biochemistry* **22,** 6054 (1983).

[12] J. R. Knutson, J. M. Beechem, and L. Brand, *Chem. Phys. Lett.* **102,** 501 (1983).

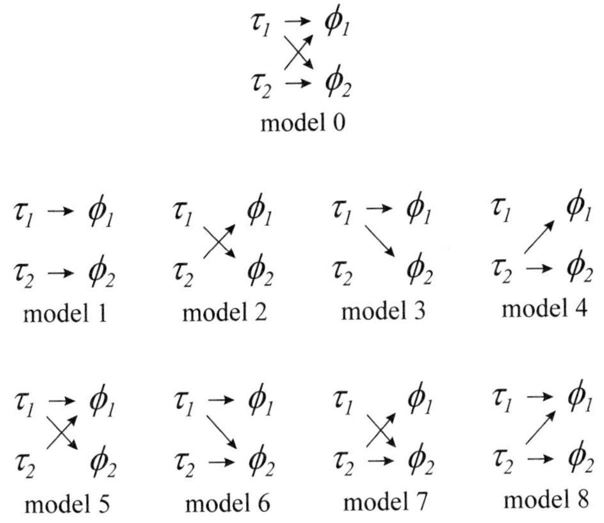

$\tau_1 \rightarrow \phi_1$, $\tau_2 \rightarrow \phi_2$ (crossed)
model 0

model 1 model 2 model 3 model 4

model 5 model 6 model 7 model 8

FIG. 1. The nine possible fluorescence intensity decay lifetime–rotational correlation time association models possible for $n = m = 2$. In analyses using these models, it is necessary to maintain an ordered numbering of lifetime and correlation times based on magnitudes. Thus, $\tau_1 < \tau_2$ and $\varphi_1 < \varphi_2$. [Reprinted from C. N. Bialik *et al., Biophys. J.* **75,** 2564 (1998) with permission of the publisher.]

procedure analogous to combining experimental data collected at multiple emission wavelengths. Because lifetimes, correlation times, and β factors should be virtually invariant with emission wavelength, each of these parameters was made common to all six data sets. By this analysis, all nine families could be assigned the correct association model. Thus, discrimination between similar association models is possible, but it requires global analysis of multiple decays collected as a function of an independent variable, such as emission wavelength, and the employment of all possible common parameters.

An additional benefit of the global analysis was that an analysis by model 0 (see Fig. 1) was sufficient to determine the actual associations between lifetimes and correlation times for each of the nine models, because the appropriate β_{ij} terms iterated to zero. This behavior was not observed in the analyses of single data sets, in which a particular model could be applied to the data only when the appropriate β_{ij} factors were fixed to zero. Therefore, it is apparent that global analysis not only increased the power to discriminate between models, but also vastly decreased the time and computational expense required. Further simulation studies will determine whether global analysis can similarly facilitate model discrimination for data with other parameter values or with n and/or $m \geq 2$.

Application to Protein Fluorescence: Liver Alcohol Dehydrogenase

To determine whether the ground-state heterogeneity model could be applied to the fluorescence of a biological macromolecule, Bialik *et al.*[8] examined the anisotropy decay of horse liver alcohol dehydrogenase (LADH). LADH is a homodimeric enzyme containing two tryptophan residues in each subunit. Trp-15 resides on the surface of the enzyme, and consequently has a red-shifted emission relative to that of Trp-314, which is situated in the hydrophobic core. Numerous investigations have concluded that each of these residues undergoes a monoexponential fluorescence intensity decay, with lifetimes of approximately 7 ns for Trp-15 and 4 ns for Trp-314.[13–16] A recent report suggested that the anisotropy decay of LADH was characterized by two correlation times, one representing the rotational diffusion of the essentially spherical enzyme and the other arising from local or segmental motion.[17] Based on the monoexponential intensity decay kinetics of each Trp residue and the evidence of local or segmental motion, it was hypothesized that the ground-state heterogeneity model would describe the anisotropy decay of this enzyme. Therefore, time-resolved fluorescence experiments were performed in an attempt to assign the local motion to one or both of the tryptophan residues.

Based on the results of the above simulations, eight LADH $r(t)$ data sets were collected as a function of emission wavelength, and analyzed globally with all lifetimes, correlation times, and β factors as common parameters. Of the nine association models in Fig. 1, only model 6 gave an acceptable analysis of these data, suggesting that the short correlation time represented the depolarization of Trp-314 by some local process. This result was surprising, because crystal structures had shown that Trp-314 is located at the subunit interface of the homodimer in a tightly packed, rigid environment.[18] Consequently, Bialik *et al.*[8] hypothesized that the depolarizing mechanism was not local dynamics, but instead might be resonance energy transfer between the Trp-314 residues of the two subunits. Consistent with this proposed homotransfer of excitation energy, the indole rings of these two residues are separated by only 7 Å, and there is significant overlap of their absorption and emission spectra.

[13] J. B. A. Ross, C. J. Schmidt, and L. Brand, *Biochemistry* **20,** 4369 (1981).

[14] M. R. Eftink and D. M. Jameson, *Biochemistry* **21,** 4443 (1982).

[15] D. R. Demmer, D. R. James, R. P. Steer, and R. E. Verrall, *Photochem. Photobiol.* **45,** 39 (1987).

[16] M. R. Eftink, Z. Wasylewski, and C. A. Ghiron, *Biochemistry* **26,** 8338 (1987).

[17] M. R. Eftink, C.-Y. Wong, D.-H. Park, G. L. Shearer, and B. V. Plapp, *Proc. SPIE* **2137,** 120 (1994).

[18] C.-I. Bränden, H. Jornvall, H. Eklund, and B. Furugren, *in* "The Enzymes" (P. D. Boyer, ed.), p. 103. Academic Press, New York, 1975.

The hypothesis that Trp-314 is depolarized by energy transfer implies that a model based on depolarization by macromolecular dynamics may not correctly describe the anisotropy decay of LADH. An alternative model (not yet developed) based explicitly on the mechanism of depolarization by energy transfer might be employed instead. Nevertheless, it was appropriate to analyze these data within the framework of ground-state heterogeneity, because the intensity decay arises from the emission of two Trp residues—Trp-15 and Trp-314—that do not interact with each other in the excited state. The proposed energy transfer mechanism of depolarization would not affect the intensity decay kinetics, because the transfer would be between two identical Trp-314 residues in different subunits. Thus, it was possible to resolve the anisotropy decay of each fluorophore, even though depolarization may have arisen from a process other than macromolecular dynamics.

Kinetic Models: Conclusions

In summary, the ground-state heterogeneity model in a global analysis of fluorescence anisotropy decay enabled assignment of depolarizing processes to particular fluorophores in both simulated and experimental data. Therefore, it is now possible to localize internal motions and other depolarizing processes to particular domains within a macromolecule.

In both the simulations and the experimental decays, discrimination between association models was critically dependent on inclusion of data sets with extreme intensity ratios. In the individual analysis of synthetic decays, most data sets with intermediate intensity ratios could be analyzed by all nine association models. However, those data sets with extreme intensity ratios could only be analyzed by a few models. For LADH, unequivocal assignment of an association model was not possible unless the decays collected at 310 and 320 nm—the high-energy edge of the emission spectrum—were included in the global analysis. The extreme α_1/α_2 ratio at these wavelengths probably restricted the acceptable ranges of the iterated parameters, enabling the rejection of several incorrect models. Therefore, it is concluded that data collected as a function of an independent variable are most informative when one or more iterated parameters are significantly perturbed by changes in that variable.

The success of the ground-state heterogeneity model suggests that equivalent benefits would accrue from the development of formally correct models of anisotropy decay kinetics in the presence of the various excited-state processes mentioned above. The opportunity exists to derive a complete set of models describing the anisotropy decay kinetics arising from each of the known mechanisms of fluorescence intensity decay. The benefit of this development would be a great increase in the utility and applicability of fluorescence anisotropy decay to the study of macromolecules.

Anisotropy Decay Data Analysis

The importance of global analysis for the application of the ground-state heterogeneity model highlights the need for powerful analytical techniques to facilitate further advances. Such techniques will become even more important as n and m increase, yielding greater numbers of possible τ–φ associations. Therefore, it is necessary to study the statistical problems inherent to time-resolved fluorescence experiments, and to develop general methods to increase the power of data analysis.

Parameter Uncertainties and Model Discrimination

Time-resolved fluorescence anisotropy data may be analyzed by several algorithms, including the method of moments,[19] the Laplace transform,[20] and nonlinear least squares (NLLS).[21] Each of these algorithms is an iterative procedure, in which the values of the parameters defined by the decay law are successively varied in an attempt to fit a physical model to the experimental data. The objective is to find the values of the iterated parameters that best fit the model to the data, and subsequently to determine the most appropriate model that describes the data.

There are two important sources of uncertainty in the kinetic parameters recovered from anisotropy decay data analysis. Although some uncertainty always arises from measurement error in experimental data, a more challenging problem is posed by the statistical cross-correlation between the parameters iterated during data analysis.[22] Cross-correlation between parameters arises from the form of the fitting functions [Eqs. (3) and (4)], which are products of two exponential terms. It effectively increases parameter uncertainities by allowing a perturbation of one parameter from its optimum value to be compensated by corresponding changes in the other parameters without reducing the goodness of fit. Consequently, it is often difficult to determine rotational correlation times with confidence, which limits the utility of anisotropy decay for studying molecular dynamics.

The cross-correlation between kinetic parameters also increases the computational expense of calculating parameter uncertainties. For uncorrelated parameters, the uncertainty can be calculated by perturbing each

[19] I. Isenberg and R. D. Dyson, *Biophys. J.* **9,** 1337 (1969).
[20] A. Gafni, R. L. Modlin, and L. Brand, *Biophys. J.* **15,** 263 (1975).
[21] A. Grinvald and I. Z. Steinberg, *Anal. Biochem.* **59,** 583 (1974).
[22] M. L. Johnson and L. M. Faunt, *Methods Enzymol.* **210,** 1 (1992).

parameter successively from its optimum value and observing the resulting deterioration in the goodness of fit. However, if parameters are highly correlated, then calculating the uncertainty in a parameter requires fixing that parameter at a perturbed value and iterating all other parameters to maximize the goodness of fit. This procedure increases the computational time required for data analysis manyfold, because each step in the uncertainty calculation is essentially a new optimization. Consequently, exact parameter uncertainties are rarely calculated.

Several approaches have been suggested to decrease the uncertainty in the iterated parameters in anisotropy decay analysis. Global analysis of decay data collected as a function of some independent variable can restrict the iterated parameters by forcing them to satisfy all data sets simultaneously.[11,12] Alternatively, methods have been proposed to constrain anisotropy decay analysis using information obtained from other physical measurements. For instance, (1) the limiting anisotropy r_0 of a fluorophore could be determined by measuring its steady-state anisotropy in a viscous solvent or a frozen glass, and used to limit the sum of the β factors; (2) approximate rotational diffusion coefficients obtained from hydrodynamic calculations could be used to estimate correlation times[23]; (3) translational diffusion coefficients could be measured by sedimentation velocity experiments, and applied as limits on rotational diffusion coefficients and correlation times[23]; and (4) the steady-state anisotropy of a macromolecule, which is equal to the time integral of the anisotropy decay, could be used to scale the observed V and H decays to each other.[6] Each of these methods attempts to apply some restriction on the iterated parameters to counter the influence of the strong cross-correlation between them.

Application of each of these constraints requires a different modification of the analysis protocol. A correlation time may simply be held constant at the value predicted by a hydrodynamic calculation. However, restriction by the limiting anisotropy or the steady-state anisotropy requires application of an algebraic relationship between iterated parameters. This type of relationship between variables must be coded directly into a data analysis program and will be different for each constraint.

Global analysis also entails modification of an analysis program. The simplest type of global analysis, in which certain iterated parameters are made common to all data sets, is fairly easy to implement. However, implementation of more complex global relationships is currently a tedious process. For example, it might be useful to apply the Stern–Volmer equation to a series of fluorescence intensity decays collected as a function of the

[23] E. Waxman, W. R. Laws, T. M. Laue, Y. Nemerson, and J. B. A. Ross, *Biochemistry* **22,** 6054 (1993).

quencher concentration. In this case, the observed intensity decay lifetimes would be functions of the quencher concentration and the quenching rate constant, k_q.[24] It would be possible to analyze these data globally by iterating k_q and calculating the lifetime at each quencher concentration from the Stern–Volmer relationship. However, this implementation would require extensive modifications to the standard algorithms used by most analysis programs. A substantially different modification would be required for each of the many global relationships that conceivably could be applied to anisotropy decay data. Consequently, more general global relationships between parameters have rarely been applied to time-resolved anisotropy data analysis.

For some forms of constraining information, no statistically rigorous method of application exists. The limits on hydrodynamic shape obtained by sedimentation velocity measurements have only been applied to anisotropy decay analysis in an *ad hoc* fashion, in which they served as a criterion for rejecting unacceptable analyses.[23] No systematic mathematical method of simultaneously analyzing time-resolved anisotropy and analytical ultracentrifugation data has been developed.

A general method of constraining NLLS analysis of fluorescence anisotropy decay has recently been developed.[25] Although this method was initially implemented to constrain an anisotropy decay analysis by the steady-state anisotropy, it could also be used to apply any of the other constraints discussed above. This algorithm enables (1) application of an arbitrary mathematical relationship between iterated parameters for a single anisotropy decay data set; (2) application of an arbitrary global relationship between iterated parameters for multiple anisotropy decay data sets collected as a function of an independent variable; (3) simultaneous analysis of anisotropy decay and any other relevant experiment (e.g., sedimentation velocity) with common iterated parameters; and (4) simultaneous application of an arbitrary number of constraints. Use of this algorithm also simplifies the modifications that must be made to an analysis program to apply any particular constraint. Thus, the effort of applying constraining information is considerably reduced. The theory and application of this novel constraint algorithm are reviewed in the following sections.

Constrained Analysis

The objective of NLLS is to minimize the reduced χ^2, which is the weighted sum of the residuals defined by the difference between the data

[24] J. R. Lakowicz, "Principles of Fluorescence Spectroscopy." Plenum, New York, 1983.

[25] E. L. Rachofsky, B. Wolf, C. N. Bialik, J. B. A. Ross, and W. R. Laws, *J. Fluoresc.* **9,** 379 (1999).

and the model at each time point t:

$$\chi^2 = \frac{1}{N-q} \sum_t \frac{1}{\sigma^2(t)} \{x_{\text{data}}(t) - x_{\text{fit}}(t)\}^2 \tag{9}$$

where N is the number of data points, q is the number of iterated parameters, and $\sigma^2(t)$ is the statistical weight of $x_{\text{data}}(t)$.[26] The weighting factors represent the experimental uncertainty of each data point. Numerous methods have been devised to minimize χ^2, the most effective of which involves calculating the gradient and the curvature of this function with respect to each of the iterated parameters.[26] For time-resolved anisotropy data analysis, a global χ^2 is defined as the cumulative reduced χ^2 for the M, V, and H curves.[23,27] This global χ^2, rather than the individual χ^2 values for each curve, is the function that is minimized. Similarly, in global analysis of multiple data sets collected as a function of an independent parameter, a global χ^2 that is the cumulative reduced χ^2 of all data sets is minimized.[11,12]

To constrain this type of minimization by an arbitrary relationship between iterated parameters, the NLLS algorithm can be modified according to the theory of Lagrange multipliers. This modified NLLS algorithm is similar to previously published methods of constrained optimization,[28] although the implementation of this technique to fluorescence decay analysis is novel. A new target function χ^2_{C} (the constrained χ^2) is defined by weighted addition of a constraining function g to χ^2:

$$\chi^2_{\text{C}} = \chi^2 + \kappa g \tag{10}$$

where κ is a Lagrange multiplier.[25] The constraining function g can be any mathematical relationship between the iterated parameters, as long as it is written to satisfy two conditions. First, g must always be greater than or equal to zero. Second, g must equal zero when the constraint is satisfied. In practice, any equation can be written to satisfy these conditions by squaring and rearranging so that all nonzero terms are on one side. For example, to constrain an $r(t)$ analysis by the steady-state anisotropy, g is defined by

$$g = (r_{\text{ss}} - \langle r \rangle)^2 \tag{11}$$

where r_{ss} is the experimentally measured value of the steady-state anisotropy, and $\langle r \rangle$ is the value calculated from the kinetic parameters of the

[26] P. R. Bevington, "Data Reduction and Error Analysis for the Physical Sciences." McGraw-Hill, New York, 1969.

[27] A. J. Cross and G. R. Fleming, *Biophys. J.* **46,** 45 (1984).

[28] R. Fletcher, "Practical Methods of Optimization." John Wiley and Sons, Chichester, United Kingdom, 1981.

anisotropy decay[24]:

$$\langle r \rangle = \frac{r_0}{1 + \langle \tau \rangle / \langle \varphi \rangle} \tag{12}$$

Here $\langle \tau \rangle$ and $\langle \varphi \rangle$ are the intensity-weighted mean decay lifetime and the harmonic mean correlation time, respectively.[29] Equation (12) is strictly valid only for the homogeneous model of anisotropy decay kinetics. In the case of ground-state heterogeneity discussed above, the expression for $\langle r \rangle$ also must account for the unique anisotropy decay of each fluorophore.[8]

Minimizing χ^2_C with respect to all iterated parameters is equivalent to finding the minimum of the function defined by the intersection of the χ^2 hypersurface with the g hypersurface in parameter space. Because the intersection of these two functions does not encompass the entire χ^2 hypersurface, the range of acceptable parameter values is diminished. Consequently, the uncertainties in the recovered parameters are decreased, and the probability of recovering correct kinetic parameters increases. Defining the constraint function g according to the two conditions described above ensures that it will be minimized when the constraint is satisfied. If the constraining function and the decay data are consistent with each other, then g will equal zero at the minimum of χ^2_C, and thus χ^2 will also be minimized.

Although a conventional Lagrange multiplier is a variable, this algorithm requires that the multiplier κ be a constant. As Eq. (10) demonstrates, it is not practical to allow κ to vary during an iterative data analysis, because χ^2_C would always decrease with decreasing κ. Thus, if κ were a variable, it would iterate to zero and the constraint would not be enforced.

The value of κ determines the weighting of the constraint function g relative to χ^2 and, consequently, the stringency with which the constraint is maintained. Just as the weighting of each data point in Eq. (9) is determined by the experimental uncertainty of that data point, so the value of κ should be determined by the experimental uncertainty in g.[30] However, the uncertainty in g depends on not only the known uncertainty in r_{ss}, but also the unknown uncertainties in the iterated parameters used to calculate $\langle r \rangle$. Therefore, it may be necessary to determine the appropriate value of κ empirically.

The effect of the constraint will depend critically on the choice of κ. If κ is too small, then the constraint will not apply. If κ is too large, then the anisotropy decay data will not be fit well, because χ^2 will not contribute

[29] For definitions of these averages, see Ref. 24.
[30] E. Di Cera, *Methods Enzymol.* **210,** 68 (1992).

significantly to χ^2_C. The value of κ must be sufficient that the approximate relationship

$$\kappa \sum_{i=1}^{q} \frac{\partial g}{\partial p_i} \approx \sum_{i=1}^{q} \frac{\partial \chi^2}{\partial p_i} \tag{13}$$

is maintained when the derivatives are evaluated near the minimum of χ^2_C. Here p_i is the ith parameter and the sum is performed over all q iterated parameters. Rachofsky *et al.*[25] have presented a general procedure for determining the acceptable range of κ for any g, and have applied this method to demonstrate that constraint by the steady-state anisotropy can be accomplished over a wide range of κ. This method is reviewed below.

Application of Lagrange Multiplier Method: Constraint by Steady-State Anisotropy

Steady-state anisotropy has been suggested as an analysis constraint for two specific cases, comprising two distinct but related mechanisms of depolarization. The first is the case where depolarization is much slower than the loss of excitation, so that there is a rotational correlation time that is more than 10 times longer than the fluorescence lifetime.[25] The subsequent poor recovery stems from the fact that most of the experimentally observable intensity decay is complete before there is any evidence of depolarization in the V and H curves. The second situation is the case in which the emission is not completely depolarized during the experimentally observable decay period, such as might occur for a fluorophore buried in a phospholipid membrane. This case can be thought of as the very long correlation time limit of the first case, but in analyzing this decay it is customary to fit for a constant r_∞ term rather than a correlation time.[6] However, it is important to realize that this r_∞ is very strongly cross-correlated with the other kinetic parameters.

Several algorithms have been employed to constrain anisotropy decay analysis by the steady-state anisotropy.[31,32] These have been critically reviewed.[6,25] The most reliable of these algorithms is to weight the areas under the V and H decay curves. To apply this method, the M, V, and H decay curves are collected to equal peak counts (to ensure equivalent experimental uncertainties for each). However, the actual emission intensities at these three polarizations are not in general equal, so the V and H

[31] W. E. Blumberg, R. E. Dale, J. Eisinger, and D. Zuckerman, *Biopolymers* **13,** 1607 (1974).

[32] J. Vanderkooi, S. Fischkoff, B. Chance, and R. A. Cooper, *Biochemistry* **13,** 1589 (1974).

curves each must be multiplied by a scalar to weight them properly relative to the M decay. The ratio of the V and H scalars can be fixed by means of the known relationship of the steady-state anisotropy to the integrals of the V and H decays:

$$\langle r \rangle = \frac{\frac{S_V}{S_H}\int_{t=0}^{\infty} V(t)\,dt - \int_{t=0}^{\infty} H(t)\,dt}{\frac{S_V}{S_H}\int_{t=0}^{\infty} V(t)\,dt + 2\int_{t=0}^{\infty} H(t)\,dt} \tag{14}$$

where S_V is the scalar for the V decay and S_H the scalar for the H decay. Note that evaluation of the integrals in Eq. (14) will yield Eq. (12). This method has been demonstrated to perform well when an r_∞ term is required.[6] However, it has the disadvantage that the experimental data, rather than the model, are weighted by the scalars. Therefore, the weighting is applied to data that include photon counting noise and convolution artifacts. In addition, this method directly constrains only two iterated parameters—the V and H scalars—and influences the other parameters only indirectly, leaving the strong cross-correlation between parameters in effect.

Application of the steady-state anisotropy constraint by the Lagrange multiplier algorithm avoids the potential problems associated with weighting scalars because this algorithm acts directly on all iterated parameters. The increment in each parameter is determined in part by the derivative of the constraint function g with respect to that parameter. Consequently, much of the parameter cross-correlation should be eliminated.

To ascertain whether the Lagrange multiplier method could effectively constrain anisotropy decay analyses by the steady-state anisotropy, Rachofsky *et al.*[25] performed a series of simulation studies. Simulated anisotropy decay data were analyzed using the r_{ss} constraint. For these synthetic data sets, both the intensity decay and the anisotropy decay had single exponential kinetics ($n = m = 1$). Data sets were generated with a variety of ratios of lifetime to correlation time (τ/φ) and a range of amplitudes (β) of depolarization. The values of r_{ss} that were used to constrain analyses were calculated from the parameters used to generate the data.

The performance of the constrained algorithm was compared to that of ordinary NLLS by two criteria. First, the uncertainties in the recovered parameters were assessed using the iterative perturbation algorithm described above. The uncertainties were represented as the increase in χ^2_C with increasing perturbation of each iterated parameter. A greater uncertainty in an iterated parameter results in a smaller increase in χ^2_C with perturbation of that parameter from its optimum value. The r_{ss} constraint significantly decreased the uncertainties of both β and φ for the data set shown in Fig. 2, for which $\varphi = 30\tau$. In contrast, the uncertainty in τ was found to be very small in the unconstrained analysis and did not change with application of the constraint. It was concluded that application of the r_{ss} constraint by the Lagrange multiplier method decreased the uncertainties in the iterated parameters. It was also noted that the cross-correlation between β and φ was decreased by the constraint.

The second comparison of the performance of constrained and unconstrained analyses was the accuracy of recovered parameters. For this experiment, the investigators formulated a cumulative measure of the accuracy of parameter recovery, the R parameter:

$$R = \prod_{i=1}^{q} \frac{1}{f_i + 1} \tag{15}$$

where the product is taken over all q iterated parameters, and f_i is the absolute fractional error in the ith parameter:

$$f_i = \left| \frac{p_i^{\text{rec}} - p_i^{\text{gen}}}{p_i^{\text{gen}}} \right| \tag{16}$$

Here p_i^{rec} and p_i^{gen} are the recovered value and the actual (generation) value, respectively, of the ith parameter. This type of assessment is only possible for simulation studies, since the true values of kinetic parameters are not known for experimental data. Figure 3 shows the R parameter for constrained and unconstrained analyses for a variety of τ/φ ratios and β values. While unconstrained analysis could only recover accurate parameters for $\varphi \leq 3\tau$, the constraint enabled recovery for $\varphi \leq 30\tau$. Therefore, the r_{ss} constraint substantially increased the maximum correlation time that could be determined for a given fluorescence lifetime. The performance of both constrained and unconstrained algorithms was predictably better for larger β factors, because the amplitude of depolarization effects on the V and H decays was greater.

To determine the optimal range of values for the Lagrange multiplier κ, these investigators formulated a benchmark assay using simulation studies. In this experiment, the goodness of fit χ^2_{C} and the parameter recovery R were determined for a single data set as a function of κ. As can be seen from the results in Fig. 4, both parameter recovery and the goodness of fit were optimized for $10^5 \leq \kappa \leq 10^{10}$. For smaller κ, the constraint was ineffective and, consequently, parameter recovery was inaccurate. For larger κ, the second term in Eq. (10) overwhelms the first, and the data cannot be fit well. The authors concluded that a default value of $\kappa = 10^8$ would be suitable for their studies on the r_{ss} constraint. This value is equal to the inverse square of the "experimental" uncertainty of r_{ss}, which in

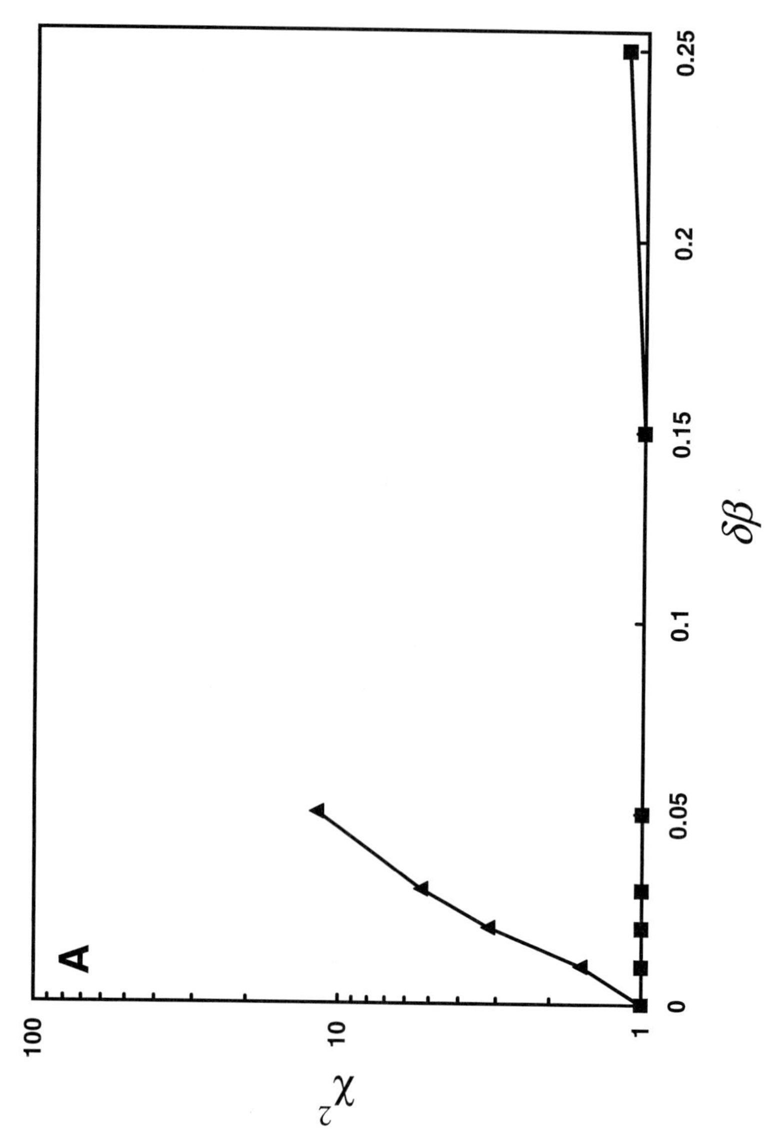

FIG. 2. Parameter uncertainties without (■) and with (▲) steady-state anisotropy constraint. The kinetic parameters used to generate the data set were $\tau = 1$ ns, $\varphi = 30$ ns, and $\beta = 0.15$, with a value of $\kappa = 10^8$ used for the analysis. For unconstrained analyses, χ^2 is plotted; for constrained analyses, χ^2_C is plotted. (A) Uncertainty in β; (B) uncertainty in φ. The lines are drawn only to connect the points, and do not represent any sort of theoretical fit to the data.

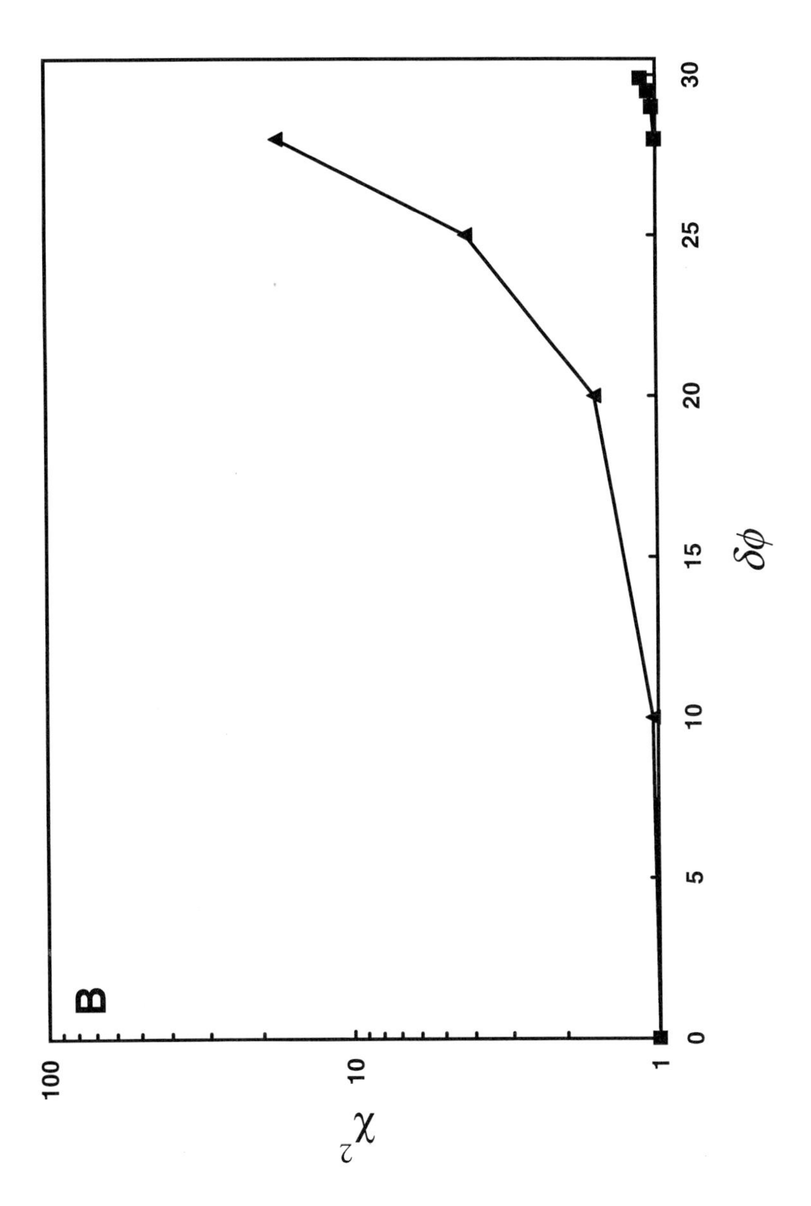
B
100
10
1
χ²
0
5
10
15
20
25
30
δφ

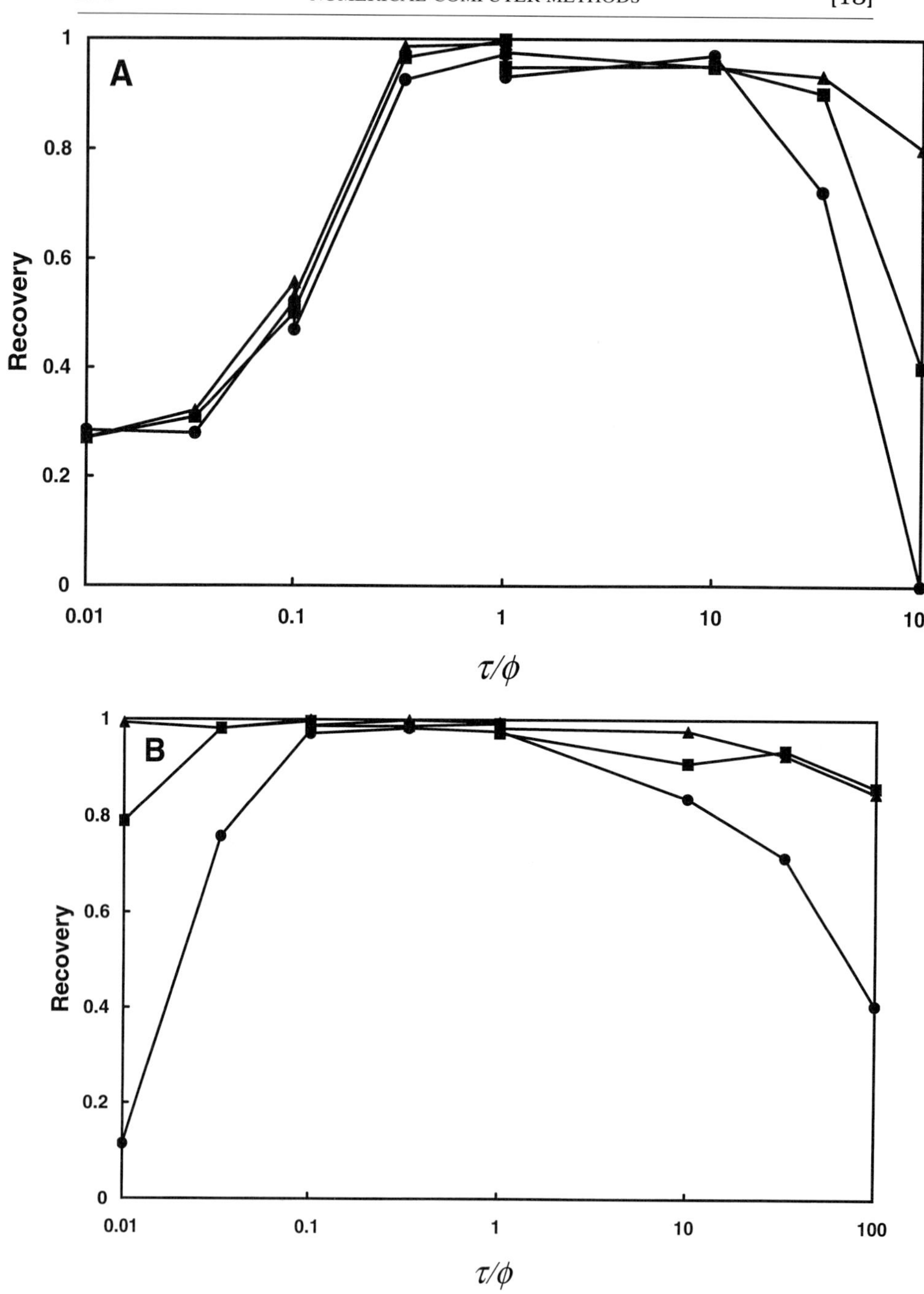
A
B
Recovery
τ/φ
1
0.8
0.6
0.4
0.2
0
0.01
0.1
1
10
100

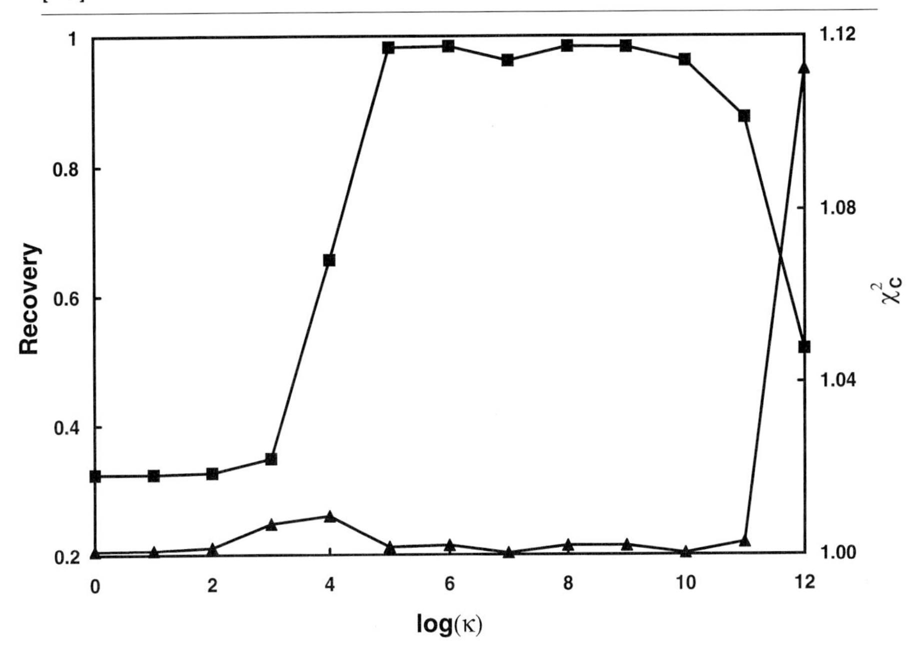

FIG. 4. Recovery of iterated parameters (■) and χ^2_C (▲) versus Lagrange multiplier. The kinetic parameters used to generate the data set were $\tau = 1$ ns, $\varphi = 30$ ns, and $\beta = 0.15$. Lines are drawn only to connect the points, and do not represent any sort of theoretical fit to the data.

these simulations was assumed to be $\pm 10^{-4}$. This equivalency suggests that the uncertainty in the value of a constraint term in general may determine the weighting of that term. However, a benchmarking procedure similar to the one described above should be performed to confirm the weighting factor for any constraint. This protocol highlights the importance of simulation studies for developing and testing new methods in data analysis.

Future Directions

Global relationships between iterated parameters for multiple data sets could be applied in the same fashion as the simple constraint on the steady-state anisotropy. In such analyses, the standard kinetic parameters α, τ, β, and φ (written without subscripts) would continue to be the ones that

---

FIG. 3. Recovery of iterated parameters as a function of the τ/φ ratio. Symbols represent $\beta = 0.05$ (●), 0.15 (■), and 0.30 (▲). (A) Unconstrained analyses; (B) constrained analyses with $\kappa = 10^8$. Lines are drawn only to connect the points, and do not represent any sort of theoretical fit to the data.

are iterated, while the parameters (such as rate or equilibrium constants) defining global relationships would be calculated secondarily. Therefore, minimal changes in program code would be required to implement a new general global constraint using the Lagrange multiplier method.

Fluorescence anisotropy decay data also may be analyzed simultaneously with other types of experimental data using this algorithm. The principal difficulty in combining different types of experimental measurements is defining an appropriate target function (e.g., a global χ^2). Each data set must be weighted by some scalar that allows it to influence the optimization without overwhelming the other data. Appropriate values for these scalars could be predicted from the experimental uncertainty in each measurement, and confirmed by the types of simulation experiments reviewed above. By simulating each type of experiment and determining the effect of changing κ on the goodness of fit and the accuracy of parameter recovery, the proper weighting factors for each data set could be determined.

Conclusions

Fluorescence anisotropy decay has the potential to provide detailed information about rotational diffusion, internal dynamics, and other fundamental processes of biological macromolecules. Historically, application of this technique to a wide variety of biochemical systems has been hindered by the technical difficulties of data collection and analysis. However, with the advent of powerful, stable solid-state lasers, efficient high-speed photodetectors, and inexpensive computer power, many of these technical obstacles have been overcome. It is now possible to rapidly collect and analyze large quantities of fluorescence anisotropy decay data, enabling the investigation of complex kinetic and dynamic processes. Kinetic models must now be formulated for each of the possible mechanisms of fluorescence intensity and anisotropy decay. Analytic methods must be developed and implemented to enable discrimination between these more complex models. The first of these new models and methods have been reviewed above. By continuing this work, it should become possible to characterize macromolecular dynamics more thoroughly by fluorescence anisotropy decay.

Acknowledgments

The authors gratefully acknowledge the contributions of co-workers Carl N. Bialik, Barnabas Wolf, and J. B. Alexander Ross to the investigations that are reviewed in this chapter. Special recognition is due Mr. Bialik for the principal contribution to the investigations of the ground-state heterogeneity models, which served as the basis of his submission to the Westinghouse Science Talent Search. This work was supported in part by NIH grants GM39750, HL29019, and CA63317, and NSF grant DBI-9724330.

[14] Analysis of Nonequilibrium Facets of Pulsatile Sex-Steroid Secretion in Presence of Plasma-Binding Proteins

By JOHANNES D. VELDHUIS and MICHAEL L. JOHNSON

Introduction

Endocrine glands include protein and steroid hormone-producing tissues that typically signal their remote target tissues by way of an intermittent rather than continuous mode of hormone delivery into the bloodstream.[1–7] Such a nonlinear burstlike mechanism of hormone release may convey important biological information to responsive cells.[8–25] Indeed, many endocrine axes never attain classical equilibrium under physiologic conditions,

[1] M. L. Hartman, A. C. Faria, M. L. Vance, M. L. Johnson, M. O. Thorner, and J. D. Veldhuis, *Am. J. Physiol.* **260**(1), E101 (1991).
[2] W. G. Rossmanith, G. A. Laughlin, J. F. Mortola, M. L. Johnson, J. D. Veldhuis, and S. S. C. Yen, *J. Clin. Endocrinol. Metab.* **70,** 990 (1990).
[3] J. D. Veldhuis, M. L. Carlson, and M. L. Johnson, *Proc. Natl. Acad. Sci. U.S.A.* **84,** 7686 (1987).
[4] J. D. Veldhuis, W. S. Evans, L. A. Kolp, A. D. Rogol, and M. L. Johnson, *J. Clin. Endocrinol. Metab.* **66,** 414 (1988).
[5] J. D. Veldhuis, A. Iranmanesh, G. Lizarralde, and M. L. Johnson, *Am. J. Physiol.* **257,** E6 (1989).
[6] J. D. Veldhuis, J. C. King, R. J. Urban, A. D. Rogol, W. S. Evans, L. A. Kolp, and M. L. Johnson, *J. Clin. Endocrinol. Metab.* **65,** 929 (1987).
[7] S. J. Winters and P. E. Troen, *J. Clin. Invest.* **78,** 870 (1986).
[8] D. J. Chase, J. A. Karle, and R. E. Fogg, *J. Reprod. Fertil.* **95,** 657 (1992).
[9] R. G. Clark and I. C. A. F. Robinson, *Nature* **314,** 281 (1985).
[10] I. J. Clarke and J. T. Cummins, *Bailliere's Clin. Endocrinol. Metab.* **1,** 1 (1987).
[11] W. S. Evans, B. J. Boykin, D. L. Kaiser, J. L. C. Borges, and M. O. Thorner, *Endocrinology* **112,** 535 (1983).
[12] M. Filicori, J. P. Butler, and W. F. Crowley, Jr., *J. Clin. Invest.* **73,** 1638 (1984).
[13] R. L. Goodman, E. L. Bittman, D. L. Foster, and F. J. Karsch, *Biol. Reprod.* **27,** 580 (1982).
[14] D. J. Handelsman and L. M. Boyland, *J. Clin. Endocrinol. Metab.* **67,** 175 (1988).
[15] J. Isgaard, L. Carlsson, O. G. P. Isaksson, and J. O. Jansson, *Endocrinology* **123,** 2605 (1988).
[16] E. L. Marut, R. F. Williams, B. D. Cowan, A. Lynch, S. P. Lerner, and G. D. Hodgen, *Endocrinology* **109,** 2270 (1981).
[17] N. Miki, M. Ono, and K. Shizume, *J. Endocrinol.* **117,** 245 (1988).
[18] M. Procknor, S. Barhir, R. E. Owens, D. E. Little, and P. G. Harms, *J. Animal Sci.* **62,** 191 (1986).

METHODS IN ENZYMOLOGY, VOL. 321
0076-6879/00 $30.00

but manifest prominent moment-to-moment ultradian and longer term circadian variations in hormone concentrations and secretion rates.[26] Although physiologic secretory input into the circulation can be viewed as highly nonlinear with bursts of hormone molecules entering the bloodstream at varying intervals over time,[3] most available metabolic studies have actually required and assumed steady-state conditions.

The nonlinear and nonequilibrium features of endocrine axes are further confounded by the presence in plasma of one or more variable affinity and capacity binding proteins. Such transport proteins are capable of associating with the secreted hormone or ligand with high or low affinity and at high or low capacity and modifying its metabolic clearance (e.g., Ref. 27). For example, in the somatotropic axis, a high-affinity growth hormone (GH)-binding protein exists physiologically in human plasma[28–31] and avidly attaches to circulating GH molecules producing a marked change in the apparent half-life of free and total GH in the circulation, while simultaneously damping the otherwise abrupt GH secretory pulse profile.[32] Recent analyses of the kinetic impact of this binding protein on the pulsatile GH axis have predicted a half-life of free GH in the absence of binding protein of only 2–7 min, whereas the half-lives of bound and total GH in the presence of the plasma high-affinity GH-binding protein are approximately threefold higher.[32,33]

[19] R. J. Santen and C. W. Bardin, *J. Clin. Invest.* **52,** 2617 (1973).

[20] M. Sato, J. Takahara, Y. Fujioka, M. Niimi, and S. Irino, *Endocrinology* **123,** 1928 (1988).

[21] M. B. Southworth, A. M. Matsumoto, K. M. Gross, M. R. Soules, and W. J. Bremner, *J. Clin. Endocrinol. Metab.* **72,** 1286 (1991).

[22] S. S. Sundseth, J. A. Alberta, and D. J. Waxman, *J. Biol. Chem.* **267**(6), 3907 (1992).

[23] M. M. Valenca, C. A. Johnston, M. Ching, and A. Negro-Vilar, *Endocrinology* **121,** 2256 (1987).

[24] J. D. Veldhuis, *The Endocrinologist* **5,** 454 (1995).

[25] T. O. F. Wagner, G. Brabant, F. Warsch, R. D. Hesch, and M. von zur Uhler, *J. Clin. Endocrinol. Metab.* **43,** 447 (1985).

[26] J. D. Veldhuis, A. Iranmanesh, M. L. Johnson, and G. Lizarralde, *J. Clin. Endocrinol. Metab.* **71,** 1616 (1990).

[27] A. Vermeulen, L. Verdonck, M. Van der Straeten, and N. Orie, *J. Clin. Endocrinol. Metab.* **29,** 1470 (1969).

[28] G. Baumann, M. W. Stolar, K. Amburn, C. P. Barsano, and B. C. De Vries, *J. Clin. Endocrinol. Metab.* **62,** 134 (1986).

[29] B. C. Cunningham, M. Ultsch, A. M. DeVos, M. G. Mulkerrin, K. R. Clauser, and J. A. Wells, *Science* **254,** 821 (1991).

[30] A. C. Herington, S. Yemer, and J. Stevenson, *J. Clin. Invest.* **77,** 1817 (1986).

[31] D. W. Leung, S. A. Spencer, G. Cachianes, R. G. Hammonds, C. Collins, W. J. Henzel, R. Barnard, M. J. Waters, and W. I. Wood, *Nature* **330,** 537 (1987).

[32] J. D. Veldhuis, M. L. Johnson, L. M. Faunt, M. Mercado, and G. Baumann, *J. Clin. Invest.* **91,** 629 (1993).

[33] A. C. S. Faria, J. D. Veldhuis, M. O. Thorner, and M. L. Vance, *J. Clin. Endocrinol. Metab.* **68,** 535 (1989).

Although a single high-affinity ligand (GH) and its corresponding plasma transport protein may describe the somatotropic axis, the gonadotropic axes in men and women are considerably more complex. Indeed, in addition to the recognized pulsatile mode of pituitary leutenizing hormone (LH) and follicle-stimulating hormone (FSH) secretion, recent direct catheterization of the human spermatic vein indicates that both testosterone and estradiol are secreted in distinct bursts occurring at a mean interval of approximately 60 min.[7] Moreover, sex steroids secreted into the bloodstream are exposed to substantial concentrations of a low-affinity high-capacity binding protein, albumin, and significant amounts of a high-affinity low-capacity glycoprotein transporter, sex-hormone binding globulin (SHBG).[27,34] Even though the steroidal secretory output of the gonad *in vivo* is pulsatile, and at least two sex-steroid hormones are secreted concurrently, and two or more distinct binding proteins exist in large quantities in plasma, available studies of the male gonadal axis to date have employed only equilibrium approximations to investigate the physiology of steroid hormone secretion, distribution, binding, and metabolic elimination. Here, we have formulated a family of coupled differential equations to investigate for the first time the impact of two or more variable-affinity and unequal-capacity binding proteins on the time structure of plasma free, bound, and total steroid hormone concentrations given three steroid hormones secreted into the circulation concurrently or at a particular lag in a physiologically pulsatile manner. This more realistic construct of burstlike hormone release in the presence of plasma transporters offers novel insights into and predictions concerning the nonequilibrium operation of the (male) gonadal–steroid axis.

Methods

Mathematical Formulation and General Assumptions

As an initial approach to describing the nonequilibrium time characteristics of free, bound, and total sex-steroid hormone concentrations in plasma, we assumed the following concerning *in vivo* physiologic steroid hormone secretion, binding, and removal:

1. Two principal sex-steroid-binding proteins exist in human plasma and contribute predominantly to steroid hormone binding in the circulation, such that contributions of any other acceptor proteins are essentially negligible by comparison.
2. The turnover of the unoccupied (steroid-) binding proteins is slow compared with that of free hormone, so that the concentration of

[34] W. M. Pardridge and E. M. Landaw, *Am. J. Physiol.* **249,** E534 (1985).

any given binding protein is essentially constant over the observation interval.

3. The intact steroid hormone-binding protein complex is removed from the bloodstream at a rate that is slow compared to the clearance of free hormone molecules.
4. The mass of hormone and binding proteins within the body is conserved over short intervals, except for irreversible removal of hormone by metabolic clearance.
5. Up to 2 mol of testosterone (T) may bind to 1 mol of SHBG, whereas albumin is not saturable at physiologic hormone concentrations in plasma.
6. The elimination function describing the removal kinetics for a steroid hormone's irreversible loss from plasma can be approximated by a single exponential, and this process acts only on free (unbound) hormone molecules within plasma, except where noted otherwise.
7. Testosterone, estradiol (E_2), and dihydrotestosterone (DHT) constitute the principal circulating gonadal steroids capable of competing for binding to albumin and SHBG in normal men.
8. The three primary steroids mentioned in item 7 are secreted in distinct and delimited secretory bursts of finite duration, specifiable amplitude (maximal rate of release achieved within the event), and quantifiable mass (integral of the secretory burst); secretory bursts can be approximated algebraically by a Gaussian or an appropriately skewed waveform (e.g., Poisson or gamma function),[35] and can be superimposed on zero or a finite positive real amount of basal (time-invariant) hormone release.
9. The possible fate of a molecule of steroid hormone secreted into the bloodstream by the gonad includes immediate entry into a free (protein unbound) compartment, reversible association with one, two, or more distinct binding proteins, dissociation without modification from the binding proteins, reassociation with the same or another acceptor protein, and/or irreversible elimination of free (but not bound unless so specified) hormone by *in vivo* metabolic mechanisms.

In addition, to the above assumptions, nominal values for the kinetic constants of the sex-steroid hormone-binding protein system must be considered. For example, see Table I, as based on available literature estimates in normal men.[34,36,37] Unless specified otherwise, we used these nominal

[35] J. D. Veldhuis, A. B. Lassiter, and M. L. Johnson, *Am. J. Physiol.* **259,** E351 (1990).
[36] R. Horton, J. Shinsako, and P. H. Forsham, *Acta Endocrinol.* (*Copenh.*) **48,** 446 (1965).
[37] F. Schaefer, J. D. Veldhuis, J. Jones, K. Scharer, and The Cooperative Study Group on Pubertal Development in Chronic Renal Failure, *J. Clin. Endocrinol. Metab.* **78,** 1298 (1994).

TABLE I

NOMINAL CHARACTERISTICS ASSUMED IN SIMULATIONS OF SEX-STEROID HORMONE SECRETION AND BINDING TO PLASMA PROTEINS IN NORMAL MALE

Parameter	Sex-steroid hormone		
	Estradiol	Testosterone	5-α-Hydrotestosterone
Daily secretion rate [mol]	40 μg [147 nmol]	7 mg [24 μmol]	0.7 mg [2.4 μmol]
Albumin			
Dissociation[a]	108 min^{-1}	210 min^{-1}	204 min^{-1}
Association[a]	$8.31 \times 10^6\ M^{-1}\ min^{-1}$	$8.40 \times 10^6\ M^{-1}\ min^{-1}$	$8.16 \times 10^6\ M^{-1}\ min^{-1}$
Sex-hormone binding globulin (SHBG)			
Dissociation[a]	4.98 min^{-1}	3.18 min^{-1}	0.96 min^{-1}
Association[a]	$2.49 \times 10^9\ M^{-1}\ min^{-1}$	$3.18 \times 10^9\ M^{-1}\ min^{-1}$	$1.9 \times 10^9\ M^{-1}\ min^{-1}$
Distribution volume of steroid[b] [protein] (L)	24 [4.9]	24 [4.9]	24 [4.9]
Half-life of free hormone[c] (min)	5	5	5

[a] Undirectional "on" (association) or "off" (dissociation) rate constants for the indicated ligand (sex-steroid hormone) at 37°. The values of k_{on} and k_{off} are related to the equilibrium dissociation constant K_d by k_{off}/k_{on}. The concentrations (and total binding capacities) of albumin and SHBG were assumed to be 680 μM and 30 nM, respectively. Data are compiled from Refs. 27, 34, 36, 37, 38, 39, 42 and 47.

[b] Apparent distribution volume of *ligands*. In constrast, *binding proteins* were assumed to be distributed in the blood volume alone (approximately 4.9 L in a 70-kg man). Irreversible elimination *of free steroid* was assumed to occur from anywhere within the (24 L) free-hormone distribution space.

[c] Various other assumed half-lives of free steroid are evaluated where indicated in the text.

values. We provisionally assumed that each theoretical steroid-hormone secretory burst contains 1/24 of the total daily pulsatile mass of steroid secreted, and occurs at a mean intersecretory burst interval of 60 min with zero or greater basal secretion. Practically, the total daily mass of steroid secreted was either considered to be exclusively pulsatile, or partitioned into 10%, 50%, or 90% pulsatile with the remainder assigned as basal. Each theoretical steroid-hormone secretory burst, whether or not superimposed on basal release, was configured as a Gaussian with a half-duration (duration at half-maximal amplitude) of 30 min centered about zero time and typically recurring at 60-min intervals. The half-life of unbound steroid hormone in plasma was assumed nominally to be 5 min, or variously to be 0.25–20 min, as indicated specifically below in individual simulations.

Initial differential equations were formulated earlier to describe the binding, dissociation, and removal of one secreted hormone from human plasma containing one binding protein.[32] Specifically, the simplest paradigm assumes a constant basal secretion rate with superimposed hormone release in physiologically delimited secretory bursts or pulses, which enter the bloodstream only by way of the free compartment. This model permits the combining of variable amounts of basal and pulsatile secretion of a single type of steroid molecule into the bloodstream containing a single binding protein, or allows exclusively burstlike hormone secretion or purely constant (basal) release as extreme models.

If the steroid hormone-binding–protein complex is removed at a significant rate (i.e., at a rate that is not inconsequential compared to that of free hormone), then an additional elimination rate constant must be interposed to operate on the concentration of the ligand-binding–protein complex at any given instant. Although some steroid-hormone-binding–protein complex is probably removed *in vivo* by certain organs,[38,39] we note that this is not a dominant pathway of hormone removal and would effectively create a higher order multicompartment model that is already made complex by the nonlinearities of pulsed hormone release into the primary sampling compartment.

Specific Mathematical Formulation for Multiple Ligands and Binding Proteins

A mathematical statement of the expected interactions among *multiple* ligands competing for two or more binding proteins is given by the following coupled differential equations:

$$\frac{d[L_i]}{dt} = \frac{\text{basal}_i}{\text{vol}} - K_{e,i}[L_i] - \sum_{j=1}^{n_{\text{pro}}} \frac{d[L_iP_j]}{dt} \tag{1}$$

$$\frac{d[L_iP_j]}{dt} = K_{a(i,j)}[L_i][P_j] - K_{d(i,j)}[L_iP_j] \tag{2}$$

where $[L_i]$ denotes the ith ligand concentration; n_{pro} the number of binding proteins; $[L_iP_j]$ the concentrations of the ith ligand bound to the jth binding protein; $K_{a(i,j)}$ the association rate constant for the ith ligand and jth binding protein; $K_{d(i,j)}$ the dissociation rate constant for the ith ligand bound to the jth protein; basal$_i$ the basal secretion rate for the ith ligand; and vol the distribution volume for the i ligands. Any given elimination rate constant $K_{e,i}$ equals ln $2/t_{1/2,i}$ for a single elimination pathway for the free ith hormone.

We have modified the above equations so that irreversible elimination of hormone can occur not only from the free compartment (as written),

[38] C. M. Mendel, *Endocrinol. Rev.* **10,** 232 (1989).
[39] W. M. Pardridge, *Endocrinol. Rev.* **2,** 103 (1981).

but also (or solely) from hormone bound to one or more specific proteins, e.g., sex-hormone binding globulin, or albumin (see below). We can also assign different distribution volumes for the binding protein(s) and the ligand(s), e.g., 4.9 L (liters) for binding proteins but, 20–24 L for free steroid in normal men.[34]

We solve the above set of coupled differential equations using a Runge–Kutta method, as suggested earlier,[32,40] when we wish to evaluate the concentration at any given instant of each free ligand and the amount of each ligand bound to each (and all) of the individual binding proteins. In implementing the simulations initially, we assumed that a Gaussian pulse of free hormone enters the circulation as a single burst that is centered at time zero. Later calculations embrace a series of recurrent 60- or 90-min pulses, and two or more hormones pulsed simultaneously or at different lags. For each hormone, the pulse amplitude (maximal rate of secretion attained within the theoretical secretory burst), frequency, and half-width (duration of the secretory peak at half-maximal amplitude) can be defined independently. Finally, we allow the amount of concurrent basal or continuous (time-invariant) hormone secretion to be varied independently for each of the multiple hormones, e.g., 0%, 10%, 50%, 90%, or 100% of total daily individual hormone secretion.

In this multiple ligand problem, we also assume these parameters: a specifiable distribution volume for each ligand and protein; a definable rate of basal secretion for each hormone; and no interactions between the binding proteins per se, or among free ligands in plasma. In this construction, we generate z differential equations, where z is the product of the number of steroid ligands and one plus the number of binding proteins. Specifically, for each ligand, there is one equation for defining its secretion into and elimination from the free compartment, and another equation denoting the binding of that ligand to and its release from (association with and dissociation from) each of the relevant binding proteins.

In a simple paradigm, the above system would involve three principal sex-steroid hormones, namely, E_2, T, and DHT. Nominal total daily secretion rates of these three steroid hormones in healthy young men are assumed as summarized in Table I. We also give the approximate unidirectional association and dissociation rate constants for binding to and release from albumin and SHBG, as well as the molar concentrations of these two binding proteins. The initially assumed distribution volume for each free ligand (24 L) is also based on earlier literature estimates. Since the undegraded binding proteins are presumably distributed predominantly in the plasma volume (4.9 L in a 70-kg man), we assume this value here for bound

[40] R. Keret, A. Pertzelan, A. Zeharia, Z. Zadik, and Z. Laron, *Isr. J. Med. Sci.* **24,** 75 (1988).

ligand and therefore allow partially overlapping distribution volumes for ligands and proteins.

Results

Using nominal literature-based kinetics and binding constants for the three sex-steroid hormones and the two principal transport proteins (SHBG and albumin; Table I), we first examined the effects of variably admixed basal and pulsatile steroid secretion as defined by three partitioning paradigms: (1) 90% of the total daily steroid production rate is partitioned as basal steroid secretion and 10% as pulsatile; (2) 50% basal secretion and 50% pulsatile; and (3) 90% pulsatile and 10% basal. Several interesting features (below) emerge from inspection and analysis of the mathematically expected time profiles of free, bound, and total sex-steroid hormone concentrations under these conditions.

For any given total daily steroid secretion rate, we find that the percentage of steroid secreted in the basal versus pulsatile mode strongly influences the mean (pseudo-steady-state) plasma concentrations of bound, free, or total hormone. In general, higher proportions of basal or continuous (versus pulsatile) secretion bring about significantly higher mean interpulse hormone concentrations and higher maximal free, bound, and total plasma hormone concentrations. Intuitively, this discrepancy occurs largely because during a pulse the free hormone level becomes higher and therefore the steroid is more rapidly cleared. In contrast, during constant (basal) hormone delivery, retention of steroid by association with binding proteins within the plasma volume delays metabolic removal. This was validated empirically in recent experiments where we measured (total) serum T concentrations over 24 hr in blood sampled every 10 min in six healthy young men, who received 8 mg/day T i.v. either via continuous infusion over 24 hr or (on a separate day) via 16 bolus pulses of 0.5 mg T each delivered over 1 min every 90 min. Endogenous T secretion was abolished by concurrent administration of ketoconazole as a potent steroidogenic enzyme inhibitor.[41] The mean 24-hr serum total T concentrations during continuous infusions were approximately twofold higher compared to values following bolus injections.

Mathematically, Eqs. (1) and (2) predicted (compared to purely pulsatile or purely continuous delivery) intermediate pseudo-steady-state (free, bound, and total) serum testosterone concentrations in a paradigm of 10% basal (Fig. 1A) or 90% basal (Fig. 1B) secretion, given an identical total daily secretion rate of T of 7 mg or 24.3 μmol. These figures also show predicted profiles of serum free, SHBG-, and albumin-bound E_2 and DHT for comparison (assuming a daily total E_2 secretion rate of 40 μg or 0.15

μmol, and a daily total DHT secretion rate of 0.7 mg or 2.4 μmol). For graphical purposes, we present the sex-steroid concentration profiles here in response to a single steroid infusion pulse.

Figure 1 demonstrates that the individual time courses and absolute excursions of free, bound, or total serum sex-steroid hormone concentrations are strongly dependent on the secretion kinetics, as assessed by our computer-assisted model. Indeed, distinctive dynamic profiles of free and bound hormone concentrations develop, and very particular time courses emerge of (1) percentage of a specific hormone that is bound and (2) percentage of a specific binding protein that is occupied, given different pulsatile versus basal partitioning assumptions. This is illustrated for 50% basal and 50% pulsatile sex-hormone secretion, assuming that bursts of E_2, T, and DHT secretion occur simultaneously. In the case of T and E_2, simultaneous secretory bursts have been demonstrated by direct sampling of the venous effluent of the human testis.[7] This allows competition to occur among the three steroids (with their different affinities) for common protein-binding sites. Assuming nominal half-lives for free steroids of 5 min each for T, E_2, and DHT, we derive the predicted nonequilibrium time profiles of free, total, and bound steroid hormone concentrations before, during, and after steroid secretory bursts, the expected percentage occupancy of each of the two binding proteins by each particular hormone, and the anticipated percentage of each hormone that is free at any given instant (Figs. 1–3).

Figure 2 shows the time profiles of (free) hormone secretion and each of the three steroids' individual percentage association with SHBG or albumin (and total protein-bound steroid). Note that momentarily high free sex-steroid concentrations during a pulse temporarily reduce the percentage of steroid bound to SHBG but not to albumin, since the latter's association rate constant is much higher (Tables II–V).

Figure 3 shows the percentage occupancy of SHBG and albumin by T, E_2, and DHT, as influenced by the relative partitioning of total steroid secretion into basal versus pulsatile; i.e., 10%, 50%, and 90% pulsatile. Note progressively greater percentage occupancy of binding proteins at higher proportions of basal (versus) pulsatile steroid delivery. This observation serves to explicate our earlier finding that constant T infusions produce higher mean serum T concentrations than pulsatile steroid delivery.[41]

The foregoing equations also allow us to vary the apparent half-life of

[41] A. D. Zwart, A. Iranmanesh, and J. D. Veldhuis, "An acute Leydig cell chemical castration model to investigate selective feedback actions of testosterone on LH release in normal men," paper presented at the 77th Annual Endocrine Society Meeting, Washington, D.C., June 14–17. #OR29-4. 1995.

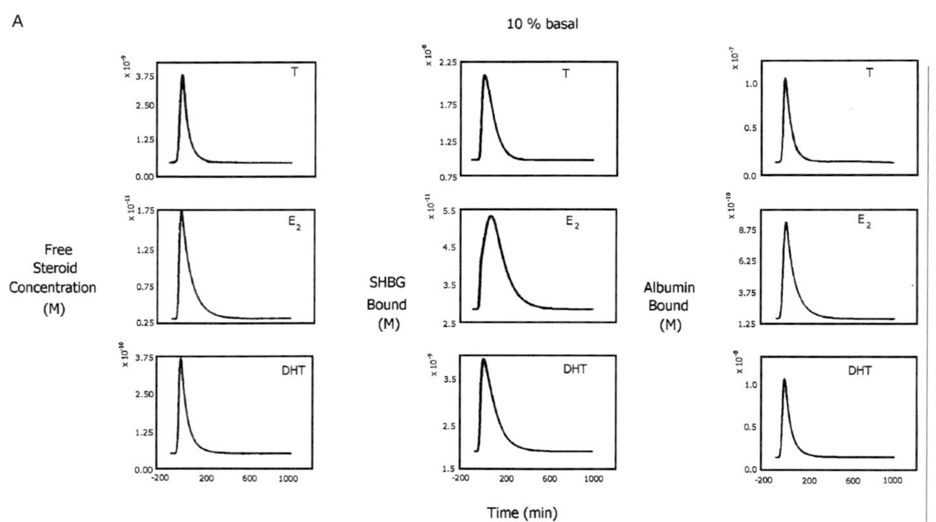

FIG. 1. Time profiles of serum concentrations of free, SHBG-bound, and albumin-bound estradiol, testosterone, and 5-α-hydrotestosterone (E_2, T, and DHT, respectively) predicted following a single concordant burst of all three steroids into a compartment containing SHBG and albumin. Kinetic parameters are summarized in Table I. We assumed a nominal half-life of each free steroid of 5 min in the absence of any protein binding, a theoretical Gaussian secretory burst half-duration of 30 min centered at time zero, and a mass of steroid of 0.5 μmol (T), 3.1 nmol (E_2), and 50 nmol (DHT) per burst superimposed on (A) 10% or (B) 90% basal secretion (see text). The steroids were assumed to distribute within 24 L and the binding proteins in 4.9 L (see Methods section).

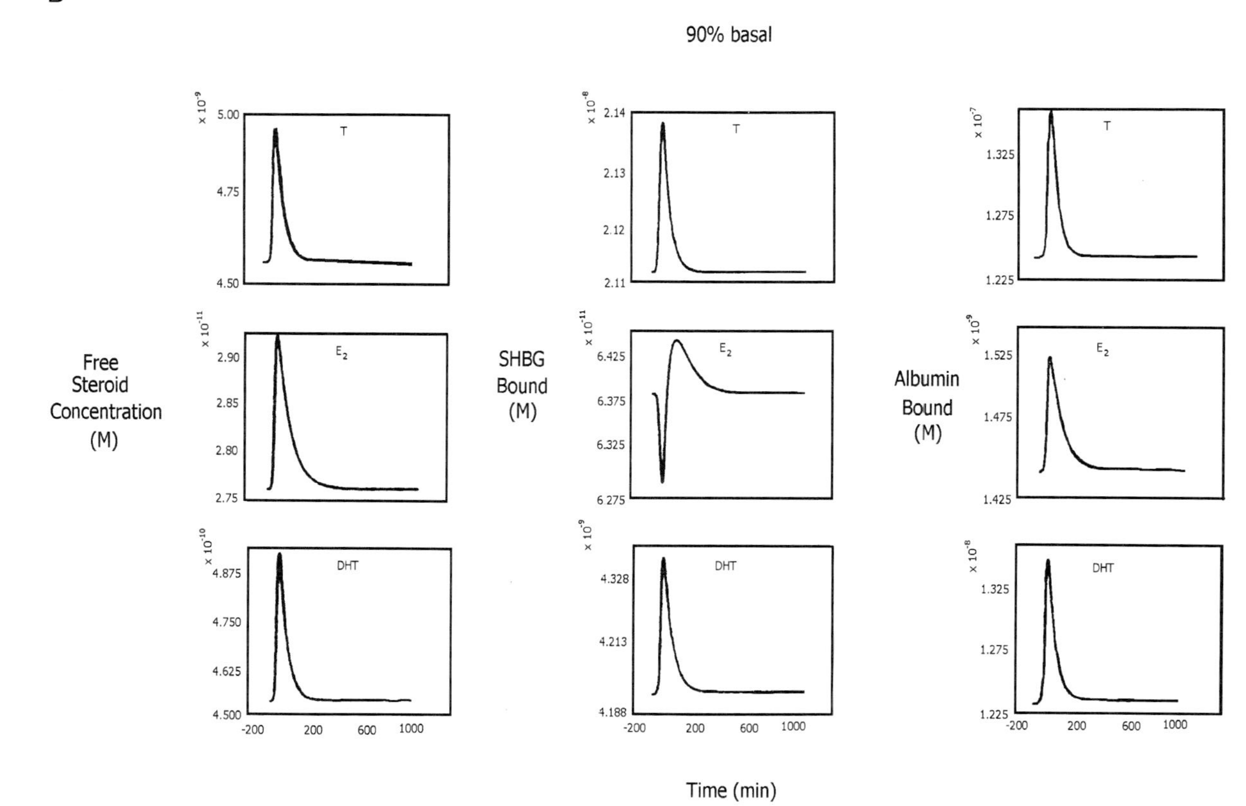
B
90% basal
Free Steroid Concentration (M)
SHBG Bound (M)
Albumin Bound (M)
T
E2
DHT
x 10-9
5.00
4.75
4.50
x 10-11
2.90
2.85
2.80
2.75
x 10-10
4.875
4.750
4.625
4.500
x 10-8
2.14
2.13
2.12
2.11
x 10-11
6.425
6.375
6.325
6.275
x 10-9
4.328
4.213
4.188
x 10-7
1.325
1.275
1.225
x 10-9
1.525
1.475
1.425
x 10-8
1.325
1.275
1.225
-200
200
600
1000
Time (min)

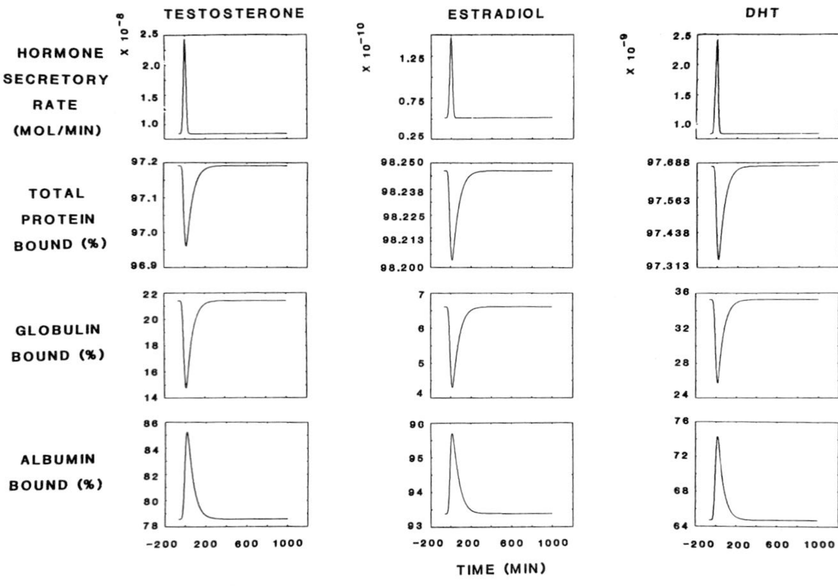

FIG. 2. Percentages of T, E_2, and DHT bound to SHBG or albumin or total protein during and following a single secretory burst of all 3 steroids. The presentation otherwise is as summarized in the legend of Figure 1, except that basal and pulsatile testosterone secretion was admixed 50%.

free sex-steroid hormone to test different physiologic premises. For example, assuming that only completely free steroid hormone (protein unbound) is irreversibly removed from the sampling compartment, and that steroids and proteins either share the same distribution volume or partially overlapping distribution volumes of 24 and 4.9 L, respectively, then the half-life of free steroid hormone must approach 0.25–1.0 min in order to create mean (pseudo-steady-state) near-physiologic serum total T, E_2, and DHT concentrations, given nominal daily steroid production rates, distributed as 50% basal and 50% in 60-min multiple secretory pulses (Table I). An alternative assumption, namely, that both free and albumin-bound steroid hormones are removed irreversibly by metabolic clearance, yields predicted serum total T, E_2, and DHT concentrations within the physiologic range, when the apparent half-lives of the free and the albumin-bound steroid available for removal are both constrained to approximately 5 min. Our simulations show that either distinct assumption (e.g., free steroid half-life of 0.25–1.0 min, or free steroid and albumin-bound steroid half-lives of 5 min) combined with the kinetic features given in Table I will yield mean serum total steroid hormone concentrations within the physiologic range. Independent physiologic experiments are required to distinguish between these two theoretical possibilities.

Figure 4 illustrates the acquisition of pseudo-steady-state by free and bound sex-steroid hormones during a repetitive series of 60-min pulses of T, E_2, and DHT, assuming that all three types of steroidal secretory bursts occur concordantly with equal free hormone half-lives of 5 min, equal (30-min) individual secretory pulse half-durations, and partitioning of total daily steroid secretion into equal pulsatile (50%) and basal (50%) components. The time-specified trajectory to pseudo-steady-state serum concentrations of *total* steroid hormone (approximates the bound data shown) is extended over about 240 min for T, 480 min for E_2, and 300 min for DHT. These values represent approximately 4 or 5 half-lives for pseudo-equilibration of serum total or bound T, E_2, and DHT concentrations. Near-equilibrium of serum total steroid-hormone concentrations occurs at different times for the three steroids, because of their different affinities largely for the high-affinity binding protein, SHBG. However, at no particular instant is true steady state achieved.

As might be predicted intuitively, following secretion or injection of one or more pulses of a steroid hormone, the apparent half-lives of free, bound, and total hormone are all longer than that of theoretical free hormone in the absence of any binding protein. This is because the steroid's association with its high-affinity binding protein imposes an additional delay on entry of steroid into the theoretically free compartment, from which it can be removed. The decay curves for free, globulin-bound, albumin-

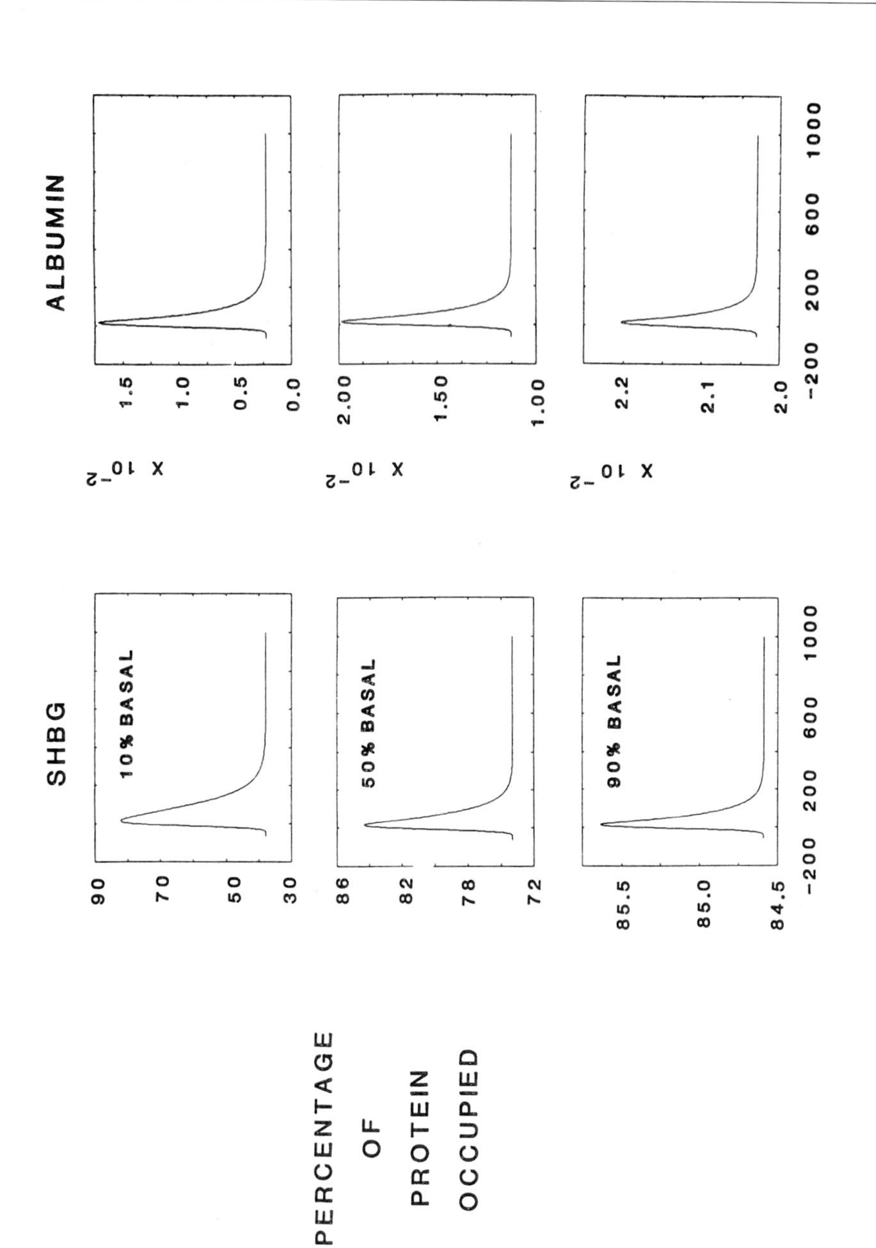
SHBG
ALBUMIN
10% BASAL
50% BASAL
90% BASAL
PERCENTAGE OF PROTEIN OCCUPIED
X 10^-2
X 10^-2
X 10^-2
90
70
50
30
86
82
78
72
85.5
85.0
84.5
1.5
1.0
0.5
0.0
2.00
1.50
1.00
2.2
2.1
2.0
-200
200
600
1000
TIME (MIN)

associated, and total ligand are illustrated in Fig. 4 for each of the three sex-steroid hormones, assuming a model in which 10 repetitive pulses of hormone are delivered concordantly and hourly for all three steroids for a total of 540 min to approach pseudo-steady-state, and then pulsatile but not basal secretion is stopped abruptly. The subpanels note the values of the calculated half-lives of free, SHBG-bound, and albumin-bound T, E_2, and DHT as estimated by monoexponential curve fitting. For example, given a *theoretical free* T, E_2, and DHT half-life of 5 min assumed in the absence of any binding proteins, then the apparent half-lives of *total* T, E_2, and DHT concentrations during decay from pseudo-steady-state in the presence of albumin and SHBG are 34, 61, and 37 min, respectively. The apparent half-lives of *free* T, E_2, and DHT when binding proteins are present are approximately 34, 60, and 36 min, respectively. Half-life values of SHBG-bound steroids are 39, 60, and 46 min for T, E_2, and DHT, and of albumin-bound steroids 34, 60, and 36 min for T, E_2, and DHT, respectively. We note that the model-predicted half-lives of free and albumin-bound steroid hormones are not distinguishable under these conditions, which indicates that dissociation of steroid-hormone molecules from albumin (unlike SHBG) is not rate limiting in the irreversible removal of free hormone.

Table II summarizes the impact of varying assumed free hormone half-lives in the presence of both SHBG and albumin. Separate half-lives are estimated for free, SHBG-bound, or albumin-bound, as well as total T, E_2, and DHT concentrations. We tested four free steroid-hormone half-lives, namely, 0.25, 1.0, 5.0, and 20 min, in each case assuming that pulsatile and basal steroid secretion rates were equal contributors to the total daily steroid production rate. Note that even the apparent half-life of *unbound* steroid is always increased when albumin and SHBG are present, compared to the expected half-life of free hormone in the complete absence of any transport proteins. In addition, the half-lives of albumin-bound and SHBG-bound T, E_2, and DHT are significantly further prolonged compared to that of unbound hormone. The half-lives of total T, E_2, and are intermediate between those of unbound and SHBG-bound hormone. The unequal affinities that T, E_2, and DHT have for albumin and especially SHBG produce differences in the predicted half-lives of these individual steroid hormones when bound to either albumin or SHBG.

FIG. 3. Percentage occupancies of SHBG and albumin during and following a single pulse of T, E_2, and DHT secretion. Data are presented otherwise as described in the legend of Fig. 1, except that the fractions of basal (versus total) hormone secretion were 0.1, 0.5, and 0.9 (10%, 50%, and 90%).

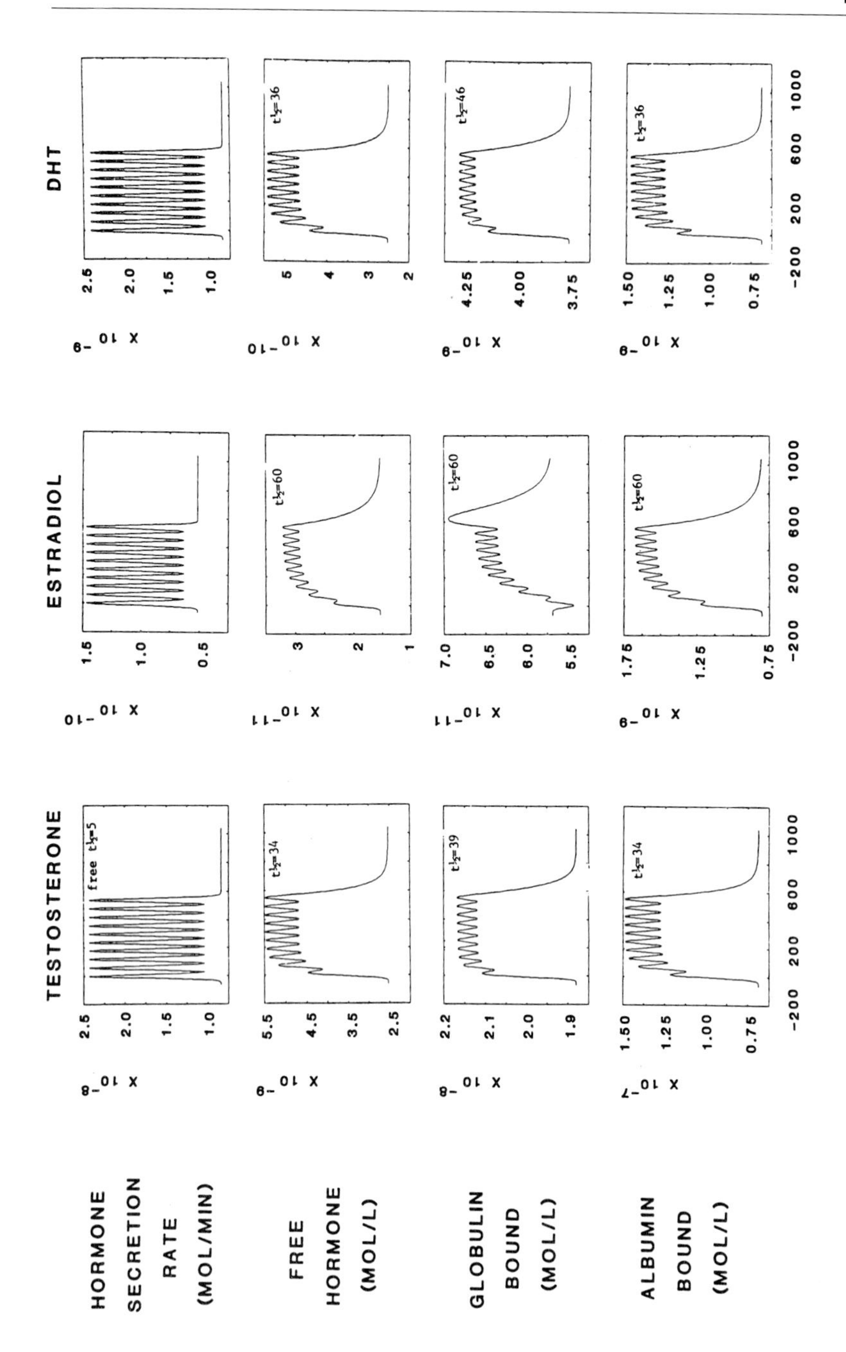
TESTOSTERONE
ESTRADIOL
DHT
HORMONE SECRETION RATE (MOL/MIN)
FREE HORMONE (MOL/L)
GLOBULIN BOUND (MOL/L)
ALBUMIN BOUND (MOL/L)

TIME (MIN)

To evaluate the specific effect of each binding protein on free, bound, and total steroid hormone half-lives, we carried out simulations with nominal transport protein concentrations (both albumin and SHBG present in approximately physiologic concentrations), as well as studies in which albumin was absent (only SHBG present), SHBG was absent (only albumin present), or both proteins were lacking. In the last circumstance the apparent and the theoretical free-steroid hormone half-lives are equal. However, when a transport protein is present, the apparent free hormone half-life is prolonged because the falling serum concentrations of free hormone are in part replenishable by dissociation of additional (free) ligand from the binding protein(s). Results from these simulations are summarized in Table III. These data show the relative importance of albumin and SHBG in controlling the apparent half-lives of unbound, bound, and total T, E_2, and DHT. Our calculations also demonstrate ligand-specific effects for any given protein.

The impact of the *degree of admixture* of basal and pulsatile hormone secretion on apparent steroid hormone half-lives was tested as well. We estimated the monocomponent half-life (single exponential) for unbound, albumin bound, SHBG bound, and total T, E_2, and DHT assuming (1) 90% basal and 10% pulsatile hormone secretion and (2) 10% pulsatile and 90% basal hormone release. The apparent half-lives of unbound, bound, and total hormone estimated from decay curves following pseudo-steady-state, here produced by a train of 10 successive pulses occurring at 60-min intervals, showed decreasing half-lives with an increasing proportion of basal secretion (Table IV). This finding reflects in large part the different baseline serum steroid concentrations existing before and after the secretory pulse, and illustrates predicted differences in steroid kinetics (despite identi-

FIG. 4. Acquisition of pseudo-steady-state by serum concentrations of free and protein-bound sex-steroid hormones during repeated pulsatile secretion of T, E_2, and DHT at 60-min intervals. Kinetic constants used and mass of steroid secreted are those given in Table I and Fig. 1, including assumed free steroid half-lives of 5 min and daily secretion rates partitioned into 50% basal and 50% pulsatile release. The plasma steroid concentrations generated by the basal rates of secretion were assumed to be at steady state at time zero, when the first pules of T, E_2, and DHT were initiated. The indicated half-lives apply to the decay of free and bound T, E_2, and DHT to baseline following abrupt cessation of pulsatile secretion after achievement of pseudo-steady-state. Note that although the theoretical half-life of each free steroid in the absence of any binding proteins was constrained to 5 min, the apparent half-lives of free and bound T, E_2, and DHT in the *presence* of nominal concentrations of SHBG and albumin are all significantly prolonged as estimated by monoexponential curve fitting of the observed decay plots. As pointed out in the text, the presence of binding protein(s) can introduce a bi- or multiexponential decay structure, even though the irreversible removal of free hormone in the absence of protein is described by a single rate constant of elimination.

TABLE II
INFLUENCE OF BINDING PROTEINS ON APPARENT MONOEXPONENTIAL HALF-LIVES FOR MULTIPLE LIGANDS

Steroid fraction	Theoretical free ligand half-life (min)[b]			
	0.25	1.0	5.0	20
Unbound[c]				
T	2.8	8.8	34	131
E_2	3.6	13.2	60	234
DHT	4.7	12.3	36	132
Albumin bound				
T	2.7	8.8	34	131
E_2	3.6	13.2	60	234
DHT	4.7	12.3	36	132
SHBG bound				
T	2.8	8.3	39	163
E_2	3.5	14.2	?[d]	?[d]
DHT	4.7	13.5	46	171
Total				
T	2.6	8.6	34	131
E_2	3.6	13.4	61	126
DHT	4.7	12.9	37	132

[a] Model consisted of 50% basal and 50% pulsatile secretion. Decay was monitored and approximated monoexponentially after the last pulse was delivered and concentrations fell from pseudo-steady-state toward values defined by continued basal secretion. V_0 albumin = 4.9-L distribution volume; V_0 SHBG = 4.9-L distribution volume; V_0 free steroid = 24-L distribution volume.

[b] Half-life of free ligand in an environment devoid of all binding proteins.

[c] Apparent half-life of free hormone given nominal binding protein concentrations and properties for SHBG and albumin (Table I).

[d] (?) Biphasic or initially ascending concentration versus time course not approximated well by a single exponential decay.

cal free steroid half-lives in the absence of binding proteins) due to relative partitioning of total sex-steroid secretion into pulsatile versus basal components. Thus, we infer that the half-life of free hormone, the amount and affinity of binding proteins present, and the relative partitioning of steroid

TABLE III

INFLUENCE OF ABSENCE OF ALBUMIN AND/OR SHBG ON APPARENT MONOEXPONENTIAL HALF-LIVES OF SEX-STEROID HORMONES IN MALE[a]

	Status of binding proteins			
Steroid compartment	Both albumin and SHBG present	SHBG alone present	Albumin alone present	No proteins
Unbound				
T	34	6.3	33	5.0
E_2	60	(Biphasic)[b]	58	5.0
DHT	36	13.1	33	5.0
Albumin bound				
T	34	—	33	—
E_2	60	—	58	—
DHT	36	—	33	—
SHBG bound				
T	39	(Biphasic)[b]	—	—
E_2	61	(Ascending)[b]	—	—
DHT	46	18.2	—	—
Total				
T	34	(Biphasic)[b]	33	—
E_2	61	(Ascending)[b]	58	—
DHT	37	17.7	33	—

[a] Data are presented as described in Table II. Secretion 50% basal and 50% pulsatile before withdrawal of pulsatile component at pseudo-steady-state after 10 successive secretory pulses. V_0 SHBG anb albumin assumed as 4.9-L distribution volume; V_0 steroid taken as 24-L distribution volume. Free half-life 5.0 min in the absence of any binding proteins. (—) irrelevant categories.

[b] Biphasic or ascending steroid-hormone concentration profile over time, which cannot be approximated readily by simple monoexponential decay model.

hormone secretion into basal and pulsatile release all influence the apparent half-lives of unbound, albumin-bound, SHBG-bound, and total steroid hormone concentrations in plasma.

The impact of delayed DHT release was evaluated, as summarized in Table V. Delaying the timing of the DHT secretory burst by 2 or 4 standard deviations (see Methods section), compared to the simultaneous T and E_2 bursts, progressively prolonged DHT half-lives only. This presumably reflects decreased competition between T and DHT for common binding sites on SHBG, when the appearance of DHT in the blood is delayed significantly compared to that of T. Thus, the *relative timing* of steroid secretory pulses will also determine the *in vivo* half-time of disappearance of the steroid, and hence influence the duration of exposure of the target cell to the hormone agonist or antagonist.

TABLE IV
INFLUENCE OF PERCENTAGE BASAL (NONPULSATILE) STEROID HORMONE SECRETION ON APPARENT HALF-LIVES OF UNBOUND, SPECIFICALLY PROTEIN-BOUND, AND TOTAL HORMONE IN HUMAN PLASMA[a]

Steroid fraction	Percentage basal secretion: 90%[b]	10%[b]
Unbound		
T	33	37
E_2	59	60
DHT	34	41
Albumin bound		
T	33	37
E_2	59	60
DHT	34	41
SHBG bound		
T	33	57
E_2	?[c]	?[c]
DHT	38	73
Total		
T	33	40
E_2	60	64
DHT	34	49

[a] V_0 steroids = 24 L; V_0 proteins = 4.9 L.

[b] Decay from pulsatile to basal after pseudo-steady-state with either 90% or 10% of total secretion assigned to the basal (time-invariant) component. Nominal free steroid half-life 5.0 min.

[c] (?) Biphasic or initially ascending concentration time course not approximated well by single exponential decay.

Discussion

We have investigated the nonequilibrium impact of two or more well-defined acceptor proteins in plasma on the time courses of multiple free, bound, and total sex-steroid hormone concentrations driven by abrupt physiologic secretory bursts superimposed on a baseline of time-invariant hormone release. This nonlinear system never attains steady state, thus realistically emulating normal physiology. Most importantly, this formalized

TABLE V
INFLUENCE OF DELAYED DHT BURST ON APPARENT HALF-LIVES OF FREE, SPECIFICALLY BOUND, AND TOTAL STEROID HORMONE IN PRESENCE OF SHBG AND ALBUMIN[a]

	Delay times		
Steroid fraction	None	2 SDs[b]	4 SDs
Unbound			
T	34	34	33
E_2	60	59	59
DHT	36	39	41
Albumin bound			
T	34	34	33
E_2	60	59	59
DHT	36	38	41
SHBG bound			
T	39	34	(Biphasic)
E_2	(Biphasic)	(Biphasic)	(Biphasic)
DHT	46	49	48
Total			
T	34	34	339
E_2	61	61	60
DHT	37	40	43

[a] Basal secretion 50% and pulsatile release 50%. Free $t_{1/2}$ 5.0 min. V_0 proteins = 4.9 L; V_0 steroids = 24 L.

[b] SD of Gaussian secretory burst, which equals its half-duration divided by 2.354. The centers of DHT secretory bursts were allowed to coincide exactly (no time delay) or follow by 2 or 4 burst SDs the centers of the concurrent T and E_2 secretory bursts.

construct of dynamic hormone delivery into the bloodstream and partitioning between transport proteins offers various new biological insights, the most salient of which are as follows.

First, the coupled differential equations indicate that the apparent half-life of the total steroid-hormone concentration in plasma is greatly influenced by the *presence, type,* and *amount of binding proteins* present, especially high-affinity transporters (e.g., SHBG). Removal of sex-steroid hormone from albumin on the other hand is so rapid that this dissociation is not rate limiting to the half-life of free steroid hormone. However, disappearance of steroid hormone from the SHBG-bound reservoir is far less rapid and therefore strongly influences apparent bound, total, and free steroid half-lives. This is also suggested by recent experiments infusing SHBG into Rhesus monkeys[42] and earlier indirect clinical studies in the human.[27]

Second, the apparent half-lives and the nonequilibrium profiles over time of free, bound, and total hormone depend on the *relative admixture* of pulsatile and basal secretion. Indeed, we showed empirically[41] as well as theoretically (present data) that a higher proportion of basal (continuous) rather than pulsatile T secretion yields significantly higher mean and integrated serum T concentrations. Thus, relative partitioning of the same total daily hormone mass secreted via pulsatile versus continuous modes provides another mechanism to control mean serum hormone levels with any given amounts of binding proteins present.

Third, lagging one of the steroid pulses compared to a companion steroid produces a biphasic time course of blood free, bound, and total ligand concentrations due to nonuniformly varying competition for shared binding sites over the time window. For example, delayed pulsatile release of DHT compared to T or E_2 promotes more rapid dissociation of previously bound steroids. Thus, the *relative timing* of the pulsatile secretion of multiple ligands also can regulate the apparent half-lives of bound and free hormones.

Fourth, in the healthy male, the gonadal-steroid axis is never at steady state if, for example, the half-life of total sex hormone is about 45 min, and testis secretory pulses occur every 60–90 min. Thus, repeated measurements are required to characterize the pseudo-steady-state with some degree of accuracy.

Fifth, removing albumin or allowing irreversible elimination of either free or albumin-bound hormone requires on mathematical grounds a free steroid half-life of 5 min in order to yield physiologic blood concentrations of T, E_2, and DHT at pseudo-steady-state. Alternatively, a free steroid half-life of approximately 0.25–0.75 min is required in the absence of these assumptions to produce physiologic plasma steroid hormone concentrations for the three ligands considered here.

Other hormones are not considered here, such as estrone, androstenedione, and DHEA, for reasons of complexity and also because of their relatively limited binding to the two proteins studied.[43] Our admixed basal and pulsatile hormone secretory system is also not so relevant for T_4 and T_3, because thyroid hormones are believed to be secreted more or less continuously. On the other hand, this general formulation is pertinent to the female reproductive axis, in which estrogen, progesterone, and testosterone serve

[42] P. H. Petra, F. Z. Stanczyk, P. C. Namkung, M. A. Fritz, and M. J. Novy, *J. Steroid. Biochem.* **22,** 739 (1985).

[43] J. D. Veldhuis, *in* "Reproductive Endocrinology" (S. C. Yen and R. B. Jaffe, eds.), p. 409. W. B. Saunders, Philadelphia, Pennsylvania, 1991.

as the three major steroids, assuming only limited protein binding by 17-hydroxyprogesterone. Although aldosterone and cortisol are also secreted in pulses,[5,44] aldosterone unlike cortisol does not compete well for cortisol-binding globulin (CBG). On the other hand, the cortisol axis can be modeled as a subset of the present formulation, with one predominant ligand. Indeed, recent steady-state experiments highlight the dominant effect of CBG *in vivo* on cortisol half-lives and plasma cortisol concentrations,[45] as predicted from the steady-state (continuous secretion only) subset of the present more broadly stated model.

The exact physiologic effects of total, bound, and free steroid concentrations on various target tissues are not known. There is uncertainty in the literature about how much protein-bound hormone is cleared, and whether protein-bound ligand or just the free moiety acts on certain tissues, etc.[38,39,46–49] At this time no clinical condition of complete absence of SHBG has been described in the human that would allow direct studies of non-SHBG-bound steroid hormone half-lives. However, a condition of ligand-binding-protein deficiency is Laron dwarfism, in which the high-affinity GH-binding protein is defective, absent, or deficient, and tissue receptors for GH also are reduced.[50] Unfortunately, the concomitant decrease in cellular GH receptors in Laron's syndrome probably confounds kinetic interpretations. Indeed, the "normal" half-life of (total) GH in some Laron patients (e.g., 18–22 min; Ref. 32) is really inappropriately long, if the plasma of such patients is truly devoid of any high-affinity GH-binding protein, under which conditions the half-life of free GH is predicted to be 2–7 min in receptor-intact individuals.[32] Thus, one can postulate that the GH half-life is inappropriately long in some Laron patients, because the tissue GH receptor itself ordinarily participates in hormone removal.

Free hormone concentrations might be important in some endocrine systems by way of feedback control. For example, in the thyrotropic axis, periodic versus constant triiodothyronine delivery exerts differential feedback on thyroid-stimulating hormone (TSH) release.[51] In the corticotropic axis, a rapid rate of increase in blood cortisol concentrations, which would

[44] H. M. Siragy, W. V. R. Vieweg, S. M. Pincus, and J. D. Veldhuis, *J. Clin. Endocrinol. Metab.* **80,** 28 (1995).

[45] G. M. Bright, *J. Clin. Endocrinol. Metab.* **80,** 770 (1995).

[46] D. A. Domassa and A. W. Gustafson, *Endocrinology* **123,** 1885 (1998).

[47] W. M. Pardridge and L. J. Mietus, *J. Clin. Invest.* **64,** 145 (1979).

[48] J. Robbins and M. L. Johnson, *in* "Free Hormones in Blood: Proceedings of the Advanced Course on Free Hormone Assays and Neuropeptides" (A. Albertini and R. P. Ekins, eds.), pp. 53–64. Elsevier, Amsterdam, 1982.

[49] W. Rosner, *Endocrinol. Metab. Clin. North Am.* **20,** 697 (1991).

[50] W. H. Daughaday and B. Trivedi, *Proc. Natl. Acad. Sci. U.S.A.* **84,** 4636 (1987).

[51] J. M. Conners and G. A. Hedge, *Endocrinology* **106,** 911 (1980).

be expected to evoke brief but remarkable elevations in circulating free cortisol concentrations, elicits fast-feedback effects not seen when the same (total) cortisol concentration is reached less rapidly.[52] However, when T is infused continuously versus in pulses in acutely androgen-deprived men[41] or short-term castrated rams,[53] suppression of episodic LH release by negative feedback is greater during continuous steroid delivery. Continuous i.v. infusion would resemble our model of purely or predominantly basal hormone release, and should (and does[41]) achieve sustainedly higher concentrations of total and free ligand at pseudo-steady-state compared to pulsatile secretion. Considerable additional study will be required to elucidate whether and how specific feedback control mechanisms depend on particular strength–duration characteristics of the time course of free versus bound and/or total hormone concentrations in plasma.

In summary, we infer prominent effects of one or more circulating sex-steroid binding proteins on the time-specified blood concentrations of unbound and total T based on computer-assisted experiments allowing omission of either albumin alone, or SHBG alone, or both proteins, from the plasma compartment. We show that binding proteins prolong the steroid hormone half-life via mechanisms that are both transporter and ligand specific. Moreover, binding proteins distinctly damp the nonlinear time courses of total, bound, and free steroid concentrations predicted in the face of physiologically episodic secretory bursts of finite nonzero duration. In addition to the nature of the binding protein and ligand, the relative timing of multiple secretory bursts also controls the half-lives of total, bound, and free steroid hormone following nonlinear secretory inputs (pulses) into the circulation. We conclude that the half-life of free hormone, the amount of sex-steroid hormone secreted in a pulsatile versus continuous manner, the amounts and types of specific transport proteins present in plasma, and the relative timing of (multiple) ligand secretory bursts all govern the temporal silhouette of free, bound, and total steroid hormone concentrations in plasma, and thereby their time-integrated concentrations, and hence regulate both the strength and duration of the hormone feedback and feed-forward signals impinging on relevant target tissues.

Acknowledgments

We thank Patsy Craig and Cindy Sites for skillful preparation of the manuscript and Paula P. Azimi for the artwork. This work was supported in part by NIH grant RR 00847 to the General Clinical Research Center of the University of Virginia, RCDA 1 KO4 HD00634 (J.D.V.), NIH grant GM-28928 (M.L.J), the Diabetes and Endocrinology Research Center

[52] M. E. Keller-Wood and F. E. Yates, *Endocrinol. Rev.* **5,** 1 (1984).
[53] T. J. Rhim, D. J. Schaeffer, and G. L. Jackson, *Endocrinology* **132,** 2399 (1993).

Grant NIH DK-38942, the Pratt Foundation, the University of Virginia Academic Enhancement Fund, and the National Science Foundation Center for Biological Timing (NSF grant DIR89-20162) and NIH P-30 Reproduction Research Center HD28934.

[15] Monte Carlo Simulations of Lateral Membrane Organization

By MADS C. SABRA and OLE G. MOURITSEN

Introduction

Biological Membranes

The biological membrane is a many-particle system consisting primarily of lipids and proteins. In a typical prokaryotic cell or in a giant lipid vesicle, the number of molecules is roughly on the order of 10^{11}.

Although the lipid bilayer is the core of the biological membrane, its role has traditionally been regarded as being that of a fairly passive two-dimensional solvent for the membrane proteins.[1,2] However, the great diversity in lipid composition that is found in membranes with different functions suggests that the chemical and hence also the physical properties of the lipid bilayer are important for membrane function.

During the last decade extensive theoretical and experimental studies have accumulated evidence indicating that the lipid bilayer is not a structureless fluid, but a very complex and heterogeneous structure with a high degree of cooperativity (see, e.g., Refs. 3–6).

[1] K. M. Merz and B. Roux, "Biological Membranes. A Molecular Perspective from Computation to Experiment." Birkhäuser, Boston, 1996.

[2] O. G. Mouritsen and P. K. J. Kinnunen, *in* "Biological Membranes. A Molecular Perspective from Computation to Experiment" (K. M. Merz and B. Roux, eds.), pp. 463–502. Birkhäuser, Boston, 1996.

[3] M. Edidin, *Comments Mol. Cell. Biophys.* **8,** 73 (1992).

[4] L. O. Bergelson, K. Gawrisch, J. A. Ferretti, and R. Blumenthal, *Mol. Membr. Biol.* **12,** 1 (1995).

[5] S. Pedersen, K. Jørgensen, T. Bækmark, and O. G. Mouritsen, *Biophys. J.* **71,** 554 (1996).

[6] J. Y. A. Lehtonen, J. M. Holopainen, and P. K. J. Kinnunen, *Biophys. J.* **70,** 1753 (1996).

0076-6879/00 $30.00

Model Membranes

Most of these studies have been carried out on model systems such as artificial lipid membranes, which are readily made by dispersing synthetic or purified natural lipids in water. Compared to native biological membranes, such model membranes are extremely simple, but they are nevertheless very useful because they allow detailed biophysical studies to be performed—studies that are often not possible on native membranes due to their complexity.

Lipid bilayers are many-particle systems with a considerable degree of cooperativity. As such, they are able to undergo several different phase transitions[7] of which the so-called main transition is the most interesting and may be of biological relevance,[8] because the cooperative nature of this phase transition implies the existence of various self-organizing lipid domains within the membrane, i.e., membrane heterogeneity,[9] which in turn may control important functions of the membrane, e.g., enzymatic activity.[10]

Computer Simulations versus Experiments

Although artificial lipid bilayers consisting of only a few different lipid components are extremely simple compared to native biological membranes, they still display complex many-particle behavior that is not always straightforward to describe. It is therefore necessary to develop theoretical models and frameworks, which can be used when trying to interpret experimental data. Without a theory or framework, such data will often be nothing more than a collection of facts, which do not by themselves provide any understanding. An alternative to analytical theories, which often fail to answer detailed questions of many-particle systems, is computer simulations on microscopic interaction models. Within this type of "theory," rather complicated models can be examined.[11] Computer simulations can be described as something in between theory and experiment. It is theory in the sense that one uses the laws of nature to define a model, and experiment in the sense that a simulation is a numerical experiment. Different parameters, e.g., the temperature or the composition, can readily be varied and the effect of this variation on the properties of the system, e.g., the molecular

[7] G. Cevc and D. Marsh, "Phospholipid Bilayers." John Wiley and Sons, New York, 1987.

[8] O. G. Mouritsen and K. Jørgensen, *Chem. Phys. Lipids* **73,** 3 (1994).

[9] O. G. Mouritsen and K. Jørgensen, *BioEssays* **14,** 129 (1992).

[10] T. Hønger, R. L. Biltonen, K. Jørgensen, and O. G. Mouritsen, *Biochemistry* **35,** 9003 (1996).

[11] O. G. Mouritsen, B. Dammann, H. C. Fogedby, J. H. Ipsen, C. Jeppesen, K. Jørgensen, J. Risbo, M. C. Sabra, M. M. Sperotto, and M. J. Zuckermann, *Biophys. Chem.* **55,** 55 (1995).

organization, can be studied. Compared to analytical theories, computer simulations have the great advantage that they are able, fully and correctly, to take into account the entropic contribution to the free energy. Analytical theories often have to approximate this contribution, which is sometimes crucial for the conclusions. Models used in computer simulations are often extremely simple, and a major criticism to statistical mechanical models is that they are too simple. However, to make progress in the understanding of how the membrane assembly works generically, it is necessary to know which physical properties and fundamental laws are governing the behavior. Furthermore, computer simulations may provide guidelines for experiments on specific systems.

The purpose of this article is to give a tutorial introduction to computer simulations on lipid membranes. In the following section, we give a brief introduction to computer simulation techniques as they can be used to study cooperative phenomena in lipid membranes. The general strategy for computer simulations is described and the reader is introduced to the basic principles behind statistical mechanics and to Monte Carlo sampling methods. In the last section we describe in detail how Monte Carlo simulations were used to propose a mechanism for the observed formation of two-dimensional "crystals" or arrays of bacteriorhodopsin, which is an integral membrane protein. The general nature of the simulations allows for specific suggestions of which experimental parameters are to be controlled in order to make two-dimensional arrays of other integral membrane proteins.

How to Perform Computer Simulations

General Strategy

The common strategy for computer simulation studies of cooperative phenomena in membrane systems basically involves three steps: (1) definition of the problem to solve, (2) formulation of a microscopic interaction model, and (3) numerical solution of the model.

Usually, the definition of the problem is obtained when the results from an experiment have given rise to a rough idea of how to explain a certain phenomenon in terms of cooperative behavior. It is now important to delimit the problem and to decide if computer simulations can be used to test this idea. The problem statement should make clear exactly which type of answer is wanted, i.e., which properties are to be calculated. Often, it is very useful to calculate properties that can be directly compared to the experimental data along with details of the spatial arrangement of the system, which is generally not available from experiments. In this way an interpretation of the experimental evidence can be obtained in terms of the spatial organization of the system.

After having defined the problem, the microscopic interaction model is formulated by writing down the energy function of the system in terms of a chosen set of variables, e.g., in terms of the spatial coordinates of the particles in the system. The formulation of the model primarily involves making the appropriate approximations, i.e., reducing the complexity (the number of variables and the number of values these variables can take) of the problem so that it becomes computationally feasible, while still preserving its basic physical properties. For example, many simulations take as a starting point the existence of the lipid membrane aggregate and only model the lateral organization of molecules within the membrane. In this way, the number of translational variables is greatly reduced from three per particle to only two per particle. Another common approximation is to let the particles be confined to a lattice, hence reducing the number of positions each particle can take from infinity to a finite number of lattice positions. The interactions between the particles are given in terms of interaction energies, which can be either attractive or repulsive. In principle, every particle of a physical system has a distance-dependent interaction with every other particle in the system. However, for most practical purposes it is a good approximation to consider only pairs of particles that are close to each other. This is particularly simple in lattice models where it is easy to reduce the calculations, e.g., by considering only pairs of particles on neighboring sites.

Many other types of approximation can be made, depending on the system and the choice of properties to be calculated. In the example described below, a few more of these are mentioned and we show in detail how to write the energy function.

We should also mention that since cooperative phenomena are a consequence of the many-particle nature of the systems under consideration, the number of molecules to be included in the model should not be too small. Indeed, one should always test to make sure that the results obtained are not dependent on the system size.

Finally, when the microsopic model has been formulated, the statistical mechanical problem posed by the energy function is to be solved using either Monte Carlo or molecular dynamics methods, and to be able to do that, only a very basic knowledge of statistical mechanics is required.

Statistical Mechanics

Statistical mechanics can be thought of as a tool that can be used to calculate the macroscopic properties of a system, once the properties of its constituents and their direct mutual interactions are known. Hence, if the direct interactions between the molecules constituting the system can be accounted for, the laws of statistical mechanics can, in principle, be used to calculate all the macroscopic properties of the system.

The statistical mechanical description of a system that contains many molecules is based on every possible microstate that the system can occupy. A microstate, or microconfiguration, of a system is a set of specific values of the mechanical variables, $\vec{\Omega}$, which describes the molecules. In the simplest situation, the mechanical variables are the spatial coordinates (x, y, z) of each of the molecules, i.e., for a system containing N molecules

$$\vec{\Omega} = \{x_1, y_1, z_1, x_2, y_2, z_2, \ldots, x_N, y_N, z_N\} \tag{1}$$

In this case, a microconfiguration is simply one of the possible ways to arrange the N molecules among each other within the system.

For each microstate, the internal energy can be calculated using the energy function (the so-called Hamiltonian), $\mathscr{H}(\vec{\Omega})$. According to the laws of statistical mechanics, the macroscopic state of a system is a weighted average over all the possible microstates which the system can possess. Each microstate is assigned a probability according to the Boltzmann probability distribution

$$\rho(\vec{\Omega}) = \frac{e^{-\mathscr{H}(\vec{\Omega})/k_{\mathrm{B}}T}}{Z} \tag{2}$$

where the normalization factor, Z, is the so-called partition function given by the sum of all the weights:

$$Z = \sum_{\{\vec{\Omega}\}} e^{-\mathscr{H}(\vec{\Omega})/k_{\mathrm{B}}T} \tag{3}$$

k_{B} is the Boltzmann constant, and T is the absolute temperature.

Having available the probability distribution, $\rho(\vec{\Omega})$, of the microstates, the thermodynamic value of any physical quantity $f(\vec{\Omega})$, which is defined in terms of the mechanical variables, can be calculated:

$$\langle f \rangle = \sum_{\{\vec{\Omega}\}} f(\vec{\Omega})\rho(\vec{\Omega}) \tag{4}$$

That is, the thermodynamic value, $\langle f \rangle$, is a weighted average of the corresponding microscopic values, f. For example, the internal energy of the system at equilibrium is given by

$$E = \langle \mathscr{H} \rangle \sum_{(\vec{\Omega})} \mathscr{H}(\vec{\Omega})\rho(\vec{\Omega}) \tag{5}$$

From the equations above follows a set of relationships between response functions and variances in the corresponding physical quantities. This is the so-called fluctuation–dissipation theorem, which for the specific heat, $C_P(T)$, takes the following form:

$$C_P(T) = \frac{\partial \langle \mathcal{H} \rangle}{\partial T} = \frac{1}{k_B T^2} \left(\langle \mathcal{H}^2 \rangle - \langle \mathcal{H} \rangle^2 \right) \tag{6}$$

The fluctuation–dissipation theorem is particularly useful when performing computer simulations because it provides the response functions (which are often quantities that can be measured directly in experiments) at single points without requiring a whole function to be calculated.

Monte Carlo Simulations

Only in very few and extremely simplified situations is it possible to obtain an analytical expression for the partition function, Z, which is necessary to calculate the thermodynamic values according to the formulas above. Another approach would be to use a computer to simply calculate the sum in Eq. (3) numerically. However, due to the huge number of possible microstates, this is not feasible either—not even on simple systems and not even when using the fastest computers available.

It is the basis of the Monte Carlo method to approximate the average in Eq. (4) by a sum over only a relatively small part of the possible microstates. In this approximation, the microstates occur in the sum according to their Boltzmann probability and hence according to their importance for the thermodynamic properties, so that

$$\langle f \rangle \simeq \sum_{i=1}^{M} f(\vec{\Omega}_i) \rho(\vec{\Omega}_i) \tag{7}$$

where M is the number of microstates included in the average. The problem is now how to choose the microstates in order to fulfill Eq. (7).

The Monte Carlo method introduces a stochastic principle, by which a sequence of microstates is generated. Each of the microstates (except for the first) is picked with a certain probability that depends on the previous microstate in the sequence. The conditional probability, $p(\vec{\Omega}_i \rightarrow \vec{\Omega}_j)$, for accepting a particular microstate, $\vec{\Omega}_j$, subsequent to the microstate, $\vec{\Omega}_i$, is chosen in a way so that the microstates are included with a frequency proportional to their Boltzmann factor, Eq. (2).

In practice, the Monte Carlo method can be realized using the following simple algorithm:

1. Choose an arbitrary initial microstate, $\vec{\Omega}_1$.
2. Pick a trial microstate, $\vec{\Omega}'_2$.
3. Calculate the energy difference between the two microstates, ΔE, using the model Hamiltonian.
4. Let the computer draw a random number, ξ.
5. If $\xi < \exp(-\Delta E/k_B T)$ the trial microstate is accepted; i.e., $\vec{\Omega}_2 = \vec{\Omega}'_2$. If not, $\vec{\Omega}_2 = \vec{\Omega}_1$; i.e., the second microstate in the sequence is the same as the first.
6. Pick a new trial microstate, $\vec{\Omega}'_3$, and so on.

The number of microstates, M, needed to make Eq. (7) a good approximation to the true thermodynamic values depends on the specific system as well as the phenomenon under consideration. It also depends on the specific values of the thermodynamic parameters such as temperature, pressure, and composition. In principle, the approximation can be made as good as desired—it is merely a question of choosing M to be sufficiently large.

Now all that remains is the question of how to choose the trial microstates. In principle, each of the trial microstates can be generated randomly from scratch. However, it is much more efficient to generate the trial microstates by randomly modifying the previous microstate slightly, e.g., by moving a single particle or by swapping the positions of two particles. Whatever method is used to generate the trial microstates, two requirements must be fulfilled. First, the sequence must be ergodic, i.e., it must be ensured that every microstate can be reached from any other microstate within the course of a finite number of steps. Second, the detailed-balance condition must be fulfilled. This simply means that the probability of picking state $\vec{\Omega}_j$ as trial state to succeed state $\vec{\Omega}_i$ must be equal to the probability of picking state $\vec{\Omega}_i$ as trial state to succeed state $\vec{\Omega}_j$. For simple models, it is usually hard not to fulfill the detailed-balance condition, but great care should be taken in the more complicated situations.

Because the canonical Monte Carlo importance sampling techniques described above operate directly on the level on the microstates, $\vec{\Omega}$, it is obvious that such techniques are extremely powerful to elucidate aspects of membrane lateral organization, as we see in the following section.

An Example: Array Formation of Membrane Proteins

In this section we work a specific example using Monte Carlo simulations in order to illustrate some of the points made above.

The study is concerned with some general conditions, which are required in order to be able to form two-dimensional (2-D) arrays of integral membrane proteins. Such arrays are important, when attempts to form three-dimensional (3-D) crystals fail, which is the case for most integral membrane proteins. Electron diffraction methods on 2-D arrays can be used as an alternative to X-ray methods on 3-D crystals, when trying to determine high-resolution protein structures.

Problem

The study is based on experimental evidence from the integral membrane protein bacteriorhodopsin from the purple bacteria *Halobacterium salinarium. In vivo,* this protein occurs as hexagonal arrays in the membrane. In a series of studies it was shown that arrays can also form *in vitro,* even in artificial membranes, provided certain charged lipids are present in the membrane.[12] This leads to the following hypothesis[13]: Certain types of lipids have a preference to the protein as compared to other lipids. Therefore, they will tend to be found at the annulus of the protein and are therefore called *annular lipids* compared to neutral lipids, which do not have this preference. The idea was now that when proteins formed arrays, and hence shared some annular lipids, some annular lipids would be set free from the protein annulus and gain translational entropy. Hence, the problem we are going to consider is the following: Is it a sufficient criterion to have some lipids with strong preference for the protein present in the membrane in order to induce the formation of arrays?

First we should consider whether it is possible to formulate an analytical theory for the problem. However, because entropy is expected to play a dominant role in this problem, it seems more feasible to use computer simulations (cf. the comment in the Introduction).

We are interested only in the lateral organization of the proteins and lipids, and do not mention the difficulties of purifying and reconstituting membrane proteins into lipid bilayers, detergents, etc., which is clearly also very important for experimental array formation, but it is not possible to simulate everything.

The following sections contain some data from simulations that have not been published before. However, the description of the off-lattice simulations is based on Ref. 13 in which more detailed information about the simulations and the results can be found.

Because we are interested in the lateral organization of molecules, the problem is two-dimensional in nature. Hence, both of the models we propose only consider two dimensions, i.e., the membrane plane. This is a simplification that speeds the calculations significantly, but at the same time

[12] A. Watts, *Biophys. Chem.* **55,** 137 (1995).
[13] M. C. Sabra, J. C. M. Uitdehag, and A. Watts, *Biophys. J.* **75,** 1180 (1998).

it precludes us from taking into account important membrane properties, like the membrane curvature, which may be important for the problem.

To illustrate how different approximations allow for computational feasibility at the expense of details of the information that can be obtained, we use two different microscopic models: a lattice model and an off-lattice model.

Lattice Model

In the lattice model, the 2-D plane to which we have confined ourselves is reduced to being discrete, i.e., the molecules are confined to sites on a lattice. As other approximations, this one increases computational speed, but at the same time it also limits the amount of information that can be obtained. We have chosen a triangular lattice since this provides for the closest packing of molecules, although a quadratic lattice, which is simpler, could also have been used. We now assume three different types of molecules: proteins, annular lipids, and neutral lipids. The lipids are chosen to occupy one lattice site each, and the proteins occupy either one or seven sites each. All of the molecules are taken to be rotationally symmetric, which leaves out the complication of having to deal with rotations. All sites on the lattice are occupied, i.e., the density of the system is fixed. Computer simulations on a more complex, but similar, model including proteins and two different lipid species have previously been performed.[14]

The central hypothesis that we wish to test assumes a particularly strong attractive interaction between the proteins and the annular lipids. Because the molecules are confined to a lattice, all distances are discrete, and for simplicity we consider only interactions between neighboring sites. Hence, we assign an attractive interaction, i.e., a negative interaction energy, between proteins and annular lipids that occupy neighboring sites. To ensure that any observed aggregation of proteins is lipid induced, we also assign a repulsive interaction (a positive interaction energy) between proteins on neighboring sites. No molecular details are included and there are no assumptions about the origin of the interaction. It is not the origin of the interaction that is important, but the fact that there is an interaction of a given relative strength. The lattice model hence contains the following parameters: the strength of the interactions, the temperature, and the relative concentrations of the three components.

We can now write down the model Hamiltonian:

$$\mathcal{H} = \sum_{\langle ij \rangle} K_{p_i p_j} \tag{8}$$

[14] T. Gil, M. C. Sabra, J. H. Ipsen, and O. G. Mouritsen, *Biophys. J.* **73,** 1728 (1997).

where $\langle ij \rangle$ denotes each pair of neighboring molecules on the lattice; and $p_i = \{\mathrm{p, a, n}\}$ is the type of molecule on the ith lattice site. The neutral lipids are denoted by n, the annular lipids by a, and the proteins by p. The interaction constants are given by (in units of $k_B T$)

$$\begin{aligned} K_{\mathrm{pp}} &= 1 \\ K_{\mathrm{pa}} &= -1 \\ K_{\mathrm{aa}} &= K_{\mathrm{nn}} = K_{\mathrm{na}} = K_{\mathrm{np}} = 0 \end{aligned} \tag{9}$$

The Hamiltonian is a prescription of how to calculate the internal energy of the system given the microstate: the interaction energies of each pair of molecules on neighboring sites are all added to obtain the total internal energy of the system.

Having constructed the model Hamiltonian, a completely random microstate is picked as the initial state. The trial states are picked in two different ways: Kawasaki exchange and long-range swaps. In Kawasaki exchange, a random molecule on the lattice is picked. Then one of its neighbors is picked, also randomly. Finally, the energy change for exchanging the positions of the two molecules is calculated, and the exchange is performed according to the Boltzmann probability as described in the previous section. Long-range swaps are similar, but instead of choosing two neighboring molecules, two random molecules are picked independently of each other. In both types of dynamics, only a very small part of the system is changed. Hence, calculating the total energies of each of the two microstates is unnecessary, and time can be saved by only considering the changing parts of the system.

When the microscopic model has been implemented into a computer program, the question of which parameter values to use for the simulations arises. Often, only a very limited set of parameter values is relevant for the problem. If some of the parameter values are known experimentally these are obvious to start with. However, since the experimental system is usually much more complicated than the model system, there may be no simple relation between the measured parameter values and the parameter of the model. One way to find the relevant parameter values is to scan different ranges of parameter values until something interesting shows up. At this point, each of the model parameters can be varied one at a time in order to systematically study the cooperative behavior of the system.

After a simulation has been started from an arbitrary microstate, it is only after a certain number of steps that the microstates in the sequence will be characteristic of thermodynamic equilibrium. It is therefore necessary to let the system equilibrate, i.e., to let the simulation run for a while, before starting to collect data. How long the equilibration time should be depends on the system in question and on the dynamics employed in the simulation.

In general, large systems need longer time per molecule to equilibrate than small systems. Furthermore, the specific choice of parameters is important—the closer the system is to a phase transition, the longer the equilibration time. In particular, when a system has to cross a phase line from a disordered to an ordered phase, equilibration times can be very long. To decide whether equilibrium has been reached, it is often useful to monitor the development of the value of one or more different properties during the simulation. When each of these properties no longer changes systematically, but rather seems to fluctuate around an average value, there is a good chance that equilibrium has been reached. However, there is always the risk that the system will get trapped in a metastable state. To make sure that true thermodynamic equilibrium has been reached, some simulations with identical parameter values should be started from different initial states, e.g., from an ordered and a disordered state. If the thermodynamic averages from these simulations are not the same, it is necessary to perform longer simulations.

When equilibrium has been reached, it is time to collect data. Which data to collect and which properties to calculate depend on which experimental data the results are going to be compared to. In the present example, the experimental evidence available is electron micrographs of the proteins in the membrane, which can be considered to be a kind of microstate. Hence, we simply need to visualize the equilibrium microstates from the model.

Figure 1A shows the random microstate from which the simulations

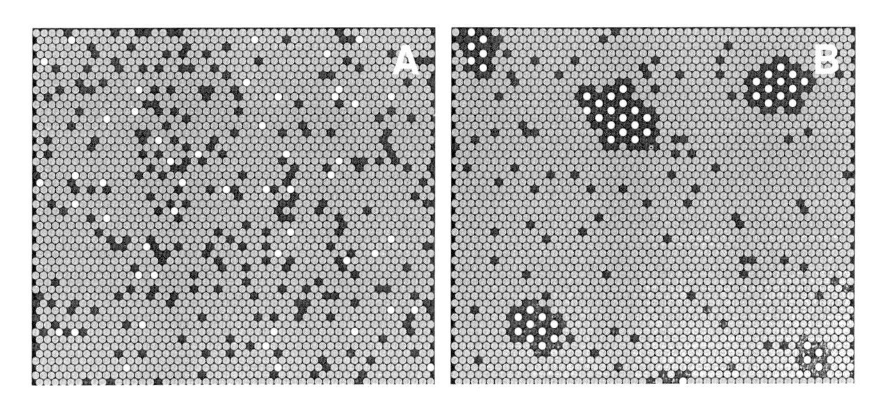

FIG. 1. Examples of microstates from simulations of the lattice model, Eqs. (8) and (9). There are 2500 molecules in each system. (A) The random microstate from which the simulations are started. (B) A microstate after 5,000,000 Monte Carlo steps per molecule. Only Kawasaki exchanges were used to generate the trial microstates. Proteins are denoted by white circles, the annular lipids by dark gray circles, and the neutral lipids are represented by light gray circles.

were started and Fig. 1B shows a microstate obtained after 5,000,000 Monte Carlo steps per molecule (MCS). Only Kawasaki exchange was used. When going from Fig. 1A to 1B, it is clear that the proteins have started aggregating and are forming two-dimensional hexagonal arrays. However, it is not possible from these microstates alone to determine whether equilibrium has been reached or not. To do so, we must monitor, e.g., the energy development of the system as a function of the number of MCS. This is done in Fig. 2, which also shows the energy development from a simulation in which long-range swaps were employed. Note that the timescale is logarithmic. It is clear from this figure that when only Kawasaki exchange is used [Fig. 2(a)], equilibrium is not reached within the course of 5,000,000 MCS, since the energy does not reach a constant level, but keeps decreasing throughout the simulation. In fact, equilibrium was not even reached after 50,000,000 MCS, at which point the simulation was stopped. When long-range swaps are used [Fig. 2(b)], we see that a constant level of the energy is reached after only about 50,000 MCS. Hence, equilibrium is reached much faster (at least 3 orders of magnitude) when long range swaps are allowed—even in this very small system. This illustrates how important it is to carefully consider how the trial microstates are to be picked in the simulation.

An equilibrium microstate from the simulation using long-range swaps is shown in Fig. 3A. There is only one protein array, and this array consists of all proteins in the system. In simulations where no annular lipids were present—or when all lipids present were annular lipids—the proteins were always randomly distributed (not shown).

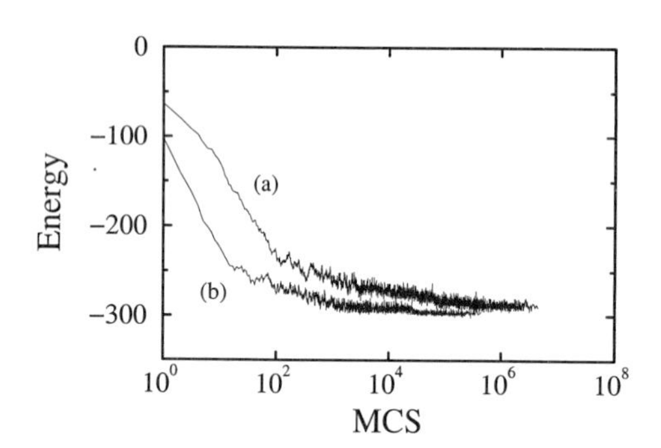

FIG. 2. The development of the total energy of the system (in units of $k_B T$) from two different simulations of the lattice model, Eqs. (8) and (9), plotted as a function of number of Monte Carlo steps per site (MCS). Note that the time axis is logarithmic. The curves correspond to (a) a simulation where only Kawasaki exchanges were employed and (b) a simulation in which long-range swaps were also used. In both simulations, the system consisted of 2500 molecules.

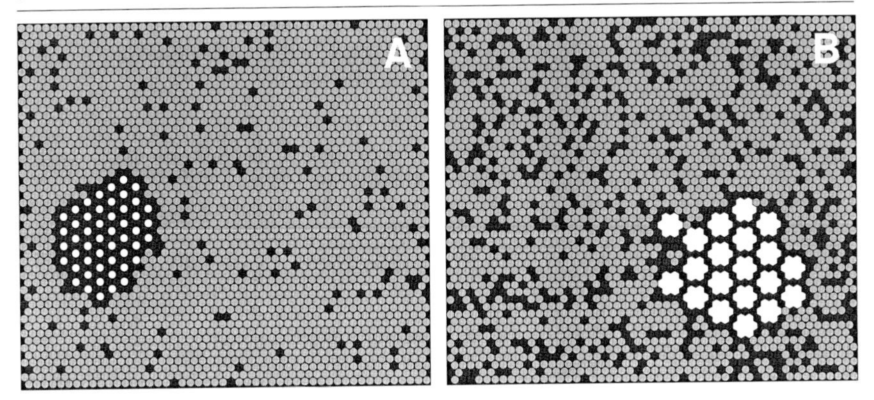

FIG. 3. Examples of microstates from simulations of the lattice model, Eqs. (8) and (9). Examples of equilibrium microstates obtained from the lattice simulations. (A) The proteins are modeled as having the same size as the lipids. There are 2500 molecules. (B) The proteins occupy seven sites each and there are 2380 molecules. In both simulations, long-range swaps were used in order to equilibrate the system quickly. Proteins are denoted by white circles, the annular lipids by dark gray circles, and the neutral lipids are represented by light gray circles.

To provide for more realism in the relative sizes of lipids and proteins, we now let each protein occupy seven sites arranged in a hexagon on the lattice. Due to the fact that the molecules are arrayed on a lattice, it is still straightforward to employ long-range swaps by swapping one protein for a hexagon of seven lipids. The microstate in Fig. 3B is an equilibrium state from such a simulation. Again, we see that an array of proteins has formed. Furthermore, we find that the proteins do not form when only one type of lipid is present (not shown).

Hence, from the simulations on the lattice model, we conclude that aggregation of small as well as of large proteins does occur due to the presence of the annular lipids.

Off-Lattice Model

The regularity of the two-dimensional arrays observed in the lattice simulations are consequences of the underlying lattice. This is clear when looking at the microstates in Figs. 1 and 3—even the lipids are ordered in hexagonal arrays. It is also clear that if the simulations had been performed on another type of lattice, e.g., on a quadratic lattice, the arrays observed would not have been hexagonal. Hence, the lattice model is only able to establish that the proteins can *aggregate* due to the annular lipids, and not whether they will form *arrays.* To be able to distinguish between aggregates and arrays, it is necessary to use a model in which there is no underlying lattice.

In the off-lattice model we describe here,[13] the molecules are confined to a square box, but they can take any position in the 2-D plane within this box. The simplest way to represent finite-sized molecules in two dimensions is by disks, which have no orientation due to their symmetry. Because there is no lattice, the concept of nearest neighbors is not well defined. Instead, we let the particles interact by a constant interaction energy when they are within a certain distance of each other. Similar to the lattice model, the proteins repel each other and there is an attractive interaction between proteins and annular lipids. As in the lattice model we make no assumptions about the origin of the various interactions.

The off-lattice model contains more parameters than the lattice model: the interaction distance, the strength of the interactions, the temperature, the density of molecules in the system, and the relative concentrations of the different types of molecules.

The Hamiltonian of the system can be written as

$$\mathcal{H} = \sum_{\langle i,j \rangle} V_{p_i p_j}(x_{ij}) \tag{10}$$

where $\langle i, j \rangle$ denotes each pair of particles in the system, x_{ij} is the distance between the particles, and $p_i = \{\mathrm{p, a, n}\}$ is the type of particle i. The neutral lipids are denoted by n, the annular lipids by a, and the proteins by p. The $V_{p_i p_j} = V_{p_j p_i}$ represent the potentials between particles and are given by (in units of $k_B T$)

$$V_{\mathrm{nn}} = V_{\mathrm{np}} = V_{\mathrm{na}} = V_{\mathrm{aa}} = \begin{cases} \infty & \text{for } x \leq 2r \\ 0 & \text{for } x > 2r \end{cases} \tag{11}$$

$$V_{\mathrm{ap}} = \begin{cases} \infty & \text{for } x \leq 2r \\ -1 & \text{for } 2r < x \leq 2r + d/r \\ 0 & \text{for } x > 2r + d/r \end{cases} \tag{12}$$

and

$$V_{\mathrm{pp}} = \begin{cases} \infty & \text{for } x \leq 2r \\ 1 & \text{for } 2r < x \leq 2r + d/r \\ 0 & \text{for } x > 2r + d/r \end{cases} \tag{13}$$

where d/r is the range of the interactions as measured from the surface of the particles, which all have equal radii, r. Hence, to calculate the internal energy of the system, all the distances between, in principle, all pairs of molecules have to be calculated, which is extremely time consuming. However, it is possible to keep track of the individual molecules during the simulation so that it becomes sufficient merely to calculate distances between particles that are in the vicinity of each other.

In this model, two different methods are used to choose the trial microstates. First, the molecules are allowed (using the Boltzmann probability condition) to move a random distance in a random direction. In principle, it would have been possible to model the lipids and proteins as being of different size, and then employ only this way of choosing the trial microstates. However, without using some kind of long-range swaps the simulations become infeasible as was seen from the lattice simulations. Hence, the second way of choosing trial microstates is to use long-range swaps that are similar to the long-range swaps in the lattice model, and they can readily be performed when the molecules are equally sized disks.

Some of the results from the off-lattice model are shown in Fig. 4. We can see that hexagonal arrays can form (Fig. 4A), as can arrays of other geometries (Fig. 4B). It is interesting to note that similar arrays of proteins have been found experimentally for bacteriorhodopsin.[12] Hence, the final conclusion from the off-lattice simulations is that two-dimensional *arrays* can actually form as a result of the presence of annular lipids. A more detailed interpretation of the results from the off-lattice model is given in Ref. 13.

Conclusion

In the preceding sections, we gave a description of the basic statistical mechanics needed to perform Monte Carlo simulations. This description

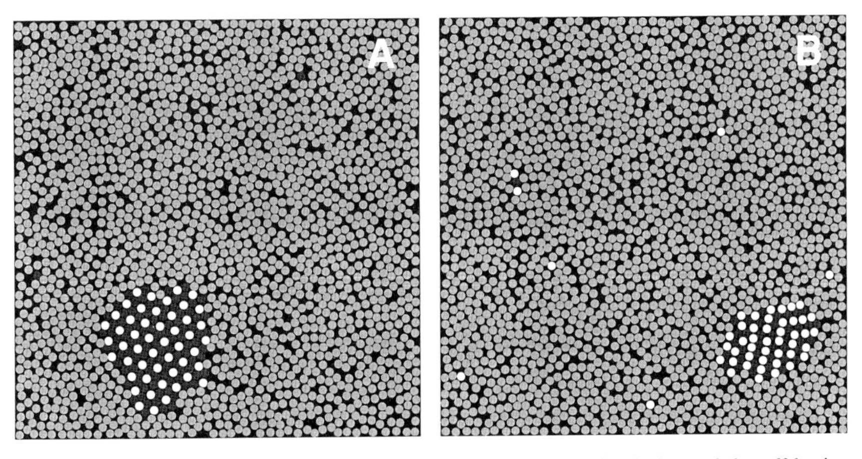

FIG. 4. Examples of equilibrium microstates obtained from simulations of the off-lattice model, Eqs. (10)–(13). (A) A hexagonal array of proteins has phase separated from the matrix of disordered lipids. (B) An example of a less ordered orthorhombic array. Each system contains 2000 molecules total. Proteins are denoted by white circles, the annular lipids by dark gray circles, and the neutral lipids are represented by light gray circles.

is very short and not complete, and those who wish to perform more complicated simulation should consult one of the many books published on the subject.[15,16] The advantage of Monte Carlo simulations of the type used here lies in their ability to demonstrate some general phenomena, and how these qualitatively depend on the parameters. Seldom can such simulations be used to predict quantitatively the behavior of cooperative systems because it is necessary to use very simple models.

The specific example of a simulation of 2-D array formation showed some of the aspects of how the complexity of a problem can be reduced (e.g., by confining molecules to a lattice) in order to gain computational feasibility at the expense of the level of detail that can be obtained.

Acknowledgments

The authors are affiliated with the Danish Centre for Drug Design and Transport, which is supported by the Danish Medical Research Council. O.G.M. is an associate fellow of the Canadian Institute for Advanced Research. Some of the work described was supported by the Danish Natural Science Research Council and the Danish Technical Research Council.

[15] M. P. Allen and D. J. Tildesley, "Computer Simulations of Liquids." Oxford University Press, United Kingdom, 1987.

[16] D. Frenkel and B. Smit, "Understanding Molecular Simulation." Academic Press, San Diego, 1996.

[16] Hydrodynamic Bead Modeling of Biological Macromolecules

By Olwyn Byron

Introduction

Although the literature is full of extremely learned and thorough works on the theory of hydrodynamic bead modeling, and papers are now beginning to accumulate on its application (especially to biological macromolecular systems), there does not appear to be a paper describing how to actually do the modeling. The aim of this chapter therefore is to fill this gap.

What is hydrodynamic bead modeling? Simply put, it is the representation of a macromolecule with an assembly of spheres for which a profile of measurable hydrodynamic parameters is then computed. As a modeling method it is complemented by whole-body modeling approaches: the use

0076-6879/00 $30.00

of ellipsoids of revolution or general triaxial ellipsoids to represent macromolecules whose shape can reasonably be approximated to such a generalized geometrical topology (see Ref. 1 for a good review of this) and for which a wide range of shape functions can be calculated (via the ELLIPS suite of computer programs[2]) and compared with hydrodynamic measurements; and the use of other more complex whole-body geometries for which small-angle scattering curves can be simulated (via the program FISH[3]). Hydrodynamic bead modeling, however, allows the user to represent essentially any shape and any level of asymmetry and has traditionally been (and still is) an approach of particular utility when high-resolution structural data are absent for a macromolecule or a complex but some idea of solution conformation is required. Reasons why the structure of a molecule at high resolution has not been determined include these: (1) there may be a very limited amount of the sample available—too little perhaps for nuclear magnetic resonance (NMR) or crystallographic studies; (2) the molecule may not crystallize, which may be due to flexibility in or between certain of its domains, high glycosylation, or an inability to obtain the molecule at sufficient purity; or (3) the species of interest may be a noncovalent complex of macromolecules in equilibrium with the monomeric species. Progress in high-resolution structural biology has provided some solutions to these problems in some cases and will happily continue to do so. But, in the case of crystallography, even if the atomic coordinates can be determined, it is not always certain that this structure is necessarily representative of the conformation adopted by the macromolecule in the more physiologically or biologically relevant solution environment. An example of a system for which this is the case (the sliding clamp of the bacteriophage T4 DNA polymerase) is given at the end of this chapter. As for NMR spectroscopy, although the coordinates obtained *are* those of the macromolecule in solution, the molecular concentration of this solution is normally very high (which may affect the shape and oligomeric state of the species in question), and pH and ionic strength are often far from physiologic. Moreover, current NMR spectroscopy studies are limited to macromolecules of mass below about 40 kDa.

Apart from offering an option for low-resolution shape determination, hydrodynamic bead modeling can be used to evaluate high-resolution homology models for macromolecules for which atomic resolution data have yet to be acquired but for which lower resolution hydrodynamic data are available. Such an example will be used to illustrate this chapter. The

[1] S. E. Harding, *Biophys. Chem.* **55,** 69 (1995).

[2] S. E. Harding, J. C. Horton, and H. Colfen, *Eur. Biophys. J.* **25,** 347 (1997).

[3] R. K. Heenan, Rutherford Appleton Laboratory Report (1989).

protein toxin, pneumolysin (PLY), a key virulence factor of the bacterium *Streptococcus pneumoniae* has a mass of 53 kDa and has been widely characterized with hydrodynamic methods,[4] electron microscopy (EM),[5] and small-angle neutron scattering.[6] It has, however, so far eluded attempts to obtain a high-resolution structure but shares a high degree of homology (48% identity, 60% sequence similarity) with perfringolysin O (PFO), a similarly functioning molecule from the bacterium *Clostridium perfringens.* The 2.7 Å resolution crystal structure for PFO was recently obtained[7] and it has been possible to generate a homology model for PLY based on these coordinates.[8] This model has been tested for its ability to reproduce experimental data[9] and the reader is referred to the original work for a proper discourse on the success of this approach.

Last but not least, there are the modular proteins in which a restricted number of functional building blocks having well-defined structures have been utilized by evolution to create a multiplicity of species, from enzymes to structural proteins (see Ref. 10 for a comprehensive review). The high-resolution structures of single and small arrays of modules are continually being solved, and the next challenge is to determine their relative spatial orientation in the whole protein, which can be composed of tens to (in some extreme cases) hundreds of copies of similar and/or different modules, and it is thus extremely difficult to derive a structure for the macromolecule in its entirety. Here, hydrodynamic bead modeling starting from the high-resolution structures of the single modules has the potential to play a major role.

The measurable hydrodynamic parameters that can currently be computed for bead models are given in Table I; the mathematical formalisms used to compute them are well described in the seminal reviews by García de la Torre and Bloomfield[11] and by García de la Torre,[12] and the more

[4] P. J. Morgan, S. C. Hyman, O. Byron, P. W. Andrew, T. J. Mitchell, and A. J. Rowe, *J. Biol. Chem.* **269,** 25315 (1994).

[5] P. J. Morgan, S. C. Hyman, A. J. Rowe, T. J. Mitchell, P. W. Andrew, and H. R. Saibil, *FEBS Letts.* **371,** 77 (1995).

[6] R. J. C. Gilbert, J. Rossjohn, M. W. Parker, R. K. Tweten, P. J. Morgan, T. J. Mitchell, N. Errington, A. J. Rowe, P. W. Andrew, and O. Byron, *J. Mol. Biol.* **284,** 1223 (1999).

[7] J. Rossjohn, S. C. Feil, W. J. McKinstry, R. K. Tweten, and M. W. Parker, *Cell* **89,** 685 (1997).

[8] J. Rossjohn, R. J. C. Gilbert, D. Crane, P. J. Morgan, T. J. Mitchell, A. J. Rowe, P. W. Andrew, J. C. Paton, R. K. Tweten, and M. W. Parker, *J. Mol. Biol.* **284,** 449 (1998).

[9] R. J. C. Gilbert, R. K. Heenan, P. A. Timmins, N. A. Gingles, T. J. Mitchell, A. J. Rowe, J. Rossjohn, M. W. Parker, P. W. Andrew, and O. Byron, *J. Mol. Biol.* **293,** 1145 (1999).

[10] P. Bork, A. K. Downing, B. Kieffer, and I. D. Campbell, *Q. Rev. Biophys.* **29,** 119 (1996).

[11] J. García de la Torre and V. A. Bloomfield, *Q. Rev. Biophys.* **14,** 81 (1981).

[12] J. García de la Torre, *Eur. Biophys. J.* **23,** 307 (1994).

TABLE I
PARAMETERS COMPUTABLE FOR HYDRODYNAMIC BEAD MODELS

Parameter[a]	Name	Units	Method of measurement[b]	Ease of measurement[c]
D_t	Translational diffusion coefficient	$cm^2\ s^{-1}$	Sedimentation equilibrium or velocity; dynamic light scattering	Easy
s	Sedimentation coefficient	S	Sedimentation velocity	Easy
R_g	Radius of gyration	nm	Small- to wide-angle static light scattering; small-angle X-ray or neutron scattering	Harder
D_r	Rotational diffusion coefficient	s^{-1}	Electric birefringence	Difficult
τ_{1-5}	Rotational relaxation times	ns	Fluorescence depolarization anisotropy decay; NMR	Difficult
$[\eta]$	Intrinsic viscosity	$ml\ g^{-1}$	Precision viscometry	Harder

[a] Other parameters may be derived from these.
[b] For reasons of space only the main methods are given for each parameter.
[c] The ease of measurement reflects a combination of the amount of sample required, its purity, the availability of appropriate equipment, its cost, and the ease of interpretation of the raw data.

recent paper by Spotorno *et al.*[13] In addition to the primary hydrodynamic parameters, certain combined and derived functions can be generated from the data of Table I that are often useful in hydrodynamic modeling (see Refs. 1 and 14 for good accounts of this). Hydrodynamic bead modeling is closely related to Debye sphere modeling with which the scattering curves acquired in small-angle X-ray, neutron (and in some instances, light) scattering studies can be interpreted. This is discussed in more depth at the end of this chapter.

Overview of Method

In this chapter three programs have been used to perform hydrodynamic bead modeling calculations and manipulations: MacBeads [written by Dr.

[13] B. Spotorno, L. Piccinini, G. Tassara, C. Ruggiero, M. Nardini, F. Molina, and M. Rocco, *Eur. Biophys. J.* **25,** 373 (1997).

Dan Thomas, then at the National Centre for Macromolecular Hydrodynamics (NCMH), University of Leicester, UK; obtainable on request from the NCMH (see http://www.nottingham.ac.uk/ncmh/unit/software.html)]; AtoB (Ref. 15; downloadable from ftp://bbri.harvard.edu/rasmb/spin/ms_dos/atob-byron/); and HYDRO (Ref. 16; downloadable from http://leonardo.fcu.um.es/macromol/software.html). These are not the only programs that have been written for this type of modeling but they are the ones routinely used by the author. HYDRO is complemented by a similar program, SOLPRO,[14] which additionally computes small- and wide-angle scattering curves and universal shape-dependent functions (similar to those generated for ellipsoids by ELLIPS[2]). SOLPRO will shortly be extended to calculate the second virial coefficient of the bead model. The process of hydrodynamic bead modeling employing HYDRO, MacBeads, and AtoB is represented by the flowchart in Fig. 1. The numbered flowchart elements shown therein are described next in more detail.

Section 1: Construction of a Bead Model in the Absence of High-Resolution Coordinates

For reasons mentioned at the start of this chapter occasions will arise when there are high-resolution structural data for neither the macromolecule of interest nor for any species with which it shares a significant level of sequence homology. In such a case some level of structural information can be obtained by careful electron microscopy (where the sample is visualized with negative staining or metal shadowing) or cryoelectron microscopy. An example of this approach is described by Tharia *et al.*[17] for the characterization of fractions of the 14S dynein from *Tetrahymena thermophila.* The structural data obtainable for this species of dynein were limited by the minute quantities in which the purified protein was available. However, it was possible for the authors to identify four distinct fractions during anion-exchange chromatography, to obtain electron micrographs of the constituent rotary-shadowed molecules, and to measure their sedimentation coefficients ($\bar{s}_{20,w}$) and masses using analytical ultracentrifugation. To test the correspondence between the conformation as visualized under the electron microscope with that in solution (in the form of $\bar{s}_{20,w}$) bead models were constructed based on the measured mass, an assumed partial specific volume of 0.73 ml/g (a typical value for protein) and the dimensions gleaned from

[14] J. García de la Torre, B. Carrasco, and S. E. Harding, *Eur. Biophys. J.* **25,** 361 (1997).

[15] O. Byron, *Biophys. J.* **72,** 408 (1997).

[16] J. García de la Torre, S. Navarro, M. C. Lopez Martinez, F. G. Díaz, and J. J. Lopez Cascales, *Biophys. J.* **67,** 530 (1994).

[17] H. Tharia, A. J. Rowe, O. Byron, and C. Wells, *J. Musc. Res. Cell Mot.* **18,** 697 (1997).

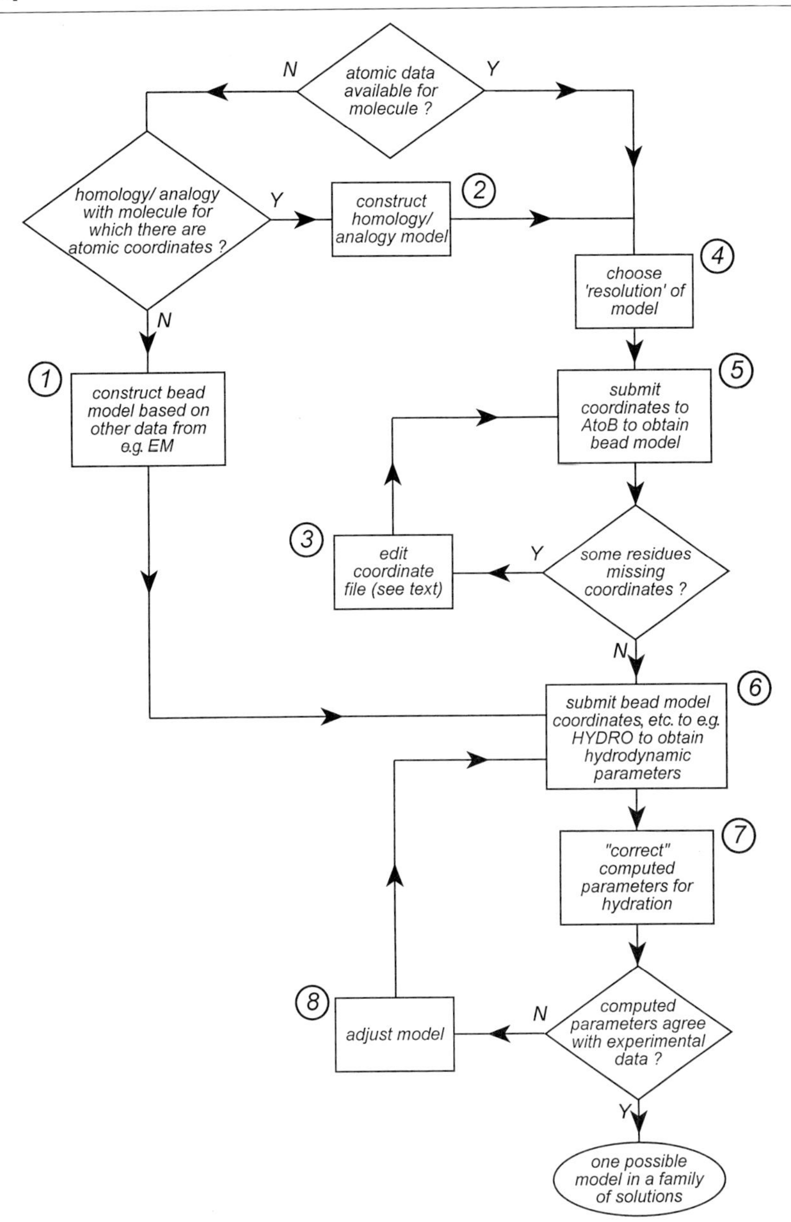

FIG. 1. Flowchart outlining a general strategy for hydrodynamic bead modeling utilizing AtoB, MacBeads, and HYDRO. The numbers refer to sections within the text in which the relevant flowchart element is discussed. A similar strategy is followed for modeling with the BEAMS suite of software.

the micrographs. Thus, for fraction 4 (Fig. 2) the mass was determined to be 527 ± 2 kDa and the model constructed was based on a mass of 524 kDa. Therefore the molecular volume [$V = M\bar{v}/N_A$, where M is the molar mass (g/mol), $\bar{v}$ is the partial specific volume (ml/g), and N_A is Avogadro's number (per mole)] was calculated to be 6.35×10^5 Å^3. This volume was subdivided so that the three heads had diameters of 65 Å and the overall molecular length was 310 Å.

Once these dimensions had been fixed, the coordinates of the constituent beads were obtained using the Macintosh personal computer program MacBeads, which allows the user graphically to create *ab initio* a model on screen and to save it as a pictorial representation (in three views: the *xy, yz,* and *zx* planes; see Fig. 2) and as a text file of the bead Cartesian coordinates and radii. This saves the user a great deal of repetitive trigonometry although for simple models the coordinates can be worked out on graph paper. The GRUMB module of the BEAMS modeling system[13] also

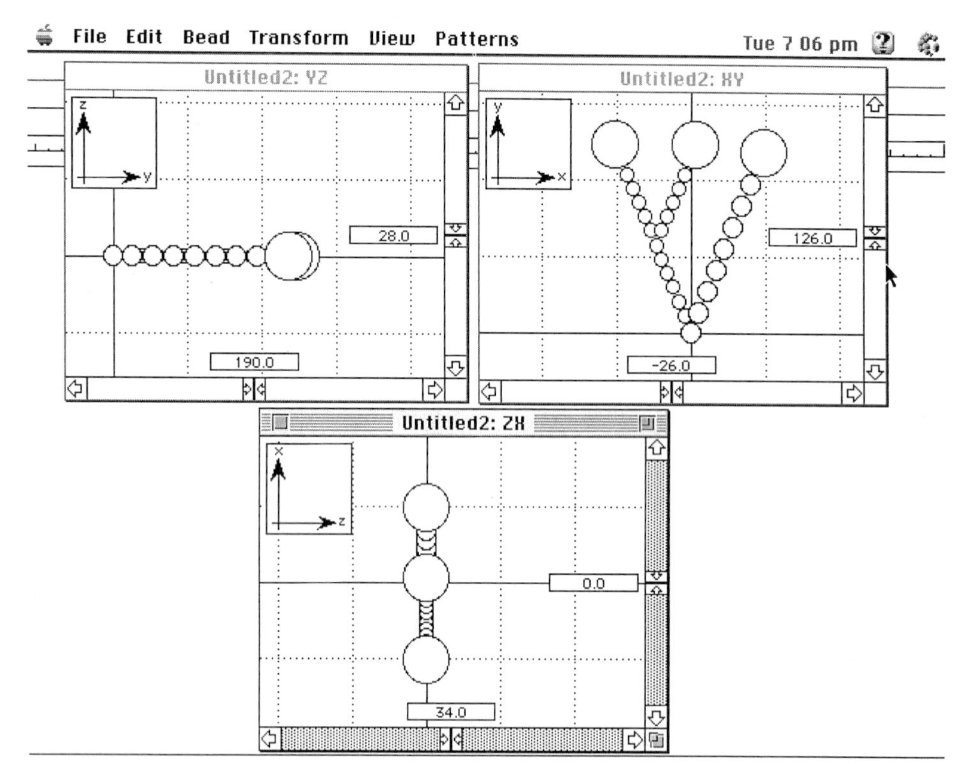

FIG. 2. A screen view of the program MacBeads in use for the construction of a model for fraction 4 of the 14S dynein from *T. thermophila.*

facilitates bead model construction, although in perhaps a less user-friendly way than MacBeads. (However, a recent addition to GRUMB makes it able to generate a script file for RASMOL, popular freeware for macromolecular visualization.[18]) Two sedimentation coefficients were recorded for fraction 4: 13.2 ± 0.6 S and 16.0 ± 0.2 S. This variability was attributed to different molecular conformations, which were represented by three models with different interstem angles (30°, 45°, and 60°). By using an interstem angle intermediate between 30° and 45° it was possible to reproduce $\bar{s}_{20,w}$ = 16.0 S for a hydration approaching a typical protein value of 0.3 g water/g protein.

A somewhat different approach to bead model building in the absence of high-resolution data has been presented by Rocco and co-workers for the modeling of fibronectin, a typical modular protein,[19] and of the integrin $\alpha_{IIb}\beta_3$.[20] In this approach, the protein is first examined at the sequence level, to identify putative domains and/or modular units, and secondary structure elements. This information is coupled with biochemical data such as proteolytic fragmentation studies and affinity labeling to subdivide the sequence into a number of segments, each of which is then represented by a bead. These operations are carried out with the help of the program PROMOLP (PROtein MOLecular Parameters),[13] which, among other things, calculates the anhydrous and hydrated molecular volume, and thus the corresponding radius, for the stretch selected. The beads are then positioned in space, utilizing GRUMB, according to EM data coupled with any biochemical information available (e.g., long-range disulfide bonds, putative noncovalent interactions between parts of the molecule, etc.). This procedure is permitted because of the direct correspondence between sequence and beads. If the EM data do not show a well-defined, constant shape, but rather a collection of different shapes (as in the case of fibronectin prepared in the presence of glycerol), it is possible to generate many different conformations using the STRINGS module of the BEAMS suite,[13,19] and to compute the properties of each conformation and the mean properties of the whole set. Moreover, this can also be done while keeping the relative structure of part(s) of the model fixed in space, and allowing the program to generate at random only the connecting segments between them, including loops. However, it is important to stress that this approach is quite different from the case of segmental flexibility that is discussed later, and rather

[18] R. A. Sayle and E. J. Milner-White, *Trends in Biochem. Sci.* **20,** 374 (1995).

[19] M. Rocco, E. Infusini, M. G. Daga, L. Gogioso, and C. Cuniberti, *EMBO J.* **6,** 2343 (1987).

[20] M. Rocco, B. Spotorno, and R. R. Hantgan, *Prot. Sci.* **2,** 2154 (1993).

applies to structures more closely related to wormlike chains or to random-coil stretches of proteins, such as the linkers sometimes present between modules in modular proteins.

Section 2: Construction of Homology and Analogy Models

When significant sequence identity is shared between two molecules it is possible to construct a high-resolution homology model for one based on the coordinates of the other. In this context "significant" means that the pairwise sequence identity for two stretches of at least 80 residues in length exceeds 25%. There are many ways in which to generate a homology model and a detailed description is outwith the scope of this chapter; however, a typical homology modeling exercise would proceed as described by Rossjohn *et al.*[8] The sequences are aligned using a computational package such as AMPS and a homology model constructed using the appropriate module (e.g., HOMOLOGY) of a commercial molecular graphics software package (e.g., Insight II, Molecular Simulations Inc., San Diego, CA). This process has two main stages. First, sizable stretches of sequence identity are identified for which coordinates for the unsolved molecule are assigned by a straightforward mapping onto the coordinates of the known molecule. Second, the remaining structure is modeled from a database of coordinates for peptide fragments. The model then has to be screened for steric clashes (using, e.g., O) and has to be energy minimized (using, e.g., X-PLOR). A measure of the success of the homology modeling process is the rms deviation for the $C\alpha$ (in the case of a protein) positions on superposition of the two structures. Further testing using a Ramachandran plot and comparison of secondary structure content with that measured for the test protein by circular dichroism give greater confidence in the validity of the final model.

A less rigorous approach is analogy modeling which is applicable when the sequence identity between two stretches of polypeptide of 80 residues or more drops to below 10%. This type of model construction relies on the existence and detection of a protein fold that has low sequence identity to the fold of interest, but which is known to be similar in structure on the basis of other data. This type of modeling is reviewed by Perkins *et al.*[21] and is particularly suited to proteins (or glycoproteins) whose domains are members of structural superfamilies.

Glycosylation of proteins often prevents their successful crystallization in an intact form. This may be because the carbohydrate moieties are flexible and inhibit efficient crystallization or because often the glycosylation is heterogeneous in a particular glycoprotein preparation. This is another

[21] S. J. Perkins, C. G. Ullman, N. C. Brissett, D. Chamberlain, and M. K. Boehm, *Immunol. Revs.* **163,** 237 (1998).

instance when hydrodynamic bead modeling enables the user to propose a set of structures consistent with lower resolution data. It is normally possible to analyze the carbohydrate content of a glycoprotein with mass spectrometry. The relatively limited set of rules that describes protein glycosylation makes their modeling more straightforward than may be initially apparent. First, there are two types of protein glycosylation: N-linked (via the amide nitrogen of an asparagine residue) and O-linked (via the hydroxyl groups of serine, threonine, or hydroxylysine). This narrows down the numbers of locations on a homology or analogy model or actual high-resolution model to which carbohydrates can be attached. Second, N-linked saccharides always have a unique core structure composed of two *N*-acetylglucosamine residues linked to a branched mannose traid. Many other sugar residues can then be added to the free mannose units to build a more complex oligosaccharide. The carbohydrate residue linked to the protein in O-linked saccharides is usually *N*-acetylgalactosamine (although mannose, galactose, and xylose residues are also found). Therefore, if the composition of these carbohydrate moieties can be determined, then they can be modeled from coordinates available on the Protein Data Bank[22] (PDB, http://www.rcsb.org/pdb/). This approach was used in a complex and elegant study of the solution conformation of human carcinoembryonic antigen (CEA, for which small-angle X-ray and neutron scattering data had been acquired) by Boehm *et al.*[23] The authors identified more than 50 glycoprotein coordinate files on the PDB and used the coordinates for the largest oligosaccharides as starting models for the moieties known to comprise the 50% carbohydrate of the CEA.

Similarly, many molecular graphics packages enable the user to construct models for *A, B,* and *Z*-conformer double-stranded DNA, based on standard base-pairing formats. These are useful as starting points for modeling nucleic acids but this aspect of hydrodynamic bead modeling is fraught with other drawbacks. Programs such as HYDRO and the COEFF strand of BEAMS treat the molecule represented by the hydrodynamic bead model as a rigid body. This is valid for double-helical DNA of fewer than 100 base pairs where the contour length of the molecule is shorter than its persistence length.[24] If a macromolecule is thought to embody

[22] E. E. Abola, F. C. Bernstein, S. H. Bryant, T. F. Koetzle, and J. Weng, *in* "Crystallographic Databases—Information Content Software Systems Scientific Applications" (F. H. Allen, G. Bergerhoff, and R. Sievers, eds.), p. 107. Data Commission of the International Union of Crystallography, Bonn, Cambridge, Chester, 1987.

[23] M. K. Boehm, M. O. Mayans, J. D. Thornton, R. H. J. Begent, P. A. Keep, and S. J. Perkins, *J. Mol. Biol.* **259,** 718 (1996).

[24] J. Gracía de la Torre, S. Navarro, and M. C. Lopez Martinez, *Biophys. J.* **66,** 1573 (1994).

domains that are connected by flexible linkers, then the bead model that successfully regenerates experimentally derived parameters represents the time-averaged conformation. Elongated double-helical DNA is not a rigid body. It can be thought of as many elements flexibly linked and is certainly best modeled as such.[25,26]

Proteins can also embody one or more points of segmental flexibility, and the computation of the hydrodynamic properties in such cases has been the subject of intensive study (reviewed by García de la Torre[12]). From the bead modeler's point of view, the problem is to generate many conformations intermediate between the limiting ones, and to then average them, choosing an appropriate weighting factor. The FLEX and SUPFLEX modules of the BEAMS suite tackle this problem in a semiautomated way for models containing just one point of segmental flexibility.[13] Clearly, this is a subset of a field where many approaches are possible, and indeed an interesting study of the solution dynamics of a two-module pair from the protein fibronectin, based on NMR data coupled to bead modeling, has recently appeared in the literature.[27]

Section 3: Editing Coordinate Files for Construction of Bead Models

Once the final high-resolution coordinate file is ready, the bead model can be generated. At least two programs are available that facilitate the conversion of atomic coordinates to bead coordinates. The ASA and TRANS modules of BEAMS are being designed for this purpose and will directly accept files in the PDB format; the program AtoB can translate PDB format coordinates for proteins, carbohydrates, and nucleic acids into bead models of selected resolution (see Section 4). AtoB is a FORTRAN program requiring little computer power and lots of development. One of its drawbacks is that the coordinate file that is used as input must follow an exact format. It must thus be edited so that there are no missing atoms from residues and so that residues contain the correct number of atoms in the correct order. The PDB uses a standard format (described comprehensively at http://www.rcsb.org/pdb/docs/format/pdbguide2.2/guide2.2_frame.html). The format of input files for AtoB is given in Fig. 3 for a model constructed from atomic coordinates (see Fig. 10 in a later section for an example of *ab initio* model construction with AtoB). In this case the program is instructed to convert the 3681 lines of atomic resolution

[25] M. L. Huertas, S. Navarro, M. C. Lopez Martinez, and J. García de la Torre, *Biophys. J.* **73,** 3142 (1997).

[26] H. Jian, A. V. Vologodskii, and T. Schlick, *J. Comput. Phys.* **136,** 168 (1997).

[27] V. Copié, Y. Tomita, S. K. Akiyama, S. Aota, K. M. Yamada, R. M. Venable, R. W. Pastor, S. Krueger, and D. A. Torchia, *J. Mol. Biol.* **277,** 663 (1998).

```
123456789*123456789*123456789*123456789*123456789*123456789*123456789*
pneumolysin homology model
'xtal'        !beads (bead) or crystallographic (xtal)?
'leave'       !manipulate final model (manip) or not (leave)?
3681          !number of lines of coordinates to be read in (ibdn)
0.727         !vbar (bead model)
464, 7.0      !number of residues, resolution (atomic)
1             !fixed (1) or mixed (2) bead radii? (atomic)
'atoms'       !atom (atoms) or resdiue (resid) coordinates? (atomic)
ATOM      1  CB  VAL     6      18.254 135.923 -18.266  1.00 25.00     6
ATOM      2  CG1 VAL     6      19.542 136.688 -17.988  1.00 25.00     6
ATOM      3  CG2 VAL     6      18.526 134.475 -18.655  1.00 25.00     6
ATOM      4  C   VAL     6      18.003 135.774 -15.717  1.00 25.00     6
ATOM      5  O   VAL     6      17.791 136.537 -14.772  1.00 25.00     8
ATOM      6  N   VAL     6      16.632 137.338 -16.973  1.00 25.00     7
ATOM      7  CA  VAL     6      17.278 135.999 -17.049  1.00 25.00     6
ATOM      8  N   ASN     7      18.842 134.743 -15.650  1.00 25.00     7
ATOM      9  CA  ASN     7      19.606 134.381 -14.451  1.00 25.00     6
ATOM     10  CB  ASN     7      20.809 133.506 -14.831  1.00 25.00     6
ATOM     11  CG  ASN     7      20.407 132.214 -15.526  1.00 25.00     6
ATOM     12  OD1 ASN     7      19.250 132.026 -15.902  1.00 25.00     8
ATOM     13  ND2 ASN     7      21.369 131.321 -15.710  1.00 25.00     7
ATOM     14  C   ASN     7      20.097 135.532 -13.569  1.00 25.00     6
ATOM     15  O   ASN     7      20.177 135.382 -12.349  1.00 25.00     8
.
.
.
ATOM   3672  N   GLU   469      13.706  97.405  56.161  1.00 25.00     7
ATOM   3673  CA  GLU   469      12.638  96.913  55.294  1.00 25.00     6
ATOM   3674  CB  GLU   469      11.530  96.248  56.122  1.00 25.00     6
ATOM   3675  CG  GLU   469      11.944  94.983  56.862  1.00 25.00     6
ATOM   3676  CD  GLU   469      10.848  94.450  57.766  1.00 25.00     6
ATOM   3677  OE1 GLU   469       9.828  93.953  57.243  1.00 25.00     8
ATOM   3678  OE2 GLU   469      11.008  94.529  59.002  1.00 25.00     8
ATOM   3679  C   GLU   469      12.051  98.094  54.515  1.00 25.00     6
ATOM   3680  O   GLU   469      12.439  99.250  54.799  1.00 25.00     8
ATOM   3681  OT  GLU   469      11.198  97.861  53.634  1.00 25.00     8
```

FIG. 3. The input for AtoB incorporating the atomic coordinates of the pneumolysin homology model. The resultant output is shown in Fig. 8. The numbers across the top are for guidance only and are not actually included as part of the input data. See text for a full explanation of the other parameters.

crystallographic coordinates (containing 464 residues) for the pneumolysin homology model[8] into a bead model composed of equally sized beads using a cubic grid of 7 Å dimension to subdivide the coordinate data set. The partial specific volume in this case is a dummy variable: AtoB will calculate the actual value from the values of $\bar{v}$ for the constituent amino acids.

As it currently stands, the program will read in coordinates for the chemical groups listed in Table II. Therefore, if for some reason part of a protein residue side chain is missing from the starting PDB file, AtoB will alert the user in the first pass of the coordinate file through the program.

TABLE II
NUMBER OF LINES OF COORDINATES AtoB IS PROGRAMMED TO READ IN FOR EACH CHEMICAL GROUP LISTED

ALA	5	GLY	4	MET	8	SER	6	FUC	10	A	21
CYS	6	HIS	10	ASN	8	THR	7	GAL	11	C	19
ASP	8	ILE	8	PRO	7	VAL	7	MAN	11	G	22
GLU	9	LYS	9	GLN	9	TRP	14	NAG	14	T	20
PHE	11	LEU	8	ARG	11	TYR	12	SIA	20	U	19

The user will then have to either remove this residue altogether [and accordingly alter the information on the fourth and sixth lines of the input file (Fig. 3)] or edit the PDB file to mimic the missing coordinates. One way in which to do this is to take a complete set of coordinates for the side chain in question from a data set where it has been successfully visualized and to translate the coordinates so that the Cα atoms coincide and the atoms of the side chain do not sterically clash with those in their immediate vicinity. An alternative is to simply assign the missing atoms the Cartesian coordinates of the Cα atom. Because the resolution of bead models tends to be considerably lower than atomic resolution the consequences of either action are usually minimal provided there are not too many occurrences of incomplete residues. For example, with the pneumolysin homology model in Fig. 4a, if the coordinate data sets for all 30 lysines are modified so that the β, γ, δ, and ε carbons and the amino nitrogen are located at the α carbon (see Fig. 4b) the resultant bead model produces only a 0.9% change in the radius of gyration (R_g) computed by HYDRO. Similarly [η] and s change by 0.3% (Fig. 5). In fact, if all the residues are represented simply by the positions on their α carbons (AtoB has an option to read in α carbon-only files) then R_g changes by 2.1%, [η] by 2.3%, and s by 1.1%, so this approach to dealing with missing atoms can be used even in extreme cases for relatively globular molecules.

AtoB does not read in hydrogen atoms and so any PDB file with these included needs to be edited accordingly. This can often easily be done by selecting a display mode on a molecular graphics package that omits hydrogens, and then saving the resultant model as a text (PDB-style) file. In AtoB the mass of the hydrogens is automatically incorporated into the masses of the atoms to which they are bonded.

Section 4: Model Resolution

The resolution of a hydrodynamic bead model is governed by a number of considerations and parameters. In HYDRO the computer time needed to solve the equation for the hydrodynamic interaction of N beads increases

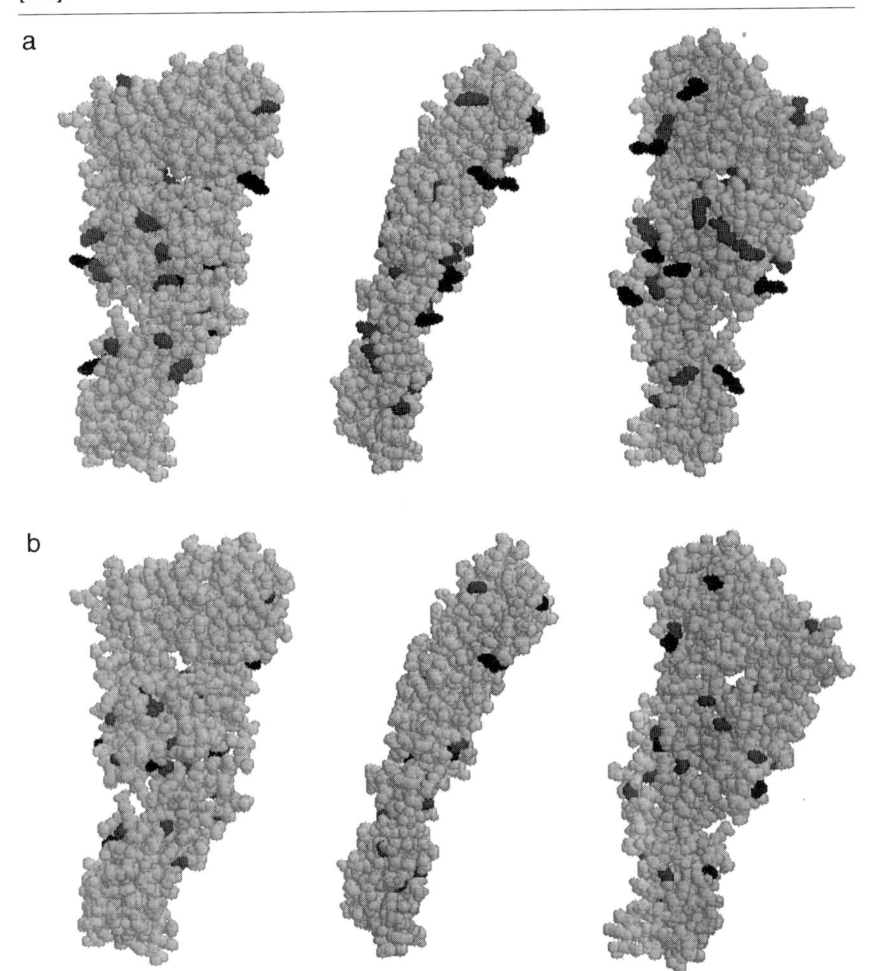

FIG. 4. Space-filling representation of the pneumolysin homology model. In (a) the lysine residues are highlighted; those that particularly extend into solution are shown in the darkest gray, while those that are less likely to remain unvisualized in a proper crystal structure are shown in medium gray. In (b) all the lysines are reduced to a single Cα atom.

at a rate close to N^3. Similarly, a table reporting the performances of the COEFF modules for the BEAMS package can be found in the paper by Spotorno *et al.*[13] As computers become faster the time needed to compute hydrodynamic parameters is becoming less of a limitation, nonetheless it is still often useful to reduce the number of elements in a file to minimize this time. In AtoB this is done by selecting the unit dimension of the

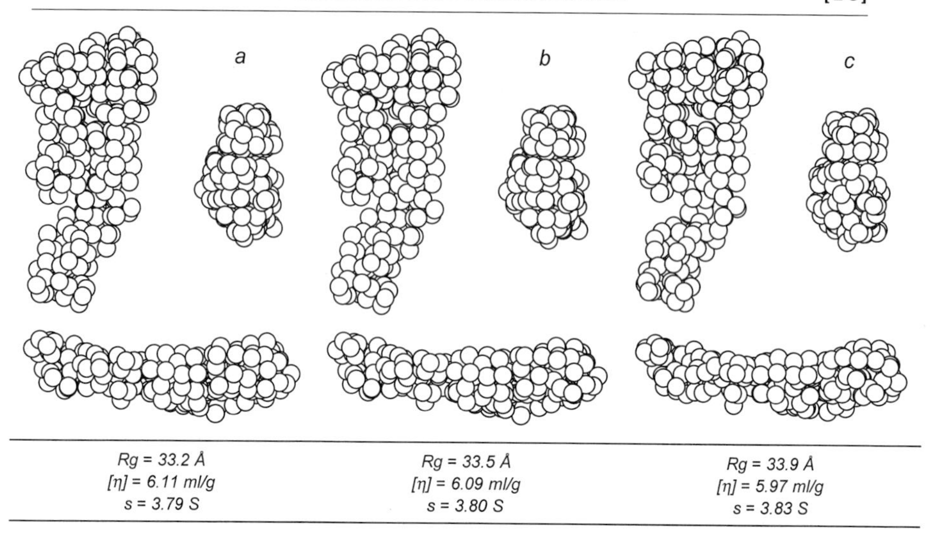

FIG. 5. Bead models (shown in three orthogonal views) constructed from the coordinates for the pneumolysin homology model using AtoB and a resolution of 7 Å. (a) Direct transformation model, using all the atoms in the original coordinate file; (b) the model generated from a coordinate file in which all the lysines are represented only by the Cα; (c) the model that results when all the residues are represented only by their Cα atoms.

cubic grid into which the constituent molecular coordinates are subdivided. Another consideration is the sensitivity of computed parameters to this resolution. From Fig. 6 it is clear that the radius of gyration (R_g) computed by HYDRO is sensitive to the resolution used to construct the bead model if the beads are all of uniform size, whereas using a mixed population of sizes the R_g generated at all resolutions tends to that computed at high resolution. In this example the resolution extends to 7 Å where the finer details of surface shape are clearly apparent (Fig. 7) and where the computation times are not preclusive. It is important not to use a resolution so low as to result in a loss of important shape information from the model and clearly, from Fig. 6, gives hydrodynamic parameters that are at odds with those obtained from a higher resolution model. In protein modeling it is advisable to aim for at least one bead per amino acid residue (a resolution of about 6–7 Å). The ASA and TRANS modules of BEAMS use two beads per amino acid residue (to represent the main chain and side chain), but a reduction in the number of beads is achieved by identifying buried beads and not using them in the subsequent hydrodynamic computations.

At this point, it is important to mention one of the current major drawbacks of bead modeling that is related to the poor packing characteristics of spheres: the mathematical formalisms used to compute the hydrody-

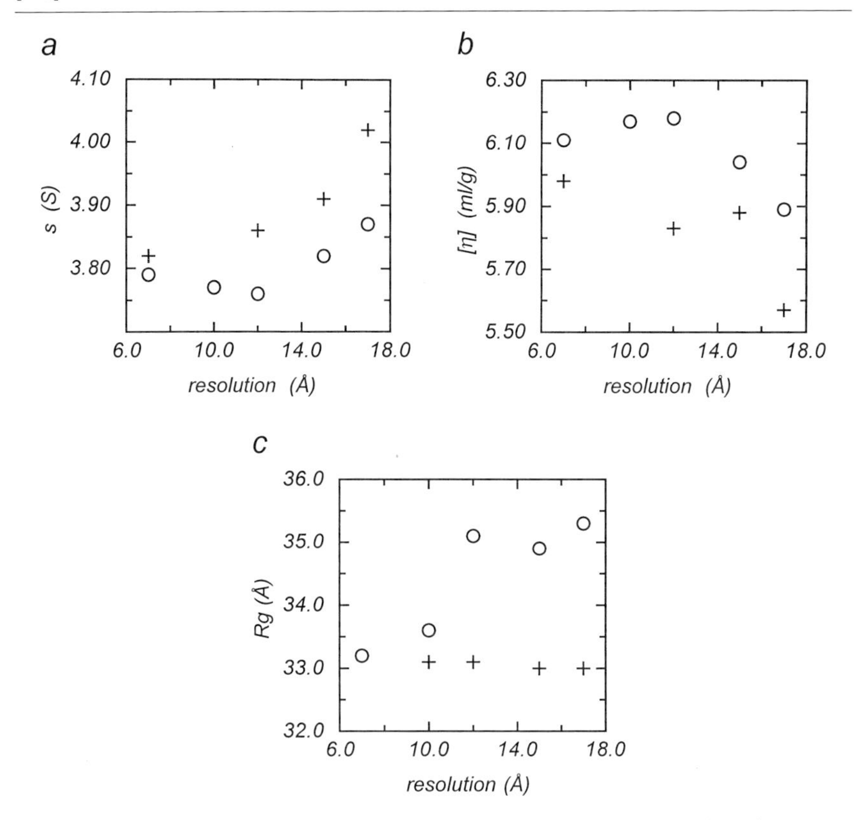

FIG. 6. The dependence on model resolution of selected computed hydrodynamic parameters: (a) sedimentation coefficient, s; (b) intrinsic viscosity, $[\eta]$; (c) radius of gyration R_g. The symbols refer to models composed to beads of equal radii (○) and of mixed radii (+).

namic properties of the ensembles can cope with overlapping beads only if all the beads are of the same size, and even in this case caution should be exerted, because multiple overlaps of the same bead(s) with many other beads are difficult to account for (this can be minimized by using many small beads). In models composed of overlapping beads of different sizes HYDRO reverts to the calculation of the unmodified Burgers–Oseen tensor to quantify the hydrodynamic interaction, thus incurring a small error. Because the utilization of beads of different size often results in a better model (see Fig. 7) it is important to have the option of both approaches.

Section 5: Generation of Bead Models

The output from AtoB takes the form of the example shown in Fig. 8 for the pneumolysin homology model. The partial specific volume is

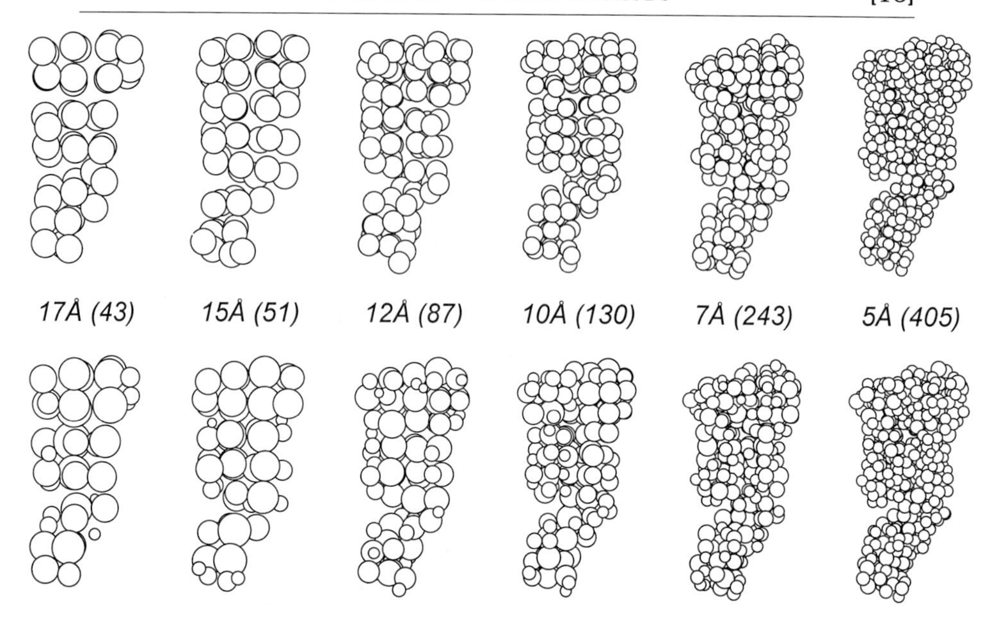

FIG. 7. Bead models at a series of resolutions constructed by AtoB from the pneumolysin homology model coordinates. The numbers in parentheses refer to the number of beads in the model. The models in the top row are composed of beads of equal sizes, those beneath are made from beads with differing sizes.

calculated from the $\bar{v}$ for the constituent amino acid residues ($\bar{v} = \Sigma \bar{v}_i w_i / \Sigma w_i$ where w is the weight fraction of each residue i), which are given in an indispensable paper by Perkins.[28] The hydration is an estimated value based on data originally reported by Cohn and Edsall[29] and later by Durchschlag[30] and should be taken only as a very rough estimate (it pertains only to amino acid residues and therefore hugely underestimates the hydration of, e.g., glycoproteins). The values that AtoB uses to calculate $\bar{v}$ for oligonucleotides are average values: for oligonucleotides, unlike proteins, $\bar{v}$ is hugely sensitive to the type and concentration of salt in the surrounding buffer (see Table III and also Ref. 31). Therefore an appro-

[28] S. J. Perkins, *Eur. J. Biochem.* **157,** 169 (1986).

[29] E. J. Cohn and J. T. Edsall, *in* "Proteins, Amino Acids and Peptides as Ions and Dipolar Ions" (E. J. Cohn and J. T. Edsall, eds.), p. 155. Reinhold Publishing Corporation, New York, 1943.

[30] H. Durchschlag, *in* "Thermodynamic Data for Biochemistry and Biotechnology" (H.-J. Hinz, ed.), p. 45. Springer-Verlag, Berlin, Heidelberg, New York and Tokyo, 1986.

[31] V. A. Bloomfield, D. M. Crothers, and J. Tinoco, "Physical Chemistry of Nucleic Acids." Harper and Row, New York and London, 1974.

```
file is called   pneumolysin homology model
number of lines  3681
there are  464 residues and resolution is   7.0
fixed (1) or mixed (2) radii? 1
reading in  464 residues
vbar calculated from atomic data is       0.736 ml/g
hydn calculated from atomic data is       0.367 g/g
mass calculated from atomic data is   52141.320 g/mole
X-ray scattering length is            78300.172 fm
neutron scattering length (2H2O) is   20684.219 fm
neutron scattering length ( H2O) is   11948.014 fm
final model of pneumolysin homology model
contains   243 beads
  n        x          y          z          r
     1   22.8297   102.7985   -15.8067     3.9716
     2   14.3104   112.6872   -17.8968     3.9716
     3    7.4193   117.4187   -17.5402     3.9716
     4   12.3469   116.3350   -20.0318     3.9716
     5    6.8998   125.0048   -18.3470     3.9716
.
.
.
   240   14.8787   100.0607    82.5806     3.9716
   241   21.0743   102.6131    84.8450     3.9716
   242   23.2822   109.3365    87.0098     3.9716
   243   26.5876   110.4680    84.1487     3.9716
```

FIG. 8. The output from AtoB for the atomic coordinates of the pneumolysin homology model in Fig. 7 which have been converted into a bead model of 243 equally sized beads (with radii of 3.97 Å). See text for a full explanation of the computed parameters.

TABLE III

PARTIAL SPECIFIC VOLUMES OF POLYNUCLEOTIDES[a]

poly(X)	0.05 *M* Tris-HCl pH 8.0	0.05 *M* HEPES–NaOH pH 7.6	0.10 *M* Sodium phosphate pH 7.0	0.04 *M* Tris–acetate pH 5.0
A	0.5219	0.5362	0.4765	0.5287
C	0.5666	0.5841	0.5286	0.5613
G	0.6193	0.6102	0.5513	0.5548
T	0.5227	0.5522	0.4527	0.5498
U	0.4061	0.4978	0.5431	0.5443

[a] Values (in ml/g) were determined via precision densimetry in a series of buffers of differing ionic strength, composition and pH. All determinations were performed at 20°.

priate $\bar{v}$ for the constituent nucleotides should ideally be used or $\bar{v}$ should be measured (either with a precision densimeter[32] or by performing sedimentation equilibrium studies in an analytical ultracentrifuge for the sample in a series of buffers which differ only in their H_2O/D_2O content and thus their density[33]). The mass and X-ray and neutron scattering lengths are calculated from the respective sums of the masses and scattering length of the constituent residues (as reported in Ref. 34). The final file of coordinates is then used as input for the bead model generation program (in this case AtoB). It is a good idea to examine the resultant model (e.g., using MacBeads or GRUMB) prior to using its coordinates as input data for the generation of hydrodynamic parameters.

Section 6: Generation of Hydrodynamic Parameters

The coordinates of the bead model are then incorporated with other data in the input file for the hydrodynamic bead modeling program that is being used. HYDRO, for example, requires the temperature, solvent viscosity, unit of length (for the input coordinates to convert them to units of centimeters), molecular weight, buoyancy factor $(1 - \bar{v}\rho)$, number of beads, and then the Cartesian coordinates and radii of the constituent beads. The COEFF programs of the BEAMS suite use a different file format (described in Ref. 13).

Section 7: Normalization for Hydration

Hydration as it pertains to the hydrodynamic situation is different from that recognized by crystallographers, by NMR spectroscopists, by small angle X-ray solution scatterers, and by small-angle neutron scatterers—and these are all different from each other. The hydration salient to hydrodynamics is the poorly defined shell of water which (in the case of the sedimentation coefficient measured by sedimentation velocity) moves with the sedimenting molecule. It has been considered in some detail in an empirical study by Squire and Himmel[35] who calculated an average hydrodynamic hydration of 0.53 ± 0.26 g H_2O/g prctein. Perkins (see, e.g., Ref. 36), uses

[32] O. Kratky, H. Leopold, and H. Stabinger, *Methods Enzymol.* **27,** 98 (1973).

[33] J. A. Reynolds and C. Tanford, *Proc. Natl. Acad. Sci. U.S.A.* **73,** 4467 (1976).

[34] S. J. Perkins, *in* "Modern Physical Methods in Biochemistry" (A. Neuberger and L. L. M. Van Deenen, eds.), Part B, p. 143. Elsevier Science Publishers B. V. (Biomedical Division), Amsterdam, New York, and Oxford, 1988.

[35] P. G. Squire and M. E. Himmel, *Arch. Biochem. Biophys.* **196,** 165 (1979).

[36] S. J. Perkins, A. W. Ashton, M. K. Boehm, and D. Chamberlain, *Int. J. Biol. Macromol.* **22,** 1 (1998).

a standard value (0.3 g H_2O/g protein) for hydrodynamic and small-angle X-ray scattering data modeling. In the latter instance, the volume of the water molecule is reduced (from 0.0299 to 0.0245 nm^3) to account for electrostriction. The uniformly hydrophilic nature of sugar residues combined with their essentially open structure points toward a likely elevated level of hydrodynamic hydration, compared with that accepted for proteins (although there is little information in the literature on an approximate level for this hydration).

Modeling particles with variable scattering density is not a problem for small-angle solution scattering data (see below) but as yet there is no mechanism for accounting for regions of differential mass density within a hydrodynamic bead model. Therefore how should hydration (with a density approaching 1.0 g/ml) be represented on an otherwise anhydrous model (with a density in the region of 1.3–1.4 g/ml)? This is discussed at some length in an earlier paper.[15] One approach is to uniformly expand the anhydrous model so that its volume encompasses the layer of surface hydration. This is a reasonable approach to adopt with globular molecules but is inappropriate for elongated particles where uniform expansion of all the beads would result in a disproportionate increase in size along the longest molecular axis. Alternatively the approach of Teller *et al.*[37] can be used where beads representing water molecules (i.e., of 3.8 Å diameter) are placed onto beads representing surface charged or polar groups. A variant of this is incorporated into the algorithm HYPRO,[38] which adds a layer of hydration spheres evenly over the protein surface. This uniform layer tends to correspond to the standard 0.3 g H_2O/g protein used by Perkins and colleagues.

A different approach has been followed by Rocco and colleagues[19,20] based on the seminal work by Kuntz and Kauzmann[39] who determined the *average* number of bound water molecules for each amino acid residue on the basis of NMR freezing experiments. Rocco *et al.* include these water molecules directly into each bead of a model generated in the absence of high-resolution data. The program PROMOLP uses the theoretical hydration values of Kuntz and Kauzmann to calculate the hydrated molecular volume. The values reside in a table and can be modified at will by the user. Putative values for carbohydrate moieties and phosphate prosthetic groups have also been calculated and are included in the PROMOLP table.

[37] D. C. Teller, E. Swanson, and C. de Haen, *Methods Enzymol.* **61,** 103 (1979).

[38] A. W. Ashton, M. K. Boehm, J. R. Gallimore, M. B. Pepys, and S. J. Perkins, *J. Mol. Biol.* **272,** 408 (1997).

[39] I. D. Kuntz and W. Kauzmann, *in* "Advances in Protein Chemistry" (C. B. Anfinsen, J. T. Edsall, and F. M. Richards, eds.), Vol. 28, p. 239. Academic Press, 1974.

This approach is currently being extended to the models generated by ASA and TRANS from *high*-resolution data, the volume occupied by the theoretically bound water of hydration being included in each bead.

A further option is to correct for hydration algebraically (see, e.g., Ref. 6). Thus for the sedimentation coefficient

$$s_\delta = s_0 \left(\frac{\bar{v}}{\bar{v} + \delta v_0^1} \right)^{1/3} \tag{1}$$

where s_δ is the hydrated (i.e., experimental) sedimentation coefficient, s_0 is the anhydrous sedimentation coefficient (as computed by, e.g., HYDRO), $\bar{v}$ is the partial specific volume, v_0^1 is the specific volume of the solvent (which approximates to its reciprocal density), and δ is the level of hydrodynamic hydration. For the intrinsic viscosity

$$[\eta]_\delta = [\eta]_0 \left(1 + \frac{\delta v_1^0}{\bar{v}} \right) \tag{2}$$

where $[\eta]_\delta$ is the hydrated (i.e., experimental) intrinsic viscosity and $[\eta]_0$ is the anhydrous intrinsic viscosity. Comparison of the data sets in Fig. 9 reveals that, at least for the sedimentation coefficient and intrinsic viscosity, there is no discernible difference in the hydrated parameter obtained either by algebraic or uniform expansion hydration for the marginally globular protein, pneumolysin. Other parameters are variously affected by hydration; for example the radius of gyration measured with small-angle X-ray scattering is increased from the anhydrous value by a layer of electrostricted water, whereas that obtained from a small-angle neutron scattering study

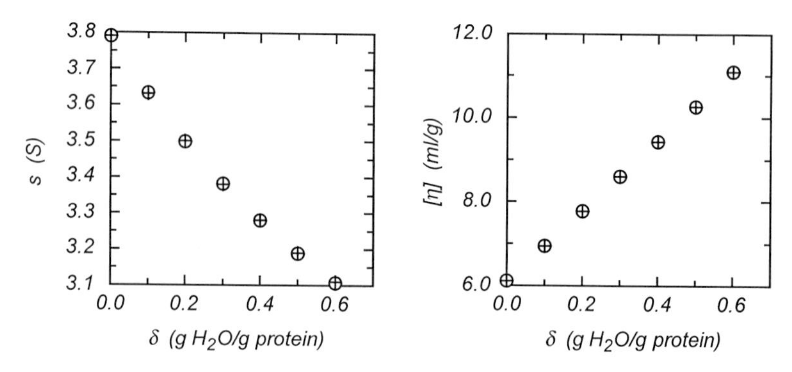

FIG. 9. The dependence of the sedimentation coefficient, s, and the intrinsic viscosity, $[\eta]$, on hydration for the pneumolysin model as implemented in the form of Eq. (1) for s and Eq. (2) for $[\eta]$ (○) and as a uniform expansion (+).

will depend crucially on the H_2O/D_2O composition of the solvent.[40] Ultimately it is vital to know how the parameter to which the hydrodynamic bead model is being constrained was derived and how precisely it depends on hydration.

Section 8: Adjustment of Model

Several aspects of a hydrodynamic bead model can be altered in order to obtain better agreement between the physical parameters predicted for it computationally and those that have been measured experimentally. Unfortunately the hydration is often a partial unknown (see above) and alteration of this can result in the convergence of modeled and experimental characteristics. Alternatively (or concurrently) points of potential or putative segmental flexibility within the macromolecule can be identified and the domains so connected can be rotated with respect to one another in a stepwise fashion. This can be done manually by generating successive bead models using, e.g., AtoB, MacBeads, or GRUMB or it can be done automatically with the type of approach described by Beavil *et al.*[41] using the program AUTOSCT (which unfortunately is not freely available but could be programmed using macros available in molecular modeling programs such as INSIGHT II). AtoB provides an option to manipulate the bead model using an unlimited series of geometrical transformations (coded in the TRIG subroutine). The options are detailed in Table IV and illustrated by an example in Fig. 10. Any beads created in a manipulation procedure are added on to the end of the current set of coordinates and as a result it is important to keep track of bead additions and deletions. The options for rotating and shifting beads actually create a new set of beads which are added to the current file of coordinates. For models with just one point of segmental flexibility, the FLEX routines of the BEAMS suite[13] can automatically generate all the intermediate models between two limiting

[40] D. I. Svergun, S. Richard, M. H. J. Koch, Z. Sayers, S. Kuprin, and G. Zaccaï, *Proc. Natl. Acad. Sci. U.S.A.* **95,** 2267 (1998).
[41] A. J. Beavil, R. J. Young, B. J. Sutton, and S. J. Perkins, *Biochemistry* **34,** 14449 (1995).
[42] S. J. Perkins, *Biochem. J.* **228,** 13 (1985).
[43] L. Gregory, K. G. Davis, B. Sheth, J. Boyd, R. Jefferis, C. Nave, and D. R. Burton, *Mol. Immunol.* **24,** 821 (1987).
[44] S. J. Perkins, *FEBS* **271,** 89 (1990).
[45] S. J. Perkins, K. F. Smith, J. M. Kilpatrick, J. E. Volanakis, and R. B. Sim, *Biochem. J.* **295,** 87 (1993).
[46] S. J. Perkins, K. F. Smith, and R. B. Sim, *Biochem. J.* **295,** 101 (1993).
[47] T. Hellweg, W. Eimer, E. Krahn, K. Schneider, and A. Müller, *Biochim. Biophys. Acta* **1337,** 311 (1997).
[48] R. D. Mullins, W. F. Stafford, and T. D. Pollard, *J. Cell Biol.* **136,** 331 (1997).

TABLE IV
GEOMETRICAL MANIPULATIONS POSSIBLE WITH TRIG SUBROUTINE OF AtoB

Letter read by AtoB	Action taken	Additional data needed[a]
a	adds bead(s)	Number of beads to be added *i, x, y, z, r*
c	calculates coordinates of a circular arrangement of touching beads, incorporates them into model	Diameter of circle *x, y, z* of circle center Number of beads in circle
d	deletes bead(s) *i*1 to *i*2 (inclusive) from model	i1, *i*2
e	"hydrates" model by expanding beads *i*1 to *i*2	*i*1, *i*2 Hydration (g H_2O/g solute)
l	lists coordinates of current model	—
r	rotates beads *i*1 to *i*2 about selected axis by required angle	Axis (*x, y,* or *z*) Angle (a real number) *i*1, *i*2
s	shifts beads *i*1, *i*2 by translational vector a defined number of times	*x, y, z* (vector) Number of translations *i*1, *i*2

[a] *i*, Bead index; *x, y, z,* Cartesian coordinates of beads; *i*1, index of start of a group of beads of which the last bead has index *i*2. See Fig. 10 for a further demonstration.

conformations, compute the average hydrodynamic parameters in the absence of restoring forces, and optionally save all the conformations for further analysis in the rigid body approximation by the SUPLEX routines with or without restoring forces.

Examples of Applied Hydrodynamic Bead Modeling

It is useful to divide published work containing applied hydrodynamic bead modeling into two broad categories: early work that was completed concurrently with the development of hydrodynamic theory and later work that uses the results of this pioneering body of literature to concentrate solely on characterizing solution macromolecular conformation as a means to an end. Good examples from the second category are given in Table V, which by no means represents an exhaustive record of the literature but should provide a comprehensive background on which to build a fuller understanding of how others have used this modeling method.

An example of a system whose X-ray structure does not fully describe its functional conformation is the siding clamp of the bacteriophage T4 DNA polymerase (gene product 45, or gp45). The high-resolution coordinates show a closed, torus-shaped trimeric complex, while analytical ultra-

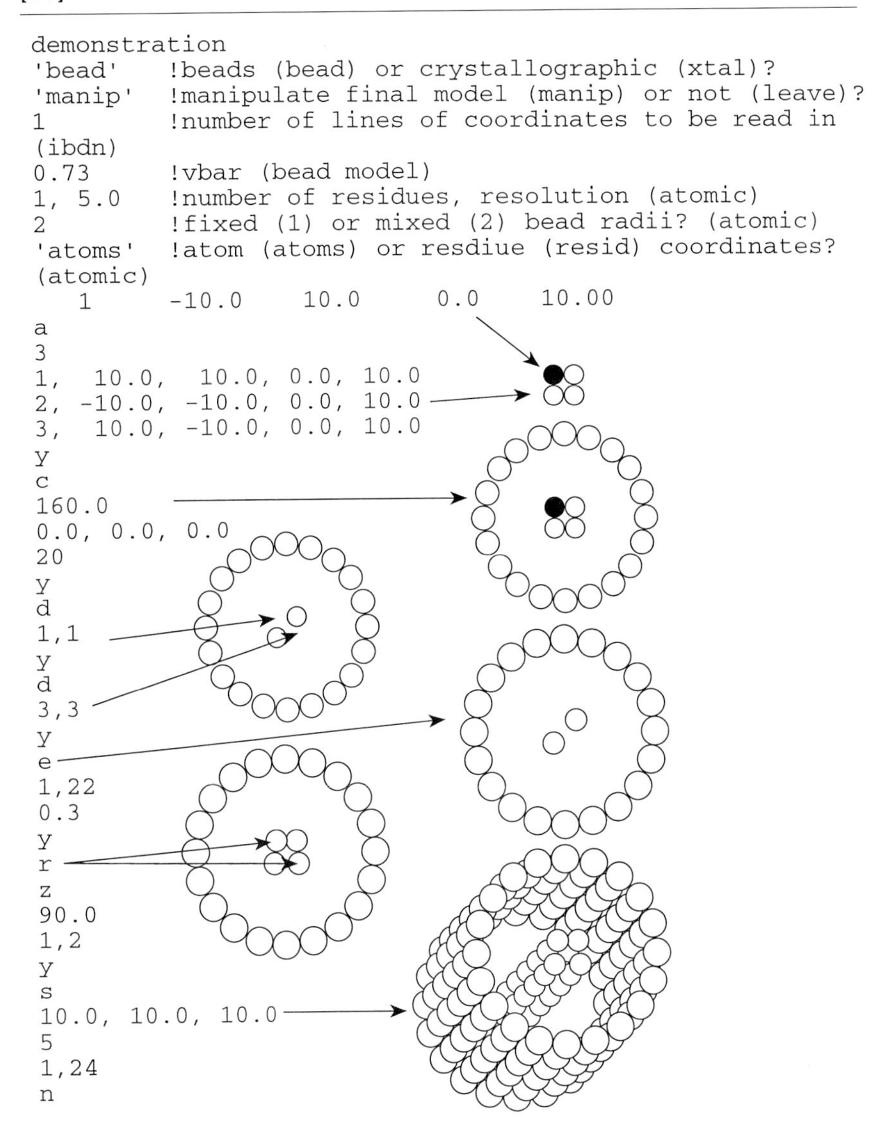

FIG. 10. An input file for AtoB to create *ab initio* a bead model. The various stages of construction are shown, related to the parts of the input file responsible for their creation.

TABLE V
PAPERS THAT HAVE MADE SIGNIFICANT USE OF HYDRODYNAMIC BEAD MODELING OF BIOLOGICAL MARCOMOLECULES

Program(s) used	System modeled	Reference
GENDIA and GENTRA, forerunners of HYDRO	C1q and its complex with $C1r_2C1s_2$	Perkins (1985)[42]
Forerunners of PROMOLP and STRINGS	Fibronectin under different solution conditions	Rocco *et al.* (1987)[19]
GENDIA and GENTRA	Subclasses of human IgG	Gregory *et al.* (1987)[43]
GENDIA	$C\bar{1}$ inhibitor and its complex with $C\bar{1}s$	Perkins (1990)[44]
GENDIA	Serine protease fold (in, e.g., β-trypsin)	Perkins *et al.* (1993)[45]
GENDIA	Factor 1 of human complement	Perkins *et al.* (1993)[46]
PROMOLP, GRUMB, and BEAMS	$\alpha_{IIb}\beta_3$ integrin	Rocco *et al.* (1993)[20]
HYDRO	Double-helical DNA	García de la Torre *et al:* (1994)[24]
TRV (a subroutine of HYDRO)	Pneumolysin bacterial toxin	Morgan *et al.* (1994)[4]
TRV	14S dynein fractions	Tharia *et al.* (1997)[17]
Modified form of TRV or HYDRO	Nitrogenase	Hellweg *et al.* (1997)[47]
None; used Riseman–Kirkwood calculation	Arp2/3 complex	Mullins *et al.* (1997)[48]
TRV?	Pair of fibronectin type III modules	Copié *et al.* (1998)[27]

centrifugation, fluorescence, and hydrodynamic modeling studies performed by Alley *et al.*[49] suggest an open, trimeric complex, with the opening occurring within the plane of the ring. Analytical ultracentrifugation of wild-type gp45 yielded a sedimentation coefficient strongly dependent on concentration, with $s^0_{20,w}$ (s at 20° in water corrected to infinite protein dilution) equal to 4.0 S. A mutant containing an intrasubunit disulfide that constrained the trimer yielded a sedimentation coefficient independent of concentration, with $s^0_{20,w}$ equal to 4.8 S. These results suggest an open complex wild-type gp45 and a closed complex for mutant gp45. Fluorescence resonance energy transfer (FRET) found that the interfaces for all three subunits were closed in the mutant gp45, while one subunit interface was open in wild-type gp45. Using the FRET distance constraints, the X-ray structure was changed to suggest solution structures consistent with the wild-type and mutant ultracentrifugation and fluorescence data. AtoB

[49] S. C. Alley, V. K. Shier, E. Abel-Santos, D. J. Sexton, P. Soumillion, and S. J. Benkovic, *Biochem.* **38,** 7696 (1999).

was used to make assemblies of beads from these new structures, and HYDRO was used to predict sedimentation coefficients. The X-ray structure yielded a predicted sedimentation coefficient of 4.72 S, whereas models in which the ring was opened (breaking one subunit interface) in the plane and out of the plane yielded 4.48 S and 4.77 S. Finally, the structure was puckered while maintaining the subunit interface contacts, yielding a predicted sedimentation coefficient of 5.33 S. When corrected for hydration (the amino acid sequence of gp45 suggests a value of 0.36 g H_2O/g protein), the model in which the ring was opened in plane was most consistent with the experimental sedimentation coefficient of wild-type gp45 (requiring 0.30 H_2O/g protein hydration), while the puckered structure was most consistent with the mutant gp45 (requiring 0.27 H_2O/g protein hydration).

Further Comments and Conclusions

This chapter has described one small branch of low-resolution modeling in some depth so that the reader should, after reading it, be in a position to attempt this type of modeling. But there are many similar types of modeling and this is a field that is developing in complexity, validity, and utility. The manual, iterative type of modeling outlined here will ultimately be superseded by automatic modeling as computers become increasingly powerful. A particularly exciting development in automated low-resolution modeling is described in a recent paper by Chacón *et al.*[50] who have combined the Debye formula (for the simulation of small-angle scattering curves from a bead model) with a genetic algorithm optimization tool so that bead models for macromolecules in solution can be generated from the experimental scattering curve alone. Briefly, their method, encoded in a FORTRAN program called DALAI_GA (GA standing for genetic algorithm), takes a family of starting bead models (e.g., beads in an ellipsoidal arrangement, hexagonally close packed), which they term as chromosomes, calculates their X-ray scattering curves (out to a scattering vector modulus, S, of 0.06 $Å^{-1}$), compares these curves with the experimental curve resulting in fitness parameters (the reciprocal of the square root of the sum of square distances between the simulated and experimental profiles), discards the half of the starting models with the lower fitness parameters, and cross-breeds and mutates the remaining 50% of the starting chromosomes (models) to yield new models, which are then processed as before. For starting models composed of 102 beads for the β_{b2}-crystallin fragment convergence

[50] P. Chacón, F. Moran, J. F. Díaz, E. Pantos, and J. M. Andreu, *Biophys. J.* **74,** 2760 (1998).

to the optimal model (containing 23 6 Å beads) was reached within about 60 generations. An additional masking process permits the conversion of the 6 Å resolution model to a 2 Å resolution starting model (or search space) with 676 beads, which results in a final model of 381 beads. The authors report that in a standard run with a starting model of 300 beads the average CPU time taken on a dedicated Silicon Graphics Indigo workstation (with an R4400 processor running at 150 MHz with 32 MB of RAM) was about 3 hr. All of the molecules used to illustrate this method were quite small (10–25 kDa)—the computing time would increase for larger molecules. The generation of a scattering curve is computationally much less intensive than the calculation of hydrodynamic parameters for a bead model. Sadly, the genetic algorithm as it stands consequently probably does not offer a realistic, more objective alternative to manual, iterative bead modeling of hydrodynamic parameters.

The work of Chacón *et al.*[50] indirectly highlights an extremely important limitation of all modeling processes but one that applies particularly to hydrodynamic bead modeling. The authors comment that even though the crystal structure of the β_{b2}-crystallin fragment has a resolution of 2.1 Å and the simulated solution scattering data they used to generate their low-resolution model extended only to 16.6 Å, this curve contains enough information to define the volume of the protein and its structure to an apparently far higher resolution; certainly by applying the masking algorithm it seems possible to obtain a highly detailed model without being guilty of overinterpretation of the experimental data. This is the converse of the situation with hydrodynamic bead modeling. Here, the aim is to obtain a model that reproduces a number of separate experimentally determined hydrodynamic parameters, and the important point is that there will be many different models that can yield a given value of a single parameter. Only by modeling several parameters simultaneously can the set of possible models be constrained to a meaningful model, but again it should be stressed that by this approach unique solutions can never be found.

Acknowledgments

I am grateful to Dawn Adkins for performing the painstaking measurements needed to compile the data in Table III and to Michael Parker for permitting me to use the coordinates of the pneumolysin homology model before details of this model are published and the coordinates are released on the PDB. I would also like to thank Stephen Alley for providing the details of the bacteriophage T4 DNA polymerase study and for kindly allowing me to use this as an illustrative example prior to full publication of his work. I am grateful to Robert Gilbert and Stephen Alley for critical reading of this chapter in draft form and for helpful comments during its composition, and I am indebted to Mattia Rocco for his significant contribution to not only this chapter but also to my understanding of the subject in a broader sense.

[17] Bayesian Hierarchical Models

By Christopher H. Schmid and Emery N. Brown

Introduction

Many random processes generating data for which statistical analyses are required involve multiple sources of variation. These sources represent randomness introduced at different levels of a nested data structure. For example, in a multicenter study, responses vary within individuals, between individuals, within sites, and between sites. When the multiple measurements taken on individuals are intended to be replicates, the variation is called measurement error.

For example, in a study of growth that we will explore in detail later, measurements of height were taken daily for 3 months on a young child. Despite careful attention to the measurement process, multiple observed readings taken consecutively in a 10-min time span were not identical. In addition to this measurement variability, each child measured varied with respect to the amount of growth and the time at which growth occurred. We will see how a hierarchical model can address these issues.

The standard statistical approach for describing processes with multiple components of variation uses mixed models of fixed and random effects. The fixed effects are quantities about which inference is to be made directly, whereas the random effects are quantities sampled from a population about which inference is desired. Typically, the variance components of the process represent the variances of these populations. An alternative method of description consists of a hierarchy of simpler models, each of which describes one component of variation with a single random effect.

Meta-analysis, a technique used to pool data from different studies in order to estimate some common parameter such as a treatment effect in medical studies, is a particular example of this type.[1–4] Assume that y_t represents the observed treatment effect from the tth study. The objective of the analysis is to synthesize the results of all of the studies, either summarizing them in a single number as an average treatment effect applica-

[1] J. B. Carlin, *Stat. Med.* **11,** 141 (1992).

[2] W. DuMouchel, *in* "Statistical Methodology in the Pharmaceutical Sciences" (D. A. Berry, ed.), p. 509. Marcel Dekker, New York, 1990.

[3] C. N. Morris and S. L. Normand, *in* "Bayesian Statistics 4" (J. M. Bernardo, J. O. Berger, A. P. Dawid, and A. F. M. Smith, eds.), p. 321. Oxford University Press, New York, 1992.

[4] T. C. Smith, D. J. Spiegelhalter, and A. Thomas, *Stat Med.* **14,** 2685 (1995).

0076-6879/00 $30.00

ble to all of the studies or as a set of effects describing different subgroups of individuals or studies. Let θ_t be the parameters describing the tth study and let these study-level parameters be expressed in terms of a set of parameters ζ common to all the studies. The hierarchy of models can then be expressed as

$$y_t | \theta_t \sim f(y_t | \theta_t) \qquad t = 1, 2, \ldots, T \tag{1}$$
$$\theta_t | \zeta \sim g(\theta_t | \zeta) \qquad t = 1, 2, \ldots, T \tag{2}$$

where the notation indicates that f is the probability distribution of y_t given θ_t and g is the distribution of θ_t given ζ. If inference is desired only about ζ, the two expressions can be combined to describe a distribution for y_t in terms of ζ, but the hierarchical structure permits inference also to be made about the study-specific parameters θ_t.

This hierarchical structure defines the dependence between the many system parameters in such a way that the model can have enough parameters to fit the data while not being overfit. Nonhierarchical models are not as effective because they must either model only the population structure or be overparameterized to describe the individual structure. To the extent that these studies are exchangeable *a priori*, the hierarchical structure also permits the use of information from one study to give a more accurate estimate of the outcome of another study.

The different levels of variation may be further characterized by ascribing causes to them based on covariates. Thus, in Eq. (2), ζ may consist of regression parameters that describe the effect of covariates, creating systematic differences between studies. Model checking may be accomplished by extending the model to incorporate nonlinearity and test for robustness to assumptions.

Maximum likelihood computation of mixed models, a standard technique for fitting hierarchical structures such as those given by Eqs. (1) and (2), assumes normality of the parameter estimates. Although adequate for the large amount of data usually available to estimate the common parameters ζ, asymptotic methods fail to give good estimates for the study-specific parameters θ_t when data are sparse in the tth study, especially if the underlying processes are not Gaussian. In such instances, external information about model parameters may be needed to inform the current data. This external information can be incorporated by setting up a Bayesian model.

Bayesian inference is an approach to statistical analysis based on Bayes rule in which model parameters θ are considered as random variables in the sense that knowledge of them is incomplete. Prior beliefs about model parameters θ are represented by a probability distribution $\pi(\theta)$ describing the degree of uncertainty with which these parameters are known. Combining this prior distribution with the data in the form of a likelihood for these parameters leads to a posterior distribution $\pi(\theta|Y)$ of the belief about the location of the unknown parameters θ conditional on the data Y. This posterior probability density of θ given Y is defined as

$$f(\theta|Y) = \frac{f(\theta)f(Y|\theta)}{f(Y)}$$

where $f(\theta)$ is the prior probability density of θ, $f(Y|\theta)$ is the likelihood of Y given θ, and

$$f(Y) = \int f(\theta)f(Y|\theta)\,d\theta$$

is the marginal distribution of the data and serves as a normalizing constant so that the posterior density integrates to one as required by the rules of probability densities. Statements of the probability of scientific hypotheses and quantification of the process parameters follow directly from this posterior distribution.

A Bayesian version of the hierarchical model can be set up by simply placing a probability distribution on the population parameter ζ in the second stage. Thus, together with Eqs. (1) and (2), we have a third stage

$$\zeta \sim h(\zeta) \tag{3}$$

where the function h represents the prior distribution of ζ. Of course, the Bayesian model is not restricted to only three stages, although in practice this number is often sufficient.

The use of Bayesian methodology has several advantages. It provides a formal framework for incorporating information gained from previous studies into the current analysis. As the data collected accrue and are analyzed, conclusions can be revised to incorporate the new information. The information from new data is combined with knowledge based on previously processed information (possibly data) to reach a new state of knowledge. The Bayesian analysis also provides estimates of the model parameters and their uncertainties in terms of complete posterior probability densities. Furthermore, inference is not restricted just to model parameters. Posterior densities of any functions of the model parameters may be simply computed by resampling methods to be discussed later.

Since its original description by Lindley and Smith,[5] the Bayesian hierarchical model has become a standard tool in Bayesian applications applied to problems in such diverse areas as corporate investment,[6] growth curves,[7,8]

[5] D. V. Lindley and A. F. M. Smith, *J. R. Stat. Soc. B* **34,** 1 (1972).

[6] A. F. M. Smith, *J. R. Stat. Soc. B* **35,** 67 (1973).

[7] R. Fearn, *Biometrika* **62,** 89 (1975).

[8] J. F. Strenio, H. I. Weisberg, and A. S. Bryk, *Biometrics* **39,** 71 (1983).

and educational testing.[9] A text of case studies in biometrical applications of Bayesian methods devoted an entire section to different hierarchical models[10] and a panel of the National Academy of Sciences advocated the hierarchical model as a general paradigm for combining information from different sources.[11]

Before describing tools for computing hierarchical models in Section III, we discuss the Gaussian model in the next section. Not only is this the most commonly employed hierarchical structure, it permits an illuminating interpretation of model parameters as averages between quantities in different levels of the structure. We will illustrate this interpretation using a state-space representation and computation of the Gaussian form of the Bayesian hierarchical model by forward and backward Kalman filter algorithms.[12]

The connection between Bayesian linear models and the Kalman filter has been previously made in several contexts. Duncan and Horn[13] and Meinhold and Singpurwalla[14] illustrated how the Kalman filter could be used to estimate the means and marginal covariance matrices of the random variables in the nonhierarchical Bayesian linear model with Gaussian error. As part of the E-step in an EM algorithm, Shumway and Stoffer[15] combined the Kalman filter and the fixed-interval smoothing algorithms[16] to compute the posterior densities of both the state vectors and the missing data in a Gaussian linear state-space model with measurement error. Schmid extended their model to a general first-order autoregressive model for continuous longitudinal data.[17] Wecker and Ansley also combined the Kalman filter and fixed-interval smoothing algorithms to compute sequentially approximate posterior densities of the state-space vectors in a polynomial spline model.[18]

We illustrate the application of these algorithms to meta-analysis with an example in Section IV. Section V describes a hierarchical model for saltatory growth and Section VI outlines Monte Carlo methods needed to compute non-Gaussian hierarchical models. Section VII presents some

[9] D. B. Rubin, *J. Educ. Stat.* **6,** 377 (1981).

[10] D. A. Berry and D. K. Stang, eds., "Bayesian Biostatistics." Marcel Dekker, New York, 1996.

[11] D. Graver, D. Draper, J. Greenhouse, L. Hedges, C. Morris, and C. Waternaux, "Combining Information: Statistical Issues and Opportunities for Research." National Academy Press, Washington, D.C., 1992.

[12] E. N. Brown and C. H. Schmid, *Methods Enzymol.* **240,** 171 (1994).

[13] D. B. Duncan and S. D. Horn, *J. Am. Stat. Assoc.* **67,** 815 (1972).

[14] R. J. Meinhold and N. D. Singpurwalla, *Am. Stat.* **37,** 123 (1983).

[15] R. H. Shumway and D. S. Stoffer, *J. Time Series Anal.* **3,** 253 (1982).

[16] C. F. Ansley and R. Kohn, *Biometrika* **69,** 486 (1982).

[17] C. H. Schmid, *J. Am. Stat. Assoc.* **91,** 1322 (1996).

[18] W. E. Wecker and C. F. Ansley, *J. Am. Stat. Assoc.* **78,** 381 (1983).

methods for model checking and we close with a brief conclusion in Section VIII.

The Gaussian Model

Assuming Gaussian probability distributions, we can express the three-stage hierarchical model as follows. The first stage describes a linear model relating the response $\mathbf{y}_t$ to a set of predictors $\mathbf{X}_t$ as

$$\mathbf{y}_t = \mathbf{X}_t\boldsymbol{\beta}_t + \boldsymbol{\varepsilon}_t \quad \boldsymbol{\varepsilon}_t \sim \mathbf{N}(0, \mathbf{V}_t) \tag{4}$$

for $t = 1, \ldots, T$ where $\mathbf{y}_t$ is an N_t-dimensional vector of data on the tth experimental unit, $\mathbf{X}_t$ is a known $N_t \times p$ design matrix, $\boldsymbol{\beta}_t$ is a p-dimensional vector of regression coefficients, and $\boldsymbol{\varepsilon}_t$ is an N_t-dimensional error vector with covariance matrix $\mathbf{V}_t$. We assume that $\boldsymbol{\varepsilon}_t$ and $\boldsymbol{\varepsilon}_u$ are uncorrelated for $t \neq u$.

The second stage describes the prior distribution of each $\boldsymbol{\beta}_t$ as

$$\boldsymbol{\beta}_t = \mathbf{Z}_t\boldsymbol{\gamma} + \boldsymbol{\omega}_t \quad \boldsymbol{\omega}_t \sim N(0, \mathbf{W}_1) \tag{5}$$

where $\mathbf{Z}_t$ is a known $p \times q$ design matrix, $\boldsymbol{\gamma}$ is a q-dimensional vector of hyperparameters, $\boldsymbol{\omega}_t$ is distributed as a p-dimensional Gaussian random vector with covariance matrix $\mathbf{W}_1$, and the random errors $\boldsymbol{\omega}_t$ and $\boldsymbol{\omega}_u$ are uncorrelated for $t \neq u$.

The third stage defines the prior distribution for $\boldsymbol{\gamma}$ as

$$\boldsymbol{\gamma} = \mathbf{G}_0\boldsymbol{\gamma}_0 + \boldsymbol{\nu} \quad \boldsymbol{\nu} \sim N(0, \mathbf{W}_2) \tag{6}$$

in terms of a known $q \times r$ matrix $\mathbf{G}_0$, r hyperparameters $\boldsymbol{\gamma}_0$, and a q-dimensional vector $\boldsymbol{\nu}$ distributed as Gaussian with covariance matrix $\mathbf{W}_2$. Prior knowledge of the hyperparameters in the third stage is often imprecise and a noninformative prior distribution such as one defined by $\mathbf{W}_2^{-1} = 0$ is frequently used.

The objective of the Bayesian analysis for the hierarchical model is to compute the marginal and joint posterior densities of the random variables $(\boldsymbol{\beta}_t|\mathbf{y}_t^*, \boldsymbol{\gamma}_0, \mathbf{W}_1, \mathbf{W}_2)$ for $t = 1, \ldots, T$ and $(\boldsymbol{\gamma}|\mathbf{y}_t^*, \boldsymbol{\gamma}_0, \mathbf{W}_1, \mathbf{W}_2)$ where $\mathbf{y}_t^* = (\mathbf{y}_1, \ldots, \mathbf{y}_t)$. Because of the linear structure of the model and its Gaussian error assumptions, it suffices to determine the first two moments of these densities.

For the present, assume the covariance matrices $\mathbf{V}_t$, $\mathbf{W}_1$, and $\mathbf{W}_2$ to be known. This is not a serious limitation because the algorithms are easily fit into an iterative scheme, such as Gibbs sampling,[19] for simultaneously estimating the mean and variance parameters.

[19] W. R. Gilks, S. Richardson and D. J. Spiegelhalter, eds., "Markov Chain Monte Carlo in Practice." Chapman and Hall, New York, 1996.

It might appear that this three-stage model has too many parameters for the amount of data collected. But note that by collapsing Eqs. (4)–(6), we can express $\mathbf{y}_t$ as a normal distribution with mean $\mathbf{X}_t\mathbf{Z}_t\mathbf{G}_0\boldsymbol{\gamma}_0$ and variance $\mathbf{V}_t + \mathbf{X}_t\mathbf{W}_1\mathbf{X}_t' + \mathbf{X}_t\mathbf{Z}_t\mathbf{W}_2\mathbf{Z}_t'\mathbf{X}_t'$. The parameters describing the data are simply γ_0, $\mathbf{V}_t$, $\mathbf{W}_1$, and $\mathbf{W}_2$. If $N_t = 1$, the model reduces to a regression with parameters $\boldsymbol{\gamma}_0$ conditional on the variance parameters.

State-Space Formulation

A state-space model typically provides a Markov representation of a system that evolves over time or space. It is defined by two equations: the state equation, which describes the temporal or spatial evolution of the system, and the observation equation, which defines the relationship between the observed data and the true state of the system. For the three-stage model in Eqs. (4)–(6), a state equation is written as

$$\boldsymbol{\theta}_t = \mathbf{F}_t\boldsymbol{\theta}_{t-1} + \boldsymbol{\omega}_t^* \tag{7}$$

for $t = 1, \ldots, T$ where $\boldsymbol{\theta}_t = \begin{pmatrix} \boldsymbol{\beta}_t \\ \boldsymbol{\gamma} \end{pmatrix}$, $\boldsymbol{\omega}_t^* = \begin{pmatrix} \boldsymbol{\omega}_t \\ 0 \end{pmatrix}$ is a $(p + q)$ vector with zero mean and covariance matrix $\mathbf{W}_1^* = \begin{pmatrix} \mathbf{W}_1 & 0 \\ 0 & 0 \end{pmatrix}$, $\mathbf{F}_t = \begin{pmatrix} 0 & \mathbf{Z}_t \\ 0 & \mathbf{I}_q \end{pmatrix}$ is the $(p + q) \times (p + q)$ transition matrix, and $\mathbf{I}_q$ is the $q \times q$ identity matrix. An observation equation is written as

$$\mathbf{y}_t = \mathbf{X}_t^*\boldsymbol{\theta}_t + \boldsymbol{\varepsilon}_t \tag{8}$$

where $\mathbf{X}_t^*$ is the $N_t \times 2p$ matrix defined as $\mathbf{X}_t^* = (\mathbf{X}_t\ 0)$.

The state-space model divides $\boldsymbol{\gamma}$ and the $\boldsymbol{\beta}_t$ among the state vectors $\boldsymbol{\theta}_t$ so that they may be estimated sequentially by the Kalman filter, fixed-interval smoothing, and state-space covariance algorithms working sequentially forward and then backward in time as each piece of data $\mathbf{y}_t$ is incorporated. Starting with initial estimates $\hat{\boldsymbol{\theta}}_0$ and $\mathbf{S}_0$ of the state vector and its covariance matrix, the forward pass uses the Kalman filter to compute the expectation and covariance of $\boldsymbol{\theta}_t$ given $\mathbf{y}_t^*$ for $t = 1, \ldots, T$. The final forward step gives the correct expectation and covariance for $\boldsymbol{\theta}_T$, but the estimated moments of $\boldsymbol{\theta}_t$ for $t < T$ are incompletely updated, using only the data up to time t. Therefore, starting with the completely updated $\boldsymbol{\theta}_T$, the fixed-interval smoothing and covariance algorithms work back from time T to time 0, updating $\hat{\boldsymbol{\theta}}_t$ and $\mathbf{S}_t$ for $\mathbf{y}_u^*$ when $u > t$, finally obtaining the fully updated moments $\hat{\boldsymbol{\theta}}_{t|T}$ and $\mathbf{S}_{t|T}$. The details of one step for each of these algorithms follow.

Kalman Filter

Starting with $\hat{\boldsymbol{\theta}}_{t-1|t-1}$, the estimate of $\boldsymbol{\theta}_{t-1}$ given $\mathbf{y}^*_{t-1}$, and its covariance matrix $\mathbf{S}_{t-1|t-1}$, the Kalman filter algorithm for this state-space model is

$$\hat{\boldsymbol{\theta}}_{t|t-1} = \mathbf{F}_t\hat{\boldsymbol{\theta}}_{t-1|t-1} \tag{9}$$

$$\mathbf{S}_{t|t-1} = \mathbf{F}_t\mathbf{S}_{t-1|t-1}\mathbf{F}_t^T + \mathbf{W}_1^* \tag{10}$$

$$\mathbf{K}_t = \mathbf{S}_{t|t-1}\mathbf{X}_t^{*T}(\mathbf{X}_t^*\mathbf{S}_{t|t-1}\mathbf{X}_t^{*T} + \mathbf{V}_t)^{-1} \tag{11}$$

$$\hat{\boldsymbol{\theta}}_{t|t} = \hat{\boldsymbol{\theta}}_{t|t-1} + \mathbf{K}_t(\mathbf{y}_t - \mathbf{X}_t^*\hat{\boldsymbol{\theta}}_{t|t-1}) \tag{12}$$

$$\mathbf{S}_{t|t} = (\mathbf{I}_{p+q} - \mathbf{K}_t\mathbf{X}_t^*)\mathbf{S}_{t|t-1} \tag{13}$$

for $t = 1, \ldots, T$ where $\hat{\boldsymbol{\theta}}_0 = \begin{pmatrix} \mathbf{Z}_t\mathbf{G}_0\boldsymbol{\gamma}_0 \\ \mathbf{G}_0\boldsymbol{\gamma}_0 \end{pmatrix}$ and $\mathbf{S}_0 = \begin{pmatrix} \mathbf{W}_1 + \mathbf{Z}_t\mathbf{W}_2\mathbf{Z}_t^T & \mathbf{Z}_t\mathbf{W}_2 \\ \mathbf{W}_2\mathbf{Z}_t^T & \mathbf{W}_2 \end{pmatrix}$ are the initial mean and covariance matrices, respectively.

Fixed-Interval Smoothing Algorithm

The associated fixed-interval smoothing algorithm is

$$\hat{\boldsymbol{\theta}}_{t|T} = \hat{\boldsymbol{\theta}}_{t|t} + \mathbf{A}_t(\hat{\boldsymbol{\theta}}_{t+1|T} - \hat{\boldsymbol{\theta}}_{t+1|t}) \tag{14}$$

$$\mathbf{A}_t = \mathbf{S}_{t|t}\mathbf{F}_{t+1}^T\mathbf{S}_{t+1|t}^{-1} \tag{15}$$

$$\mathbf{S}_{t|T} = \mathbf{S}_{t|t} + \mathbf{A}_t(\mathbf{S}_{t+1|T} - \mathbf{S}_{t+1|t})\mathbf{A}_t^T \tag{16}$$

recursively computed from $t = T-1, \ldots, 1$ where $\hat{\boldsymbol{\theta}}_{t|T}$ and $\mathbf{S}_{t|T}$ are, respectively, the estimate of $\boldsymbol{\theta}_t$ and its covariance matrix given $\mathbf{Y}_T^*$. The initial conditions, $\hat{\boldsymbol{\theta}}_{T|T}$ and $\mathbf{S}_{T|T}$, are obtained from the last step of the Kalman filter. Thus, computation of $\hat{\boldsymbol{\theta}}_{t|T}$ and $\mathbf{S}_{t|T}$ requires only the output of $\hat{\boldsymbol{\theta}}_{t|t}$ and $\mathbf{S}_{t|t}$ from the tth forward step and the output of $\hat{\boldsymbol{\theta}}_{t+1|T}$ and $\mathbf{S}_{t+1|T}$ from the previous backward step.

State-Space Covariance Algorithm

The state-space covariance algorithm[20] gives the covariance between $\hat{\boldsymbol{\theta}}_{t|T}$ and $\hat{\boldsymbol{\theta}}_{u|T}$ as

$$\mathbf{S}_{t,u|T} = \mathbf{A}_t\mathbf{S}_{t+1,u|T} \qquad t < u \leq T. \tag{17}$$

By the definitions of the Kalman filter and the fixed-interval smoothing algorithm, the probability density of $(\boldsymbol{\beta}_t|\mathbf{y}_T^*, \boldsymbol{\gamma}_0, \mathbf{W}_1, \mathbf{W}_2)$ is the Gaussian density whose mean vector is the first p components of $\hat{\boldsymbol{\theta}}_{t|T}$ and whose covariance matrix is the left upper $p \times p$ submatrix of $\mathbf{S}_t|_T$. The probability density of $(\boldsymbol{\gamma}|\mathbf{y}_T^*, \boldsymbol{\gamma}_0, \mathbf{W}_1, \mathbf{W}_2)$ is the Gaussian density whose mean is the

[20] P. DeJong and M. J. Mackinnon, Biometrika **75,** 601 (1988).

second q components of any $\hat{\boldsymbol{\theta}}_{t|T}$ and whose covariance matrix is the right lower $q \times q$ submatrix of any $\mathbf{S}_{t|T}$. The posterior covariance between $\boldsymbol{\beta}_t$ and $\boldsymbol{\gamma}$ is given by the off-diagonal blocks of $\mathbf{S}_{t|T}$ for $t = 1, \ldots, T$, whereas the posterior covariance between $\boldsymbol{\beta}_t$ and $\boldsymbol{\beta}_u$ for $t \neq u$ is given by the left upper $p \times p$ submatrix of $\mathbf{S}_{t,u|T}$. The posterior covariance between $\boldsymbol{\beta}_t$ and $\boldsymbol{\gamma}$ can be used to assess the proximity of the tth experimental unit to the mean of the second stage of the hierarchy.

When no order dependence is assumed *a priori* among the $\mathbf{y}_t$ values, the estimates of the posterior densities are independent of the sequence in which the data enter the algorithm. The sequence chosen determines only the point in the algorithm at which a particular posterior density is estimated. As we show next, the Kalman filter updating makes the effect on the posterior densities of adding new data simple to understand and simple and quick to compute.

Quantifying Contribution of New Information

Consider the simplest form of the model where each unit has one response (i.e., $N_t = 1$) and a simple mean is estimated at each stage so that

$$y_t = \beta_t + \varepsilon_t \tag{18}$$
$$\beta_t = \gamma + \omega_t \tag{19}$$
$$\gamma = \gamma_0 + \nu \tag{20}$$

Because only the Kalman (forward) filter is needed to compute the posterior of γ, we only need to examine Eqs. (9)–(13) to understand how the population mean γ is computed.

Let us write $\gamma^{(t-1)}$ for the expectation $E(\gamma|\mathbf{y}_{t-1}^*)$ of γ given $\mathbf{y}_{t-1}^*$. Application of Eqs. (11)–(13) gives the expectation of γ after the new datum y_t is observed as the weighted average

$$\gamma^{(t)} = (1 - \omega_1)\gamma^{(t-1)} + \omega_1 y_t \tag{21}$$

with posterior variance

$$V(\gamma|\mathbf{y}_t^*) = V_{\gamma^{(t)}} = (1 - \omega_1)V_{\gamma^{(t-1)}} \tag{22}$$

where the weight $\omega_1 = V_{\gamma^{(t-1)}}/(V_{\gamma^{(t-1)}} + W_1 + V_t)$.

The updated posterior mean for γ is a compromise between the previous estimate of the mean, $\gamma^{(t-1)}$, and the new data, y_t. If the new observation, y_t, is larger than expected, then the expectation of γ will increase; if it is smaller than expected, the expected value of γ will decrease. Large measurement variability of the new data (large V_t), large between-unit variation (large W_1), or precise prior information about γ (small $V_{\gamma^{(t-1)}}$) lead to small changes in γ. Conversely, substantial change in the posterior of γ will occur when y_t is precisely measured, experimental units are homogeneous, or γ is not well estimated.

The posterior variance is reduced by an amount related to the previous estimate of the variance $V_{\gamma^{(t-1)}}$, the variance of the new data V_t, and the between-unit variance W_1. Reduction is greatest when V_t and W_1 are small relative to $V_{\gamma^{(t-1)}}$ because then the tth unit is providing a lot of information relative to that provided by previous units.

These conditions are intuitively reasonable. If the new data are imprecisely measured, we should not have much confidence in them and therefore would not want them to affect our model estimates substantially. Likewise, large between-unit variation implies only a loose connection between the units and so a new observation will have less effect on the existing structure. Finally, new data should be less influential if we already have a good estimate of the population mean.

To consider information about β_t, we note that from Eq. (9), the best estimate of β_t before seeing y_t is the population mean $\gamma^{(t-1)}$ and the estimated precision is $V_{\gamma^{(t-1)}} + W_1$, namely, the variance of the population mean plus the variance of the random effect. Applying Eqs. (11)–(13) gives

$$E(\beta_t|\mathbf{y}_t^*) = \beta_t^{(t)} = (1 - \omega_2)\gamma^{(t-1)} + \omega_2 y_t \tag{23}$$

and

$$V(\beta_t|\mathbf{y}_t^*) = V_{\beta_t^{(t)}} = \omega_2 V_t = (1 - \omega_2)(V_{\gamma^{(t-1)}} + W_1) \tag{24}$$

where $\omega_2 = (V_{\gamma^{(t-1)}} + W_1)/(V_{\gamma^{(t-1)}} + W_1 + V_t)$.

The updated posterior mean for β_t is a compromise between the prior mean, $\gamma^{(t-1)}$ and the new data, y_t. A precisely estimated data value (small V_t), substantial prior between-unit heterogeneity (large W_1), or a poor prior estimate of γ [large $V_{\gamma^{(t-1)}}$] will give more weight to y_t. The last condition follows from the idea that if prior knowledge of β_t, as expressed by the current knowledge $\gamma^{(t-1)}$ of γ, is imprecise, the previous observations will have little information to give about β_t.

The posterior variance is also closely related to the precision of the new data. If y_t is well estimated so that V_t is small, then its associated random effect β_t will also be well estimated. Furthermore, even if V_t is not small, but the prior variance $V_{\gamma^{(t-1)}} + W_1$ is small relative to V_t, then β_t can be precisely estimated using information from previous units.

In general, data on new units have the greatest effect on current posterior estimates if the new data are precisely measured and the new units and current unit are closely related. The filtering equations, therefore, quantify the amount of information provided by new data and the effect of the complete hierarchical structure on the posterior distributions of model parameters. Knowledge about a unit-specific parameter depends not just on the data gathered on that unit, but through the model, on information gathered from other units. As data collection proceeds, the influence of new units on population parameters decreases, but the influence of the population parameters on the new units increases.

Simplified Computations for Incorporating Additional Data

The sequential form of these computations also leads to efficient incorporation of new data. To update the posterior densities of β_t and γ when data from several new experimental units $y_{T+1}, \ldots, y_{T+k}$ are collected, we could combine the original and new data into a single sequence and replace T with $T + k$ in the Kalman filter algorithm, but this approach ignores the previous processing of the original data and is therefore not computationally efficient.

A better method uses the stored $\hat{\boldsymbol{\theta}}_{t|t}$ and $\mathbf{S}_{t|t}$ from the original calculations as inputs to a Kalman filter, which proceeds forward from $t = T + 1$ to $T + k$. A backward pass from $t = T$ to $t = 1$ with the fixed-interval smoothing and state-space covariance algorithms then completes updating the posterior density. This process saves the first T steps of the forward filter compared to repeating the full algorithm.

When new studies are added incrementally and the second-stage regression model is of primary interest, as it might be in meta-analysis, further efficiency can be realized by skipping the backward pass. Rather, we can update the second-stage regression estimates by running the forward algorithm one step further as each new unit is added. When the regression model is sufficiently precise, we can make one pass back through the data to update all the individual studies.

Another form of data accumulation occurs when new observations arrive for experimental units y_t for which some data have already been collected. In this case, the structure of the problem is different because the updated β_t are no longer partially exchangeable given $\mathbf{y}_t^*$. Hence, the Kalman filter, fixed-interval smoothing algorithm, and state-space covariance algorithms as stated here cannot be applied. In meta-analysis, at least, this type of accumulating data is uncommon because new data usually represent new studies. Nevertheless, interim analysis of several concurrent studies might contribute new data of this type.

Computation

The complete posterior distribution of $\boldsymbol{\gamma}$ and $\boldsymbol{\beta} = (\beta_1, \beta_2, \ldots, \beta_T)$ may be expressed algebraically as

$$(\boldsymbol{\beta}, \boldsymbol{\gamma})|\mathbf{y} \sim N(\mathbf{B}\boldsymbol{\eta}, \mathbf{B})$$

with

$$\mathbf{B}^{-1} = \begin{pmatrix} \mathbf{X}^T\mathbf{V}^{-1}\mathbf{X} + \boldsymbol{\Omega}_1^{-1} & -\boldsymbol{\Omega}_1^{-1}\mathbf{Z} \\ -\mathbf{Z}^T\boldsymbol{\Omega}_1^{-1} & \mathbf{Z}^T\boldsymbol{\Omega}_1^{-1}\mathbf{Z} + \mathbf{W}_2^{-1} \end{pmatrix} \quad \text{and} \quad \boldsymbol{\eta} = \begin{pmatrix} \mathbf{X}^T\mathbf{V}^{-1}\mathbf{y} \\ \mathbf{W}_2^{-1}\mathbf{G}_0\boldsymbol{\gamma}_0 \end{pmatrix}$$

where $\mathbf{y}$ is the vector of T observed responses,

$$\mathbf{X} = \begin{pmatrix} \mathbf{X}_1 & 0 & \cdot & 0 \\ 0 & \mathbf{X}_2 & \cdot & 0 \\ \cdot & \cdot & \cdot & \cdot \\ 0 & 0 & \cdot & \mathbf{X}_T \end{pmatrix}$$

$$\mathbf{V} = \begin{pmatrix} \mathbf{V}_1 & 0 & \cdot & 0 \\ 0 & \mathbf{V}_2 & \cdot & 0 \\ \cdot & \cdot & \cdot & \cdot \\ 0 & 0 & \cdot & \mathbf{V}_T \end{pmatrix}$$

$$\boldsymbol{\Omega}_1 = \begin{pmatrix} \mathbf{W}_1 & 0 & \cdot & 0 \\ 0 & \mathbf{W}_1 & \cdot & 0 \\ \cdot & \cdot & \cdot & \cdot \\ 0 & 0 & \blacklozenge\text{c} & \mathbf{W}_1 \end{pmatrix}$$

and

$$\mathbf{Z} = \begin{pmatrix} \mathbf{Z}_1 \\ \mathbf{Z}_2 \\ \cdot \\ \mathbf{Z}_T \end{pmatrix}$$

Inverting $\mathbf{B}$ is the major computational task in evaluating this joint posterior distribution. The Kalman filter algorithm described in Section II is generally much faster than brute force inversion as the number of observations and regression parameters increases, but the SWEEP algorithm as described in Carlin[21] is even quicker.

To compute by SWEEP, first express the hierarchical model as a multivariate Gaussian density

$$\begin{pmatrix} \mathbf{y} \\ \boldsymbol{\beta} \\ \boldsymbol{\gamma} \end{pmatrix} \sim N\left[\begin{pmatrix} \boldsymbol{\mu}_{\mathbf{y}} \\ \boldsymbol{\mu}_{\beta} \\ \boldsymbol{\mu}_{\gamma} \end{pmatrix}, \begin{pmatrix} \Sigma_{\mathbf{y}} & \Sigma_{\mathbf{y},\beta} & \Sigma_{\mathbf{y},\gamma} \\ \Sigma_{\beta,\mathbf{y}} & \Sigma_{\beta} & \Sigma_{\beta,\gamma} \\ \Sigma_{\gamma,\mathbf{y}} & \Sigma_{\gamma,\beta} & \Sigma_{\gamma} \end{pmatrix}\right]$$

[21] J. B. Carlin, *Aust. J. Stat.* **32,** 29 (1990).

and put this in the following symmetric tableau (omitting the lower triangular portion for ease of presentation)

$$\begin{array}{c|cc} (\mathbf{y} - \boldsymbol{\mu}_y)^T & -\boldsymbol{\mu}_\beta^T & -\boldsymbol{\mu}_\gamma^T \\ \hline \boldsymbol{\Sigma}_\mathbf{y} & \boldsymbol{\Sigma}_{\mathbf{y},\beta} & \boldsymbol{\Sigma}_{\mathbf{y},\gamma} \\ & \boldsymbol{\Sigma}_\beta & \boldsymbol{\Sigma}_{\beta,\gamma} \\ & & \boldsymbol{\Sigma}_\gamma \end{array}$$

For a general block matrix $\begin{pmatrix} A & B \\ C & D \end{pmatrix}$, sweeping on D gives $\begin{pmatrix} A - BD^{-1}C & BD^{-1} \\ D^{-1}C & -D^{-1} \end{pmatrix}$. The reverse operation, called reverse sweeping, that undoes the sweeping operation changes $\begin{pmatrix} A & B \\ C & D \end{pmatrix}$ into $\begin{pmatrix} A - BD^{-1}C & -BD^{-1} \\ -D^{-1}C & -D^{-1} \end{pmatrix}$. Thus sweeping on $\mathbf{y}$ in the tableau above gives

$$\begin{array}{c|cc} (\mathbf{y} - \boldsymbol{\mu}_\mathbf{y})^T \boldsymbol{\Sigma}_\mathbf{y}^{-1} & -\boldsymbol{\mu}_\beta^T - (\mathbf{y} - \mu_\mathbf{y})^T \boldsymbol{\Sigma}_\mathbf{y}^{-1} \boldsymbol{\Sigma}_{\mathbf{y},\beta} & -\boldsymbol{\mu}_\gamma^T - (\mathbf{y} - \mu_\mathbf{y})^T \boldsymbol{\Sigma}_\mathbf{y}^{-1} \boldsymbol{\Sigma}_{\mathbf{y},\gamma} \\ \hline -\boldsymbol{\Sigma}_\mathbf{y}^{-1} & \boldsymbol{\Sigma}_\mathbf{y}^{-1} \boldsymbol{\Sigma}_{\mathbf{y},\beta} & \boldsymbol{\Sigma}_\mathbf{y}^{-1} \boldsymbol{\Sigma}_{\mathbf{y},\gamma} \\ & \boldsymbol{\Sigma}_\beta - \boldsymbol{\Sigma}_{\beta,\mathbf{y}} \boldsymbol{\Sigma}_\mathbf{y}^{-1} \boldsymbol{\Sigma}_{\mathbf{y},\beta} & \boldsymbol{\Sigma}_{\beta,\gamma} - \boldsymbol{\Sigma}_{\beta,\mathbf{y}} \boldsymbol{\Sigma}_\mathbf{y}^{-1} \boldsymbol{\Sigma}_{\mathbf{y},\gamma} \\ & & \boldsymbol{\Sigma}_\gamma - \boldsymbol{\Sigma}_{\gamma,\mathbf{y}} \boldsymbol{\Sigma}_\mathbf{y}^{-1} \boldsymbol{\Sigma}_{\mathbf{y},\gamma} \end{array}$$

from which the joint posterior probability density $[\boldsymbol{\beta}, \boldsymbol{\gamma}|\mathbf{y}]$ may be read directly. The means of $\boldsymbol{\beta}$ and $\boldsymbol{\gamma}$ are given by the negatives of the entries in the second and third columns, respectively, of the first row. The entries in the last two rows give the posterior covariance matrix of $\boldsymbol{\beta}$ and $\boldsymbol{\gamma}$.

Rather than sweeping $\mathbf{y}$ directly, it is actually more efficient to first sweep $\boldsymbol{\gamma}$ and $\boldsymbol{\beta}$ before sweeping $\mathbf{y}$ and then undo the sweeps on $\boldsymbol{\gamma}$ and $\boldsymbol{\beta}$ by reverse sweeps to reproduce the effect of sweeping on $\mathbf{y}$ alone. Though this modified SWEEP algorithm involves five sets of computations, efficiency is gained because the initial sweeps on $\boldsymbol{\gamma}$ and $\boldsymbol{\beta}$ diagonalize $\boldsymbol{\Sigma}_\mathbf{y}$ to the simple form $\sigma_\varepsilon^2 \mathbf{I}$ so that $\mathbf{y}$ may be swept quite simply analytically. This leaves only the smaller dimensional reverse sweeps of $\boldsymbol{\gamma}$ and $\boldsymbol{\beta}$ for numerical computation.

While the Kalman filter and SWEEP are each efficient algorithms for computing the posteriors of the mean parameters $\boldsymbol{\gamma}$ and $\boldsymbol{\beta}$ in Gaussian models, they assume fixed variance parameters $\mathbf{V}_t$, $\mathbf{W}_1$, and $\mathbf{W}_2$. Typically, variances will be unknown and so these algorithms must be embedded inside a larger algorithm in which the variances are first estimated and then the means are computed conditional on the variances.

One algorithm that can be used for meta-analysis uses the result that the within-study covariance matrices $\mathbf{V}_t$ are usually so well estimated from each study that they may be assumed known.[2] To incorporate the uncertainty arising from the between-study variance $\mathbf{W}_1$, average the conditional posterior distribution of $\boldsymbol{\beta}_t$ and $\boldsymbol{\gamma}$ given $\mathbf{W}_1$ over the posterior distribution of $\mathbf{W}_1$. Assuming a noninformative prior for $\mathbf{W}_2$ so that $\mathbf{W}_2^{-1} \to 0$ gives

$$[\gamma|\mathbf{y}_T^*, \mathbf{W}_1][\mathbf{W}_1|\mathbf{y}_T^*] \propto [\mathbf{W}_1][\gamma|\mathbf{W}_1][\mathbf{y}_T^*|\gamma, \mathbf{W}_1]$$

where the notation $[\mathbf{Y}|\mathbf{X}]$ represents the conditional density of the random variable $\mathbf{Y}$ given the random variable $\mathbf{X}$. Letting the prior distribution for $[\gamma, \mathbf{W}_1]$ be constant, suggested in Berger[22] as an appropriate form of a noninformative prior distribution for hierarchical models, gives

$$[\mathbf{W}_1|\mathbf{y}_T^*] \propto [\mathbf{y}_T^*|\gamma, \mathbf{W}_1]/[\gamma|\mathbf{y}_T^*, \mathbf{W}_1]$$

Combining Eqs. (4) and (5), we have

$$\mathbf{y}_t^*|\gamma, \mathbf{W}_1 \sim N(\mathbf{X}_t\mathbf{Z}_t\gamma, \mathbf{V}_t + \mathbf{X}_t\mathbf{W}_1\mathbf{X}_t^T)$$

and standard calculations show that

$$\gamma|\mathbf{y}_T^*, \mathbf{W}_1 \sim N(\mathbf{m}, \Sigma)$$

where $\mathbf{m} = (\mathbf{Z}'\Sigma^{-1}\mathbf{Z})^{-1}\mathbf{Z}'\Sigma^{-1}\mathbf{y}_T^*$, Σ is the diagonal matrix with diagonal elements $\mathbf{V}_t + \mathbf{W}_1$, and $\mathbf{Z}$ is the matrix defined previously. If V_t and W_1 are scalars, it can then be shown using these last two expressions that the log posterior density for W_1 above is proportional to

$$g = \sum_{t=1}^{T} \log[(V_t + W_1)]^{-1} - \log|\mathbf{Z}'\Sigma^{-1}\mathbf{Z}|$$
$$- \sum_{t=1}^{T} y_t^2/(V_t + W_1) + (\mathbf{Z}'\Sigma^{-1}\mathbf{y}_T^*)^T(\mathbf{Z}'\Sigma^{-1}\mathbf{Z})^{-1}(\mathbf{Z}'\Sigma^{-1}\mathbf{y}_T^*)$$

An estimate of the posterior for $\boldsymbol{\beta}$ and $\boldsymbol{\gamma}$ may then be obtained using Monte Carlo integration over this log posterior by setting up a grid of, say, 100 points that span the range of the distribution of $\mathbf{W}_1$ and then computing the ergodic average $\frac{1}{100}\sum_{k=1}^{100} f_k(\boldsymbol{\beta}, \gamma|W_1) * g_k(W_1)$ where the f_k and g_k are evaluated at each of the 100 points. Another technique for performing this numerical integration uses Markov chain Monte Carlo and is discussed in Section VI.

[22] J. O. Berger, "Statistical Decision Theory and Bayesian Analysis," 2nd Ed. Springer-Verlag, New York, 1985.

Example: Meta-Regression

To illustrate application of the Bayesian hierarchical linear model to meta-analysis, we look at the effect of the time from chest pain onset until treatment on the relative risk of mortality following thrombolytic therapy for acute myocardial infarction. Eight large (>1000 patients) randomized control trials of the thrombolytic agents streptokinase (SK), urokinase (UK), recombinant tissue plasminogen activator (tPA), and anistreplase (APSAC) have reported outcomes by subgroups of patients according to time-to-treatment.[23–30]

The Bayesian hierarchical model may be used to perform a meta-analysis combining the results from the subgroups in these studies. For the first stage, let y_t be the observed log relative risk in the tth subgroup having a Gaussian distribution centered about the true subgroup relative risk β_t with random error ε_t having variance σ_ε^2. The second stage describes the representation of β_t as a simple linear regression on time-to-treatment Time_t with Gaussian error ω_t having variance W_1, i.e., $\beta_t = \delta_0 + \delta_1 * \text{Time}_t + \omega_t$. Thus $\boldsymbol{\gamma} = \begin{pmatrix} \delta_0 \\ \delta_1 \end{pmatrix}$ and $\mathbf{Z}_t = (1 \ \text{Time}_t)$. Finally, the third stage describes a Gaussian prior distribution for $\boldsymbol{\gamma}$ with mean $\boldsymbol{\gamma}_0$ and variance $\mathbf{W}_2$. This hierarchical structure may be put into the state-space form of Eqs. (7) and (8) by setting $\boldsymbol{\theta}_t = \begin{pmatrix} \beta_t \\ \delta_0 \\ \delta_1 \end{pmatrix}$, $\mathbf{F}_t = \begin{pmatrix} 0 & 1 & \text{Time}_t \\ 0 & 1 & 0 \\ 0 & 0 & 1 \end{pmatrix}$,

$$\boldsymbol{\theta}_{t-1} = \begin{pmatrix} \beta_{t-1} \\ \delta_0 \\ \delta_1 \end{pmatrix}, \ \mathrm{X}_t^* = (1 \quad 0 \quad 0), \ \mathbf{W}_t^* = \begin{pmatrix} W_1 & 0 & 0 \\ 0 & 0 & 0 \\ 0 & 0 & 0 \end{pmatrix}, \text{ and } V_t = \sigma_\varepsilon^2.$$

Our objective is to compute the joint posterior density of β_t and $\boldsymbol{\gamma}$. This may be accomplished by the Kalman filter algorithms conditional on the variance components $\mathbf{V}_t$, W_1, and $\mathbf{W}_2$ together with the Monte Carlo integration as presented above.

Though averaging over the variance component makes the posterior strictly non-Gaussian, the use of a uniform prior and the substantial amount

[23] GISSI Study Group, *Lancet* **i,** 397 (1986).
[24] ISAM Study Group, *N. Eng. J. Med.* **314,** 1465 (1986).
[25] ISIS-2 Collaborative Group, *Lancet* **ii,** 349 (1988).
[26] ASSET Trial Study Group, *Lancet,* **ii,** 525 (1988).
[27] AIMS Trial Study Group, *Lancet* **i,** 545 (1988).
[28] USIM Collaborative Group, *Am. J. Card.* **68,** 585 (1991).
[29] LATE Study Group, *Lancet* **342,** 759 (1993).
[30] EMERAS Collaborative Group, *Lancet,* **342,** 767 (1993).

of data available make an asymptotic Gaussian posterior a good approximation. Figure 1 shows the posterior density of W_1 under both a constant regression model, $\beta_t = \delta_0$, and the model linear in time, $\beta_t = \delta_0 + \delta_1 \text{Time}_t + \omega_t$. The introduction of time-to-treatment reduces the between-study variance so that the posterior mode for W_1 is essentially zero. Nevertheless, the skewness of the posterior indicates that it is quite probable that $W_1 > 0$.

Table I shows the raw data (in terms of the percent risk reduction) for the subgroups together with the point estimates and 95% confidence intervals based on (1) the observed data, y_t; (2) the posteriors for the β_t (averaging over the likelihood for W_1); and (3) the regression line $\delta_0 + \delta_1 \text{Time}_t$. Figure 2 plots the fitted regression line versus time-to-treatment superimposed on the observed point estimates and Bayes estimates for the subgroups.

Table I shows that the Bayes estimate of each true study effect is a weighted average of the observed effect y_t and the regression estimate, $\delta_0 + \delta_1 \text{Time}_t$. Figure 2 shows this weighting graphically with the Bayes estimates (represented by + signs) falling between the observed effects (squares)

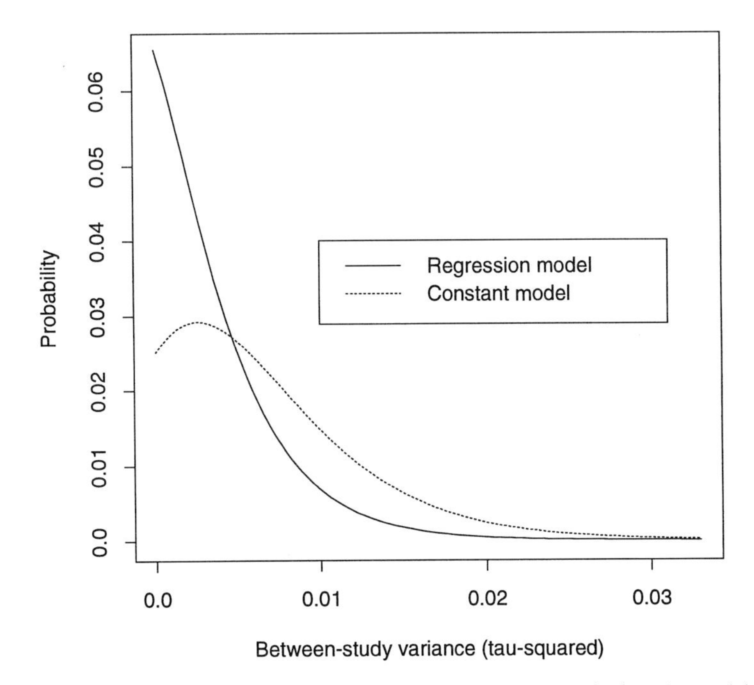

FIG. 1. Unnormalized posterior density of between-study variance (W_1) under models with and without time-to-treatment as a regression variable.

TABLE I

MORTALITY DATA FROM NINE LARGE PLACEBO-CONTROL STUDIES OF THROMBOLYTIC AGENTS AFTER MYOCARDIAL INFARCTION[a]

			Treated		Control		y_t			β_t			$\delta_0 + \delta_1$ *Time		
Study	Year	Time (hrs)	Deaths	Total	Deaths	Total	Lower	Mean	Upper	Lower	Mean	Upper	Lower	Mean	Upper
GISSI-1	1986	0.7	52	635	99	642	27	47	61	17	28	38	16	25	34
ISIS-2	1988	1.0	29	357	48	357	6	40	61	15	26	36	14	25	35
USIM	1991	1.2	45	596	42	538	−45	3	35	12	22	33	14	25	35
ISAM	1986	1.8	25	477	30	463	−37	18	51	13	24	34	11	24	35
ISIS-2	1988	2.0	72	951	111	957	13	35	51	15	26	35	10	24	36
GISSI-1	1986	2.0	226	2381	270	2436	−1	14	28	12	21	30	10	24	36
ASSET	1988	2.1	81	992	107	979	2	25	43	14	24	33	10	24	36
AIMS	1988	2.7	18	334	30	326	−3	41	67	13	24	34	8	23	36
USIM	1991	3.0	48	532	47	535	−51	−3	30	10	21	31	7	23	36
ISIS-2	1988	3.0	106	1243	152	1243	12	30	45	15	24	33	7	23	36
EMERAS	1993	3.2	51	336	56	327	−25	11	37	11	22	31	6	23	37
ISIS-2	1988	4.0	100	1178	147	1181	13	32	46	14	24	32	3	22	37
ASSET	1988	4.1	99	1504	129	1488	2	24	41	13	22	31	3	22	37
GISSI-1	1986	4.5	217	1849	254	1800	2	17	30	12	20	28	1	21	37
ISAM	1986	4.5	25	365	31	405	−49	11	46	10	21	31	1	21	37
AIMS	1988	5.0	14	168	31	176	14	53	74	11	22	32	0	21	38
ISIS-2	1988	5.5	164	1621	190	1622	−5	14	29	10	19	27	−2	20	38
GISSI-1	1986	7.5	87	693	93	659	−17	11	32	7	17	26	−9	18	39
LATE	1993	9.0	93	1047	123	1028	4	26	42	8	18	27	−14	17	39
EMERAS	1993	9.5	133	1046	152	1034	−7	14	30	6	15	24	−16	16	39
ISIS-2	1988	9.5	214	2018	249	2008	−2	14	28	7	16	24	−16	16	39
GISSI-1	1986	10.5	46	292	41	302	−71	−16	21	0	13	24	−20	15	39
LATE	1993	18.0	154	1776	168	1835	−17	5	23	−9	6	18	−49	6	40
ISIS-2	1988	18.5	106	1224	132	1227	−3	20	37	−8	8	21	−51	5	40
EMERAS	1993	18.5	114	875	119	916	−27	0	21	−11	4	17	−51	5	40

[a] The estimated true percent risk reduction (with 95% confidence limits) for each time subgroup is shown for the observed data and results from three regression models: the fixed effects model, the Bayes model, and the estimate that would be predicted if a new study were performed with the same mean time to treatment.

and the regression line. Larger studies (denoted by bigger squares) get more weight; their Bayes estimates are proportionately closer to the observed effect than to the regression line. Conversely, smaller studies carry little weight; their Bayes estimates are pulled almost completely into the regression line. The small amount of between-study variance, however, shrinks all estimates close to the regression line.

Intuitively, if we believe the model, the exchangeability of the study means lets us use information from the other studies to better estimate each treatment effect. Thus, for the first subgroup, the GISSI-1 study patients with mean time-to-treatment of 0.7 hr, the Bayes estimate is a 28% reduction, between the 47% observed reduction and the 25% average reduction at 0.7 hr from the regression model. Because the evidence from the other studies indicates less benefit, we downweight the estimate from GISSI-1 toward the regression average. If we could discover other factors that explained the additional benefit observed in GISSI-1, we could incorporate

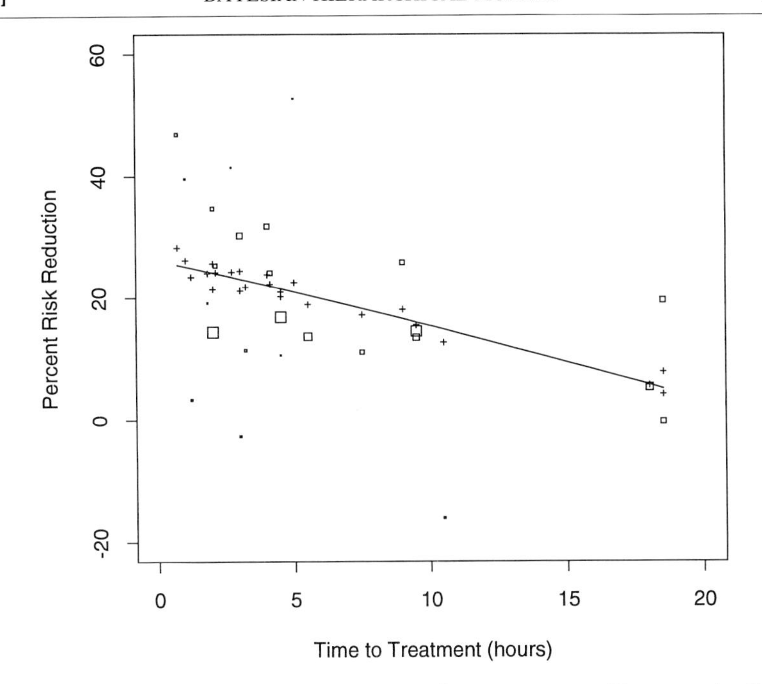

FIG. 2. Estimated percent risk reduction versus time-to-treatment. The regression line is shown together with the Bayes estimates (+ signs) and the observed estimates (squares). The size of the squares is inversely proportional to the study variance.

them into the regression model, thus increasing the regression estimate (and so the Bayes estimate) toward the observed relative risk for GISSI-1. The Bayes confidence intervals are much narrower than those based solely on the observed data because they have borrowed strength from the other studies to increase the precision available from a single study.

An important clinical objective for cardiologists is determining the latest time at which treatment still shows a benefit. All of the models demonstrate a benefit of treatment in all of the studies, at least up until 18.5 hr. The lower bound on the confidence intervals are not as sanguine, of course, but the Bayes model shows some benefit until at least 10.5 hr even in the USIM study, which by itself showed little benefit. The lower limits on the intervals about the regression estimates show that we could reasonably expect benefit in any study for which the average time-to-treatment was 5 hr or less, but that random variation would make benefit more uncertain if the time were greater than 5 hr. Nevertheless, if we are concerned not with the results from a single trial, but rather with pooled results, we can reasonably conclude that thrombolytic treatment is beneficial for at least the first 10.5 hr after the

TABLE II
CHANGING ESTIMATES OF SLOPE (δ_1) AND INTERCEPT (δ_0) AS NEW STUDIES ARE ADDED

	Intercept		Slope	
Study added	Estimate	SE	Estimate	SE
GISSI-1	−0.360	0.093	0.0428	0.0192
ISAM	−0.353	0.090	0.0421	0.0190
AIMS	−0.373	0.089	0.0413	0.0189
Wilcox	−0.380	0.082	0.0410	0.0183
ISIS-2	−0.315	0.050	0.0136	0.0072
USIM	−0.291	0.048	0.0114	0.0071
EMERAS	−0.293	0.044	0.0134	0.0056
LATE	0.296	0.043	0.0130	0.0048

onset of chest pain and perhaps for several hours after that also. These conclusions are similar to those drawn by the Fibrinolytic Therapy Trialists' Collaborative Group.[31]

To illustrate the updating features of the Kalman filter, Table II shows the change in the estimates of the regression coefficients δ_0 and δ_1 (expressed on the log relative risk scale) from the regression on time-to-treatment as the studies are added in chronological order. Because δ_0 and δ_1 are second-stage parameters common to each study, computing their posterior distribution requires only the Kalman (forward) filter.

The dominating effect of the two largest studies, GISSI-1 and ISIS-2, is apparent. These two studies reduce the standard errors substantially and also change estimates of the mean and variance. The USIM study also shifts the estimates substantially because its small treatment effects contrast with the larger ones from other short time-to-treatment studies.

Table III shows that an individual study's estimates can also be followed as the algorithm progresses. Based on data from the subgroup of patients treated within 1 hr in GISSI-1, the estimate of the true effect of treatment for patients treated within 1 hr is a reduction in risk of 47% [= exp(−0.633)]. When data from the rest of the GISSI-1 study are incorporated into a regression model, the reduction is estimated to be only 28%. This reduction is shrunk to 25% based on data from all eight studies. Conversely, the estimated effect for patients treated after 9 hr changes from a 9% increased risk (16% if only data from that subgroup are used) to a 15% decreased risk with the information from the other eight studies.

[31] Fibrinolytic Therapy Trialists' (FTT) Collaborative Group, *Lancet* **343,** 311 (1994).

TABLE III
THREE DIFFERENT GISSI-1 TIME SUBGROUP ESTIMATES (WITH STANDARD ERRORS)[a]

Time subgroup (hr)	Subgroup only	All GISSI-1 data	All studies
<1	−0.633 (0.162)	−0.330 (0.082)	−0.287 (0.040)
1–3	−0.155 (0.085)	−0.274 (0.064)	−0.270 (0.036)
3–6	−0.184 (0.086)	−0.167 (0.052)	−0.237 (0.029)
6–9	−0.117 (0.139)	−0.039 (0.084)	−0.198 (0.026)
>9	0.149 (0.198)	0.089 (0.134)	−0.159 (0.031)

[a] The first are the observed estimates from each subgroup. The second are estimates for the subgroup from a hierarchical model fit to all the GISSI-1 data. The third are estimates from a hierarchical model fit to data from all eight studies.

This readjustment is not solely of academic interest. The results of the GISSI-1 study were so influential that not only did they establish streptokinase as a standard treatment for myocardial infarction, but they also convinced physicians that effectiveness decreased as time since symptom onset increased. Based on this study, treatment was advocated for patients who arrived within 6 hr after symptom onset.[23] The evaluation of this study by a Bayesian hierarchical linear model derived from data that also include the other large clinical trials of thrombolytic therapy suggests now that treatment is also beneficial for patients arriving much later than 6 hr after symptom onset.

Example: Saltatory Model of Infant Growth

Lampl *et al.*[32] reported daily measurements of infants for whom whole-body length appears to increase at discrete jumps, termed *saltations,* with much longer intervening periods of no growth, termed *stasis.* Figure 3 shows daily measurements of total recumbent length (height) of one infant taken over a 4-month period (from day 321 to day 443 of age) using a specially designed infant measuring board. The pattern of saltation and stasis is apparent in the figure, although measurement error does introduce some uncertainty. Sources of this error include the equipment, measurement technique, and the cooperation of the individual subjects.

The within-individual variation may involve the random nature of the growth events or could involve more systematic changes in growth patterns related to the age of the child or individual genetic, physiologic, or environmental factors modifying the growth rates. Between-individual variation

[32] M. Lampl, J. D. Veldhuis, and M. L. Johnson, *Science* **258,** 801 (1992).

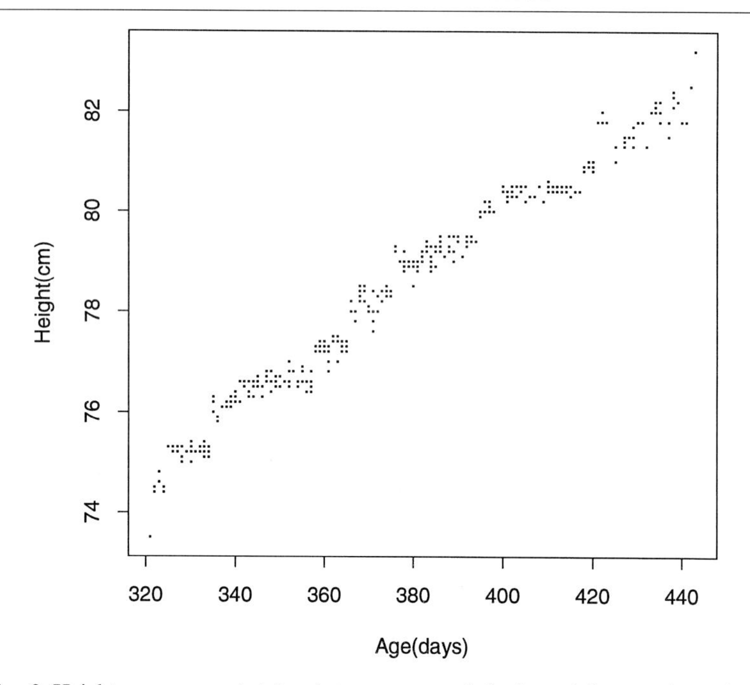

FIG. 3. Height measurements taken between one and six times daily over 4 months on a single infant.

would describe broad characteristics of populations and subpopulations of individuals, for instance, the average height at a given age or the distribution of the mean amplitude or mean stasis interval.

These requirements can be described with a three-level Bayesian hierarchical model. The first level represents an individual's height as the sum of the true height and the error of measurement. The second level represents this true height in terms of components that relate to the average height of similar individuals in the population and to the specific growth characteristics of this individual. The third level provides a probabilistic specification of growth in the population. This hierarchical structure permits the growth measurements from one individual to inform the growth distribution of another individual through the population components in the model.

We can represent such saltatory growth as a stochastic process as follows. First, let y_t be the measured height and H_t be the true height at time t for $t = 1, \ldots, T$ time points measured on one individual. Then, take $u = (u_1, u_2, \ldots, u_K)^T$ to be the set of times at which the saltations occur with h_{u_k}, the saltation at time u_k. Then

$$H_t = H_{t-1} + \sum_{t-1<u_k\le t} h_{u_k}$$

describes the true growth process and we observe

$$y_t = H_t + \varepsilon_t$$

where ε_t is measurement error distributed as a Gaussian random variable with mean 0 and variance σ_ε^2. We assume that the distribution of each saltation h_{u_k} is Gaussian with mean μ_h and variance σ_h^2. This can be expressed by the equation $h_{u_k} = \mu_h + \varepsilon_{u_k}$ where $\varepsilon_{u_k} \sim N(0, \sigma_h^2)$. Conditional on the set of saltation times, u, the problem may then be set up as a Bayesian hierarchical model as follows.

Let J_t represent the number of saltations that occur during the interval $(t - 1, t]$, let the prior distribution for the average amplitude, μ_h, be Gaussian with mean μ_{h_0} and variance $\sigma_{h_0}^2$ and let the prior for the initial height, H_0, be Gaussian with mean μ_0 and variance σ_0^2. We can write this in the state-space form of Eqs. (7) and (8) as

$$\begin{pmatrix} H_t \\ \mu_h \end{pmatrix} = \begin{pmatrix} 1 & J_t \\ 0 & 1 \end{pmatrix}\begin{pmatrix} H_{t-1} \\ \mu_h \end{pmatrix} + \begin{pmatrix} \sum_{t-1<u_k\le t} \varepsilon_{u_k} \\ 0 \end{pmatrix}$$

and

$$y_t = (1 \quad 0)\begin{pmatrix} H_t \\ \mu_h \end{pmatrix} + \varepsilon_t$$

so that

$$\boldsymbol{\theta}_t = \begin{pmatrix} H_t \\ \mu_h \end{pmatrix}, \quad \mathbf{F}_t = \begin{pmatrix} 1 & J_t \\ 0 & 1 \end{pmatrix}, \quad \mathbf{X}_t^* = (1 \quad 0), \quad \mathbf{W}_1^* = \begin{pmatrix} J_t\sigma_h^2 & 0 \\ 0 & 0 \end{pmatrix}$$

and $V_t = \sigma_\varepsilon^2$. Given the variances σ_ε^2 and σ_h^2 and the prior distributions for μ_h and H_0, the computations for the Kalman filter are started by setting $\boldsymbol{\theta}_0 = \begin{pmatrix} \mu_0 \\ \mu_{h_0} \end{pmatrix}$ and $\mathbf{S}_0 = \begin{pmatrix} \sigma_0^2 & 0 \\ 0 & \sigma_{h_0}^2 \end{pmatrix}$. For simplicity, we assume that the variance components σ_ε^2 and σ_h^2 are fixed at $\hat{\sigma}_\varepsilon^2 = 0.059$ and $\hat{\sigma}_h^2 = 0.038$ from a prior analysis using the EM algorithm,[33] and that $\boldsymbol{\theta}_0 = \begin{pmatrix} 73 \\ 0.94 \end{pmatrix}$ and $\mathbf{S}_0 =$

[33] R. J. A. Little and D. B. Rubin, "Statistical Analysis with Missing Data." John Wiley and Sons, New York, 1987.

$\begin{pmatrix} 4.7 & 0 \\ 0 & 4.9 \end{pmatrix}$, based on values estimated from data from another infant. In fact, the results turn out to be robust even to extreme changes in these prior values. Using Fig. 3, we fix 12 saltations occurring on days 322, 325, 335, 341, 358, 366, 376, 395, 400, 418, 421, and 443. The assumptions of fixed variances and growth times are made for pedagogical reasons in order to illustrate use of the Kalman filter. We will discuss removal of these restrictions presently.

Table IV gives estimates and standard errors for the 12 saltations $\mathbf{h} = (h_1, h_2, \ldots, h_{12})$, the mean saltation μ_h, and the initial height H_0 (Table IV). Figure 4 shows the fitted growth curve overlaid on the data. The infant's initial height was estimated to be 73.6 ($\pm$0.3) cm (mean $\pm$ 2 standard errors) and the estimated final height was 83.0 cm, a total growth of 9.4 cm. This corresponds to an average growth of 0.788 cm per growth event with individual growth amplitudes varying between 0.42 and 1.19 cm. These estimates are fairly precise with standard errors ranging from 0.05 to 0.14 cm. All of the amplitudes are significantly different from zero, indicating our choice of saltation times was reasonable.

Incorporation of Variance Components

Computation of the full model with random variances and growth times requires embedding the Kalman filter inside a larger Markov chain Monte

TABLE IV
POSTERIOR MEANS AND STANDARD ERRORS OF MODEL PARAMETERS FOR GROWTH AMPLITUDES $(h_1, h_2, \ldots, h_{12})$, MEAN GROWTH (μ_h), AND INITIAL HEIGHT (H_0)

Parameter	Mean	SE
h_1	0.904	0.139
h_2	0.734	0.089
h_3	0.876	0.059
h_4	0.486	0.054
h_5	0.704	0.048
h_6	0.922	0.056
h_7	0.963	0.050
h_8	0.813	0.062
h_9	0.421	0.062
h_{10}	0.550	0.070
h_{11}	0.900	0.071
h_{12}	1.187	0.107
μ_h	0.788	0.058
H_0	73.586	0.137

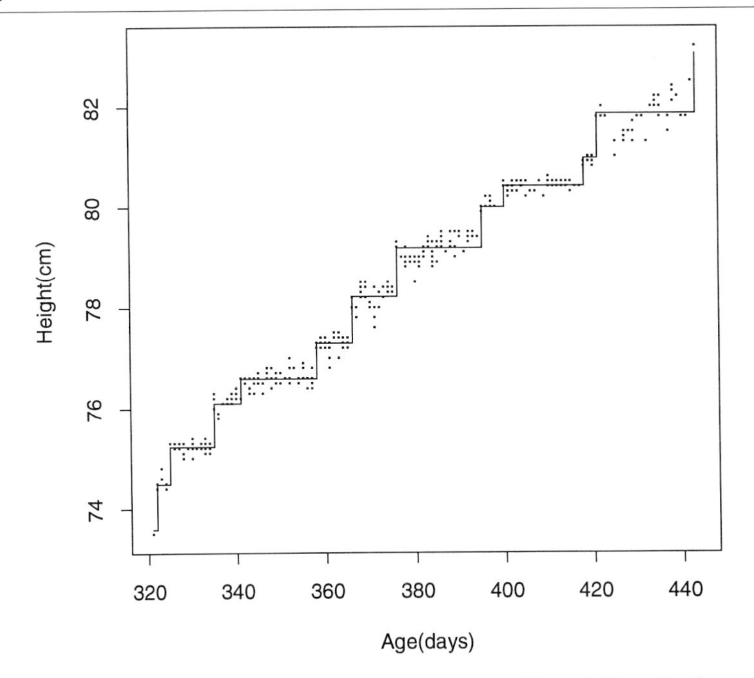

FIG. 4. Fitted curve from the hierarchical growth model overlaid on the data.

Carlo (MCMC) algorithm that could incorporate the non-Gaussian structure of these components. A brief outline of MCMC is therefore warranted.

Return to the basic Bayesian formula describing the posterior density of a parameter θ as

$$f(\theta|Y) = \frac{f(\theta)f(Y|\theta)}{f(Y)}$$

If we wish to calculate the posterior mean (or in fact any posterior moment), we must evaluate the expectation of θ with respect to this density, $E(\theta) = \int \theta f(\theta|Y)\, d\theta$. This involves an integration which, except in very simple problems, must be performed numerically. In complex problems with many parameters, computing the marginal density requires integrating over all parameters but one. This is simply impossible for many real problems.

MCMC methods work by simulating a series of N draws from the correct posterior distribution for each parameter in the model. With these simulated draws, moments may be calculated simply by Monte Carlo integration. For instance, the mean of θ_i is calculated simply from the average of all the drawn values of the θ_i. Moreover, since draws from the complete joint posterior are available, we can compute the posterior of an arbitrary function g of the parameters by simply averaging the values obtained by applying this function to the simulated parameters, obtaining $E[g(\theta)] \approx \frac{1}{N} \sum_{i=1}^{N} g(\tilde{\theta}_i)$ for the N draws $\tilde{\theta}_i$ from $f(\theta|Y)$.

The theory behind this method is too complex to describe other than briefly here. Interested readers may refer to Gilks *et al.*[19] and the many papers cited therein for details. Consider starting a chain such that having drawn $\theta^{(t-1)}$ we can draw from $h(\theta^{(t)}|\theta^{(t-1)}, y)$. Because this draw depends only on the previous state, it is a Markov chain. Eventually, the chain will forget its starting value and it can be shown that the chain will converge to its stationary distribution, which is the correct posterior $f(\theta|y)$. Thus, to simulate a Markov chain from the correct posterior we need only be able to draw from $h(\theta^{(t)}|\theta^{(t-1)}, y)$. It is often simpler to proceed one parameter at a time, simulating from $h_i(\theta_i^{(t)}|\theta_{[i]}^{(t-1)}, y)$ where the notation $\theta_{[i]}$ represents all members of θ except the ith. If all of these full conditional distributions can be sampled, the MCMC algorithm is described as the Gibbs sampler. Except when conjugate distributions have been used, however, not all of these full conditonals will be able to be represented in terms of known densities that can be simulated.

When the conditional density can be written down but not simulated, the Metropolis–Hastings form of MCMC can be used instead. This consists of drawing $\theta_i^{(t)}$ from a transition density $q_i(\theta_i^{(t)}|\theta_{[i]}^{(t-1)})$ that is constructed to resemble $h_i(\theta_i^{(t)}|\theta_{[i]}^{(t-1)})$ and then accepting this draw with probability

$$R = \min\left(1, \frac{h_i(\theta_i^{(t)}|Y, \theta_{[i]}^{(t-1)})q_i(\theta_{[i]}^{(t-1)}|\theta_i^{(t)})}{h_i(\theta_i^{(t-1)}|Y, \theta_{[i]}^{(t-1)})q_i(\theta_i^{(t)}|\theta_{[i]}^{(t-1)})}\right)$$

This acceptance ratio can be seen as the ratio of the posterior probability that $\theta_i = \theta_i^{(t)}$ conditional on the current draws of the other parameters to the conditional posterior probability that $\theta_i = \theta_i^{(t-1)}$ weighted by the importance ratio $q_i(\theta_{[i]}^{(t-1)}|\theta_i^{(t)})/q_i(\theta_i^{(t)}|\theta_{[i]}^{(t-1)})$. If the new drawn value has substantially higher posterior probability than the previous value, the acceptance rate will be high. In the Gibbs sampler, the transition density is $q_i(\theta_i^{(t)}|\theta_{[i]}^{(t-1)}) = h_i(\theta_i^{(t)}|Y, \theta_{[i]}^{(t-1)})$ so that the ratio is always one and the draw is always accepted. Gibbs sampling therefore uses the optimal transition density if it can be found and if not uses a Metropolis–Hastings step instead.

In the growth problem, the parameters for which posterior inference are desired include those describing the growth amplitude, $\{\mathbf{h}, H_0, \mu_h, \sigma_h^2\}$, the measurement error variance σ_e^2, and the parameters ϕ involved in describing the saltation times $\mathbf{u}$. The Kalman filter set out above describes the distribution of $\mathbf{h}, H_0, \mu_h|Y, \sigma_e^2, \mathbf{u}, \phi$. MCMC is required to compute the full posterior. This is the subject of current research.

Model Checking

Once models have been fit to data, it is important that they be checked against the data to ensure that they describe the important features. Plots can help. Here we describe the idea of using posterior predictive checks to evaluate the accuracy of a Bayesian model.[34] The basic idea is to simulate draws of the outcomes from the model posterior and then to check these simulated responses against the actual responses.

For a given realization of the parameters from an iteration of the MCMC algorithm, we calculate the test statistic $T(y_k^{\text{rep}}, \theta_k)$, a measure of discrepancy where y_k^{rep} is a sample from the predictive distribution $[y|\theta_k]$. $T(y_k^{\text{rep}}, \theta_k)$ is then compared with $T(y^{\text{obs}}, \theta_k)$ computed using the observed data y^{obs}. A Bayesian p value may then be computed from $\Pr[T(y_k^{\text{rep}}, \theta_k) > T(y^{\text{obs}}, \theta_k)]$. As in the classical hypothesis test, this p value will be small when the observed disrepancy is most often higher than would be expected if the data were generated from the model.

A variety of test statistics may be employed to validate the model. Two standard ones are:

1. An overall-goodness-of-fit test: $T_1(y, \theta) = \Sigma_{i=1}^{N} [y_i - E(y_i|\theta)]/[\text{Var}(y_i|\theta)]$
2. The test of maximal deviation: $T_2(y, \theta) = \max|y_i - E(y_i|\theta)|$.

We can also tailor statistics to the problem at hand. For example, to evaluate the growth model, we could examine the following test statistics:

1. Number of sign changes (a rough test of serial correlation): $T_3(y, \theta) = \{\text{sgn}(y_i - E(y_i|\theta)) \neq \text{sgn}(y_{i+1} - E(y_{i+1}|\theta))\}$
2. Height at the end of observation: $T_4(y, \theta) = y_N$
3. Largest growth increment: $T_5(y, \theta) = \max|\bar{y}_i - \bar{y}_{i-1}|$
4. Largest stasis interval: $T_6(y, \theta) = \{$longest consecutive run such that $\bar{y}_i - \bar{y}_{i-1} < k \forall i\}$
5. Number of growth events: $T_7(y, \theta) = \Sigma I\{(\bar{y}_i - \bar{y}_{i-1}) > k\}$ for some constant k where I is the indicator function.

All of these checks use the entire set of data. A second type of posterior check examines how a model predicts on new data. For this purpose, we can split each data series into two parts using the first two-thirds to develop the model and the last third to test the model.[35] For each realization θ_k, we can compute y_k^{pred} from $[y^{\text{new}}|\theta_k, y^{\text{old}}]$ where y^{new} represents the last

[34] A. Gelman, J. B. Carlin, H. S. Stern, and D. B. Rubin, "Bayesian Data Analysis." Chapman and Hall, New York, 1995.

[35] B. P. Carlin and T. A. Louis, "Bayes and Empirical Bayes Methods for Data Analysis." Chapman and Hall, New York, 1996.

third and y^{old} the first two-thirds of y^{obs}. Bayesian p values are then computed from $\Pr[T(y_k^{pred}, \theta_k) > T(y^{new}, \theta_k)]$.

Conclusion

Hierarchical models are becomig increasingly popular for representing probability structures with multiple components of variance. Bayesian forms of these models allow a much more flexible description of model features and explicit representation of prior information that may be combined with the data to improve scientific inference. The computation of Bayesian models has been significantly simplified by MCMC techniques, which now allow computation of complex probability models. Representation of a Bayesian hierarchical linear model in state-space form and sequential computation by linear filters provides a useful way to understand the model structure and the interrelationships among model parameters, as well as to appreciate how new data update the posterior densities of the parameters. In problems with an inherent temporal or ordered spatial structure, this representation can also facilitate model construction.

Acknowledgments

This research was supported by grants 19122 and 23397 from the Robert Wood Johnson Foundation, by NAGW 4061 from NASA, and by R01-HS08532 and R01-HS07782 from the Agency for Health Care Policy and Research.

[18] Monte Carlo Applications to Thermal and Chemical Denaturation Experiments of Nucleic Acids and Proteins

By D. JEREMY WILLIAMS and KATHLEEN B. HALL

Introduction

Information about the states of systems must often be extracted indirectly from measurements of properties considered characteristic of these states. However, even for systems with stable, well-defined states, the ability to determine the *true* value of any property is complicated by errors in the measurement process. Statistical fluctuations due to processes on the molecular scale may further obfuscate attempts to determine the most likely (probable) value of the property under consideration. As a consequence of these error-generating processes, repeat measurements under nominally

0076-6879/00 $30.00

constant environmental conditions give a distribution of values for the property instead of a single value. If we hypothesize that there exists a parental probability distribution that contains complete information on the possible values and expected frequencies of occurrence of the property of interest, then an approximate parental distribution can be constructed by binning a large number of measurements to produce a frequency histogram. From this probability distribution all statistical quantities of interest can then be calculated, including the most probable (*true*) value of the parameter, its mean value, and the variance of the distribution—a measure of the spread of the distribution around the mean.

As an example of this process we consider the measurement of the absorbance of an RNA solution under constant solution conditions (constant temperature, pH, salt concentration, and nucleic acid concentration). Because the extinction coefficient of the molecule is constant under these conditions we expect a single absorbance value A_0. Instead, individual A_i values are measured, each differing from A_0 by some amount ε_i:

$$A_i = A_0 + \varepsilon_i \tag{1}$$

Let us assume that we can infer from a large number of absorbance measurements (N) that the A_i distribution is symmetric about some value A_m. In this case, an estimate of the most probable absorbance value can be obtained from the mean or the median of the distribution since for symmetric unimodal distributions these three parameters are equal.[1] The sample mean A_m is particularly easy to calculate and can be used to estimate the true absorbance value A_0:

$$A_0 \cong A_m = \frac{1}{N} \sum_N A_i \tag{2}$$

The sample variance (s^2) is also easily calculated and can be used as an estimate of the true variance of the distribution, which in turn furnishes information on the precision with which A_0 can be known from a finite number of absorbance measurements.

$$\sigma^2 \cong s^2 = \frac{1}{N-1} \sum_N (A_i - A_m)^2 \tag{3}$$

Observables like absorbance are often of little intrinsic interest but serve rather as reporters of other properties and system parameters that cannot be measured easily. We are particularly interested in changes in

[1] P. R. Bevington, "Data Reduction and Error Analysis for the Physical Sciences," p. 15. McGraw-Hill, New York, 1969.

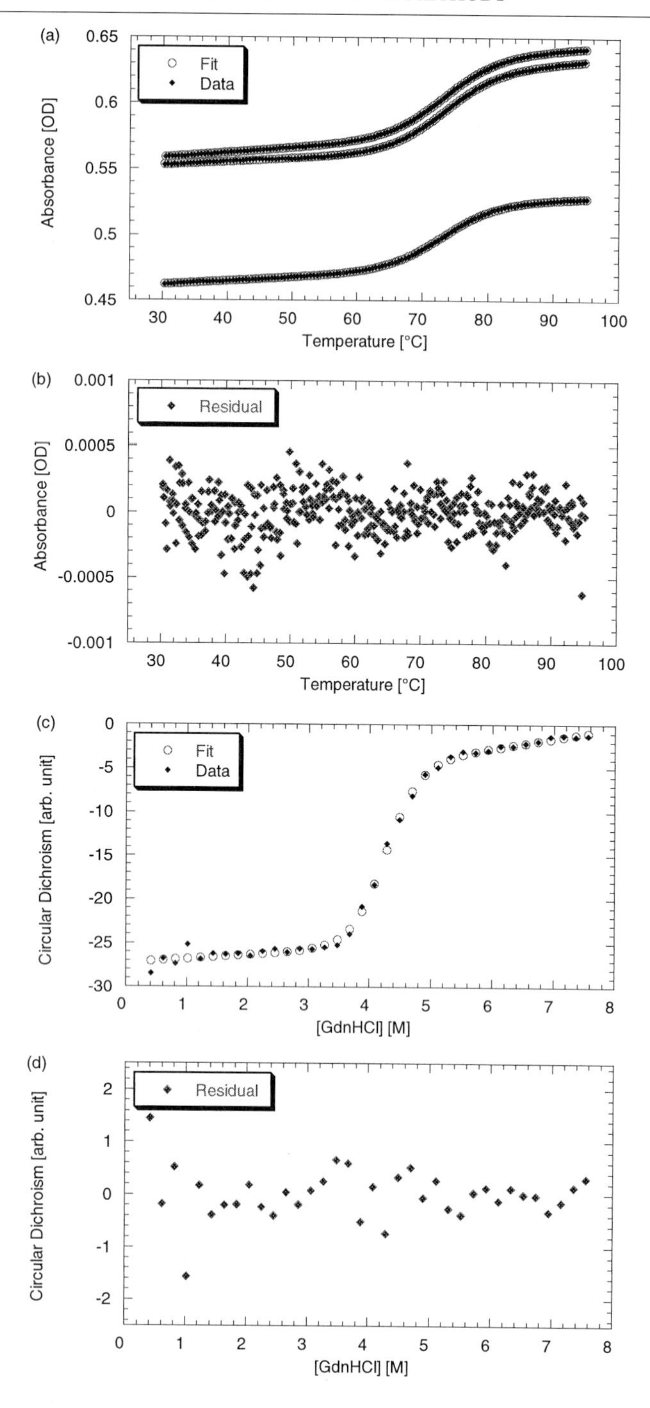
(a)
Fit
Data
Absorbance [OD]
0.65
0.6
0.55
0.5
0.45
30 40 50 60 70 80 90 100
Temperature [°C]
(b)
Residual
Absorbance [OD]
0.001
0.0005
0
-0.0005
-0.001
30 40 50 60 70 80 90 100
Temperature [°C]
(c)
Fit
Data
Circular Dichroism [arb. unit]
0
-5
-10
-15
-20
-25
-30
0 1 2 3 4 5 6 7 8
[GdnHCl] [M]
(d)
Residual
Circular Dichroism [arb. unit]
2
1
0
-1
-2
0 1 2 3 4 5 6 7 8
[GdnHCl] [M]

the values of these parameters in response to perturbations, which reflect underlying changes in other properties that cannot be monitored directly. The determination of the free-energy change of unfolding is an example of this methodology. Here the fraction of folded (or unfolded) protein or RNA is measured as a function of increasing denaturant concentration or temperature and then related to a free-energy change for the process. Nucleic acid structural transitions are commonly monitored by absorbance measurements in the ultraviolet region of the spectrum from which changes in species fractions can be extracted and enthalpy, entropy, and free-energy changes of unfolding calculated. Circular dichroism is often used in an analogous fashion to monitor chemically or thermally induced folding processes in proteins and peptides.

Examples of the data from UV/vis melting studies on an RNA hairpin and from chemical denaturation studies on a small globular protein are shown in Fig. 1. The fits to the data were generated by assuming a model involving only two species in equilibrium (single strand and native)—the two-state assumption. This model is valid in the limit of high positive cooperativity where intermediate states are greatly suppressed. The extinction coefficients (or molar circular dichroism) of the species are assumed to depend linearly on the perturbant (temperature in the case of the RNA and denaturant for the protein). It is further assumed that the enthalpy and entropy change of unfolding is independent of temperature and that the free-energy change depends linearly on denaturant concentration allowing free energies to be extrapolated to other temperatures and denaturant concentrations.[2,3]

Parameter Estimation

Given a model such as the one described previously, how is it then applied to the data to produce the results shown in Fig. 1? As implied

[2] S. M. Freier, D. D. Albergo, and D. H. Turner, *Biopolymers* **22,** 1107 (1983).

[3] C. N. Pace, *Methods Enzymol.* **131,** 266 (1986).

Fig. 1. (a) Global analysis of three independent absorbance (260 nm) versus temperature melting experiments for an RNA hairpin. Fits to two-state model with linear baselines as described in text and by D. J. Williams and K. B. Hall, *Biochemistry* **35,** 14,665 (1996). Experimental conditions: 0.5 m*M* EDTA, 10 m*M* sodium cacodylate, pH 7. (b) Residuals (difference between fit and experimental data) versus temperature for UV/vis melting experiment in (a). (c) Guanidine hydrochloride (GdnHCl) denaturation experiment of a small globular protein monitored by CD at 220 nm. Data fit to two-state model with linear baselines as described in text and by J. Lu and K. B. Hall, *J. Mol. Biol.* **247,** 739 (1995). Experimental conditions: 50 m*M* NaCl, 10 m*M* sodium cacodylate, pH 7.0. (d) Residuals versus [GdnHCl] for the data and fit shown in (c).

earlier, this process involves what is known as data fitting. For this purpose we assert that the dependent variable can be described as a function of the independent variables and parameters of interest: that is, $y(x_i) \cong f(x_i, P)$. The task is then to determine parameter values capable of "explaining" the experimental observations. Despite our best efforts, we know the dependent variable to be contaminated with experimental errors and so even given knowledge of the true parameter values P_T, we expect the function $f(x_i, P_T)$ to differ from $y(x_i)$ by some finite amount d_i. We might intuit that as P approaches P_T the difference between the experiment and its predicted value would decrease until at P_T it would be some minimum value. The remaining (or residual) error (d_i) could then serve as an estimate of the experimental error (ε_i). As an estimate of P_T we might seek parameters P_M that minimize the differences between the dependent variable $y(x_i)$ and its predicted value $f(x_i, P)$. But how do we measure the "difference" between the measurement and the prediction? In other words, what distance metric do we minimize? Two possible metrics are the sum of the average deviations $\Sigma|y(x_i) - f(x_i, P)|$ and the sum of the square deviations $\Sigma[y(x_i) - f(x_i, P)]^2$. However, the choice of functional form to minimize cannot be arbitrary if the goal is to obtain parameters that have the maximal probability of being correct (maximum likelihood parameters). The choice must be dictated by the nature of the error distribution. If we assume that each experimental measurement is independent, that the errors are entirely in the dependent variable, and that these errors are normally distributed with zero means and standard deviations of σ [$N(0, \sigma^2)$], then least squares minimization will yield maximum likelihood parameter estimates.[4,5] In this case, the chi-square (χ^2), which is simply the weighted sum of the squares deviation, is minimized:

$$\chi^2 \equiv \sum \frac{[y_i(x_i) - f(x_i, P)]^2}{\sigma_i^2} \tag{4}$$

Where there are significant errors in the independent variable or where these errors are correlated with those in the dependent variable, alternative analyses are necessary in order to extract maximum likelihood estimates from the data.[6] Additionally, a least squares analysis is not appropriate in situations where the experimental errors are known to be highly asymmetric, as happens in some counting processes with low numbers of events or in experiments where the value of a parameter is closely bounded on

[4] N. R. Draper and H. Smith, "Applied Regression Analysis," p. 60. Wiley Series in Probability and Mathematical Statistics, John Wiley and Sons, New York, 1966.

[5] M. L. Johnson and L. M. Faunt, *Methods Enzymol.* **210,** 4 (1992).

[6] M. L. Johnson, *Anal. Biochem.* **148,** 471 (1985).

one side. Also problematic are cases where the error distribution contains significant contributions from higher order moments (i.e., the data show large or frequent outliers).[7,8] In many cases, however, experimental errors are likely to be the result of several processes acting in concert, and it is commonly assumed from the central limit theorem, that the total error distribution will be approximately normal.[9] Furthermore, least squares analysis seems to provide reasonably accurate parameter estimates for some nonnormal error distributions provided that outliers are not large or frequent.[8]

We do not discuss the mechanics of how to perform least squares minimization of a function to a set of data. Interested readers are referred to the excellent discussions by Johnson and Faunt[10] and by Bevington.[11]

Parameter Confidence

Having obtained an estimate of the parameters from a least squares analysis, the next step is to evaluate parameter confidence. Imagine sitting at the center of a hemispherical depression corresponding to the maximum likelihood parameters of a two-parameter least squares fit. The space extending away and up from this central position is the variance space of the fit. As we walk away from the center in a radial direction, the surface starts to rise as a result of the increase in the χ^2 value. If the surface rises very quickly, we can assume that we have left the minimum—the minimum is well determined. If the surface more closely resembles the proverbial pancake, we might be less confident that we had previously determined the minimum. Because the curvature of the surface depends both on the function $f(x_i, P)$ and on the error in the data, we might expect intuitively that if the surface rises very steeply away from the minimum value, the parameters are well determined by the data and vice versa. At this point some readers might object that one either has or has not previously determined the minimum, and that the flatness or steepness of the surface should not matter. The objection is valid, but we view this particular bowl and its associated minimum as only one of a large number of possible bowls—each bowl having a different set of minimum parameter values—that could have

[7] P. R. Bevington, "Data Reduction and Error Analysis for the Physical Sciences," p. 40. McGraw-Hill, New York, 1969.

[8] E. H. Henry, *Biophys. J.* **72,** 652 (1997).

[9] S. Ghahramani, "Fundamentals of Probability," p. 390. Prentice Hall, Upper Saddle River, New Jersey, 1996.

[10] M. L. Johnson and L. M. Faunt, *Methods Enzymol.* **210,** 9 (1992).

[11] P. R. Bevington, "Data Reduction and Error Analysis for the Physical Sciences," p. 204. McGraw-Hill, New York, 1969.

been realized had the experiment been repeated. We seek statistical tests to relate increases in the chi-square value away from the minimum to probabilities that other bowls taken from the same parental distribution would actually have minima at these new positions. The variance space could then be divided into a region where the parameters would lie with a certain probability given the fitting function and the error in the data. Thus we could determine confidence intervals for the parameters of the fit. The mathematical detail of this treatment of parameter confidence is given by Bard.[12]

Figure 2 shows four possible two-parameter variance spaces from the point of view of looking down onto the minimum along the chi-square (z) axis. These surfaces are depicted as circular or elliptical, which is only true when the fitting parameters are linear [i.e., all second and higher order derivatives of the $f(x_i, P)$ with respect to P are zero].[5] In the general nonlinear case, the chi-square surface is not a P-dimensional ellipse as shown (P is the number of parameters) and is not necessarily symmetric. In both panels the parameter axes are oriented along the directions of the sides of the squares (N $\leftrightarrow$ S and E $\leftrightarrow$ W). In Fig. 2a, the axes of the variance spaces (axis of the ellipse) are aligned with the parameter axes, whereas in Fig. 2b they are at an angle. This seemingly small difference has important consequences when determining parameter confidences or comparing parameter distributions to determine if they are consistent with the same parental distribution.

In Fig. 2a we can walk away from the minimum (1) in the direction of one of the parameter axes to the maximum allowed value of that parameter (the circular or elliptical boundary) consistent with the given data and the stated probability level *without* changing the value of the other parameter. Naturally, the distance of the circular (elliptical) boundary from the minimum at the center depends on how quickly the surface rises and on how stringent we want to make the test of whether to accept or reject a parameter value as being consistent with the fit (the confidence level). For example, the barrier that divides parameters consistent with the fit from those that are inconsistent is farther away at the 95% confidence level than at the 67% confidence level. In other words, to be more certain that we have considered all parameter values consistent with the data and fit we must include more of the surface of possible parameter values. We must move the barrier farther away (make the ellipse larger) and so include parameters that have higher χ^2 values (lower probabilities).

The important feature of the variance spaces in Fig. 2a is that the direction of *minimum increase* of the chi-square value—movement along

[12] Y. Bard, "Nonlinear Parameter Estimation," p. 184. Academic Press, New York, 1974.

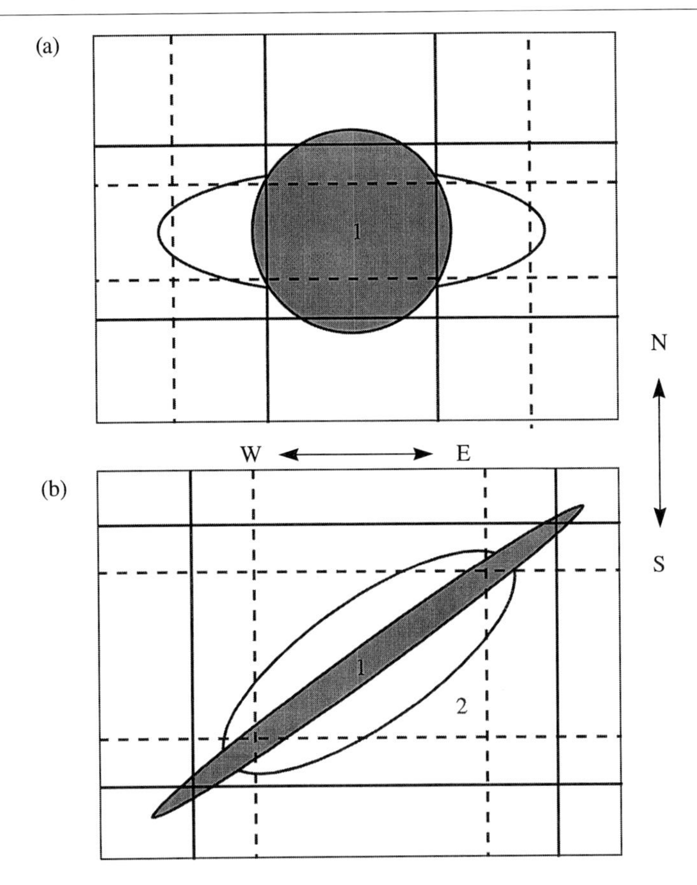

FIG. 2. (a) Two-dimensional representation of the variance space for a two-parameter fit involving linear *uncorrelated* parameters. Shaded circle (unshaded ellipse) encloses hypothetical joint 95% confidence interval for parameters of equal (different) variances. Parallel lines of the same type bracket individual parameter 95% confidence intervals. (b) Two-dimensional representation of the variance space for a two-parameter fit involving linear *correlated* parameters. Shaded ellipse (unshaded ellipse) encloses hypothetical joint 95% confidence interval for highly (moderately) correlated parameters. Parallel lines of the same type bracket individual parameter 95% confidence intervals.

the axes of the elliptical variance space—is in the direction of the parameter axes. As a consequence, the value of a parameter that minimizes χ^2 is independent of the value of the other parameter. In this case the two parameters are said to be uncorrelated.

The situation is quite different in the correlated case shown in Fig. 2b. Here the directions of minimum χ^2 increase (the axes of the ellipse) are not aligned with the parameter axes. In this case, the value of a parameter that minimizes χ^2 is highly dependent on the value of its correlated partner. Now, to reach the maximum value of either parameter consistent with a specified confidence level requires movement along both parameter axes. Attempts to determine parameter confidence intervals by movement away from the minimum (1) to the dividing line, along a single parameter axis, results in an underestimation of the range of parameter values consistent with the fit to the data and chosen probability level. The value of a parameter at the dividing line, obtained by movement along a single parameter axis away from the minimum, is not the maximum (minimum) value of the parameter consistent with the fit to the data at the stated probability level. With increasing parameter correlation the concomitant elongation and narrowing of the parameter distribution results in serious underestimation of parameter confidence intervals if the parameters are assumed to be uncorrelated. A natural consequence of parameter correlation is that changes in the value of one parameter leading to an increase in the sum of the square residuals can be reduced (compensated) by changes in a correlated parameter. An example of this is the antagonistic effects of the slope and intercept on the chi-square value in a linear fit. Increases in χ^2 due to changes of the slope away from its best fit value can be partially compensated by opposing changes in the intercept. Consequently, attempts to determine parameter confidences for linear fits that do not take this correlation into account will seriously underestimate the confidence intervals of these two parameters.

A second critical difference between the case of correlated and uncorrelated parameters concerns the ability to generate the joint parameter distribution from knowledge of the individual distributions. In Fig. 2, the circles and ellipses map joint confidence probabilities or the probability that a given x,y pair of parameter values is consistent with the fit to the data. Two parallel lines of the same type give the range of parameter values consistent with the fit to the data at the stated probability level for the parameter axis they intersect. (*Note*: The dashed lines are associated with the unshaded ellipse and the solid lines with the shaded ellipse.) These individual parameter confidence intervals spanned by the parallel lines define confidence intervals for single parameters without knowledge of (or concern for) the value of the other parameter. Whereas the circles and ellipses allow the evaluation of the probability that a given x,y parameter pair is consistent with the fit, the parallel lines enclose the probability that a given x value is consistent in the absence of knowledge of the value of y and vice versa. As shown in Fig. 2a (the uncorrelated case), the square defined by the intersection of parallel lines of the same type maps an area very similar to that enclosed by the associated ellipse or circle. Under these circumstances, joint confidence intervals can be calculated quite reliably from knowledge of the individual (single-parameter) confidence intervals. This is not true in the case of correlated parameters where the areas mapped by the squares from individual parameter confidences overestimate the true joint confidence intervals. The more correlated the parameters, the more elongated the true joint distribution and the more severe the problem becomes. The point labeled 2 in the lower panel lies within the individual confidence intervals of both parameters and yet is clearly outside the joint confidence region. Consequently, when comparing parameter distributions involving two or more highly correlated variables it is safer to use the joint confidence intervals than to compare the one-dimensional single-paramcter confidence intervals. Experimental cases of interest involving two or three highly correlated parameters are particularly amenable to graphical comparisons. We will discuss applications to two common situations involving highly correlated parameters.

Confidence Interval Estimation

Several methods differing in severity of assumptions and computational expense are available for calculating parameter confidence intervals.[13] The simplest method estimates parameter confidences from the variance–covariance matrix, which is a reflection of the curvature of the χ^2 surface at the minimum. The matrix is assumed to be diagonal, which is tantamount to assuming that the parameters are uncorrelated. This method yields asymptotic standard errors and is only valid for linear least squares of uncorrelated parameters. It will underestimate the confidence interval—sometimes severely—when these assumptions are not satisfied.

A more rigorous method due to Box[14] also estimates parameter confidence from the variance–covariance matrix but takes parameter correlation into account. Single-parameter confidence intervals are found by converting the correlation matrix (derived from the variance–covariance matrix) into a suitable diagonal form to find its eigenvalues and eigenvectors. The eigenvectors are closely related to the axes of the hyperdimensional ellipse (the directions of minimum χ^2 increase) and the curvature at the minimum is again used to predict confidence intervals. Conceptually this process is equivalent to finding a new set of parameter axes that are aligned with the axes of the variance space from linear combinations of the old ones. This method assumes that the parameter space is elliptical and uses the *F*-test

[13] M. L. Johnson and L. M. Faunt, *Methods Enzymol.* **210,** 20 (1992).
[14] G. E. P. Box, *Ann. N.Y. Acad. Sci.* **86,** 792 (1960).

to associate probabilities with a given confidence interval; it is strictly valid only for the linear least squares case where the variance space is quadratic.

The prediction of symmetric confidence intervals is one consequence of the linear approximation employed in the two methods discussed previously. However, in the general nonlinear case, the variance space may be nonelliptical and/or asymmetric. A third method popularized by Johnson[15] relaxes the assumption that the variance space is quadratic. Instead of predicting confidence intervals from the curvature at the minimum, the parameter space is searched for a critical χ^2 value using the F-test to determine the probability level. To avoid the expense of searching the entire variance space, search directions along the parameter axes and the axes of the hyperdimensional ellipse are chosen. These additional axes are simply the eigenvectors of the correlation matrix multiplied by the square root of the corresponding diagonal element of the variance–covariance matrix. In several applications that we have examined, this method gives similar individual parameter confidence intervals to more rigorous techniques.

The grid search technique is a logical extension of the preceding method. Instead of searching in specified directions, the entire χ^2 surface is mapped. The variance space is divided into a P-dimensional grid (P is the number of parameters) and the value of the weighted sum of the squares is evaluated at each grid point.[13,16] The F-test is again commonly applied to determine confidence levels at a given probability, with the caveat that the F-test gives the stated probability rigorously for the linear case only. The grid search is expensive computationally because it scales as the product of the sampling density per parameter. It becomes exorbitant for any reasonable number of parameters or sampling density. Fortunately, in many cases only a subset of the total number of parameters is of interest and the grid search can be performed on these parameters with optimization (fitting) of all other parameters at each grid point. The grid search method shares a critical advantage with the Monte Carlo technique in that it provides the complete joint parameter distributions, which allows easy comparison of highly correlated parameters and facilitates simple and accurate propagation of errors into derived parameters.

Of the methods to calculate confidence intervals, the Monte Carlo technique makes the least number of assumptions.[17,18] The Monte Carlo method involves creating perfect data using the best fit parameter values, the func-

[15] M. L. Johnson, *Biophys. J.* **44,** 101 (1983).

[16] J. M. Beecham, *Methods Enzymol.* **210,** 37 (1992).

[17] W. H. Press, B. P. Flannery, S. A. Teukolsky, and W. T. Vetterling, "Numerical Recipes in C: The Art of Scientific Computing," p. 689. Cambridge University Press, Cambridge, Massachusetts, 1992.

[18] M. Straume and M. L. Johnson, *Methods Enzymol.* **210,** 117 (1992).

tion, and the independent variables used initially to determine these parameters. Error is then added to the noise-free simulated data and fits carried out on the new pseudo-data. A reasonable estimate of the error—from which to create fictitious experiments—is the main technical requirement of the Monte Carlo method. This error can be synthesized *ab initio* given knowledge of the form of the error distribution (standard Monte Carlo), or the error of the original fit (the residuals) can be resampled (bootstrap).[17,18] In the case applicable to least squares fitting, synthetic data are generated from the least-squares determined minimum values, at the value of the independent variables used in the original fit and using the same fitting function. Normally distributed errors (Gaussian errors) with zero mean and standard deviation $\sigma[N(0, \sigma^2)$ distribution] are then added to the noise-free data. The variance of the fit is often used as an estimate of the true experimental error ($s^2 \cong \sigma^2$). In the bootstrap version the noise-free synthetic data are constructed in the same manner, but errors are obtained by randomly sampling from the residuals of the original fit with replacement (i.e., no memory in the selection process; a residual is selected at random from the collection of residuals without regard to whether it was selected before). In both cases errors must be properly weighted if necessary to provide maximum likelihood parameter estimates. For certain technical reasons neither the error estimate from the sample variance ($s^2 = \Sigma d_i^2/n - p$) nor the variance of the bootstrap ($b^2 = \Sigma d_i^2/n$) is an unbiased estimate of the true variance (σ^2) in the nonlinear case.[19,20] Furthermore, the variance of the bootstrap is downward biased because the number of degrees of freedom due to independent observations (*n*, the number of data points) is not corrected for the degrees of freedom used to fit the data (*p*, the number of parameters).[19] Nevertheless, both s^2 and b^2 are often considered reasonable surrogates for the true experimental error. Pure error estimates of $\sigma^2(x_i)$ can be determined by performing replicate measurements of the dependent variable at fixed values of the independent variable.[20]

Experimental Applications

Protein Stability by Chemical Denaturation

We have applied Monte Carlo analyses to the calculation of joint confidence intervals to aid in the comparison of the effects of mutations on

[19] B. Efron, "The Jacknife, the Bootstrap and Other Resampling Plans," p. 36. SIAM, Philadelphia, 1982.

[20] N. R. Draper and H. Smith, "Applied Regression Analysis," p. 282. Wiley Series in Probability and Mathematical Statistics, John Wiley and Sons, New York, 1966.

correlated variables of interest. The first example involves a thermodynamic study of several proteins constructed by swapping homologous β strands between the N- (RBD1) and C- (RBD2) terminal domains of the human U1A RNA binding domain. These two RBDs have an $\alpha\beta$ sandwich tertiary fold ($\beta\alpha\beta\beta\alpha\beta$ secondary structure) common to RNA binding domains, a 25% amino acid identity (40% amino acid similarity), and are structurally homologous.[21] Functionally, the two RBDs are very different. The N-terminal domain binds to stemloop II of the U1 snRNA whereas RBD2 does not stably associate with any RNA.[22,23] Chimeric proteins consisting of swaps of regions of the two protein domains and other point mutations were constructed to test the effects on the RNA binding properties of the two RBDs. As part of this study, circular dichroism was used to monitor guanidine hydrochloride (GdnHCl) unfolding to assess the stability of these mutants. Stability measurements were carried out and analyzed as previously described,[23] except in this study, global analysis of three independent data sets was performed for each mutant using the program Conlin. [Conlin is a C implementation of Gauss–Newton optimization with the Aitken δ^2 method to improve convergence.[10,11] The program was inspired by the data analysis software NONLIN.[15] Conlin incorporates methods for confidence interval estimation such as the Monte Carlo (standard and bootstrap) and grid search, in addition to semianalytical methods of searching the variance space.]

A Monte Carlo analysis (1000 synthetic data sets) of the denaturation data is summarized in Fig. 3. The linear approximation used to calculate the free-energy change at an arbitrary denaturant concentration assumes that $\Delta G = \Delta G^\circ - m$[denaturant]. At the midpoint of the transition (C_m), the free-energy change is assumed to be zero and the standard free-energy change (ΔG°) can be calculated from m and C_m. Not surprisingly, the ΔG° determined in this analysis is highly correlated with the value of m. The highly elongated appearance of the m versus ΔG° distributions shown in Fig. 3 (correlation coefficient > 0.99) indicates that joint confidence intervals for the two parameters cannot be accurately estimated from the single-parameter confidence intervals. The individual parameter confidence intervals formed by projecting the joint distributions onto the parameters axes would show extensive overlap for the different mutants although the joint distributions are in almost every case distinct. This is the crux of the problem with comparing individual parameter confidence intervals for highly corre-

[21] J. Lu and K. B. Hall, *Biochemistry* **36,** 10393 (1997).

[22] D. Scherly, W. Boelens, W. J. van Venrooij, N. A. Dathan, J. Hamm, and I. Mattaj, *EMBO J.* **8,** 4163 (1989).

[23] J. Lu and K. B. Hall, *J. Mol. Biol.* **247,** 739 (1995).

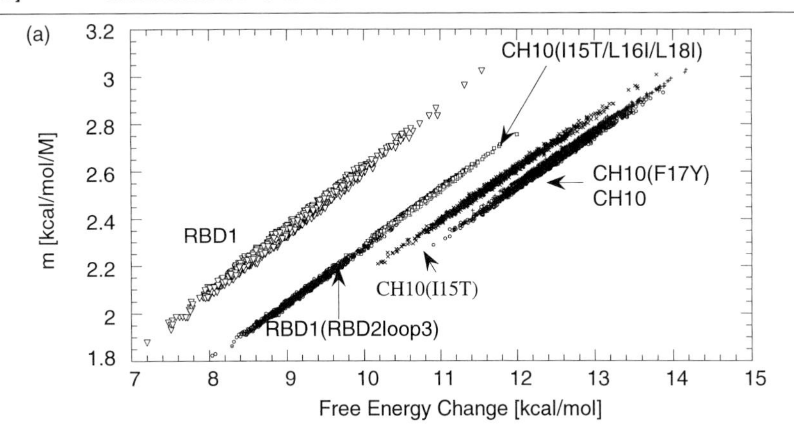

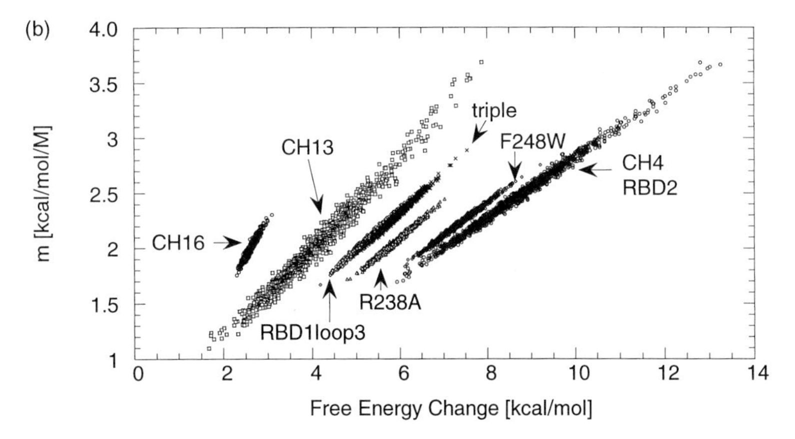

FIG. 3. Plots of m versus ΔG° for the two parent and mutant RBDs generated from Monte Carlo simulations (1000 replicates) using the sample variance as an estimate of the experimental error. (a) RBD1 and closely related chimeras and mutants. (b) RBD2 and closely related chimeras and mutants.

lated parameters. The joint distributions allow us to distinguish mutations that have no net thermodynamic effect (overlapping joint distributions) from those that only appear to have no net effect because of overlap in the individual dimensions. In addition, the 2-D comparison increases resolution, making the trends more easily visible (e.g., compare the information in Fig. 3 to Table I).

From the graphical display of the parameter distributions, it is apparent

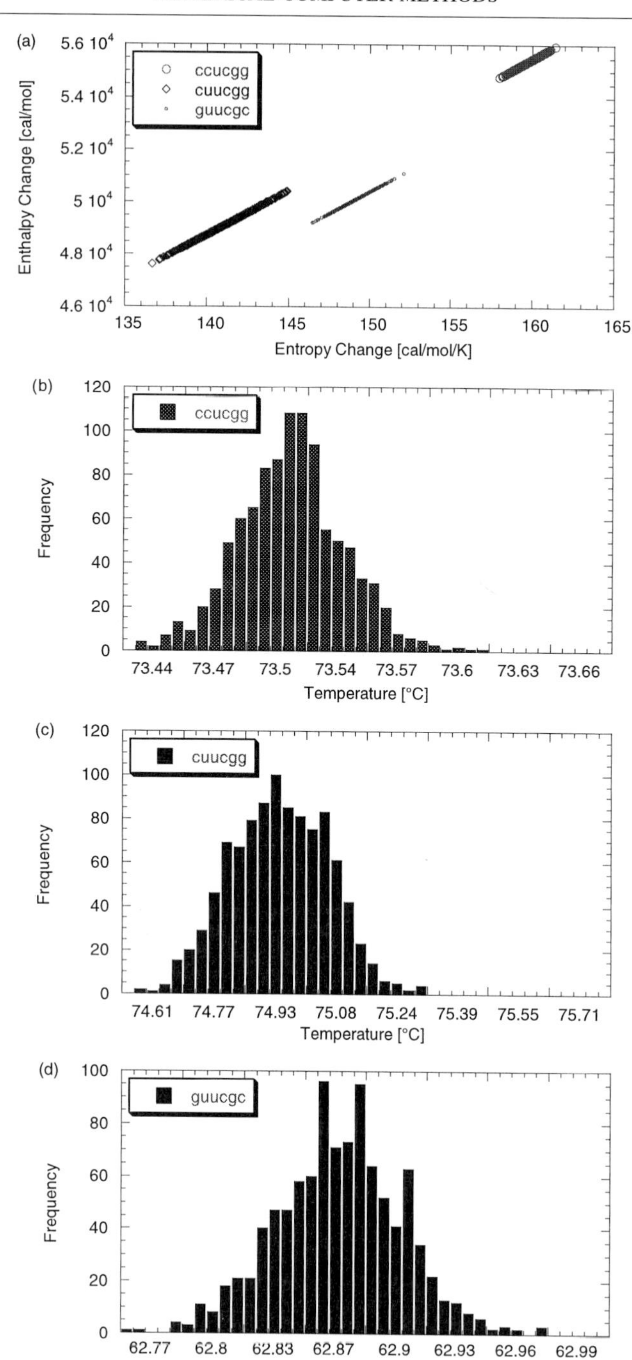
(a)
ccucgg
cuucgg
guucgc
Enthalpy Change [cal/mol]
Entropy Change [cal/mol/K]
(b)
ccucgg
Frequency
Temperature [°C]
(c)
cuucgg
Frequency
Temperature [°C]
(d)
guucgc
Frequency
Temperature [°C]

TABLE I
THERMODYNAMIC PARAMETERS OF WILD-TYPE AND MUTANT U1A PPROTEINS

Protein[a]	m (kcal/mol/M)	ΔG° (kcal/mol)	C_m (M)
RBD1	−2.35	9.0	3.8
RBD1(RBD2loop3)	−2.09	9.2	4.4
Chimera 10	−2.70	13.0	4.8
Chimera 10 (I15T)	−2.49	10.7	4.3
Chimera 10 (I15T/L16I/L18I)	−2.59	11.9	4.6
Chimera 10 (F17Y)	−2.80	13.0	4.6
RBD2	−2.41	8.5	3.5
Chimera 4	−2.42	8.6	3.5
Chimera 13	−1.90	3.9	2.0
Chimera 16	−2.00	2.6	1.3
RBD2(I209T/F211Y/T213N)	−2.09	5.3	2.5
RBD2(RBD1loop3)	−2.27	5.9	2.6
RBD2(F248W)	−2.20	7.3	3.7
RBD2(R238A)	−2.03	5.8	2.9

[a] RBD1 and related mutants upper half; RBD2 and related mutants lower half.

that there is significantly more overlap of the distributions in the m dimension than in ΔG°. One interpretation of these results is that the mutations predominantly affect the value of ΔG° not m. The m value is thought to reflect differences in the accessible surface areas between the native and denatured states. Evidently, swapping homologous regions between these two very similar proteins does not significantly affect these surface area differences.

RNA Stability from Thermal Denaturation

Figures 4 and 5 and Table II summarize the least squares fits and Monte Carlo analyses of a series of model RNA hairpins. These results are part of a larger study of a class of ultrastable four-membered RNA hairpin loops known as UNCG tetraloops. These tetraloops are found frequently in complex cellular RNA such as ribosomal RNAs and are thought to be

FIG. 4. (a) Plots of ΔH° versus ΔS° generated from Monte Carlo simulations (1000 replicates) of three RNA hairpins (ggac[CUCG]gucc*uau*, ggac[UUCG]gucc*uau*, and ggag[UUCG] cucc*uau*) using the sample variance as an estimate of the experimental error. (b)–(d) T_m distributions calculated from the joint $\Delta H^\circ/\Delta S^\circ$ distributions [$T_m(^\circ C) = \Delta H^\circ(cal/mol)/\Delta S^\circ(cal/mol/K) - 273.15$].

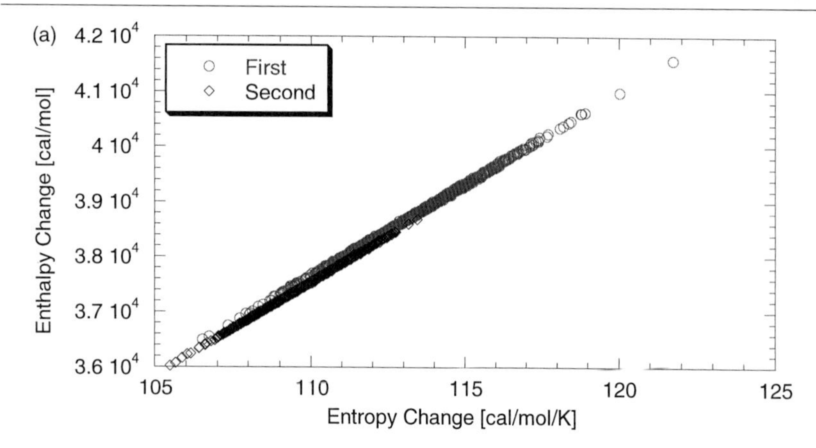

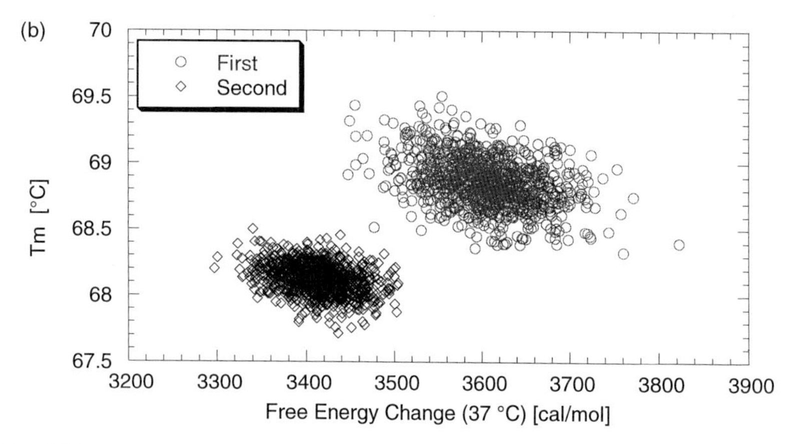

FIG. 5. (a) Plots of ΔH° versus ΔS° generated from Monte Carlo simulations (1000 replicates) for melting experiments on the RNA hairpin (cgc[UUCG]gcg) and a second experiment 1 month later. Sample variance used as an estimate of the experimental error. (b) T_m versus ΔG° distributions calculated from the joint $\Delta H^\circ/\Delta S^\circ$ distributions [T_m(°C) = ΔH°(cal/mol)/ΔS°(cal/mol/K) − 273.15].

important for the proper folding of these molecules.[24,25] The study aims to examine in detail the sequence determinants of this unusual stability and the effects of various nucleotide substitutions on the structure, dynamics,

[24] C. Turek, P. Gauss, C. Thermes, D. R. Groebe, M. Gayle, N. Guild, G. Stormo, Y. Aubenton-Carafa, O. C. Uhlenbeck, I. Tinoco, Jr, E. N. Brody, and L. Gold, *Proc. Natl. Acad. Sci. U.S.A.* **85,** 1364 (1988).

[25] C. R. Woese, S. Winker, and R. R. Guttell, *Proc. Natl. Acad. Sci. U.S.A.* **87,** 8467 (1990).

TABLE II
THERMODYNAMIC PARAMETERS: LEAST SQUARES FITS AND MONTE CARLO ANALYSES OF RNA HAIRPINS

Parameter	ΔH° (kcal/mol)	ΔS° (cal/mol/K)	T_m (°C)	ΔG° (37°) (kcal/mol)
Best fit value[e]	55.37[a]	159.72	73.51	5.83
	49.09[b]	141.01	74.95	5.35
	50.09[c]	149.07	62.87	3.86
	38.68[d]	113.10	68.86	3.60
	37.42[d]	109.63	68.12	3.41
Mean[f]	55.37[a]	159.71	73.52	5.83
	49.07[b]	140.96	74.95	5.35
	50.09[c]	149.06	62.87	3.86
	38.70[d]	113.15	68.86	3.61
	37.43[d]	109.69	68.12	3.41
Standard deviation[f]	0.18[a]	0.53	0.03	0.02
	0.50[b]	1.47	0.13	0.04
	0.27[c]	0.80	0.03	0.02
	0.65[d]	1.94	0.18	0.05
	0.44[d]	1.30	0.11	0.03

[a] ggac[CUCG]gucc*uau* hairspin.
[b] ggac[UUCG]gucc*uau* hairpin.
[c] ggag[UUCG]cucc*uau* hairpin.
[d] Replicate experiments of the cgc[UUCG]gcg hairpin.
[e] Value from global least squares fit of three independent UV/vis melting experiments.
[f] Analysis of distributions from Monte Carlo simulations of least squares data (1000 replicates).

and stability of the motif. The UV/vis thermal denaturation studies were carried out as previously described[26] except that all analyses were performed with Conlin.

In these studies, ΔH° and ΔS° are analogous to ΔG° and m in the protein chemical denaturation studies, and the T_m is analogous to the C_m. The free-energy change for the reaction is the difference of the enthalpy change and the product of the temperature and entropy change ($\Delta G = \Delta H - T\Delta S$), and at the T_m (the transition midpoint temperature) the free-energy change is zero [$\Delta G(T_m) = 0$]. As a consequence, enthalpy and entropy changes determined for these van't Hoff analyses are highly correlated (correlation coefficient >0.999), leading to large individual (single) parameter confidence intervals. However, the joint confidence interval is well defined and derived parameters such as ΔG° and T_m that are composites of ΔH and ΔS are known with much higher confidence.

[26] D. J. Williams and K. B. Hall, *Biochemistry* **35,** 14665 (1996).

Reliability of Monte Carlo Analyses

The central assumption of the Monte Carlo technique is that the distribution of parameter values about the fitted value (P_F) is the same as the distribution about the true value (P_T).[17] This does not imply that P_F is equal to P_T, a fact that can have interesting consequences. An instructive example is provided by comparing the $\Delta H°$ values for the cUUCGg and gUUCGc RNA mutants. This particular mutation has been compared in different labs under a variety of solution conditions, and generally the cUUCGg mutant has a more favorable enthalpy change by at least 5 kcal/mol, contrary to what would be concluded from Fig. 4 and Table II. The fitted values of $\Delta H°$ and $\Delta S°$ may fluctuate greatly from one experiment to another as a result of the high correlation between the two parameters. Consequently, partitioning the effects of a molecular perturbation between these two parameters is very problematic unless one has a large number of experiments and a good sense of the true parameters for the two distributions. However, the fluctuations in the enthalpy and entropy are compensatory, with the consequence that the errors in the value of the free energy are much less. This particular mutation is predicted to have a $\Delta G°$ (37°) difference of 1.7–2.3 kcal/mol. Although the enthalpy and entropy fluctuations are large, pairs of $\Delta H°$ and $\Delta S°$ values fall essentially on the same lines resulting in similar $\Delta G°$ and T_m values.

Monte Carlo analyses give estimates of parameter precision with very few assumptions. It is important to keep in mind, that these are estimates of parameter precision not accuracy, and only if the error distribution of the real data has been correctly simulated. Critical to the accurate application of the Monte Carlo technique is the creation of simulated data sets with realistic error. In this regard both the T_m plots in Fig. 4 and the standard deviations summarized in Table II suggest suspiciously narrow confidence intervals for the measured thermodynamic parameters. Replicate experiments in our lab and several others suggest that the standard deviations for $\Delta H°$, $\Delta S°$, $\Delta G°$, and T_m should be much larger.[27,28] The melting temperature in particular is known to ± 0.5°, a standard deviation considerably greater than the values listed in Table II.[29] These unusually small confidence intervals are not specific to the standard Monte Carlo analysis as can be inferred from the results summarized in Table III. The problem lies with the assumption that the sample variance (s^2) is a reasonable estimate of the true experimental error (σ^2). This problem arises because not all sources

[27] M. Petersheim and D. H. Turner, *Biochemistry* **22,** 256 (1983).
[28] H. T. Allawi and J. Santa Lucia, Jr., *Biochemistry* **36,** 10581 (1997).
[29] A. Lesnik and S. M. Freier, *Biochemistry* **37,** 6991 (1998).

TABLE III
CONFIDENCE INTERVALS[a] FOR REPLICATE EXPERIMENTS OF RNA HAIRPIN CGC[UUCG]GCG

Analysis	ΔH° (kcal/mol)	ΔS° (cal/mol/K)	T_m (°C)	ΔG° (37°) (kcal/mol)
Semianalytical[b]	35.6–41.7	104.0–122.2	68.0–69.7	3.4–3.8
	35.1–39.7	102.8–116.4	67.6–68.7	3.2–3.6
Grid[c]	35.8–41.5	104.3–121.7	67.7–70.2	3.4–3.8
	35.1–39.6	102.8–116.1	67.5–68.7	3.3–3.6
Monte Carlo[d]	36.5–41.6	106.5–121.7	68.3–69.5	3.4–3.8
	36.0–38.7	105.5–113.5	67.7–68.5	3.3–3.5
Bootstrap[e]	36.6–40.8	107.0–119.2	68.3–69.4	3.4–3.8
	36.2–38.8	106.0–113.8	67.7–68.5	3.3–3.5

[a] Variance ratio (F-statistic) chosen to give approximately 99.99% confidence ($\pm 4\sigma$).
[b] Variance space searched along parameter axes and axes of the hyperdimensional ellipse. [From M. L. Johnson, *Biophys. J.* **44,** 101 (1983)].
[c] 64×64 grid search on ΔH° and ΔS° with all other parameters optimized.
[d] Monte Carlo analysis on best fit values using 1000 replicates and the square root of the variance at the minimum as an estimate of the experimental error.
[e] Monte Carlo analysis of best fit values using errors generated by resampling the residuals with replacement.

of error *equilibrate* on the timescale of the experiment. Setup errors resulting from small differences in sample preparation and handling are only sampled extensively over multiple experiments. Consequently, the errors estimated from individual experiments or global analysis of experiments with similar setup conditions are an underestimate of the true error. This is particularly problematic when the *rapidly equilibrating* errors sampled by the experiment are small relative to the other sources of error, as is the case with high-precision techniques like UV/vis monitored thermal denaturation. In these cases, if the sample variance is used as an estimate of the experimental error, replicate experiments will often give nonoverlapping (or minimally overlapping) parameter distributions as seen in Fig. 5. The underestimation of the experimental error in these experiments is reminiscent of sampling problems encountered in molecular dynamics simulations when one attempts to generate ensemble averages from the time averages of individual trajectories. Unless the total simulation time is very long or sevral experiments are performed from different starting points, the calculated thermodynamics will reflect the features of the local surface and not the global phase space.

Ideally, good estimates of the true experimental error can be obtained from replicate experiments that properly sample the total error distribu-

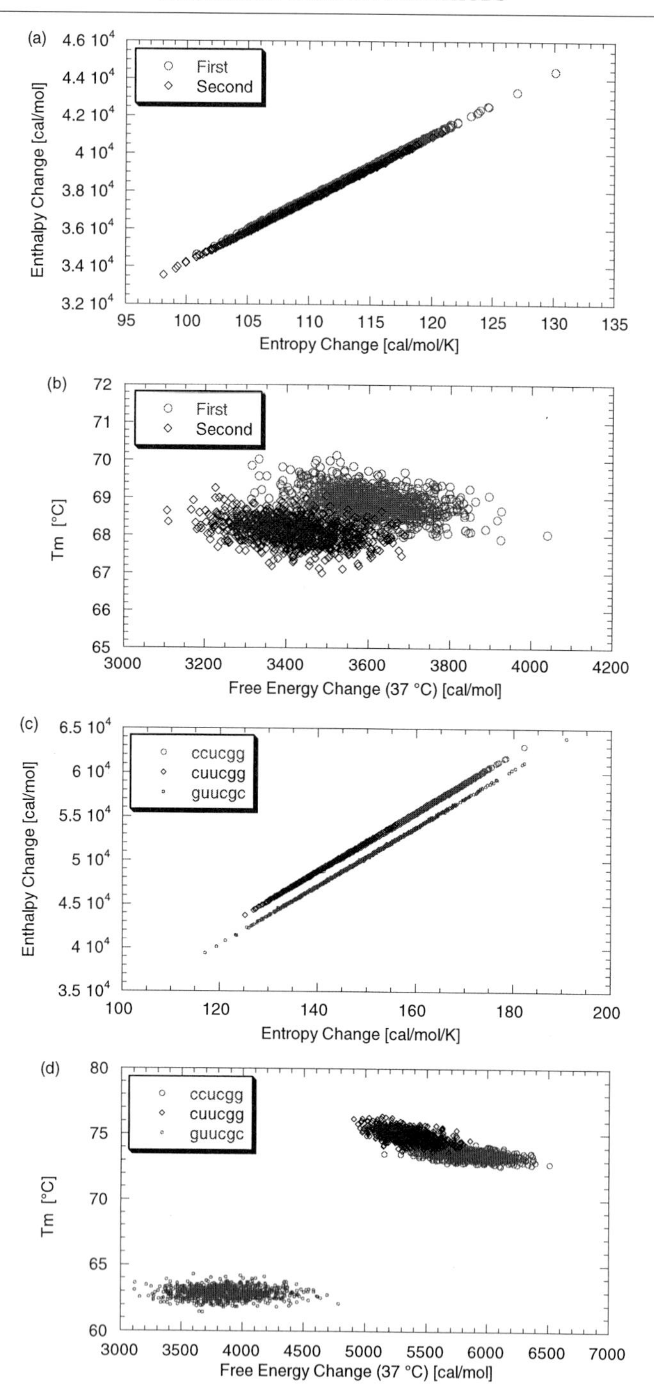
(a)
First
Second
Enthalpy Change [cal/mol]
Entropy Change [cal/mol/K]
(b)
First
Second
Tm [°C]
Free Energy Change (37 °C) [cal/mol]
(c)
ccucgg
cuucgg
guucgc
Enthalpy Change [cal/mol]
Entropy Change [cal/mol/K]
(d)
ccucgg
cuucgg
guucgc
Tm [°C]
Free Energy Change (37 °C) [cal/mol]

TABLE IV
THERMODYNAMIC PARAMETERS: MONTE CARLO ANALYSES OF LEAST SQUARES FITS OF RNA HAIRPINS

Parameter	ΔH° (kcal/mol)	ΔS° (cal/mol/K)	T_m (°C)	ΔG° (37°) (kcal/mol)
Mean[e]	55.43[a]	159.89	73.54	5.84
	49.06[b]	140.95	74.96	5.35
	50.21[c]	149.43	62.86	3.86
	38.73[d]	113.25	68.87	3.61
	37.48[d]	109.84	68.12	3.42
Standard deviation[e]	2.23[a]	6.52	0.33	0.21
	1.84[b]	5.44	0.46	0.16
	3.49[c]	10.43	0.42	0.26
	1.26[d]	3.76	0.36	0.10
	1.24[d]	3.70	0.32	0.10

[a] ggac[CUCG]gucc*uau* hairpin.
[b] ggac[UUCG]gucc*uau* hairpin.
[c] ggag[UUCG]cucc*uau* hairpin.
[d] Replicate experiments of the cgc[UUCG]gcg hairpin.
[e] Analysis of distributions from Monte Carlo simulations of least squares data (1000 replicates).

tion.[30] Not knowing $\sigma(x_i)^2$ exactly we simply adjusted the error used to generate the synthetic data sets to give standard deviations in the T_m that correspond to what is known experimentally. The results of these additional simulations are summarized in Fig. 6 and Table IV. The joint distributions of the replicate experiments now show much more reasonable overlap. Conversely, the joint distributions for the cUUCGg and gUUCGc motifs are still clearly distinct and there is almost no overlap of the free energy or T_m distributions, although the uncertainty in either ΔH° or ΔS° is now considerable. In contrast, the cUUCGg and cCUCGg motifs now show overlapping joint distributions. This is a particularly satisfying result as this substitution is predicted to be isosteric, and preliminary nuclear magnetic

[30] N. R. Draper and H. Smith, "Applied Regression Analysis," p. 28. Wiley Series in Probability and Mathematical Statistics, John Wiley and Sons, New York, 1966.

FIG. 6. Monte Carlo analysis (1000 replicates) of least squares fits of UV/vis thermal denaturation experiments for four RNA hairpins with the experimental error chosen to give standard deviations in T_m of approximately 0.5° [$\sigma(T_m) \approx \pm 0.5°$]. (a) and (b) Duplicate experiments on hairpin cgc[UUCG] gcg 1 month apart. (c) and (d) Experiments on ggac[CUCG]gucc*uau,* ggac[UUCG]gucc*uau* and ggag[UUCG]cucc*uau* hairpins.

resonance studies suggest very little structural change as a result of the mutation (data not shown).

Conclusions

With a minimum number of assumptions, Monte Carlo techniques provide complete joint parameter probability distributions, which allow graphical comparisons of parameters and also facilitate the accurate propagation of errors into derived parameters. Unlike grid search analyses, additional tests to assign probabilities to the confidence intervals are not needed. Furthermore, knowledge of the orientation (bias) of the variance space with respcct to the parameter axes is not necessary in order to efficiently sample the parameter space. However, if a Monte Carlo analysis is to yield meaningful results, careful attention must be given to the experimental error estimate used to generate pseudo-data.

With realistic errors, the complete joint distributions generated by Monte Carlo analyses can be indispensable for the analysis of highly correlated parameter pairs. In particular, knowledge of the joint parameter distributions is of critical importance when attempting to assess the effects of mutations in systems where the parameters of interest are highly correlated. (Slopes and intercepts in straight line fits, m and ΔG° values in chemical denaturation experiments, and ΔH° and ΔS° in van't Hoff studies are all examples of highly correlated parameter pairs.) Due to the large fluctuations in the values of correlated parameters, the effects of mutations on a system can often be obscured by overlap in both single-parameter dimensions. Graphical comparison of the joint parameter distributions allows mutations with no net effect on either parameter to be easily distinguished from those which have no *apparent* effect because of overlap in the individual parameter dimensions.

Acknowledgments

The design and synthesis of the chimeric proteins and the chemical denaturation study was performed by Dr. Jirong Lu. We also thank her for many lively and stimulating discussions over the course of this work. We are grateful to Dr. Janid Ali, Dr. Michael L. Johnson, Dr. Alex Klinger, and Nasib K. Maluf for helpful comments during various stages of the project. Conlin is available on the Web at http://www.biochem.wustl.edu/kbhlab. This work was supported in part by NIH grant GM46318 to K.B.H. and a Gerty T. Cori Sigma Chemical Company Predoctoral Fellowship to D.J.W.

[19] Analysis of Drug–DNA Binding Data

By XIAOGANG QU and JONATHAN B. CHAIRES

Introduction

The rational design of new DNA binding agents requires a thorough understanding of the thermodynamics of the DNA binding of existing drugs.[1,2] Fundamental to any thermodynamic characterization of drug–DNA interactions is the determination of binding constants. Because many DNA binding drugs exhibit large changes in absorbance or fluorescence on binding, these changes are commonly used to determine the distribution of free and bound drug in solution to construct binding isotherms that may be used to obtain binding constants.[3,4] For example, for an absorbance change on binding, one can calculate the concentration of bound drug (C_b) from the difference in absorbance at a given wavelength (ΔA_λ) between the drug alone and the drug in the presence of DNA from the equation

$$C_b = (\Delta A_\lambda / \Delta \varepsilon_\lambda) \tag{1}$$

where $\Delta\varepsilon_\lambda$ is the difference in molar extinction coefficients between the free and bound species, $\Delta\varepsilon_\lambda = \varepsilon_0 - \varepsilon_b$ (Ref. 5). Equation (1) assumes that the drug exists in only two states, free and bound. Once the concentration of bound drug is determined, the concentration of free drug may be calculated from the conservation equation, $C_f = C_t - C_b$, where C_t is the known total drug concentration. In the case of fluorescence, there are two common useful situations, one in which drug emission is quenched on DNA binding, another when emission is enhanced. If the drug is quenched on binding,

$$C_f = C_t(F/F_0 - P)/1 - P \tag{2}$$

where F is the observed fluorescence emission under a particular set of DNA and drug concentrations, F_0 is the emission of the same drug concentration in the absence of DNA, and P is the ratio of the fluorescence of the completely bound drug to the free drug, $P = F_b/F_0$ (Ref. 5). If drug

[1] J. B. Chaires, *Biopolymers* **44,** 201 (1997).
[2] I. Haq, T. C. Jenkins, B. Z. Chowdhry, J. Ren, and J. B. Chaires, *Methods Enzymol.* **323,** [16], 2000.
[3] H. Porumb, *Prog. Biophys. Mol. Biol.* **34,** 175 (1978).
[4] G. Dougherty and W. J. Pigram, *CRC Crit. Rev. Biochem.* **12,** 103 (1982).
[5] A. Blake and A. R. Peackocke, *Biopolymers* **6,** 1225 (1968).

0076-6879/00 $30.00

fluorescence emission is enhanced, a useful equation is

$$C_b = C_t(F - F_0)/(P - 1)F_0 \qquad (3)$$

where the definitions of the symbols are the same as for Eq. (2) (Ref. 6).

Equations (1)–(3) are all based on the assumption that the binding process is two state, and that the observed spectral response is a linear combination of only two spectral species, corresponding to the free and bound forms. These assumptions require verification. Practical use of Eqs. (1)–(3) requires knowledge of the extinction coefficient or relative fluorescence of the bound species. These values are often obtained by linear transformation and extrapolation of primary absorbance or fluorescence emission data.[7,8] Such transformations, particularly the often recommended double-reciprocal plot, are fraught with error, and their use can lead to considerable uncertainty in the estimates of the optical properties of the bound form that are so crucial to the accurate determination of the distribution of free and bound drug concentration.[9–12]

The purpose of this chapter is to describe the protocols for the numerical analysis of primary fluorescence and absorbance titration data that have evolved during the last decade in our laboratory. These protocols were developed to rigorously test the two-state assumption and to provide a reliable means of obtaining the limiting spectral properties of the free and bound drug forms with appropriate error estimates. Singular value decomposition[13–17] is used in the first case, and nonlinear least squares fitting of primary titration data,[18] along with Monte Carlo error analysis,[19] is used in the latter case.

Several reviews have appeared that discuss the analysis of fluorescence

[6] J. B. LePecq and C. Paoletti, *J. Mol. Biol.* **27,** 87 (1967).

[7] V. A. Bloomfiled, D. M. Crothers, and I. Tinoco, Jr., "Physical Chemistry of Nucleic Acids," pp. 408–410. Harper & Row, New York, 1974.

[8] D. J. Winzor and W. H. Sawyer, "Quantitative Characterization of Ligand Binding," pp. 35–37. Wiley-Liss, New York, 1995.

[9] K. Zierler, *Biophys. Struct. Mech.* **3,** 275 (1977).

[10] M. R. Panjehshahin, C. J. Bowmer, and M. S. Yates, *Biochem. Pharm.* **38,** 155 (1989).

[11] K. M. Rajkowski, *Biochem. Pharm.* **39,** 895 (1990).

[12] R. B. Martin, *J. Chem. Ed.* **74,** 1238 (1997).

[13] E. R. Henry and J. Hofrichter, *Methods Enzymol.* **210,** 129 (1992).

[14] R. W. Hendler and R. I. Shrager, *J. Biochem. Biophys. Meth.* **28,** 1 (1994).

[15] K. D. Vandegriff and R. I. Shrager, *Methods Enzymol.* **232,** 460 (1994).

[16] I. Haq, B. Z. Chowdhry, and J. B. Chaires, *Eur. J. Biophys.* **26,** 419 (1997).

[17] A. K. Dioumaev, *Biophys. Chem.* **67,** 1 (1997).

[18] M. L. Johnson and L. M. Faunt, *Methods Enzymol.* **210,** 1 (1992).

[19] M. Straume and M. L. Johnson, *Methods Enzymol.* **210,** 117 (1992).

titration data, usually with emphasis on protein–ligand interactions.[20–22] The general treatment of such primary data is similar for both protein and nucleic acid interactions, although the ultimate treatment of binding isotherms may differ because of the need to incorporate possible lattice effects (neighbor exclusion) in models for ligand binding to long DNA or RNA molecules.[23] The approaches described here, while intended for the analysis of drug binding to a DNA lattice, are of general utility.

Overview of Procedures

The overall approach used is as follows. A fixed concentration of drug (ligand) is titrated with increasing concentration of DNA (receptor). At each point in the titration, complete absorbance or fluorescence emission spectra are recorded. The accumulated families of spectra are then analyzed by singular value decomposition (SVD) to enumerate, by a model-free analytical procedure, the number of significant spectral species that comprise the family. This provides a rigorous test of the two-state assumption. If the two-state assumption is valid, nonlinear least squares fitting is done on the primary spectral data using an appropriate binding model to obtain estimates of the binding constant and the limiting optical properties of the free and bound species. Monte Carlo methods are employed to examine the error in these estimates. Each of these steps is discussed and illustrated, using primary data obtained for ethidium bromide, the prototypical DNA intercalator.[6]

Gathering Primary Data

It is not generally appreciated that an enormous concentration range must be covered in order to define a complete binding isotherm. For simple, noncooperative binding, it takes 1.8 logarithmic (base 10) units in concentration to go from 10% to 90% saturation, about a 100-fold change in concentration.[24] If one is trying to add a sufficient concentration of sites to completely bind a fixed concentration of ligand to measure the optical properties of the bound form, a large excess of site concentration is need. In this laboratory, we have found it expedient to proceed in the following way. Titrations

[20] L. D. Ward, *Methods Enzymol.* **117,** 400 (1985).

[21] M. R. Eftink, *Methods Enzymol.* **278,** 221 (1997).

[22] C. R. Bagshaw and D. A. Harris, *in* "Spectrophotometry and Spectrofluorimetry: A Practical Approach" (D. A. Harris and C. L. Bashford, eds.), p. 91. IRL Press, Oxford, United Kingdom, 1987.

[23] J. D. McGhee and P. H. von Hippel, *J. Mol. Biol.* **86,** 469 (1974).

[24] G. Weber, *Adv. Prot. Chem.* **29,** 1 (1975).

are first done at a fixed drug concentration, using the lowest concentration possible that still yields a good absorbance or fluorescence signal. Typically this is 1–10 μM for DNA binding drugs. Spectra are then measured over a wide range of DNA concentrations, from 1 nM to 1 mM. Rather than attempt to incrementally add DNA from a stock solution directly to the cuvette to cover this concentration range, we have found it most efficient to prepare a series of separate tubes containing increasing DNA concentrations, and then to add the desired fixed drug concentration to each tube. The spectrum of each sample is then recorded and stored for analysis. It is convenient to prepare separate DNA stock solutions that cover the desired range. We typically prepare DNA solutions starting with 1 nM base pairs (bp), and increase concentration in 0.25 $\log_{10}$ units up to 1 mM bp. This allows for the construction of an isotherm with 4 points per decade in DNA concentration, ranging from 1 nM to 1 mM bp. About 25 points define the complete isotherm. Such a range is required to define the limiting value of free and bound ligand.

Figure 1 shows primary spectral data obtained by this approach for ethidium bromide binding to calf thymus DNA. The visible absorbance of ethidium is red shifted on binding with a marked hypochromic effect, while its fluorescence emission is significantly enhanced. These families of spectra form the primary data set for subsequent analysis. In our laboratory, absorbance spectra are recorded on a Cary 3E UV/vis spectrophometer with a thermoelectric temperature controller. Fluorescence emission spectra are recorded on an ISS, Inc. Greg 200 spectrofluorometer, connected to a Neslab circulating water bath for temperature control. Both instruments acquire spectra in digital form. We typically measure absorbance or fluorescence at 1-nm intervals over a specified range. The digital data are cast into the form of a matrix with elements in the matrix corresponding to an absorbance (or fluorescence) value, with rows corresponding to wavelengths and columns corresponding to DNA concentrations.

Singular Value Decomposition

The first step in the numerical analysis of the drug–DNA binding data is to use singular value decomposition to enumerate the number of significant spectral species in spectra such as shown in Fig. 1. The data matrix **M** constructed from the primary spectral data is imported into the program MATLAB (The MathWorks, Inc., Natick, MA), which contains suitable routines for the SVD computation and for easy graphical display of the results. Several excellent and detailed discussions of the SVD technique exist,[13–15] so the method will be only briefly reviewed here.

The data matrix **M** is decomposed by the SVD method into the product

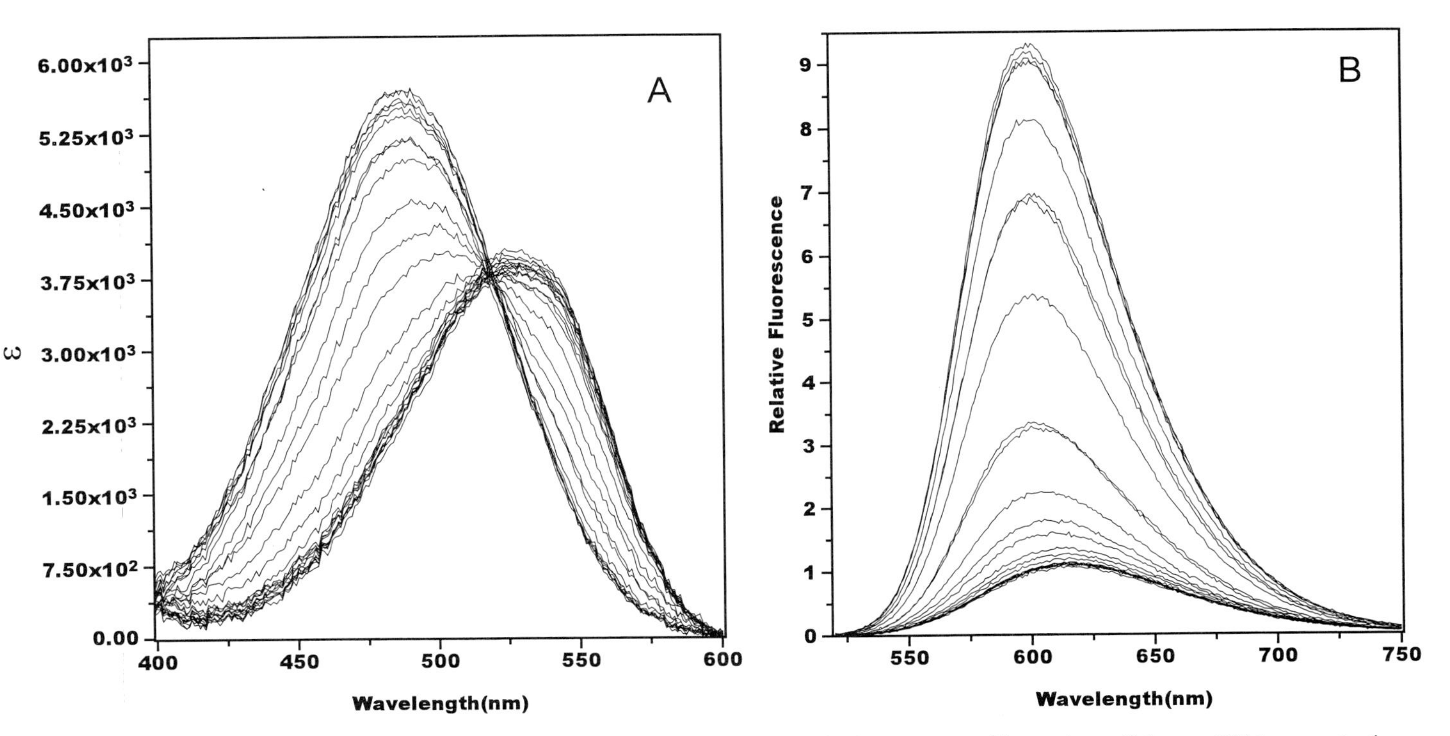

FIG. 1. (A) Absorbance and (B) fluorescence emission spectra of ethidium bromide in the presence of increasing calf thymus DNA concentrations. (A) Spectra for ethidium bromide at a concentration of 11.4 μM are shown for DNA concentrations ranging from 1 nM (bp) to 1 mM (bp). On binding to DNA, the visible absorbance spectrum of ethidium is red-shifted, with a pronounced hypochromic effect. (B) Fluorescence emission spectra for 3.1 μM ethidium in the presence of 1 nM to 1 mM DNA. The excitation wavelength was 480 nm. On binding to DNA, ethidium fluorescence is greatly enhanced.

of three matrices, $\mathbf{M} = \mathbf{USV}^t$. The matrix **U** contains the so-called basis spectra; **S** is a diagonal matrix that contains the singular values; **V** is a matrix that contains amplitude vectors. The challenge in the use of SVD to enumerate the number of significant spectral species in a family of spectra is to rationally decide on the number of significant singular values. In the protocol that we have adopted, this is done in a three-step process, each of which is described.

In the first step, the magnitudes of the singular values are tabulated and inspected. Table I shows the first 10 singular values that result from SVD analysis of the data shown in Fig. 1. For both the absorbance and fluorescence data, the magnitudes of the first two singular values are clearly larger than the remaining values, suggesting that there are in fact only two significant spectral species and that binding of ethidium to DNA is truly a two-state process. The problem is to evaluate the statistical significance of the third and higher singular values.

In the second step, values for the first-order autocorrelation of the

TABLE I
SINGULAR VALUE DECOMPOSITION ANALYSIS OF ETHIDIUM BINDING TO CALF THYMUS DNA

Method	Singular value	Autocorrelation analysis	
		U matrix	**V** matrix
Absorbance	2.0587	0.9997	0.9582
	0.6080	0.9993	0.9383
	0.0229	0.9681	−0.1557
	0.0065	0.7947	0.0047
	0.0044	0.4695	0.1117
	0.0040	0.3644	0.0160
	0.0039	0.3255	−0.2296
	0.0037	0.1115	−0.0713
	0.0035	0.1330	−0.2630
	0.0033	0.0234	0.2607
Fluorescence	8.3446	0.9997	0.9151
	0.4704	0.9989	0.9496
	0.0251	0.7972	0.7572
	0.0178	0.1245	−0.6100
	0.0154	0.0679	−0.0434
	0.0147	0.0260	−0.5838
	0.0132	0.0794	0.3051
	0.0127	−0.1319	−0.1929
	0.0114	−0.0538	−0.3853
	0.0104	−0.1003	−0.1738

columns of the **U** and the **V** matrices are calculated. The autocorrelation function is:

$$C(X_i) = \Sigma(X_{j,i})(X_{j+1,i})$$

where $X_{j,i}$ and $X_{j+1,i}$ are the jth and jth + 1 row elements of column I from either the **U** or **V** matrix. The value of $C(X_i)$ is a measure of the smoothness between adjacent row elements. Because the column vectors of **U** and **V** are normalized to unity, $C(X_i)$ values can vary between -1 and 1. Values near -1 indicate rapid row-to-row variation, or "noise," and indicate random behavior in the column vector. Significant singular values are associated with nonrandom column vectors in the **U** and **V** matrices. A $C(X_i)$ of 0.8 corresponds to a column vector with a signal-to-noise ratio of 1.0 (Ref. 13). We therefore have selected a cutoff of $C(X_i) \geq 0.8$ for columns in both the **U** and **V** matrices as a criterion for accepting a singular value as significant. Autocorrelation values are listed in Table I. For the absorbance data, the autocorrelation value for the **V** matrix column associated with the third singular value is negative, from which we conclude that only the first two singular values are significant. For the fluorescence data autocorrelation values for both the **U** and **V** matrices fall below the cutoff value of 0.8 for the third singular value, again suggesting that only the first two singular values are significant. From the magnitude of the singular values and from the values of the autocorrelation values for columns of the **U** and **V** matrices, we can conclude that there are only two significant spectral species for both the absorbance and fluorescence data. Ethidium binding to DNA seems to be a strictly two-state process.

The third step provides an additional test of the conclusions drawn from steps 1 and 2. The difference between the original data matrix and a computed matrix is calculated. The computed matrix is calculated by creating an **S** matrix in which the singular values are set to zero for all diagonal elements except those that have been judged to be significant:

$$\mathbf{D} = \mathbf{M} - \mathbf{U}\mathbf{S}_c\mathbf{V}^t$$

Where $\mathbf{S}_c$ is the modified **S** matrix. If the correct number of singular values has been selected, a contour plot of the **D** matrix ought to be completely random, since it represents the random noise in the original data matrix **M**. If a significant singular value has been neglected, the contour plot of the **D** matrix will clearly show ordered regions. Figure 2 shows contour plots obtained for different assumed numbers of significant singular values for fluorescence data and the values shown in Table I. For ethidium binding to DNA, random difference contour plots were observed for fluorescence data after inclusion of only two singular values. Inclusion of three or four singular values did not significantly improve the difference plots, verifying

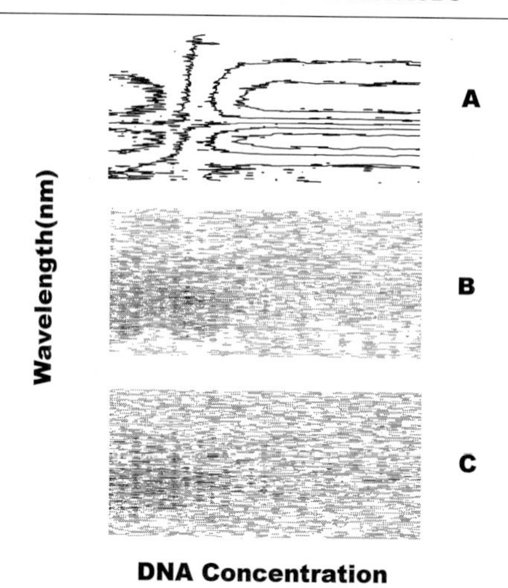

FIG. 2. Contour plots showing the difference between the experimental data matrix of fluorescence emission spectra as a function of DNA concentration (as shown in Fig. 1B) and calculated matrices obtained for different numbers of singular values assumed to be significant. In these contour plots, the abcissa corresponds to DNA concentrations and the ordinate to fluorescence emission wavelengths. Each element in the matrix represents the difference between the experimental and calculated fluorescence emission intensity. The differences result from calculations with (A) one, (B) two, and (C) three singular values.

the conclusion that there are but two significant spectral species. This point can be confirmed more quantitatively. The elements of the **D** matrices can be squared and summed to obtain the sum of the squares of the residuals (SS). These SS values can be used to compute F values to evaluate whether or not the inclusion of the next highest singular value is warranted.[25] Addition of the second singular value resulted in $F = 167.5$, with $P = 0.0065$, indicating unambiguously that its inclusion was warranted. Addition of the third singular value, however, resulted in $F = 0.94$ and $P = 0.438$, indicating that there is no strong statistical basis for including it. Similarly, addition of the fourth singular value resulted in $F = 0.75$, $P = 0.46$, again arguing against its inclusion.

It is now useful to inspect the basis spectra for the ethidium data that emerge from the SVD analysis. Figure 3 shows the weighted basis spectra (the product **US**) corresponding to the first three singular values. For both absorbance and fluorescence data, two prominent basis spectra have emerged. The third basis spectrum in each case is small in magnitude and featureless in shape and barely, if at all, differs from a flat baseline. The

[25] H. J. Motulsky and L. A. Ransnas, *FASEB J.* **1,** 365 (1987).

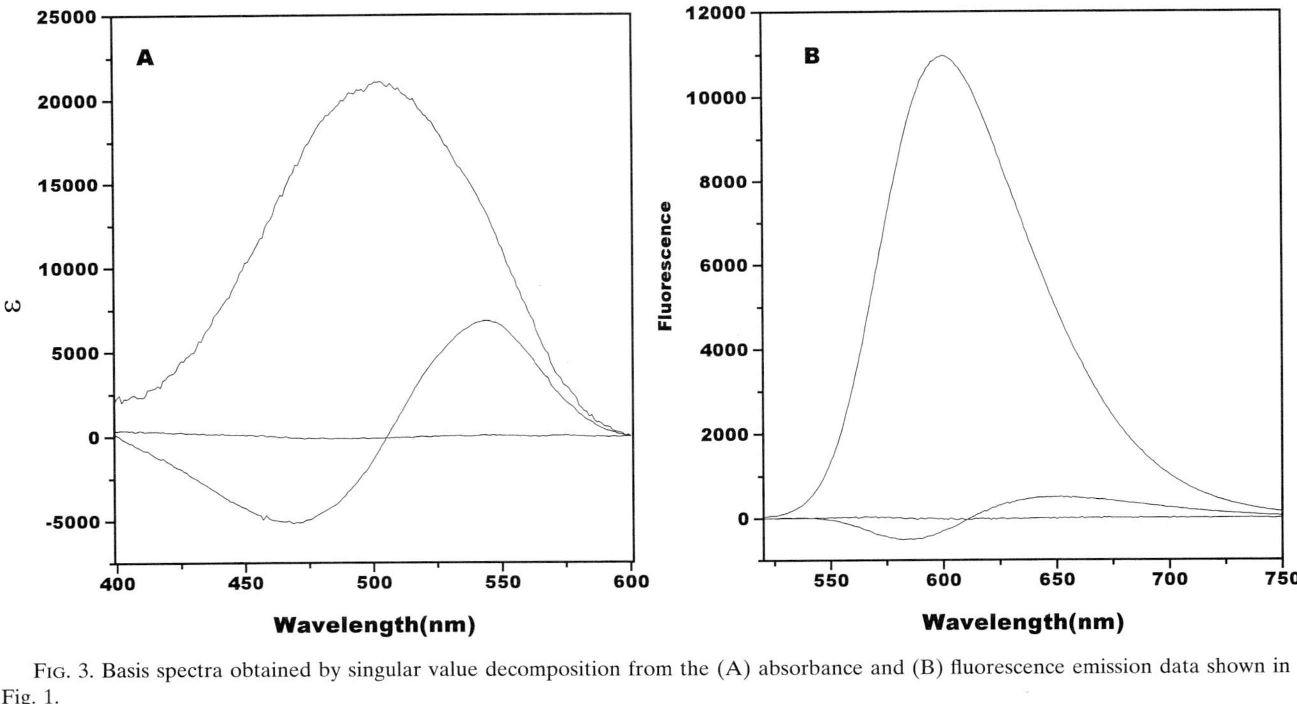

FIG. 3. Basis spectra obtained by singular value decomposition from the (A) absorbance and (B) fluorescence emission data shown in Fig. 1.

two significant basis spectra are the ones that may be linearly combined to construct the whole family of spectra observed in the primary data. It is important to caution that these basis spectra are mathematical constructs, and do not necessarily represent directly the spectra of the free and bound forms.

For ethidium binding to DNA, we can now state with confidence that the process is two state, and proceed with further analysis. What if SVD had revealed more than two significant basis spectra, signifying that the two-state assumption was not valid? One is then faced with a more complicated problem, and must use an appropriately more complicated analysis. The main complication is that there is no guarantee that the spectral response is linear. A variety of "model-free" analytical methods for the determination of binding isotherms have been described that must be used if the simple two-state model is proven not to be true.[26–28] These methods require considerable more effort, but are unavoidable if the simple equations, Eqs. (1)–(3), cannot be applied.

Nonlinear Fitting of Primary Data

The next step in the quantitative analysis of the primary data is to construct graphs of the titration curves by plotting the spectral response (absorbance or fluorescence emission) at a selected wavelength against the logarithm of the DNA concentration. Alternatively, for fluorescence emission, the integrated area of the emission band could be plotted to reduce noise. Figure 4 shows titration curves constructed for the absorbance and emission spectral data shown in Fig. 1. These titration curves are sigmoidal in shape, and are symmetrical around a midpoint. The large span of DNA concentrions required to define these complete titration curves should be noted.

The data shown in Fig. 4 may be fit to a simple binding model. Over most of the titration curve, the concentration of DNA sites is in large excess over the total ligand concentration. Neighbor exclusion effects can be neglected. Because SVD has rigorously established that binding is two state, possible intermediate or multiple species may be neglected. The simple equilibrium

$$D + S \rightleftharpoons C$$

describes the binding process, where D is the ligand, S the DNA binding

[26] W. Bujalowski and T. M. Lohman, *Biochemistry* **26,** 3099 (1987).

[27] H. Fritzsche and A. Walter, *in* "Chemistry and Physics of DNA–Ligand Interactions" (N. R. Kallenbach, ed.), p. 1. Adenine Press, Schenectady, New York, 1990.

[28] T. M. Lohman and W. Bojalowski, *Methods Enzymol.* **208,** 258 (1991).

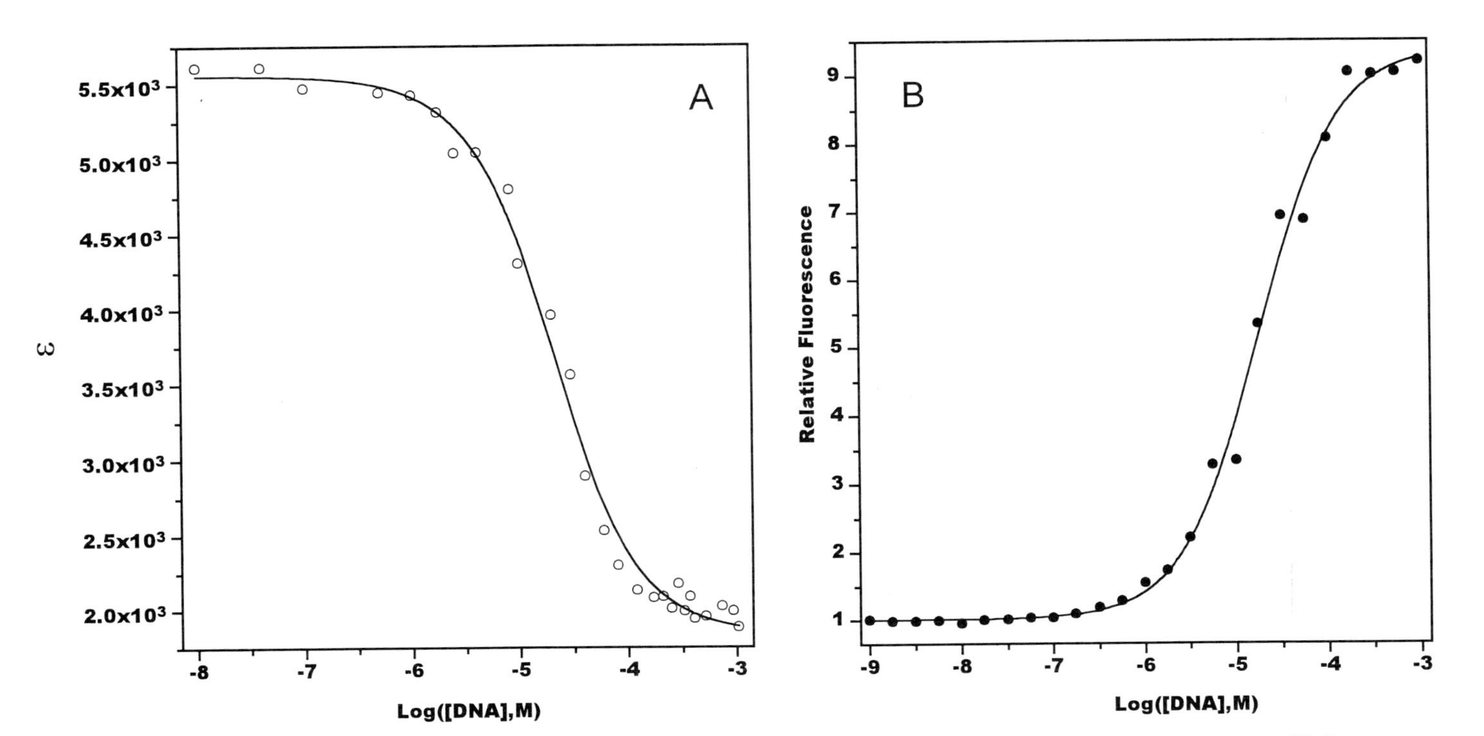

FIG. 4. Titration curves for the interaction of ethidium with calf thymus DNA as monitored by (A) visible absorbance at 480 nm or (B) fluorescence emission. Data from the spectra in Fig. 1 were used to construct these titration curves. The solid curves are the best fits to the experimental data obtained by nonlinear least squares analysis with the parameters shown in Table II.

site, and C the complex; S is taken to be a DNA base pair. The observed fluorescence (or absorbance) is assumed to be the sum of the fluorescence of the free and bound species, weighted by their respective concentrations:

$$F = F^0(C_t - C_b) + F^b C_b \tag{4}$$

where F is the apparent fluorescence at each DNA concentration, F^0 is the fluorescence intensity of free ligand, and F^b is the fluorescence intensity of the bound species. For the interaction of a ligand D with a DNA site S, it may be easily shown that:

$$Kx^2 - x(KS_0 + KD_0 + 1) + KS_0 D_0 = 0 \tag{5}$$

where x equals C_b, K is the association constant, S_0 is the total concentration, and D_0 is the total ligand concentration. Equation (5) is readily solved using the quadratic formula. Data in the form of fluorescence response F as a function of total DNA site concentration at fixed concentration of ligand can then be fit by nonlinear least squares methods to get K, F_0, and F_b. The same approach can be used to obtain a fitting function for absorbance data to obtain K, ε_0, and ε_b. Fitting functions have been incorporated into the program FitAll (MTR Software, Toronto). Fits to the data are shown as solid lines in Fig. 4, with the parameters of the fit listed in Table II.

It is vital to evaluate carefully the error in the parameters resulting from the nonlinear least squares fitting procedure. A variety of approaches can be used for this purpose.[18] Our preference is to apply a Monte Carlo analysis, for which a routine has recently been added to the FitAll package and which is now available as part of the standard research edition of the software. The Monte Carlo method has previously been described in detail.[19] A brief description follows. Once the best fit parameters have been obtained, a synthetic data set is calculated from those parameters that matches exactly the number of points and span of the original experimental

TABLE II
ETHIDIUM DNA BINDING PARAMETERS OBTAINED BY ABSORBANCE AND FLUORESCENCE TITRATIONS

Parameter	Absorbance titration	Fluorescence titration
$K/10^4$ (M^{-1})	6.52 ± 0.63	6.15 ± 0.55
ε_0 or F_0	5,560 ± 81 M^{-1} cm^{-1}	466 ± 37 a.u.[a]
ε_b or F_b	1837 ± 79 M^{-1} cm^{-1}	4317 ± 108 a.u.

[a] a.u., arbitrary units.

data set. To this "perfect," noise-free data, pseudo-random noise is added, with a standard deviation equal to that obtained by the original nonlinear fit. The perturbed synthetic data set is then fit, and the fitting parameters are stored. The process is repeated for a large number of different sets of noise. The distribution of fitted parameters is then examined, and their statistics computed. Figure 5 shows examples of such Monte Carlo simulations for the fits shown in Fig. 4. In these cases, the fitting is very well behaved, and the distribution of estimated parameters is symmetrical and Gaussian in shape. Such need not be the case, which is when the Monte Carlo method is most valuable. The method provides a reliable evaluation of the error in each parameter, and for possible cross-correlations in parameter estimates. Error estimates for the parameters obtained in Fig. 4 are shown in Table II.

Knowledge of the error in parameter estimates allows for a proper propagation of the error in the derived quantities needed in Eqs. (1)–(3). The error in the difference in molar extinction coefficients, $\Delta\varepsilon = \varepsilon_0 - \varepsilon_b$, is given by

$$\delta\Delta\varepsilon = [(\delta\varepsilon_0)^2 + (\delta\varepsilon_b)^2]^{1/2}$$

where δ indicates the error in the parameter estimate. For the parameter P in Eqs. (2) and (3),

$$\delta P/P = [(\delta F_0/F_0)^2 + (\delta F_b/F_b)^2]^{1/2}$$

Using these formulas along with the fitted parameters and their errors presented in Table II, we calculate $\Delta\varepsilon = 3723 \pm 113$ and $P = F_b/F_0 = 9.26 \pm 0.76$. With these appropriate error estimates in the key parameters used to calculate the amount of bound ligand using Eqs. (1)–(3), the error in the concentration of free and bound ligand can be accurately determined by propagation.

Titration of DNA with Ligand

While the titration of a fixed concentration of ligand with DNA yields a reliable estimate for the binding constant, that protocol is not useful for the determination of binding stoichiometry. Other protocols must be used for that purpose, either the method of continuous variations[29] or the titration of a fixed concentration of DNA with increasing concentrations of ligand. We describe the latter. Figure 6 shows an example of the titration of calf thymus DNA with ethidium, monitored by fluorescence emission. Figure 6A shows the primary fluorescence data, in which the enhanced

[29] C. Y. Huang, *Methods Enzymol.* **87,** 509 (1982).

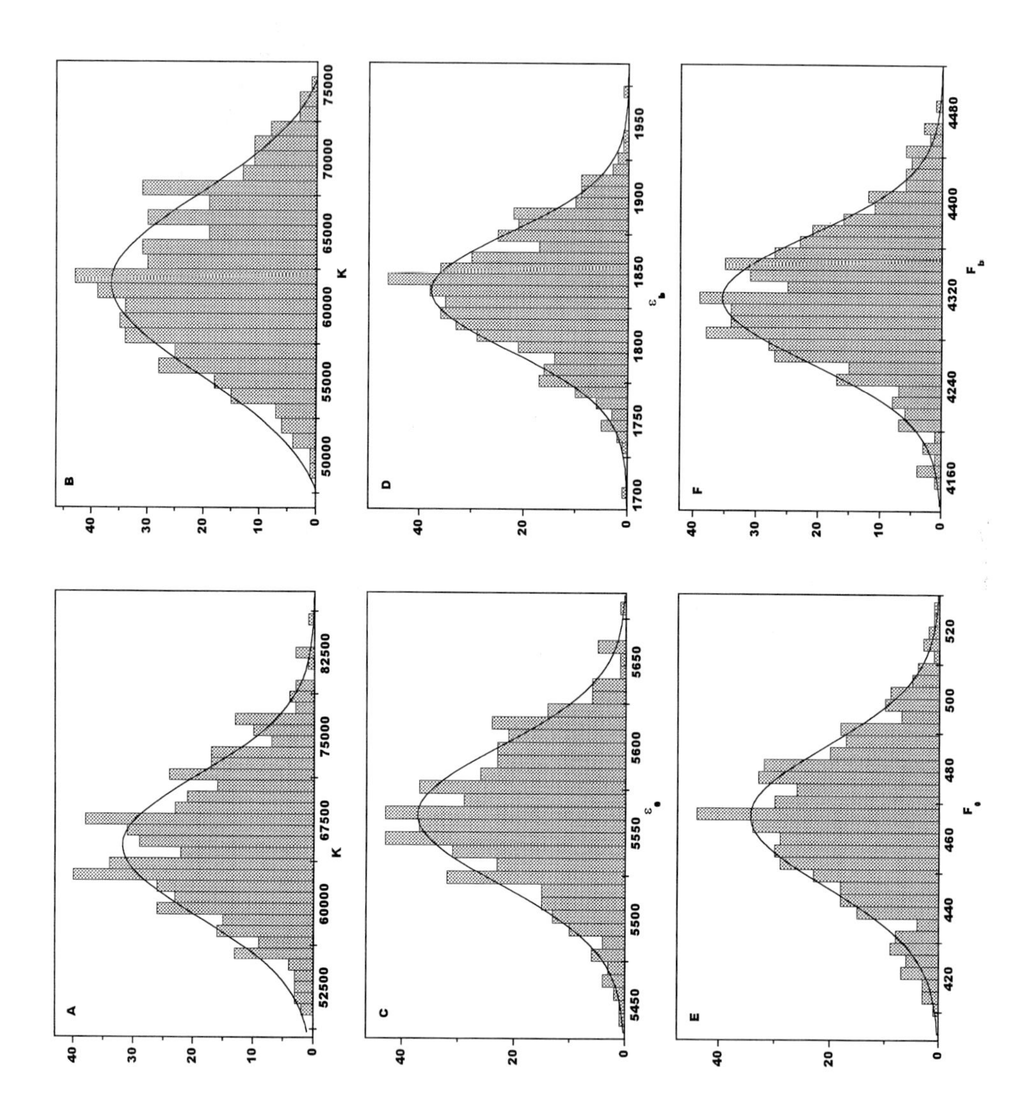
A
B
C
D
E
F
K
K
ε_e
ε_b
F_e
F_b
0
10
20
30
40
52500
60000
67500
75000
82500
50000
55000
60000
65000
70000
75000
5450
5500
5550
5600
5650
1700
1750
1800
1850
1900
1950
420
440
460
480
500
520
4160
4240
4320
4400
4480

fluorescence emission of ethidium on binding is evident. These primary data can be used to calculate the distribution of free and bound ethidium at each point in the titration by use of Eq. (3), along with the derived value of P obtained as described above. Free and bound ethidium concentrations may then be used to construct a Scatchard plot[30] of r/C_f versus r, where r is the binding ratio, $r = C_b/[DNA]_{total}$. Figure 6B shows the Scatchard plot for the ethidium–DNA interaction, which can be analyzed to obtain estimates for the binding constant and stoichiometry. In this case, the neighbor exclusion model of McGhee and von Hippel[23] provides the most appropriate description of the binding data. The simple neighbor exclusion model (neglecting cooperative interaction) is expressed in a closed-form equation:

$$r/C_f = K_{NE}(1 - nr)\{[1 - nr]/[1 - (n - 1)r]\}^{(n-1)}$$

where K_{NE} is the binding constant to an isolated site and n is the exclusion parameter, the number of base pairs in a ligand binding site. Application of this model to the analysis of drug–DNA binding data has been fully described, along with possible limitations in parameter estimation.[31] Non-linear least squares fits to the data shown in Fig. 6B yield estimates of $K_{NE} = 1.04\ (\pm 0.02) \times 10^5\ M^{-1}$ and $n = 1.83\ (\pm 0.04)$ bp. The error estimates were obtained by Monte Carlo analysis. K_{NE} is the binding constant for drug binding to an isolated site (n base pairs long) along the DNA lattice. The binding constant K obtained from fits of titration data such as those shown in Fig. 4 is related to K_{NE} by the relationship $K = K_{NE}/n$. Since $K_{NE}/n = 5.2 \times 10^4\ M^{-1}$, the binding constants obtained by fits to the neighbor exclusion model are in excellent agreement with the values presented in Table II.

[30] G. Scatchard, *Ann. N.Y. Acad. Sci.* **51,** 660 (1949).
[31] J. J. Correia and J. B. Chaires, *Methods Enzymol.* **240,** 593 (1992).

FIG. 5. Results from Monte Carlo simulations to explore the error in parameter estimates obtained by nonlinear least squares analysis. The parameters shown in Table II were used to generate perfect data using the same range and number of points as in the experimental data sets shown in Fig. 4. Pseudo-random noise was added with the same standard deviation as obtained in the fit, and the data were then fit to obtain parameter estimates that were stored for analysis. Five hundred cases were examined. Histograms of the parameter estimates are shown, along with fits to a Gaussian distribution. Simulations corresponding to fits of absorbance data are shown in (A), (C), and (D). (B), (E), and (F) show the results from simulations corresponding to fits to fluorescence data.

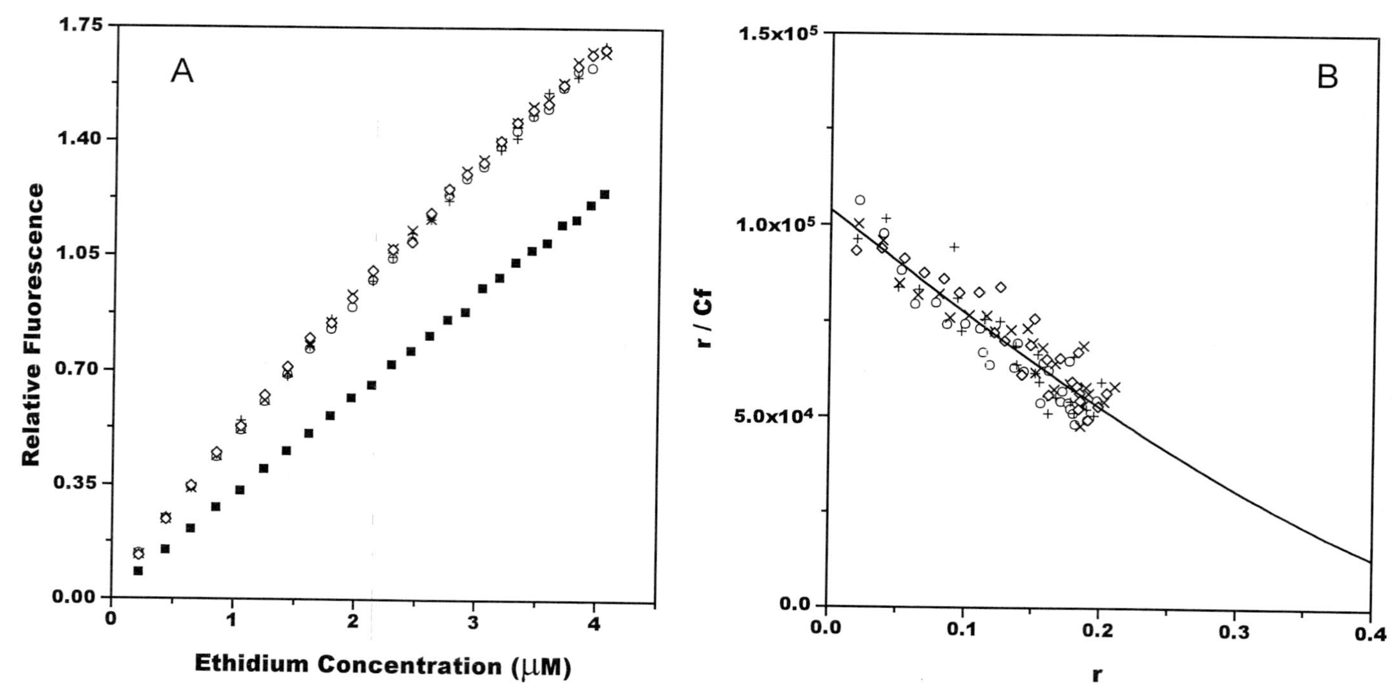

FIG. 6. Ethidium binding to calf thymus DNA. (A) Fluorescence response of ethidium alone (closed squares) and in the presence of a fixed DNA concentration. (B). Scatchard plot derived from the data shown in (A). The solid line is the best fit to the neighbor exclusion model.

Conclusion

Fluorescence and absorbance spectroscopies provide a powerful means of determining drug–DNA binding constants. We have described protocols that have evolved in our laboratory that are designed to evaluate the number of significant spectral species that contribute to changes in fluorescence or absorbance spectra on drug binding. This provides a rigorous test of the two-state assumption normally invoked in the analysis of spectroscopic data to obtain the distribution of free and bound drug. In addition, we have described the application of nonlinear least squares fitting methods, coupled with Monte Carlo analysis, to reliably estimate limiting optical parameters and their errors, which are necessary to compute the distribution of free and bound ligand from spectroscopic data. Although we used ethidium binding to calf thymus DNA to illustrate these methods, the approaches are general, and could be used to analyze drug binding to DNA oligonucleotides, synthetic polynucleotides, or alternate structures (triplexes, quadraplexes). Once the methods are established in the laboratory, rigorous analysis can proceed quickly. The complete acquisition and analysis of a titration curve by the methods described is now done in our laboratory within one working day.

Acknowledgments

Supported by grant CA35635 from the National Cancer Institute.

[20] Neural Network Techniques for Informatics of Cancer Drug Discovery

By William W. van Osdol, Timothy G. Myers, and John N. Weinstein

Introduction

Neural computing is a relatively recent development in the information sciences, a product of artificial intelligence research in the 1950s and 1960s.[1,2] Artificial network models were constructed as aids to the understanding of processes such as perception, motor function, learning, memory, and

[1] T. Khanna, "Foundations of Neural Networks." Addison-Wesley, New York, 1991.

[2] M. H. Hassoun, "Fundamentals of Artificial Neural Networks." MIT Press, Cambridge, Massachusetts, 1995.

0076-6879/00 $30.00

consciousness. Later, it become clear that such networks could perform interesting types of calculations and pattern mappings, whether or not they did it in a way pertinent to the biology. Artificial neural networks differ from the usual computer programs in that information is coded in distributed form in terms of the strength of "synaptic" connections between neurons, also called processing elements, that are nonlinear threshold gates. Most types learn from a set of examples, rather than being programmed to get the right answer. Neural networks are being used in the biomedical sciences for pattern recognition and/or decision making in diverse fields (reviewed in Refs. 3–5), including nucleic acid sequence analysis, protein sequence analysis, quantitative cytology, diagnostic imaging, neural organization, speech recognition, and clinical diagnosis. They have also been used in drug discovery, the subject of this chapter.

Back-Propagation Neural Networks for Prediction of Drug Mechanism

To illustrate how neural networks can be used to predict mechanism of action and related properties of potential drugs, we give here an example from the cancer drug discovery program of the National Cancer Institute (NCI).[6] Since 1990, the NCI Developmental Therapeutics Program has screened more than 65,000 chemical compounds and a larger number of natural product extracts against a panel of 60 different tumor cell lines *in vitro*. We focus here on the chemically defined compounds. At various points in the drug discovery–development pipeline, it is necessary to select which compounds to send forward—to *in vivo* studies in animals, to pharmacological and toxicological characterization, and to the clinic. One early observation was that the pattern of cell response to a compound in the 60-cell line screen appeared to correlate with chemical structure and with presumed mechanism of action. The first evidence was provided by the COMPARE program of Paull *et al.*[7] For a designated "seed" compound, COMPARE searches the database of already-tested agents for those with the most similar patterns of activity against the 60-cell line panel, as indi-

[3] J. N. Weinstein K. W. Kohn, M. R. Grever, V. N. Viswanadhan, L. V. Rubinstein, A. P. Monks, D. A. Scudiero, L. Weich, A. D. Koutsoukos, A. J. Chiausa, and K. D. Paull, *Science* **258,** 447 (1992).

[4] J. N. Weinstein, J. J. Casciari, K. Raghavan, J. Buolamwini, and T. G. Myers, *Symp. World Congr. Neural Networks* **1,** 121 (1994).

[5] W. W. van Osdol, T. G. Myers, K. D. Paull, K. W. Kohn, and J. N. Weinstein, *J. Natl. Cancer Inst.* **86,** 1853 (1994).

[6] M. R. Boyd. *Princ. Pract. Oncol. Updates* **3,** 1 (1989).

[7] K. D. Paull, R. H. Shoemaker, L. Hodes, A. Monks, D. A. Scudiero, L. V. Rubinstein, J. Plowman, and M. R. Boyd. *J. Natl. Canc. Inst.* **81,** 1088 (1989).

cated in pairwise comparisons by Pearson's correlation coefficient. Compounds matched by pattern of activity were, indeed, found very often to have similar chemical structures, and the matches also correlated with *in vitro* biochemical mechanism of action (whether known with certainty or putatively).[8–10] Subsequent studies and analyses have delineated structure–function–target relationships for the tested compounds.[11–20]

Another line of evidence for the information content of the cell sensitivity patterns came from neural network analyses. We first used back-propagation neural networks to address quantitatively the following question: Can patterns of activity against cell lines in the panel be used to classify new agents by their mechanisms of action?[3] As described in subsequent sections of this chapter, we then developed two-dimensional Kohonen self-organizing maps to explore in greater detail the relationships between pattern of activity and mechanism.[5]

[8] R. Bai, K. D. Paull, C. L. Herald, L. Malspeis, G. R. Pettit, and E. Hamel, *J. Biol. Chem.* **266,** 15882 (1991).

[9] K. D. Paull, L. Hodes, J. Plowman *et al., Proc. Am. Assoc. Cancer Res.* **29,** 488 (1988).

[10] L. Hodes, K. D. Paull, A. D. Koutsoukos, and L. V. Rubinstein, *J. Biopharm. Statist.* **2,** 31 (1992).

[11] J. N. Weinstein, T. G. Myers, J. Buolamwini, K. Raghavan, V. N. Viswanadhan, J. Licht, L. V. Rubinstein, A. D. Koutsoukos, K. W. Kohn, D. Zaharevitz, M. R. Grever, A. Monks, D. A. Scudiero, B. A. Chabner, N. L. Anderson, and K. D. Paull, *Stem Cells* **12,** 13 (1994).

[12] M. Alvarez, K. D. Paull, C. Hose, J.-S. Lee, J. N. Weinstein, M. Grever, S. Bates, and T. Fojo, *J. Clinic. Invest.* **95,** 2205 (1995).

[13] J. N. Weinstein, T. G. Myers, P. M. O'Connor, S. H. Friend, A. J. Fornace, K. W. Kohn, T. Fojo, S. E. Bates, L. V. Rubinstein, N. L. Anderson, J. K. Buolamwini, W. W. van Osdol, A. P. Monks, D. A. Scudiero, E. A. Sausville, D. W. Zaharevitz, B. Bunow, V. N. Viswanadhan, G. S. Johnson, R. E. Wittes, and K. D. Paull, *Science* **275,** 343 (1997).

[14] T. G. Myers, M. Waltham, G. Li, J. K. Buolamwini, D. A. Scudiero, L. V. Rubinstein, K. D. Paull, E. A. Sausville, N. L. Anderson, and J. N. Weinstein, *Electrophoresis* **18,** 647 (1997).

[15] L. M. Shi, Y. Fan, T. G. Myers, M. Waltham, K. D. Paull, and J. N. Weinstein, *Math. Model. Sci. Comput.* **38,** 189 (1998).

[16] L. Wu, A. M. Smythe, S. F. Stinson, L. A. Mullendore, A. P. Monks, D. A. Scudiero, K. D. Paull, A. D. Koutsoukos, L. V. Rubinstein, M. R. Boyd, and R. H. Shoemaker, *Cancer Res.* **52,** 3029 (1992).

[17] A. D. Koutsoukos, L. V. Rubinstein, D. Faraggi, S. Kalyandrug, J. N. Weinstein, K. D. Paull, K. W. Kohn, and R. M. Simon, *Statis. Med.,* in press.

[18] K. Wosikowski, D. Schuurhuis, K. Johnson, K. D. Paull, T. G. Myers, J. N. Weinstein, and S. E. Bates, *J. Natl. Cancer Inst.* **89,** 1505 (1997).

[19] J. M. Freije, J. A. Lawrence, M. G. Hollingshead, A. De la Rosa, V. Narayanan, M. Grever, M. Sausville, K. D. Paull, and P. S. Steeg, *Nature Med.* **3,** 395 (1997).

[20] H. M. Koo, A. P. Monks, A. Mikheev, L. V. Rubinstein, M. Gray-Goodrich, M. J. McWilliams, W. G. Alvord, H. K. Oie, A. F. Gazdar, K. D. Paull, H. Zarbl, and G. F. Vande Woude, *Cancer Res.* **56,** 5211 (1996).

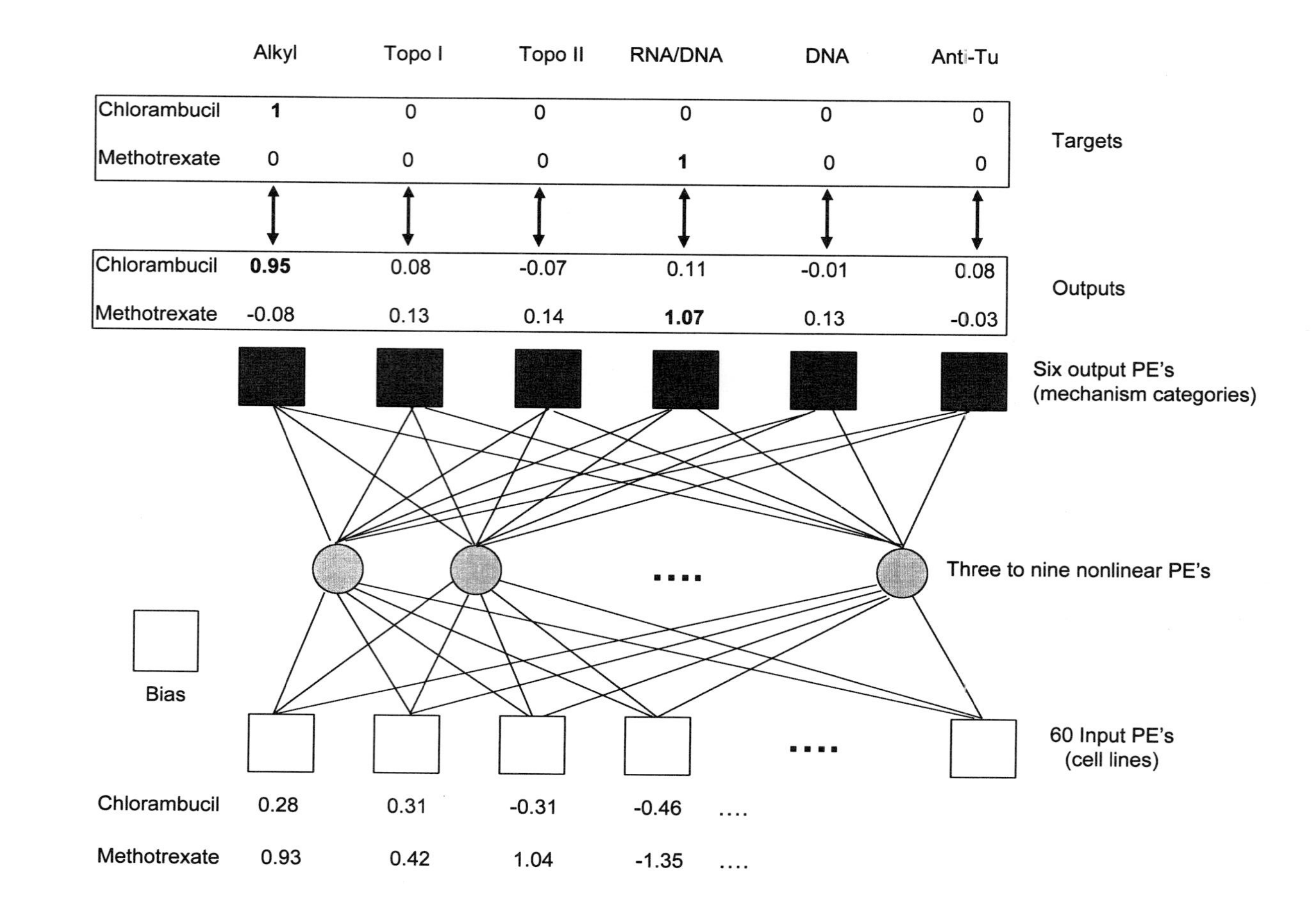
Alkyl
Topo I
Topo II
RNA/DNA
DNA
Anti-Tu
Chlorambucil 1 0 0 0 0 0
Methotrexate 0 0 0 1 0 0
Targets
Chlorambucil 0.95 0.08 -0.07 0.11 -0.01 0.08
Methotrexate -0.08 0.13 0.14 1.07 0.13 -0.03
Outputs
Six output PE's (mechanism categories)
Three to nine nonlinear PE's
Bias
60 Input PE's (cell lines)
Chlorambucil 0.28 0.31 -0.31 -0.46 ....
Methotrexate 0.93 0.42 1.04 -1.35 ....

Methods for Back-Propagation Neural Network Analysis

To predict mechanism of action, we developed neural networks, such as that illustrated in Fig. 1, using the Neural Works Professional II/Plus program package (NeuralWare, Pittsburgh, PA), with 60 input processing elements, one for each cell line, and six output processing elements, one for each of the mechanism categories under consideration.[3] Between the inputs and outputs, we included a hidden layer whose number of processing elements (PEs) could be varied. The more hidden layer elements, the more complex and nonlinear the patterns that could be learned. Other techniques, linear or nonlinear, could be used to explore the information content of the screening data, but neural networks appeared to be the most flexible method available, especially in the absence of any established functional relationship between cell line responses to chemotherapeutic agents, *in vitro,* and the mechanisms of action of those agents.

Using a set of drugs (Table I) whose mechanisms were reasonably well established, we trained the network to predict mechanism of action. This training was done by presenting iteratively the activity pattern of each agent at the input layer, as illustrated in Fig. 1 for two standard drugs, methotrexate and chlorambucil. Each output processing element corresponded to a mechanism category; the target value for an output was 1 if the drug input belonged to that category, and 0 if it did not. The synaptic weights were updated progressively during training. That is, they were reinforced or penalized, based on the difference between the desired and actual outputs, using the "back-propagation" algorithm[21,22] so that the network outputs came to approximate the target values (that is, 1s and 0s for the different classes). The drug was then assigned to the category having the largest value as its output.

[21] P. Werbos, Ph.D. Thesis, Harvard University, 1974.
[22] D. E. Rumelhart, G. E. Hinton, and R. J. Williams, *Nature* **323,** 533 (1986).

FIG. 1. Neural networks with three to nine hidden layers of PEs used to classify chemotherapeutic agents according to their mechanism of action. Sample inputs, outputs, and targets for the trained network are shown for two standard agents, chlorambucil and methotrexate. Solid interconnections schematically represent different weights in the trained network. The output patterns closely match the target patterns for the two drugs (note values in boxes), indicating that their mechanisms of action are correctly predicted. The bias is connected to all hidden layer and output processing elements, but the connections have been omitted for clarity. It represents offset values in the nonlinear functions that comprise the processing elements.

TABLE I

SET OF "STANDARD" ANTICANCER AGENTS AND SET OF 60 CELL LINES USED IN NCI PRIMARY CANCER SCREEN[a]

Number and type	Compound
35 Alkylating agents	Asaley (167780), AZQ (182986), carmustine (409962), busulfan (750), carboxyphthalatoplatinum (271674), carboplatin (241240), lomustine (79037), iproplatin (256927), chlorambucil (3088), chlorozotocin (178248), cisplatinum (119875), clomesone (338947), cyanomorpholinodoxorubicin (357704), cyclodisone (348948), dianhydrogalactitol (132313), fluorodopan (73754), hepsulfan (329680), hycanthone (142982), melphalan (8806), semustine (95441), mitomycin C (26980), mitozolamide (353451), mechlorethamine (762), PCNU (95466), piperazine alkylator (344007), piperazinedione (135758), pipobroman (25154), porfiromycin (56410), spiromustine (172112), tetraplatin (363812), triethylenemelamine (9706), thio-tepa (6396), teroxirone (296934), uramustine (34462), yoshi-864 (102627)
32 Topoisomerase I inhibitors	Camptothecin (94600), camptothecin Na salt (100880), camptothecin derivatives (95382, 107124, 643833, 629971, 295500, 249910, 606985, 374028, 603071, 176323, 295501, 606172, 606173, 610458, 618939, 610457, 610459, 606499, 610456, 364830, 606497, and 9 confidential camptothecin derivatives)
16 Topoisomerase II inhibitors	Doxorubicin (123127), amonafide (308847), amsacrine (249992), anthrapyrazole derivative (355644), pyrazoloacridine (366140), bisantrene hydrochloride (337766), daunorubicin (82151), deoxydoxorubicin (267469), mitoxanthrone (301739), menogaril (269148), morpholinodoxorubicin (354646), *N*,*N*-dibenzyldaunomycin (268242), oxanthrazole (349174), rubidazone (164011), teniposide (122819), etoposide (141540)
19 RNA/DNA antimetabolites	L-Alanosine (153353), 5-azacytidine (102816), 5-fluorouracil (19893), acivicin (163501), aminopterin (132483), aminopterin derivative (184692), aminopterin derivative (134033), ornithine derivative (633713), ornithine derivative (623017), Baker's soluble antifol (139105), dichloroallyl lawsone (126771), brequinar (368390), ftorafur (148958), 5,6-dihydro-5-azacytidine (264880), methotrexate (740), methotrexate derivative (174121), PALA (224131), pyrazofurin (143095), trimetrexate (352122)

As inputs to the network we used the quantities Δ (Ref. 7), defined for each drug–cell line pair as

$$\Delta = \log_{10}(1/GI_{50}) - \mathrm{mean}[\log_{10}(1/GI_{50})] \quad (1)$$

where GI_{50} is the concentration of a drug required to decrease the growth of a cell line by 50% in the standard assay. Assay conditions and technical

TABLE I (*continued*)

Number and type	Compound
16 DNA antimetabolites	3-HP (95678), 2′-deoxy-5-fluorouridine (5-FUdR) (27640), 5-HP (107392), α-TGDR (71851), aphidicolin glycinate (303812), cytarabine (63878), 5-Aza-2′-deoxycytidine (127716), β-TGDR (71261), cyclocytidine (145668), guanazole (1895), hydroxyurea (32065), inosine glycodialdehyde (118994), geldanamycin (330500), pyrazoloimidazole (51143), thioguanine (752), thiopurine (755)
13 Antimitotic agents	Allocolchicine (406042), halichondron B (609395), colchicine (757), colchicine derivative (33410), dolastatin 10 (376128), maytansine (153858), rhizoxin (332598), paclitaxel (125973), paclitaxel derivative (608832), thiocolchicine (361792), tritylcysteine (83265), vinblastine sulfate (49842), vincristine sulfate (67574)
Source	Cell line
Non-small-cell lung cancer	NCl-H23, NCl-H522, A549/ATCC, EKVX, NCl-H226, NCl-H322M, NCl-H460, HOP-62, HOP-18, HOP-92, LXFL 529
Small-cell lung cancer	DMS 114, DMS 273
Colon cancer	HT29, HCC-2998, HCT-116, SW-620, COLO 205, DLD-1, HCT-15, KM12, KM20L2
Ovarian cancer	OVCAR-3, OVCAR-4, OVCAR-5, OVCAR-8, IGROV1, SK-OV-3
Leukemia	CCRF-CEM, K-562, MOLT-4, HL-60(TB), RPMI-8226, SR
Renal cell carcinoma	UO-31, SN12C, A498, CAKI-1, RXF-393, RXF-631, 786-O, ACHN, TK-10
Melanoma	LOX IMVI, MALME-3M, SK-MEL-2, SK-MEL-5, M14, SK-MEL-28, M19-MEL, UACC-62, UACC-257
CNS tumor	SNB-19, SNB-75, SNB-78, U251, SF-268, SF-295, SF-539, XF 498

[a] Numbers in parentheses are NSC numbers. Modified from Refs. 3 and 5.

details have been described previously.[9,23–26] The value subtracted in Eq. (1) for a given drug relates to the mean of its cytotoxic potencies over the entire panel of cells. Because of this correction, differences in *overall* potency of the agents were nulled out, and their patterns of *differential activity* emphasized.

[23] M. C. Alley, D. A. Scudiero, A. P. Monks *et al., Cancer Res.* **48,** 589 (1988).

[24] D. A. Scudiero, R. H. Shoemaker, K. D. Paull *et al., Cancer Res.* **48,** 4827 (1988).

[25] R. H. Shoemaker, A. P. Monks, M. C. Alley *et al., in* "Prediction of Response to Cancer Chemotherapy" (T. Hall, ed.). Alan R. Liss, New York, 1988.

[26] A. P. Monks, D. A. Scudiero, P. Skehan, R. H. Shoemaker, K. D. Paull, D. Vistica, C. Hose, J. Langley, P. Cronise, A. Vaigro-Wolff, M. Gray-Goodrich, H. Campbell, J. Mayo, and M. R. Boyd, *J. Natl. Cancer Inst.* **83,** 757 (1991).

The database for training (Table I) was formulated from a list of "standard agents," supplemented by additional drugs for which sufficient mechanistic information was available. Mechanisms of action were established for a set of 131 agents using literature sources, structural homologies, and data from experiments on mechanism but independent of results from the 60-cell line screen.

In almost all cases, the screening data represented arithmetic averages over multiple experiments run at different times. In the matrix of values, there were 604 (7.1%) missing values (that is, drug/cell line combinations not tested or else eliminated at the time of the experiment for quality control reasons). For each missing value, we inserted the mean obtained for the drug over all cell lines in the tested set.[3] Insofar as possible, construction of the database was kept independent of prior knowledge of results in the screen. However, screen results were used in a number of cases for the limited purpose of deciding which dose-range vector for a given drug captured the dynamic range of activity.

The control strategy used for the network was as follows: Learning was achieved by the normal cumulative delta algorithm with epoch size 30. Input vectors were presented in randomized order during training to prevent cyclic behavior. The initial learning coefficients and momentum were 0.3 and 0.45 for the hidden layer and 0.15 and 0.4 for the output layer (all four values decreasing by one-half after 10,000 presentations). Each vector was presented a total of 15,000 times. This number of iterations was determined in preliminary experiments to fall within a broad optimal range with respect to the root mean square error of prediction in the 10 test sets. The classification rule used was "plurality wins." For each pair of training and test sets, the network was trained and then tested seven times from seven different initial random number starting points. The resulting output values were then averaged, a procedure that decreased considerably the number of incorrect predictions (for example, from a mean of 15.4 to 12 in the case of seven hidden layer elements).

Specification of mechanism is, to some degree, arbitrary and uncertain. A molecule may inhibit cell growth in multiple ways, depending on culture conditions and cell type. To the extent that a mechanism could be unambiguously assigned, it was also not clear that mechanistic category would actually correlate well with functional pattern. For all of these reasons, and given inevitable experimental noise in the data, we were not sure beforehand whether reasonably predictive results could be obtained. However, the findings were unexpectedly good; as indicated in Tables II and III, for a total of 141 drug activity patterns over the 131 agents, the network predicted 12 (9%) incorrectly, a 67-fold improvement in the odds ratio over random assignment of categories. It is important to note that these results were

TABLE II

NEURAL NET PREDICTIONS OF MECHANISM FROM THE NATIONAL CANCER INSTITUTE'S CANCER DRUG SCREEN DATA[a]

No. hidden layer PEs	Incorrect predictions	Correct predictions	Comparison with LDA (p value)[b]
3	20 (14.2%)	121	1.00
5	13 (9.2%)	128	0.07
7	12 (8.5%)	129	0.02
9	12 (8.5%)	129	0.04
Linear discrimination analysis (LDA)	20 (14.2%)	121	—

[a] Summary of results as a function of number of hidden layer PEs and in comparison with results from linear discriminant analysis. From Ref. 3.

[b] Two-tail p value from McNemar test[25] for comparison with LDA. The p values for seven and nine hidden layer PEs differ because there were a total of 10 discrepancies in the case of seven PEs and 12 discrepancies in that of nine PEs.

TABLE III

PREDICTION OF MECHANISM CATEGORY FROM CANCER DRUG SCREEN DATA USING A BACK-PROPAGATION NEURAL NETWORK WITH SEVEN HIDDEN LAYER ELEMENTS[a]

	Predicted category[b]						
Actual category	Alkyl 1	Topo I 2	Topo II 3	R/D 4	DNA 5	Antimitotic 6	None 7
1. Alkylating	**33**		2				
2. Topo I		**35**					
3. Topo II	1		**18**				
4. RNA/DNA	1		1	**17**		1	
5. DNA	1		1	1	**12**	1	
6. Antimitotic	1	1				**14**	
Correct predictions: 129 (91.5%)							
Incorrect predictions: 12 (8.5%)							

[a] Left-hand column gives putative mechanism. Numbers in boldface are correct predictions. R/D indicates RNA/DNA antimetabolites; DNA indicates DNA antimetabolites. From Ref. 3.

[b] Note that the table reflects results for 141 drug concentration vectors representing 131 different drugs. For 10 of the drugs, two different dose–response profiles were included because it was not clear which best indicated the properties of the agent.

predictive rather than simply correlative, because a "cross-validation" protocol was used. That is, a network was trained on 90% of the agents and then tested on the remaining 10% (in all 10 permutations).

For comparison with the neural network results, linear discriminant analysis (LDA) was performed using the SAS program package (SAS Institute, Inc., Cary, NC). LDA categorizes vectors of numbers (in this case, the Δ values) according to known examples, where the vectors are assumed to follow a multivariate normal distribution with known or (as in this case) estimated parameters. LDA can, thus, be considered a purely linear equivalent of the network. LDA (with equal prior probabilities) produced 20 wrong classifications (p value $= 0.02$ versus the neural network in the two-tailed McNemar test, a binomial calculation based on the discrepant predictions of the two methods that asks how likely it is that the difference in number of incorrect predictions arose by chance).[27]

These findings substantiated the hypothesis that patterns of differential growth inhibition in the screening panel contain useful information on mechanism and that the neural network can recognize that information. We did not expect such clear-cut answers. We expected, for instance, that more of these agents would prove to express mixed mechanisms in their patterns of activity.[3] To complicate matters further, the subcellular distribution of drugs can be influenced by a variety of factors, including transmembrane potentials, transmembrane pH gradients, structure-selective transporters, and P-glycoprotein-mediated multidrug resistance. These factors may be viewed as confounding variables in the system and may explain some of the incorrect classifications.[3,11]

As often happens, some of the most interesting insights have come from data that did not fit. To cite one example, the network with three hidden layer PEs classified porfiromycin and mitomycin as antimitotics instead of alkylating agents, whereas the net with five hidden layer PEs classified them correctly.[3] This ambiguity may have resulted from the fact that porfiromycin and mitomycin bind in the minor groove of DNA, whereas the other alkylators in the data set bind in the major groove. Testing other types of minor groove binders, such as tomamycin, the anthramycins, and the pyrrolo-1,4-benzodiazepines, may resolve this question. The classification network thus appears to be a good source of clues as to the fine structure of mechanistic categories.

Application of Kohonen Self-Organizing Maps

We now describe the use of another neural network paradigm, the Kohonen self-organizing map (SOM), to illuminate more detailed relation-

[27] Q. NcNemar, *Psychometrika* **12,** 151 (1947).

TABLE IV
MECHANISMS OF ACTION REPRESENTED IN DATA SET FOR SOM

Agent	Mechanism
Alkylating agents (36)	Alkylating at N2 position of guanine—A2
	Alkyltransferase-dependent cross-linkers—AC
	Alkylating at N7 position of guanine—A7
	DNA intercalators—AI
DNA synthesis inhibitors (16)	Compounds incorporated into DNA—DI
	DNA polymerase inhibitors—DP
	Ribonuclease reductase inhibitors—DR
Nucleotide synthesis inhibitors (19)	Antifolates—RF
	Irreversible inhibitors—RI
	Anti-other precursors—RO
	Unknown site of inhibition—R
Topoisomerase I inhibitors (32)	T1
Topoisomerase II inhibitors (16)	T2
Tubulin-active antimitotics (13)	TU

ships among a set of chemotherapeutic agents, as reflected by their patterns of differential activity in the NCI cancer screen.[5,28] The SOM accomplishes this by capturing the salient topological features of the distribution of high-dimensional response patterns and representing them in a low-dimensional lattice. The SOM is an example of a topology-preserving map obtained from an unsupervised, competitive learning algorithm. As such, it is especially well suited for our purpose: exploratory data analysis, the prospective search for structure in collections of data. These studies were performed about 2 years after the classification experiments presented above.

Materials and Methods

Patterns of Differential Response. As input data we used differential response patterns from a group of 131 drugs identical to those described above, representing six broadly defined mechanisms of action, several of which were further divided into submechanisms, as listed in Table IV. For this study we assembled data from the panel of 45 cell lines that have been included in the cancer screen since its inception. These cell lines are listed

[28] W. W. van Osdol, T. G. Myers, K. D. Paull, K. W. Kohn, and J. N. Weinstein, *in* "Proceedings of the World Congress on Neural Networks," Vol. 2, p. 762. INNS, 1995.

in Table V and comprise a subset of the 60 cell lines used in the earlier neural network study (see Table I). Thus, the data set consisted of 131 input vectors, each having 45 real-valued components (Δ values) derived from the dose–response curves of the cell lines. The values for an agent were calculated from Eq. (1), and, for most agents, the biological response patterns reflected averages over multiple experiments performed at different times. As a result, although 45 cell lines were identical in both neural network studies, the activity patterns for a given agent differed between studies because of rescreening during the intervening time. For example, there were far fewer missing Δ values (<1%) in this second study (and they were filled as described above).

Each response pattern was then normalized as follows:

$$\Delta' = \Delta/(\sigma\sqrt{2N}) \tag{2}$$

where σ is the standard deviation over all N cell lines in the response pattern. This had the effect of transforming the Euclidean distance between two patterns, $\mathbf{x}$ and $\mathbf{y}$, to $\sqrt{1-r}$, where r is their Pearson correlation coefficient,

$$r = N^{-1} \sum_{i=1}^{N} (x_i - \bar{x})(y_i - \bar{y})/(\sigma_x \sigma_y) \tag{3}$$

and was done so that the SOMs might be readily comparable to the calculations of the DISCOVERY programs[13,14] and pairwise correlation in COM-

TABLE V
TUMOR CELL LINES PROVIDING RESPONSE PATTERNS FOR INPUTS TO THE SOM

Source	Cell line
Non-small-cell lung cancer (NSCLC)	A549/ATC, EKVX, HOP-62, HOP-92, NCI-H226, NCI-H23, NCI-H322, NCI-H460, NCI-H522
Colon	COLO 205, HCC-2998, HCT-116, HCT-15, HT29, KM12, SW-620
Ovarian	IGROV1, OVCAR-3, OVCAR-4, OVCAR-5, OVCAR-8, SK-OV-3
Leukemia	CCRF-CEM, HL-60TB, MOLT-4, RPMI-822, SR
Renal	A498, CAKI-1, RXF-393, SN12C, UO-31
Melanoma	LOX IMVI, MALME-3M, SK-MEL-2, SK-MEL-5, SK-MEL-28, UACC-257, UACC-62
Central nervous system (CNS)	SF-268, SF-295, SF-539, SNB-19, SNB-75, U251

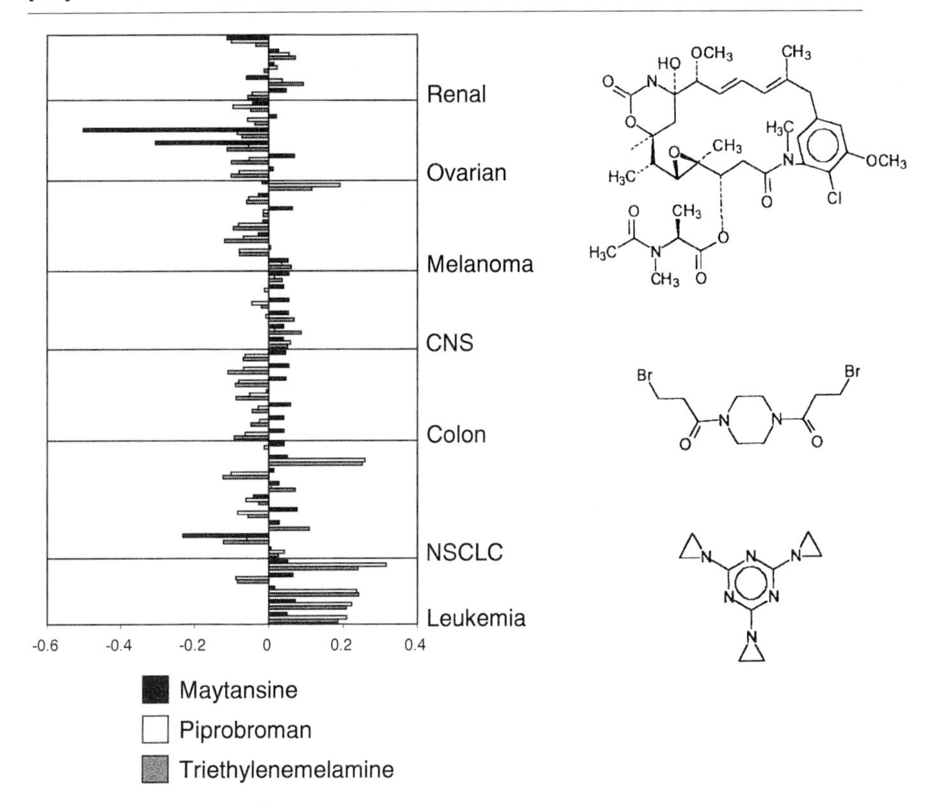

FIG. 2. "Mean graph"[7] of normalized (Δ') response patterns for three chemotherapeutic agents included in the data set under study. Clearly, there is substantial similarity between the patterns for piprobroman and triethylenemelamine, and dissimilarity between those patterns and the pattern for maytansine. The first two agents alkylate the N-7 position of guanine, in the major groove of DNA, while the latter agent is a tubulin-active antimitotic. The Kohonen self-organizing map is used to reflect such pattern differences quantitatively in two dimensions. Structures for the three compounds are shown at right.

PARE.[7-10,29] The Δ' response patterns for three chemotherapeutic agents are plotted in Fig. 2.

Although the 131 vectors of the input set were nominally 45 dimensional, they might have occupied, at least approximately, a space of considerably lower dimension. This possibility was examined via principal component analysis (PCA), a linear technique for finding low-dimensional representations of a set of vectors.[30] This analysis showed that the input set had 12

[29] K. D. Paull, C. M. Lin, L. Malspeis, and E. Hamel, *Cancer Res.* **52,** 3892 (1992).

[30] K. V. Mardia, J. T. Kent, and J. M. Bibby, "Multivariate Analysis." Academic, London, 1979.

eigenvalues greater than unity, but the associated eigenvectors accounted for only 67% of the variance in the data. To account for 90% of the variance, 23 eigenvectors would have been required. As a result, we chose to work with the original data set.

Kohonen SOM Algorithm. Kohonen's algorithm[31] is an example of a topology-preserving map obtained via unsupervised learning. It is typically applied to N continuous-valued inputs, $x_1, \ldots, x_N$, defining a vector (pattern), $\mathbf{x}$, in an N-dimensional space. M output units are arranged in a one-, two-, or three-dimensional array and are fully connected via weights, $w_{i,j}$, to the input units. Thus, each output unit, i, has associated with it an N-dimensional weight vector, $\mathbf{w}_i$. An example of this architecture in two dimensions is illustrated in Fig. 3.

After selection of the discrete output array and initial randomization of the weights, a step in the development of a map proceeds as follows. An input vector of activity values is chosen, and the output unit, i^*, is found such that its weight vector minimizes the Euclidean distance to the current input vector:

$$i^* = \text{argmin}\{d(\mathbf{x}, \mathbf{w}_i)\}, \qquad i = 1, \ldots, M \tag{4}$$

where $d(\mathbf{x}, \mathbf{w}_i)$ is the Euclidean distance between vectors. The components of that weight vector and the weight vectors of neighboring output units are then modified according to the learning rule

$$\Delta w_{i,j} = \alpha\chi(i, i^*)(x_j - w_{i,j}) \tag{5}$$

where α is a learning rate parameter, $\chi(i, i^*)$ is a neighborhood function, and j denotes the components of the vectors $\mathbf{w}_i$ and $\mathbf{x}$. The learning rate governs the extent to which weight vector $\mathbf{w}_i$ is modified to approximate $\mathbf{x}$ more closely, and the neighborhood function is the mechanism that captures topological features of the input set. The value of $\chi(i, i^*)$ is unity for $i = i^*$, but decreases monotonically with increasing distance (measured in the discrete geometry of the output array) between output units i^* and i. Output units close to unit i^* have their weight vectors drawn appreciably toward $\mathbf{x}$, whereas more distant output units, for which χ is small, are affected only slightly. As a result, nearby output units receive similar updates. Input vectors are presented to the map and compete for the weight vectors (and thereby the output units), continually attracting the nearest and its neighbors in N-dimensional space. Ideally, over many presentations of the entire input set, output units proximal in the array come to represent input vectors proximal in the input space.

Training usually begins with a wide range for χ (so that the neighborhood

[31] T. Kohonen, "Self-Organization Maps." Springer Verlag, Berlin, 1995.

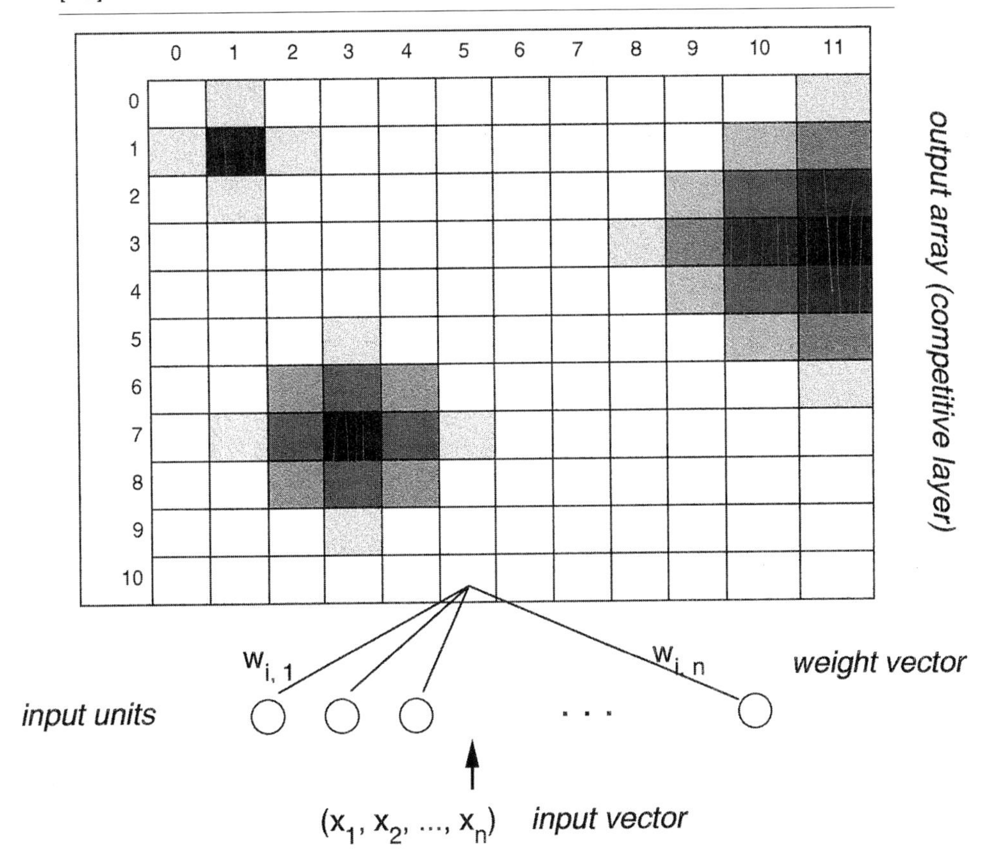

FIG. 3. Schematic representations of the network architecture for the Kohonen self-organizing map; input vectors, input units, the two-dimensional array of output units and weight vector associated with output unit (10, 5) are shown. Also indicated are three neighborhoods, defined by $\lambda = 1, 2$, and 3. The central cell (shaded black) indicates the activated output unit, i^*. The value of the neighborhood function declines monotonically (indicated by progressively lighter shading) with increasing distance from i^*, reaching a value of 0.607 ($= 1/\sqrt{e}$) at distance λ. Note that the neighborhood corresponding to $\lambda = 3$ is truncated by the edge of the output array. (Reprinted from van Osdol *et al.*[5])

of any output unit covers almost the entire output array) and a large value for α (e.g., 0.8). It is essential for convergence that α and the range of the neighborhood decrease as training proceeds.[31,32] Thus, α and χ must be $\alpha(t)$ and $\chi(t)$, where t denotes the extent of the training.

[32] J. Hertz, A. Krogh, and R. G. Palmer, "Introduction to the Theory of Neural Computation." Addision-Wesley, Reading, Massachusetts, 1991.

Application of SOM. Using the SOM-PAK software package, version 1.2 (Laboratory of Computer and Information Science, Helsinki University of Technology, Espoo, Finland), we mapped the set of 131 biological activity patterns from 45 dimensions to a two-dimensional array. Our goal was to extract detailed topological information from the input set and to represent this information visually and quantitatively.

Training was done in two phases: a short initial organization of the inputs, followed by a longer training period in which the organization was refined. There are many possible functional forms for $\alpha(t)$; in SOM-PAK during each phase of training, the learning rate declines linearly from its initial value to a final value of zero. The neighborhood function took the form of a Gaussian:

$$\chi(i, i^*, t) = \exp[-\|\mathbf{r}_i - \mathbf{r}_{i^*}\|^2/2\lambda^2(t)] \tag{6}$$

where $\mathbf{r}_i$ and $\mathbf{r}_{i^*}$ are two-dimensional vectors representing output units i and i^*, and $\lambda(t)$ declines linearly from its initial value to a final value of one during each phase of training.

In SOM-PAK the quantization error, ε, measures how accurately the map has come to represent a particular input vector:

$$\varepsilon_i(t) = d(\mathbf{x}, \mathbf{w}_{i^*}) \tag{7}$$

where $\mathbf{w}_{i^*}(t)$ is the closest weight vector to input vector $\mathbf{x}_i$ at stage t of training. In an output array of M units, with choices of $\alpha(t)$ and $\chi(t)$, there are many possible representations of an input set, and training can be viewed as a process of minimizing $\bar{\varepsilon}$, the mean quantization error.[32] Thus, it is common practice to select from a large number of maps, each involving the same training parameters but a different random initialization of the weight vectors, the map(s) with the smallest value of $\bar{\varepsilon}$ at the end of training.

The value of $\bar{\varepsilon}$ also depends on the form of $\chi(t)$, the initial values of $\alpha(t)$ and $\lambda(t)$, and the number of presentations of the input set during each phase of training. Also note that $\bar{\varepsilon}$ does not reflect the topological fidelity of the output representation to the input set. In fact, we observed that minimizing $\bar{\varepsilon}$ (representing individual input vectors well) and capturing the topological features of the input set (representing collective features of the input set) are conflicting goals of the SOM. Therefore, we developed empirically the following training protocol to balance these opposing tendencies.

A two-dimensional competitive array size of 11×12 output units was initially trained on 100 presentations of the input set, with learning rate and Gaussian neighborhood radius decreasing linearly from 0.8 and 7, respectively. A refinement phase of training followed, consisting of 1250 presentations of the input set, with learning rate and neighborhood radius declining linearly from 0.6 and 4, respectively. The optimal maps were defined as those that minimized the mean quantization error over 50 maps generated from different random initial distributions of weights. The arrangement and number, M, of units in the output array are important aspects of the map. We investigated a broad range in M (4×4 to 12×14), and focused primarily on arrays of 11×12 units in order to permit substantial resolution of the input set. This array is essentially square; rather higher quantization errors were obtained with more rectangular arrays (10×13, 8×16).

The last issue that needs to be addressed is how to measure performance of the SOM algorithm. Cross-validation, a procedure generally applicable for feed-forward, back-propagation networks, is not appropriate, because the SOM does not partition input sets into preexisting categories or map vectors to a dependent variable. Nonetheless, there are several other possibilities. First, the correlation matrix for the input set of response patterns can be calculated and compared with the distribution of the input set over the output array. Highly correlated response patterns should be mapped to the same or adjacent output units, and poorly correlated response patterns should not. Second, cluster analysis,[33] a deterministic technique, can be applied to represent distance relationships among the response patterns, and the representation compared with that derived from the SOM. Cluster methods fall into two broad categories, partitioning and hierarchical, within each of which are available many different algorithms. We used a hierarchical, agglomerative algorithm, based on Euclidean distances between response patterns, and a group-average definition of distance between clusters,[33] to order the rows and columns of the correlation matrix of the input set.[3,13] This cluster-ordered correlation matrix for the input set allows one to see the collective relationships that exist among groups of highly correlated response patterns, although it is cumbersome for large data sets. Third, one can compare SOM results to random projection of the response patterns into the output array. Such a projection can be obtained by choosing randomly a row and column for each input vector, or by using an initial neighborhood radius of zero [$\lambda(0) = 0$] in both training phases of the SOM. In this latter case, the map reflects the initial random distribution of weight vectors. Comparison of a random map with one properly trained makes readily apparent the degree of organization achieved by the SOM, but is not quantitative at the level of individual input vectors.[5] Thus, we have

[33] L. Kaufman and P. J. Rousseeuw, "Finding Groups in Data: An Introduction to Cluster Analysis." John Wiley & Sons, New York, 1990.

relied on the clustered correlation matrix as a reference standard for the SOM results.

Results

The training protocol produced a minimum $\bar{\varepsilon}$ consistent with good topological organization as corroborated by agglomerative cluster analysis and calculation of correlation matrices. All maps calculated via this schedule (but from different random initial distributions of weight vectors) had the same final value of $\bar{\varepsilon}$ (0.277235) and were identical up to a symmetry operation (rotation, reflection, or inversion). One such map, shown in Fig. 4, reveals that agents sharing the same broadly defined mechanism of action tended to be projected to a particular region of the output array. The tubulin-active antimitotics and topoisomerase I inhibitors mapped most cohesively; the topoisomerase II inhibitors and inhibitors of nucleotide synthesis less so, followed by the alkylating agents and inhibitors of DNA

Kohonen self-organizing map of 131 chemotherapeutic agents:
45 cell lines, gaussian neighborhood, 100 cycles initial phase,
1250 cycles refinement phase; mean ε = 0.277235

	0	1	2	3	4	5	6	7	8	9	10	11
0	2R.	RO	RO	AI	DI	3DI	DR		3A7	2AC	AC	5AC
1			A7									
2	3RF		A7		DI	A7	A7	A7		2A7		DR
3	2RF	RO		2T2	T2			2A7	2A7			DR
4	RF, RI	RO	T2				A7	3A7	A7			3DR
5	RF, RO	RO			2DP		A7	3A7				DP
6	RF, RO			AI	2A2						2T1	T1
7			DP				3T1	2T1	3T1	T1		T2
8	3TU				2T1		2T1	T1	2T1			3T2
9		TU				T1		T1	2T1			T2
10	4TU	2TU	3TU, DP	RF		3T1	2T1	4T1		2T2	T2	4T2

FIG. 4. Final state of the self-organizing map for the input set of 131 normalized response patterns. Each cell represents a unit in the competitive array of output units. Cells are labeled by the two-letter code indicating the mechanism of action (and submechanism, if available) of the agent(s) mapped to it. If a unit is activated by the agents of different mechanisms, all relevant codes are indicated. Blank cells indicate output units that are inactive in the final state of the map. (Reprinted from van Osdol *et al.*[28] with permission.)

synthesis. The coordinates of the 131 agents in the SOM output array are listed in the Appendix.

Examination of the input set by means of its clustered correlation matrix indicates several characteristics of the map, as illustrated for the tubulin-active antimitotic agents (TU) in Fig. 5. Differential response patterns that mapped to the same output unit are well correlated (>70%); colchicine and its derivatives, for example. However, even well-correlated response patterns can be distributed over more than one output unit—compare vincristine sulfate, maytansine and rhizoxin, [8, 0], to the colchicines, [10, 0]—due to the large number of output units available. The maps are also sensitive to rather low correlation: geldanamycin, an inhibitor of DNA polymerase, maps with paclitaxel and a paclitaxel derivative because it correlates best with these agents, although only at a 60% level. Thus, several factors influence the range of topological relationships captured by the map, and external measures of correlation or distance among the input vectors can help to elucidate them.

The maps have some success distinguishing among the submechanisms that comprise several of the general categories. The clearest example is provided by the DNA alkylating agents. The acyltransferase-dependent cross-linking agents (AC) and agents alkylating at position N-7 of guanidine but not dependent on the transferase (A7) are well resolved, in accord with COMPARE analysis, and the latter are partially resolved into several small subgroups. The submechanisms of the nucleotide synthesis inhibitors (RF, RO) are partially resolved, but there are some fairly strong correlations between agents having different submechanisms. The inhibitors of DNA synthesis (DI, DP, DR) tend to be as well correlated with agents having other mechanisms as they are among themselves, although the agents mapped to [2, 11], [3, 11], and [4, 11] constitute an exception. These aspects are captured by the self-organizing map. The input set is rather small, and it remains to be seen whether the resolution seen here can be retained with larger sets of agents encompassing such detailed mechanisms.

Discussion

We used Kohonen's algorithm to project vectors nonlinearly from 45-dimensional space to a discrete, two-dimensional array. The input vectors were the patterns of differential response of 45 cultured cancer cell lines to 131 agents in the National Cancer Institute drug discovery program. These agents represent six broadly defined mechanisms of action. Empirically, we developed training schedules that produced close approximations to an equilibrium map. We examined an apparently optimal map for the information it could provide about the input set.

NAME	MAP	MOA	DP	RF	TU	TU	TU	TU	TU	TU	TU	TU	TU	TU	TU	TU	TU
Geldanamycin	10,2	DP		57	62	58	41	44	53	32	40	34	62	32	44	49	41
Baker's sol antifolate	10,3	RF	38		76	72	56	38	47	40	44	42	64	50	49	56	52
Paclitaxel	10,2	TU	62	76		77	54	53	58	36	48	38	77	49	55	61	55
Paclitaxel derivative	10,2	TU	58	72	77		62	49	46	28	44	38	75	51	49	61	65
S-Trityl-L-cysteine	10,2	TU	41	56	54	62		56	54	62	55	56	69	49	50	50	44
Dolastatin	10,1	TU	44	38	53	49	56		75	48	48	50	71	43	46	51	46
Halichondrin B	9,1	TU	53	47	58	46	54	75		55	60	65	77	53	60	57	56
Vincristin sulfate	8,0	TU	32	40	36	28	62	48	55		85	82	71	76	76	72	61
Maytansine	8,0	TU	40	44	48	44	55	48	60	85		88	79	77	86	81	82
Rhizoxin	8,0	TU	34	42	38	38	56	50	65	82	88		71	68	76	70	72
Vinblastin sulfate	10,1	TU	62	64	77	75	69	71	77	71	79	71		72	79	82	78
Colchicine derivative	10,0	TU	32	50	49	51	49	43	53	76	77	68	72		88	92	87
Colchicine	10,0	TU	44	49	55	49	50	46	60	76	86	76	79	88		91	89
3-Demethyl Thiocolchicine	10,0	TU	49	56	61	61	50	51	57	72	81	70	82	92	91		94
Allocolchicine	10,0	TU	41	52	55	65	44	46	56	61	82	72	78	87	89	94	

FIG. 5. Correlation matrix for the tubulin-active antimitotics and two agents that map with them under the training schedules described in the text. The i, jth entry is the Pearson correlation (expressed as a percentage) between the differential response patterns of the compounds in the ith row and jth column. Thus, the matrix is symmetric about the main diagonal. Positions of the agents in the maps are also listed. Correlations between the antimitotics and all other compounds in the data set are less than 60% and are not shown. (Reprinted from van Osdol *et al.*[28] with permission.)

Agents having the same mechanism tend to map to the same region of the output array. Exceptions to this can be rationalized on the basis of correlation coefficients obtained from the COMPARE computer program of Paull *et al.*[7–10,29] or from the corresponding full correlation matrix. In the following paragraphs, we suggest plausible interpretations of the mapping of the compounds. These should not be regarded as statistically validated conclusions, but, rather, as clues to guide further analysis and experiment.

The topoisomerase I inhibitors and the tubulin-active antimitotic agents (TU) are mapped quite cohesively, reflecting a combination of high internal correlation (similar patterns of differential response) and low external correlation (dissimilar patterns for agents that differ in mechanism). The topoisomerase I inhibitors (T1) in the data set are all variations on the structure of camptothecin; hence, their coherence cannot, on the basis of this analysis alone, be assigned unequivocally to an action on this enzyme. The antimitotics, however, do include a wide variety of structures, suggesting that the coherence of these drugs on the map is directly related to anti-tubulin activity.[29] Note that the colchicines, which promote disaggregation of tubulin, mapped to [10, 0], and that paclitaxel and a derivative, which promote aggregation of tubulin, mapped to [10, 2].

The topoisomerase II inhibitors (T2) behave somewhat less cohesively in that several of them map with the inhibitors of nucleotide synthesis. Currently, we cannot rationalize this split on the basis of chemistry or mechanism. The inhibitors of nucleotide synthesis map rather loosely together, and several agents with other mechanisms map consistently with them. The inhibitors of DNA synthesis show a low tendency to cohere, consistent with their correlation with other mechanisms of action.

The nitrogen mustards (mechlorethamine, chlorambucil, melphalan, uramustine) and their chemical relatives, the aziridines (triethylenemelamine, thiotepa, diaziridinylbenzoquinone), occupy adjacent regions of the output layer ([2, 6], [3, 7], [3, 8], and [4, 7], [5, 7], respectively). All of these drugs react at guanine N-7 positions in the major groove of DNA and are able to cross-link the paired DNA strands.[34,35] An exception, spiromustine, [0, 8], is unusually lipophilic and may have unique pharmacological properties, but it is not clear why the other two compounds, asaley and fluorodopan, map with it. Partially overlapping the aziridines are six weakly reactive alkylating agents, chemically defined as bifunctional epoxides (dianhydrogalactitol and teroxirone) and methanesulfonates (hepsulfam and yoshi-864), ([4, 6], [4, 7], and [5, 6], [5, 7]). In DNA, the only sites sufficiently nucleophilic to react extensively with these weak alkylators

[34] K. W. Kohn, *Prog. Cancer Res. Ther.* **28,** 181 (1984).
[35] K. W. Kohn, *Anticancer Drugs* **191,** 77 (1989).

are guanine N-7 positions in the major groove of the helix. Busulfan, a bismethane sulfonate, having little or no ability to form DNA interstrand cross-links, mapped to [2, 7], one position away from the other weak alkylators. Also interspersed among these agents are three piperazine derivatives (piprobroman, piperazine alkylator, and piperazinedione) that represent variations on nitrogen mustard structure.

A second mechanistically defined subgroup of alkylating agents includes the chloroethylating agents: carmustine and lomustine, [0, 9]; semustine, [0, 10]; and chlorozotocin, clomesone, cyclodisone, mitozolamide, and PCNU [0, 11]. These compounds differ from the nitrogen mustards and aziridines in that they react strongly at guanine O-6 positions in the DNA major groove. The resulting monoadducts can form lethal interstrand cross-links unless they are rapidly removed by an alkyltransferase DNA repair enzyme that is deficient in some tumor cell types.[36,37] Carmustine, lomustine, and, to a lesser extent, semustine generate reactive isocyanates that can produce a variety of cytotoxic effects by mechanisms that do not contribute to antitumor activity. The response patterns apparently manifest these differences, and the SOM is sufficiently sensitive to reflect them as well.

A different subcategory of alkylating agent is represented by mitomycin and porfiromycin, which cross-link DNA by reacting with sites in the minor groove. These two chemically similar drugs map as a distinct subgroup at position [6, 4]. Platinum complexes constitute another category of drugs able to cross-link DNA.[38-40] These fall in at least three widely separated locations on the map: cisplatin and carboplatin, mapped to [2, 9], lie within the area occupied by most alkylating agents, as does iproplatin at [2, 5], but tetraplatin and diaminocyclohexyl-Pt(II) mapped to positions [1, 2] and [2, 2]. The latter two agents are known to retain activity against cisplatin-resistant tumor cells, apparently because of the presence of the diaminocyclohexyl ligand.[38]

The projection of the input set obtained from the SOM reflects mechanistic categories and chemical and mechanistic subcategories. It also gives intriguing mechanistic clues. Thus, the response patterns obtained from the cancer screen contain detailed information about mechanism of action, and efforts to develop a detailed understanding of the reasons for this are in

[36] K. W. Kohn, L. C. Erickson, G. Laurent *et al., in* "Nitrosoureas: Current Status and New Developments" (A. Prestayko, ed.), p. 69. Academic, New York, 1981.

[37] K. W. Kohn, *Recent Results Cancer Res.* **76,** 141 (1981).

[38] E. Reed and K. W. Kohn, *in* "Cancer Chemotherapy: Principles and Practice" (B. A. Chabner and J. M. Collins, eds.), p. 465. Lippincott, Philadelphia, 1990.

[39] M. C. Christian, *Semin Oncol.* **19,** 720 (1992).

[40] T. Hamilton, P. O'Dwyer, and R. Ozols, *Curr. Opin. Oncol.* **5,** 1010 (1993).

progress.[14,18,41] We note that the utility of the maps depends crucially on the accuracy of the activity patterns and assignment of mechanism of action. Compounds of interest are generally tested more than once, so that their activity patterns represent averages over multiple measurements. The automated screen also involves a number of quality control procedures. Nevertheless, the measurement of GI_{50} is subject to errors, which we have neither assessed quantitatively nor treated qualitatively in the present calculations. In addition, despite the apparent success of the self-organizing map with our chosen input set, it is impossible generally to project from high to low dimension without losing topological information. The loss becomes increasingly severe as the dimensionality of the input space increases and can be only partly ameliorated by using a three-dimensional output array.

We have shown that the SOM can produce results similar to those of hierachical agglomerative cluster analysis. This should not be interpreted to mean that the techniques are redundant. By varying the choice of training parameters, the SOM can be directed to weight more heavily the individual or collective aspects of the input set. In addition, both the two-dimensional representation in the output array and the weight vector representation can be used for further processing. The many hierarchical agglomerative cluster algorithms available all produce different results (because they all define distances among clusters differently), but it is not obvious how their results can be used for further processing, largely because the notion of cluster is difficult to define satisfactorily.

Several applications of the SOM can be envisioned. First, optimal map(s) for a set of selected input vectors of known mechanism and for a particular output array could be developed. A test set of differential response patterns for agents of unknown mechanism can then be projected by the SOM into the output array to indicate the relationships of the unknown to the known agents and to suggest mechanisms of action. The quantization errors would indicate the closeness of the relationship. For example, the set of 131 compounds treated here might be useful for this purpose. Second, the SOM can be used to develop exemplars, as weighted averages of weight vectors, to represent groups of highly correlated input vectors. These can subsequently be used either in the SOM or cluster analysis to classify sets of newly tested compounds according to known mechanisms of action. This is essentially a more sophisticated version of the first application suggested. Third, the SOM might serve as a preprocessing step for feed-forward back-propagation networks, to reduce the number of inputs from the dimensionality of the response patterns to the dimensionality of the output array.

[41] S. E. Bates, A. T. Fojo, J. N. Weinstein, T. G. Myers, M. Alvarez, K. D. Paull, and B. A. Chabner, *J. Cancer Res. Clin. Oncol.* **121,** 495 (1995).

Conclusion

Fundamental work continues to be done on the theory of the SOM, focusing primarily on what self-organization means, and trying to prove or disprove that the SOM always converges to an organized state. The latter has been proved rigorously in the one-dimensional case (mapping from the real line to a one-dimensional lattice), but not for the general case of projecting from high-dimensional spaces to two- or three-dimensional lattices.[32,42–45]

Mechanism of action assignments and chemical structures for the set of 122 agents used in this work (proprietary compounds excluded) can be found at the following World Wide Web site: http://epnsw1.ncifcrf.gov:2345/dis3d/cancer_screen/stdmech.html. Activity patterns for these agents can be found at http://helix.nih.gov/ncidata//canscr/stdagnt.tar.Z. Mean graph[7] representations and COMPARE lists for approximately 20,000 nonconfidential compounds can be retrieved by NSC number at the site http://epnws1.ncifcrf.gov:2345/dis3d/cancer_screen/nsc4.html. Other World Wide Web sites are given in Ref. 13.

Appendix

Below are listed the final positions of the input vectors in the SOM output array that resulted from the training schedule presented in the text.

NSC	Name	MoA	Row	Col	Error
19893	5-Fluorouracil	R.	0	0	0.32
148958	Ftorafur	R.	0	0	0.44
102816	5-Azacytidine	RO	0	1	0.39
264880	5,6-Dihydro-5-azacytidine	RO	0	2	0.46
142982	Hycanthone	AI	0	3	0.44
71851	α-2′-Deoxythioguanosine	DI	0	4	0.31
752	Thioguanine	DI	0	5	0.24
71261	β-2′-Deoxythioguanosine	DI	0	5	0.28
755	Thiopurine	DI	0	5	0.35

[42] M. Cottrell and J-C. Fort, *Annales de l'Institut Henri Poincaré* **23**(1), 1 (1987).
[43] Th. Villmann, R. Der, and T. Martinez, *in* "Proceedings of the ICANN 94, Sorrento" (M. Marinaro and P. G. Morasso, eds.), p. 298. Springer-Verlag, Berlin, 1994.
[44] J. A. Flanagan, *Neural Networks* **9,** 1185 (1996).
[45] J.-C. Fort and G. Pagès, *Neural Networks* **9,** 773 (1996).

NSC	Name	MoA	Row	Col	Error
118994	Inosine glycodialdehyde	DR	0	6	0.45
167780	Asaley	A7	0	8	0.3
73754	Fluorodopan	A7	0	8	0.31
172112	Spiromustine	A7	0	8	0.39
409962	Carmustine	AC	0	9	0.24
79037	Lomustine	AC	0	9	0.25
95441	Semustine	AC	0	10	0.31
353451	Mitozolamide	AC	0	11	0.22
338947	Clomesone	AC	0	11	0.26
178248	Chlorzotocin	AC	0	11	0.27
95466	*N*-(2-Chloroethyl)-*N*′-2,6-dioxo-3-piperidinyl-*N*-nitrosourea	AC	0	11	0.28
348948	Cyclodisone	AC	0	11	0.43
363812	Tetraplatin	A7	1	2	0.3
174121	Methotrexate derivative	RF	2	0	0.26
184692	Amiopterin derivative	RF	2	0	0.26
633713	*N*-8-(4-Amino-4-deoxypteroyl)-*N*δ-*hemiphthaloyl*-L-*ornithine*	*RF*	*2*	*0*	*0.36*
271674	*Diaminocyclohexyl-Pt(II)*	*A7*	*2*	*2*	*0.3*
127716	*5-Aza-2′-deoxycytidine*	*DI*	*2*	*4*	*0.42*
256927	*Iproplatin*	*A7*	*2*	*5*	*0.38*
762	*Mechlorethamine*	*A7*	*2*	*6*	*0.28*
750	*Busulfan*	*A7*	*2*	*7*	*0.4*
119875	*Cisplatin*	*A7*	*2*	*9*	*0.27*
241240	*Carboplatin*	*A7*	*2*	*9*	*0.34*
107392	*5-Hydroxypicolinaldehyde thiosemicarbazone*	*DR*	*2*	*11*	*0.35*
740	*Methotrexate*	*RF*	*3*	*0*	*0.26*
352122	*Trimetrexate*	*RF*	*3*	*0*	*0.27*
126771	*Dichloroallyl lawsone*	*RO*	*3*	*1*	*0.35*
268242	*N,N*-Dibenzyldaunorubicin	T2	3	3	0.25
354646	Morpholinodoxorubicin	T2	3	3	0.41
8806	Melphalan	A7	3	7	0.17
25154	Pipobroman	A7	3	7	0.19
34462	Uramustine	A7	3	8	0.13
3088	Chlorambucil	A7	3	8	0.15
95678	3-Hydroxypicolinaldehyde thiosemicarbazone	DR	3	11	0.34
163501	Acivicin	RI	4	0	0.32
134033	Aminopterin derivative	RF	4	0	0.38
368390	Brequinar	RO	4	1	0.25
366140	Pyrazoloacridine	T2	4	2	0.39
102627	Yoshi-864	A7	4	6	0.17
6396	Thiotepa	A7	4	7	0.11
329680	Hepsulfan	A7	4	7	0.2
296934	Teroxirone	A7	4	7	0.21

NSC	Name	MoA	Row	Col	Error
344007	Piperazine alkylator	A7	4	8	0.16
51143	Pyrazoloimidazole	DR	4	11	0.22
1895	Guanazole	DR	4	11	0.27
32065	Hydroxyurea	DR	4	11	0.27
132483	Aminopterin	RF	5	0	0.34
153353	L-Alanosine	RO	5	0	0.41
143095	Pyrazofuran	RO	5	1	0.36
63878	Cytarabine	DP	5	4	0.24
145668	Cyclocytidine	DP	5	4	0.29
135758	Piperazinedione	A7	5	6	0.16
9706	Triethylenemelamine	A7	5	7	0.11
182986	Diaziridinylbenzoquinone	A7	5	7	0.25
132313	Dianhydrogalactitol	A7	5	7	0.27
303812	Aphidicolin glycinate	DP	5	11	0.22
224131	PALA	RO	6	0	0.36
623017	*N*-8-(4-Amino-4-deoxypteroyl)-*N*γ-*hemiphthaloyl*-L-ornithine	RF	6	0	0.44
357704	Cyanomorpholinodoxorubicin	Al	6	3	0.34
26980	Mitomycin C	A2	6	4	0.24
56410	Porfiromycin	A2	6	4	0.26
606985	Camptothecin derivative	T1	6	10	0.23
610458	Camptothecin derivative	T1	6	10	0.28
618939	Camptothecin derivative	T1	6	11	0.27
27640	2′-Deoxy-5-fluorouridine	DP	7	2	0.3
610456	Camptothecin derivative	T1	7	6	0.13
proprietary	Camptothecin derivative	T1	7	6	0.19
proprietary	Camptothecin derivative	T1	7	6	0.24
proprietary	Camptothecin derivative	T1	7	7	0.16
proprietary	Camptothecin derivative	T1	7	7	0.17
proprietary	Camptothecin derivative	T1	7	8	0.13
606173	Camptothecin derivative	T1	7	8	0.23
606172	Camptothecin derivative	T1	7	8	0.27
643833	Camptothecin derivative	T1	7	9	0.26
249992	Amsacrine	T2	7	11	0.19
153858	Maytansine	TU	8	0	0.23
332598	Rhizoxin	TU	8	0	0.26
67574	Vincristin sulfate	TU	8	0	0.3
proprietary	Camptothecin derivative	T1	8	4	0.22
propietary	Camptothecin derivative	T1	8	4	0.26
95382	Camptothecin derivative	T1	8	6	0.21
364830	Camptothecin derivative	T1	8	6	0.22
610459	Camptothecin derivative	T1	8	7	0.17
295500	Camptothecin derivative	T1	8	8	0.16
374028	Camptothecin derivative	T1	8	8	0.33
122819	Teniposide	T2	8	11	0.18
141540	Etoposide	T2	8	11	0.21

NSC	Name	MoA	Row	Col	Error
269148	Menogaril	T2	8	11	0.21
609395	Halichondron	TU	9	1	0.39
proprietary	Camptothecin derivative	T1	9	5	0.22
610457	Camptothecin derivative	T1	9	7	0.2
606497	Camptothecin derivative	T1	9	8	0.16
606499	Camptothecin derivative	T1	9	8	0.19
267469	Deoxydoxorubicin	T2	9	11	0.24
361792	Thiocolchicine	TU	10	0	0.22
757	Colchicine	TU	10	0	0.26
406042	Allocolchicine	TU	10	0	0.27
33410	Colchicine derivative	TU	10	0	0.3
49842	Vinblastin sulfate	TU	10	1	0.22
376128	Dolastatin 10	TU	10	1	0.47
125973	Paclitaxel	TU	10	2	0.35
608832	Paclitaxel derivative	TU	10	2	0.36
83265	Tritylcysteine	TU	10	2	0.45
330500	Geldanamycin	DP	10	2	0.47
139105	Baker's soluble antifolate	RF	10	3	0.32
proprietary	Camptothecin derivative	T1	10	5	0.39
100880	Camptothecin Na salt	T1	10	5	0.39
603071	Camptothecin derivative	T1	10	5	0.52
94600	Camptothecin	T1	10	6	0.21
107124	Camptothecin derivative	T1	10	6	0.33
249910	Camptothecin derivative	T1	10	7	0.17
176323	Camptothecin derivative	T1	10	7	0.18
629971	Camptothecin derivative	T1	10	7	0.2
295501	Camptothecin derivative	T1	10	7	0.23
301739	Mitoxantrone	T2	10	9	0.22
355644	Anthrapyrazole derivative	T2	10	9	0.24
349174	Oxanthrazole	T2	10	10	0.22
82151	Daunorubicin	T2	10	11	0.18
164011	Rubidazone	T2	10	11	0.2
123127	Doxorubicin	T2	10	11	0.2
337766	Bisantrene	T2	10	11	0.44

Acknowledgments

The authors would like to thank the members of the SOM Programming Team of the Laboratory of Computer and Information Science, Helsinki University of Technology, for providing the software and technical support necessary to perform these calculations. We would also like to acknowledge Dr. Kenneth D. Paull, who initiated efforts at analysis of the NCI cell screen when he developed the COMPARE program package. He died in January 1998, a great loss to us personally and professionally.

[21] Dynamic Network Model of Glucose Counterregulation in Subjects with Insulin-Requiring Diabetes

By BORIS P. KOVATCHEV, MARTIN STRAUME, LEON S. FARHY, and DANIEL J. COX

Introduction

Recent studies demonstrate that the most effective treatment for insulin-dependent diabetes mellitus (IDDM) is the strict control of blood glucose (BG) levels approximating a normal range. Chronic high BG levels are proven to cause many complications in multiple body systems over time. The only established alternative for patients with IDDM is intensive therapy that can prevent or delay the development of long-term diabetes complications. The latest results are described by the Diabetes Control and Complications Trial[1] and its European counterpart.[2]

Because patients with IDDM are unable to produce insulin, the standard daily regimen involves multiple insulin injections that lower BG, thus balancing the elevation of BG caused by digestion of carbohydrates. In general, the functional consequences of insulin injection, food intake, and physical activity are well known to depress, elevate, and depress, respectively, BG levels. As such, they historically have been implemented as primary factors in treatment protocols designed to maintain good glycemic control. However, multiple insulin injections required by intensive therapy increase patient's risk of hypoglycemia. Too much insulin can even result in severe hypoglycemia (SH), defined as a low BG resulting in stupor, seizure, or unconsciousness that precludes self-treatment.[3] Although most SH episodes are not fatal, there remain numerous negative sequelae leading to compromised occupational and scholastic functioning, social embarrassment, poor judgment, serious accidents, and possible permanent cognitive dysfunction.[4–6] The more distressing the SH episode, the greater the psychological

[1] Diabetes Control and Complications Trial (DCCT) Study Group, *N. Engl. J. Med.* **329,** 977 (1993).

[2] P. Reichard and M. Phil, *Diabetes* **43,** 313 (1994).

[3] The Diabetes Control and Complications Trial Research Group, *Diabetes* **46,** 271 (1997).

[4] A. E. Gold, I. J. Deary, B. M. Frier, *Diabetic Med.* **10,** 503 (1993).

[5] I. J. Deary, J. R. Crawford, D. A. Hepburn, S. J. Langan, L. M. Blackmore, and B. M. Frier, *Diabetes* **42,** 341 (1993).

0076-6879/00 $30.00

fear of hypoglycemia,[7] and this fear can lead to avoidance behaviors that result in poor metabolic control.[8] In other words, the threat and fear of SH can significantly discourage patients and health care providers from pursuing intensive therapy, thus increasing their vulnerability to long-term diabetes complications. Consequently, hypoglycemia has been identified as the major barrier to improved metabolic control.[9]

In addition to these external interventions, in most individuals with IDDM, low BG levels trigger the release of counterregulatory hormones. This in turn prompts the release of stored glucose into the bloodstream to restore euglycemia. However, in IDDM subjects, the ability to counterregulate is frequently impaired by factors such as long-standing diabetes, autonomic neuropathy, and intensive therapy.[10–12] Insufficient or absent counterregulatory responses allow BG to fall further until SH occurs. Consequently, the risk for severe hypoglycemia in IDDM has been attributed not only to relative insulin excess, but also to impaired glucose counterregulation.[10,13,14]

In summary, the struggle for tight metabolic control is influenced by many external and internal interacting factors that cause, often dramatic, blood glucose fluctuations over time. Glucose metabolism has been studied with isotopic tracer methods in animal and human studies[15,16] and described by multicompartment models.[17,18] Investigation of counterregulation through insulin infusion has typically been pursued in two ways: (1) during the induction of hypoglycemia, various hormones are sampled to determine whether their levels increase as BG falls, (2) whether or not BG spontane-

---

[6] N. B. Lincoln, R. M. Faleiro, C. Kelly, B. A. Kirk, and W. J. Jeffcoate, *Diabetes Care* **19,** 656 (1996).

[7] A. A. Irvine, D. J. Cox, and L. A. Gonder-Frederick, *Health Psych.* **11,** 135 (1992).

[8] A. A. Irvine, D. J. Cox, and L. A. Gonder-Frederick, *Diabetes Care* **14,** 76 (1991).

[9] P. E. Cryer, J. N. Fisher, and H. Shamoon, *Diabetes Care* **17,** 734 (1994).

[10] P. E. Cryer and J. E. Gerich, *N. Engl. J. Med.* **313,** 232 (1985).

[11] S. A. Amiel, R. S. Sherwin, D. C. Simonson, and W. V. Tamborlane, *Diabetes* **37,** 901 (1988).

[12] S. A. Amiel, W. V. Tamborlane, D. C. Simonson, and R. S. Sherwin, *N. Engl. J. Med.* **316,** 1376 (1987).

[13] N. H. White, D. A. Skor, P. E. Cryer, D. M. Bier, L. Levandoski, and J. V. Santiago, *N. Engl. J. Med.* **308,** 485 (1983).

[14] J. E. Gerich, *Diabetes* **37,** 1608 (1988).

[15] G. Hetenyi, R. Ninomiya, and G. A. Wrenshall, *J. Nucl. Med.* **7,** 454 (1966).

[16] D. M. Brier, K. J. Arnold, W. R. Sherman, W. H. Holland, W. F. Holmes, and D. M. Kipnis, *Diabetes* **26,** 1005 (1977).

[17] P. A. Insel, J. E. Liljenquist, J. D. Tobin, R. S. Sherwin, P. Watkins, R. Andres, and M. Berman, *J. Clinic. Invest.* **55,** 1057 (1975).

[18] R. Steele, H. Rostami, and N. Altszuler, *Federation Proc.* **33,** 1869 (1974).

ously rises or plateaus despite the continual infusion of regular insulin.[19] While these basic approaches to describing glucose counterregulation include quantifying plasma hormonal concentrations, they have not described the dynamic interaction of insulin and glucose nor yielded a precise mathematical model of the sequential components of the glucose counterregulation.

In this manuscript we present a dynamical network model of insulin–glucose–physical activity–counterregulation interactions. A subset of this model is then applied to describe the dynamics of this process during a euglycemic hyperinsulinemic clamp and subsequent reduction of BG to hypoglycemic levels. Using this model, we approximated the temporal pattern of each subject's BG fluctuations and evaluated parameters of insulin and glucose sensitivity. On that basis we computed dynamic estimates for the time of onset, rate, and sequential components of these subject's counterregulatory responses. These results were verified by correlating for each subject the model-estimated counterregulatory dynamics with subsequently analyzed plasma epinephrine concentrations.

Insulin–Glucose Counterregulation Network

Figure 1 presents the general scheme of BG dynamics as a function of three principal temporal behavioral variables known to exert regulatory control on BG level, (PB1) insulin injection, (PB2) food intake, and (PB3) physical activity, as well as one principal temporal physiological variable, CR, a counterregulatory process through which the available liver stores replenish low BG levels. The latter cannot be measured directly, but rather will be estimated by the model, given its interactions. Thus, the system is defined in terms of five time-dependent state variables: (1) BG level, (2) insulin injections, (3) food intake, (4) physical activity, and (5) liver stores for replenishing low BG levels. A network of processes provides the functional regulatory interactions responsible for BG control. BG levels will be positively affected by food intake (process PB2), as well as by the potential for replenishment when BG is low from available liver stores (process CR). The negative effect on BG level by insulin injections is denoted by process PB1, whereas the reduction in BG brought about by physical activity is referred to by process PB3. The counterregulatory loop between liver stores and BG is implemented by way of (1) inhibition of process CR by elevated BG levels (process Reg) (i.e., a release from inhibition of process CR below some threshold BG level, thus providing

[19] G. B. Bolli, P. DeFeo, S. DeCosmo, G. Perriello, M. M. Ventura, M. Massi Benedetti, F. Sateusario, H. E. Gerich, and P. Brunetti, *Diabetes* **33,** 732 (1984).

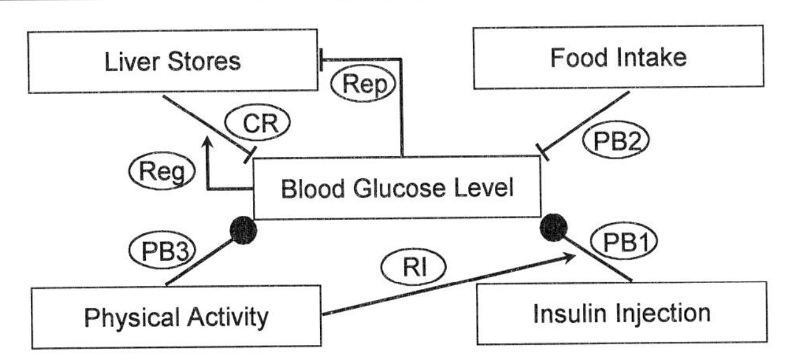

FIG. 1. Insulin-glucose counterregulation network. System state variables are enclosed in boxes. System processes are denoted by bold lines terminating in either "T" or "ball" endings, and are referred to with oval-enclosed labels. Processes terminating in "T" endings are positively affecting their target variable or process, whereas those terminating in "ball" endings are negatively affecting their target variable or process. Regulatory processes have arrow endings. PB1, principal behavioral process 1, model process by which injected insulin reduces BG; PB2, principal behavioral process 2, model process by which consumed food elevates BG; PB3, principal behavioral process 3, model process by which physical activity reduces BG; CR, counterregulatory process, model process by which liver stores elevate BG; Reg, model process by which BG levels inhibit process CR (i.e., when BG falls below the counterregulatory threshold, process CR is disinhibited thus permitting liver stores to promote recovery from low BG levels); Rep, replenishment process, model process by which BG levels act to replenish liver stores; RI, regulatory interaction, model process by which a history of physical activity acts to promote process PB1 (i.e., a history of physical activity enhances insulin's effectiveness at reducing BG levels).

for counterregulatory recovery from low BG levels by recruitment of available liver stores), and (2) replenishment of liver stores under conditions during which BG levels are not low (by process Rep). One additional regulatory interaction (process RI) makes reference to the physical activity-based enhancement of insulin's effectiveness in down-regulating BG via PB1.

A simplified version of this dynamic network emerges during controlled hyperinsulinemic clamp. During such a procedure (described in detail below) the influence of the subjects' physical activity is negligible (processes PB3 and RI) and the replenishment process Rep can be ignored since there is no time for liver reserves to be restored. Thus, neither of these parameters was assumed to influence BG during the study. Instead of insulin injections (PB1) there is a continuous insulin infusion at a constant rate and direct variable dextrose infusion replaces food intake (PB2). Thus, the insulin infusion (process I) replaces the behavioral process PB1 while the dextrose infusion (process D) replaces the behavioral process PB2. As described above, the counterregulation loop (CR-Reg) is anticipated only at lower BG levels during the study. The resulting simplified network of functional interactions is shown in Fig. 2.

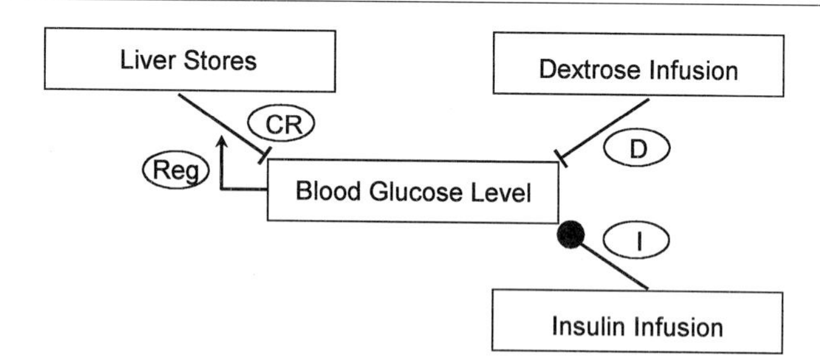

FIG. 2. Network of functional interactions during hyperinsulinemic clamp. BG levels are positively affected by the dextrose infusion (D) and by the potential for replenishment when BG is low (CR), and negatively affected by the insulin infusion (I). A regulatory loop (Reg) between liver stores and BG inhibits the process CR at elevated BG levels.

Thus, during hyperinsulinemic clamp the system was defined in terms of three time-dependent state variables: (1) BG level, (2) insulin infusion, and (3) dextrose infusion. A network of processes provides the functional regulatory interactions responsible for BG control: BG level was positively affected by the dextrose infusion (process D), as well as by the potential for replenishment when BG is low from available liver stores (process CR). The negative effect on BG level by the insulin infusion is denoted by process I. A regulatory loop (Reg) between liver stores and BG is implemented by way of inhibition of process CR by elevated BG levels (i.e., a release from inhibition of process CR below some threshold BG level, thus providing for counterregulatory recovery from low BG levels by recruitment of available liver stores). Time rates of change of BG (i.e., $d[\mathrm{BG}]/dt$) were described by a nonlinear ordinary differential equation. The counterregulatory response was modeled as a release of glucose from a multicompartment storage pool. It was assumed that during euglycemia no counterregulatory response occurred, while during descent into hypoglycemia, the model expected onset of counterregulation.

Experimental Design and Methods

This model was applied to data collected for 40 subjects during standardized hyperinsulinemic clamping. The subjects were recruited through newsletters, notices posted in diabetes clinics, and direct physician referral. All subjects had to have IDDM for at least 2 years and have taken insulin since the time of diagnosis. There were 16 males and 24 females, with mean age 35.5 years (SD = 8.1), mean duration of disease 16.9 years (SD = 9.6), mean inulin units/day 42.0 (SD = 15.5), and mean glycosylated hemoglobin

8.5% (SD = 1.7). The normal range for the glycosylated hemoglobin assay was 4.4 to 6.9%.

The procedure begins with orientation meetings where the subjects signed consent forms. To ensure that subjects' BGs are not in a low range (<3.9 m*M*) for 72 hr prior to the study, their insulin dose is reduced by 10% and long-acting insulin is discontinued 36 hr prior to the study. Subjects are instructed to eat prophylactically 10 g of glucose whenever BG is ≤5.6 m*M* and are required to test their BG five times a day (1 hr before each meal, at bedtime, and 4 hr into their sleep). If low BG occurs, the study is rescheduled. Subjects are admitted to the University of Virginia General Clinical Research Center. On admission, subjects are given a physical exam, including an assessment for autonomic neuropathy. BG is maintained overnight between 5.6 and 8.3 m*M* with intravenous regular human insulin as per a previously published insulin infusion protocol.[19] Subjects are given dinner and a bedtime snack the evening before the study, but remain fasting on the morning of the study. No caffeinated beverages are consumed after hospital admission. On the morning of the study, IVs are placed in the nondominant forearm. Insulin is continuously infused at a constant rate of 1.0 mU/kg/min and a 20% dextrose solution is infused at a variable rate to maintain BG at 5.6 m*M*. During phase 1 (euglycemia) BG is maintained between 5.6 and 8.3 m*M*. During phase 2 (BG reduction), BG is steadily lowered to a target level of 2.2 m*M* by varying the dextrose infusion. Adjustments in dextrose infusion are made every 5 min. Arterialized blood (achieved by warming the hand in a heated glove to 50°) is sampled for glucose concentration every 5 min and for plasma epinephrine concentration every 10 min.

Algorithm and Computation

Phase 1: Maintained Euglycemia

It was assumed that (1) the dextrose infusion influenced BG positively through an unknown dextrose conversion parameter a, and (2) BG decay rate was inversely proportional to the BG level, through an unknown insulin utilization parameter b. This led to the following nonlinear differential equation for the BG time rate of change:

$$\frac{d\mathrm{BG}(t)}{dt} = aD(t) - b\frac{\mathrm{BG}(t)}{\varepsilon + \mathrm{BG}(t)^2}I \tag{1}$$

where $D(t)$ is the variable dextrose infusion rate (mg/kg/min) and I is the constant insulin infusion. The inverse proportion $\mathrm{BG}(t)/[\varepsilon + \mathrm{BG}(t)^2]$, where

ε is a small constant, was used instead of standard I/BG(t) for better computational stability.

Phase 2: BG Reduction

During the second phase of the test counterregulation was anticipated and Eq. (1) was expanded by an additional term:

$$\frac{d\mathrm{BG}(t)}{dt} = aD(t) - b\frac{\mathrm{BG}(t)}{\varepsilon + \mathrm{BG}(t)^2} I + \mathrm{CR}(t) \tag{2}$$

We allowed the counterregulation term CR(t) to be a uni- or bimodal function, corresponding to one- or two-compartment modeling. We would not impose a specific analytical form on the counterregulation function. In general, it needs to be a positive pulsatile function that, depending on the subject's data, has one, two, or possibly more additive components: $\mathrm{CR}(t) = \mathrm{CR}_1(t)$, or $\mathrm{CR}(t) = \mathrm{CR}_1(t) + \mathrm{CR}_2(t)$. For this particular application in order to be able to approximate our data, we suggest each counterregulation component be defined by

$$\mathrm{CR}_i(t) = \begin{cases} 0 & \text{when } t < T_i \\ C_i r_i^2 (t - T_i) \exp[-r_i(t - T_i)] & \text{when } t \geq T_i, i = 1, 2 \end{cases}$$

This way the function $\mathrm{CR}_1(t)$, would be zero when $t < T_1$, and would increase at time T_1. Thus, the parameter T_1 would be interpreted as time of onset of counterregulation, while the product $C_1 r_1^2$ would be the counterregulation slope at onset (the derivative at time $t = T_1$). The same would be valid for the second component $\mathrm{CR}_2(t)$. The analytical form of CR(t) was carefully selected to allow for this physiologic interpretation of its parameters; however, it is not restricted and other functions that meet certain mathematical requirements would be suitable descriptors of counterregulatory responses.

Parameter Estimation

The input data for the model were each subject's dextrose infusion records and corresponding BG levels, each of which were recorded every 5 min. An automated algorithm for analysis of these data was developed as follows: Prior to identification by the algorithm of onset of counterregulation, each subject's parameters a and b were estimated, along with a maximum-likelihood estimate of their initial BG level. This was accomplished by a modified Gauss–Newton nonlinear least squares parameter estimation algorithm[20,21] in which differential Eq. (1) was integrated numerically for BG(t) by a fourth-order Runge–Kutta method.[22] Applied to each subjects' data set, this procedure successfully evaluated individually for each subject these characteristics of the dynamics of dextrose utilization and BG elimination during euglycemia and descent into hypoglycemia prior to onset of counterregulation.

The algorithm was initialized to consider first only those time points in which there was clearly no potential for counterregulation (i.e., those points comprising the euglycemic phase 1 of the experiment). The parameters a and b and the initial BG level were nonlinear least squares estimated to this subset of data, followed by evaluation of the standard deviation of fit to the BG data. At this point, the difference between the observed BG level for the next time point and the BG level predicted by the model in the absence of counterregulation was evaluated. If this difference was less than two standard deviations of fit, this next observed BG level was considered to be prior to onset of counterregulation and was included as an additional point for estimation of parameters a and b and an initial BG level by Eq. (1). This process was repeated iteratively until a BG level was identified that differed from the predicted value by more than two standard deviations of fit, indicating onset of counterregulation. From this point onward, parameters a and b and the estimated initial BG level were fixed, onset of counterregulation (parameter T_1) was defined as the time of the previous time point, and the model began fitting to differential Eq. (2). The process again proceeded iteratively one point at a time until either the remainder of the data set was successfully considered or until the need for a second component of the counterregulatory response was identified (in the same manner as above).

Results

Goodness-of-Fit of Model

The algorithm was applied to the data sets of each of the 40 subjects. The accuracy of the data fit was tested by the coefficient of determination, usually interpreted as the percentage of the total variation of the dependent

[20] M. L. Johnson, S. G. Frasier, *Methods Enzymol* **117,** 301 (1985).

[21] M. Straume, S. G. Frasier-Cadoret, and M. L. Johnson, *in* "Topics in Fluorescence Spectroscopy, Vol. 2: Principles" (J. R. Lakiwicz, ed.), pp. 177–241. Plenum, New York, 1991.

[22] W. H. Press, B. P. Flannery, S. A. Teukolsky, and W. T. Vetterling, "Numerical Recipes: The Art of Scientific Computing (Fortran Version)," pp. 551–552. Cambridge University Press, Cambridge, Massachusetts, 1989.

TABLE I
COEFFICIENTS OF DETERMINATION OF NONLINEAR MODEL FIT

Coefficient of determination	Number of subjects	% of subjects
Above 99%	7	17.5
95–99%	24	60
90–95%	7	17.5
86–90%	2	5

variable around its mean that is explained by the fitted model.[23] Since the model is nonlinear and standard ANOVA p value cannot be computed, the goodness-of-fit of each model was evaluated by the closeness of its coefficient of determination to 100%. (Note that our model is intrinsically nonlinear, and therefore the usual F-statistic and its significance level cannot be used.) The average coefficient of determination across subjects was 96.6% (SD = 3.1), with a maximum of 99.7% and a minimum of 86.5%. This indicates an extremely good model fit for all subjects. Seven subjects had coefficients of determination above 99%, while only two subjects had coefficients of determination below 90% (see Table I).

Counterregulation

The rate of counterregulation of each subject was estimated in units equivalent to mg/kg/min dextrose infusion, on the basis of the difference between Eqs. (2) and (1), as explained above (Parameter Estimation section). Our model indicated that 3 subjects did not counterregulate, 3 subjects had a counterregulatory response that was not statistically significant (i.e., not different from zero at p level 0.05), and 34 subjects had significant counterregulation, $p < 0.05$. For 11 of these 34 subjects a unimodal counterregulatory function was sufficient for the data fit. For the remaining 23 subjects a bimodal function was needed.

Figures 3, 4, and 5 present data and model fits for selected subjects with bimodal, unimodal, and nonsignificant counterregulation, respectively. The x axis of each figure is the elapsed time in minutes. The left y axis [BG (mM)] refers to three variables: (1) the BG data plotted by circles; (2) the full model fit, based on Eq. (2) (thick black line); and (3) the model-predicted decay of BG if counterregulation did not occur, based on Eq. (1) (thin black line). Two other variables are plotted along the right y-axis [dextrose infusion/counterregulation (mg/kg/min)]: (1) the dextrose

[23] T. O. Kvalseth, *Am. Statistic.* **39,** 279 (1985).

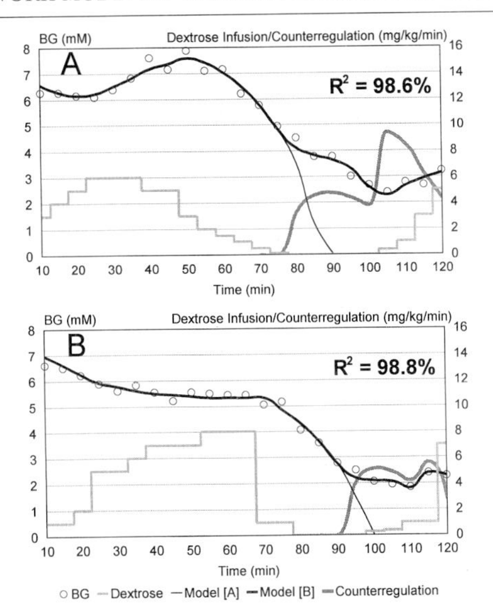

FIG. 3. Bimodal counterregulation. (A) Model indicates an onset of counterregulation at minute 77 (BG level of 4.3 m*M*) and an acceleration of the counterregulatory response at minute 105. (B) Counterregulation onset is at minute 90 (BG level of 2.6 m*M*), followed by a second component at minute 110. The coefficients of determination of the model fit are 98.6% and 98.8%, respectively.

infusion rate, represented by a stepwise gray line, is a constant within each 5-minute interval and is adjusted every 5 min. (2) The counterregulatory rate in units equivalent to mg/kg/min dextrose infusion is plotted by a thick gray curve. All figures include the coefficients of determination R^2 of the nonlinear model fit.

Figures 3A and 3B present subjects who demonstrated bimodal counterregulation. In Fig. 3A, dextrose infusion decreased after minute 40 and was discontinued at minute 75, thus the BG level was expected to fall. However, at minute 80 (BG = 4.3 m*M*,) the measured BG was higher than the expected BG, indicating an onset of counterregulation a few minutes before that, which reduced the BG decline until minute 100. At minute 105 the BG level started rising from its nadir value of 2.3 m*M* indicating an acceleration of the counterregulatory response, since it occurred *before* the renewed dextrose infusion. This prompted the algorithm to switch to a bimodal counterregulatory mode, which is depicted by the higher second part of the counterregulation curve. Figure 3B presents a similar bimodal counterregulation, but its onset at minute 90 was at a lower BG level of 2.6 m*M*. The second peak of the counterregulation rate was not as well defined and the total area under the curve was smaller.

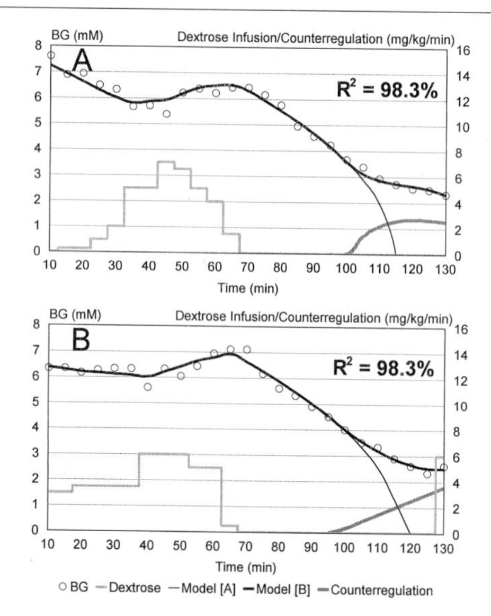

FIG. 4. Unimodal counterregulation. (A) Model indicates an onset of counterregulation at minute 100 (BG level of 3.3 m*M*) that plateaus 10 min later. (B) Counterregulation onset is at minute 100 (BG level of 4 m*M*) and the counterregulation rate steadily increases over the next 35 min. Both coefficients of determination are 98.3%.

Figures 4A and 4B present subjects whose data were fitted by a unimodal model of counterregulation. In Fig. 4A the onset of counterregulation was at a BG level of 3.3 m*M* and after 10 min the counterregulation rate plateaus. In Fig. 4B the onset was at a BG level of 4 m*M* and the counterregulation rate steadily increased over the following 30 min.

Figure 5A presents data for a subject whose BG nadir was 2.8 m*M*, but who did not counterregulate. This was confirmed by this subject's subsequent plasma epinephrine analyses, which showed no increase (see also Table II). Figure 5B presents a subject whose counterregulation was not statistically different from zero at $p = 0.05$, although his BG nadir was below 3 m*M*. The model indicated a minor response at minute 122 when BG was 3.5 m*M*.

External Verification of Model

The predicted counterregulation rate was correlated with the corresponding epinephrine data for each subject. To illustrate these correlations, Table II presents epinephrine concentrations, recorded every 10 min, and model-evaluated counterregulation rates at corresponding 10-min intervals

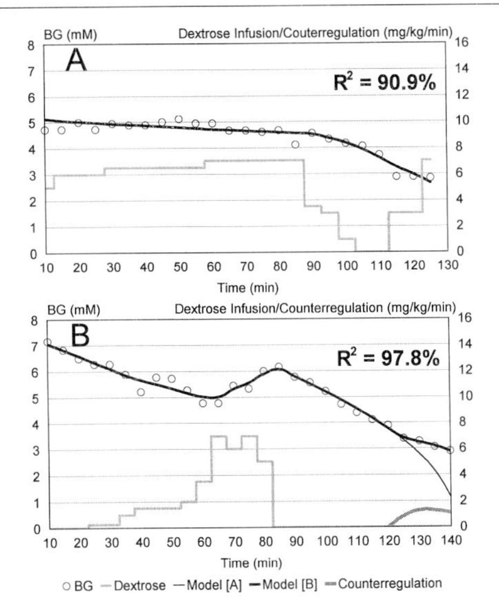

FIG. 5. Nonsignificant counterregulation. (A) No counterregulatory response was observed despite a BG nadir of 2.8 m*M*. (B) Statistically insignificant counterregulation—a minor response at minute 122 (BG = 3.5 m*M*) that declined 15 min later. The coefficients of determination of the model fit are 90.9% and 97.8%, respectively.

for the subjects' data shown in Figs. 3A, 4A, and 5A. The correlation coefficients and their p levels are listed at the bottom of each column.

In summary, for all but the six subjects who did not have significant counterregulation responses, the correlations were above 0.5 and statistically significant at $p = 0.05$. These results are summarized in Table III.

Discussion

The deterministic differential equation model developed in the present study accounts, in a highly reliable manner, for the dynamics of both exogenous dextrose infusion rate-dependent changes in blood glucose levels as well as endogenous physiologic counterregulation during euglycemic hyperinsulinemic clamping of patients with IDDM. During euglycemia in the absence of counterregulatory response, the model is mechanistically parameterized, individually and separately for each patient considered, in terms of two physiologic processes: (1) the dextrose-to-blood glucose conversion efficiency and (2) the insulin utilization efficiency for elimination of blood glucose. The model was implemented during analysis to use an objective criterion for identifying the time of onset of physiologic counterregulation

TABLE II

MODEL-PREDICTED COUNTERREGULATION RATES[a] AND CORRESPONDING PLASMA CONCENTRATION[b]

	Subject 3A		Subject 4A		Subject 5A	
Time (min)	Plasma epinephrine	Counter-regulation	Plasma epinephrine	Counter-regulation	Plasma epinephrine	Counter-regulation
10	51	0	34	0		0
20	38	0	41	0	104	0
30	47	0	38	0	90	0
40	56	0	22	0	100	0
50	47	0	62	0	125	0
60	39	0	34	0	96	0
70	53	0	25	0	92	0
80	26	3.3	31	0	85	0
90	29	4.7	34	0	93	0
100	167	3.8	32	0.01	99	0
110	289	9.4	91	2.2	98	0
120	266	8.5	211	2.7	112	0
130			504	2.5		
Correlation	$r = 0.85$		$r = 0.78$		n/a	
P level	$P < 0.0005$		$p < 0.001$		n/a	

[a] In units equivalent to mg/kg/min dextrose infusion.
[b] pg/ml.

at low blood glucose levels. Counterregulation was then parameterized, again, individually and separately for each patient considered, in terms of the rate and volume of counterregulatory response, as well as for the potential for bimodal counterregulation.

The objective analysis performed by this implementation of the model was successful in all 40 of the blood glucose–dextrose infusion data sets of

TABLE III

CORRELATION BETWEEN MODEL-PREDICTED COUNTERREGULATION RATES AND SUBSEQUENTLY ANALYZED PLASMA EPINEPHRINE CONCENTRATIONS[a]

Correlation coefficient	Significance	Number (%) of subjects
$r > 0.9$	$p < 0.0005$	12 (30%)
r between 0.8 and 0.9	$p < 0.001$	12 (30%)
r between 0.5 and 0.8	$p < 0.05$	9 (22.5%)
No counterregulation and no epinephrine response		6 (15%)

[a] No epinephrine data were available for one subject.

IDDM patients considered in this study, typically accounting for greater than 95% of the observed variance in blood glucose time series. Additionally, the counterregulatory responses predicted by the model are consistent with observed plasma epinephrine concentrations, as indicated by the high correlation between the two. Interestingly, the modeling results suggest the previously unrecognized possibility that blood glucose counterregulation may be a multicomponent process (because a bimodal counterregulatory response was predicted for 23 of the 40 individuals examined).

Whereas insulin infusion in the present study was constant throughout (at 1.0 mU/kg/min), the mathematical formulation of blood glucose elimination kinetics is such that variable insulin infusion rates, or effects of changes in insulin action over time (as, for example, in the case of bolus infusions of insulin), can be accommodated by imposing a time dependence to the insulin-effect term [i.e., $I \rightarrow I(t)$]. Implementing a time-dependent insulin-effect term may require using standard insulin action curves, but the ability to readily do so points out the potential widespread utility of this model for describing the dynamics of blood glucose levels in a variety of clinical or experimental settings.

The demonstrated ability of the present model to highly reliably describe the dynamics of blood glucose changes in both euglycemia and during descent into hypoglycemia with subsequent initiation of (a sequence of) counterregulatory response(s) will promote development of a more detailed understanding of the mechanisms underlying the physiology of blood glucose regulation, in general, and hypoglycemia, in particular, in IDDM.

In purely practical clinical terms, the mathematical evaluation of subjects' counterregulatory responses in interpretable units, equivalent to mg/kg/min dextrose infusion, could be of a great advantage to various clamp studies. This utility is enhanced by the fact that it is extremely difficult to directly measure epinephrine concentrations. This assay is available in only a few laboratories across the United States and is very expensive.

Theoretically, the ability to objectively identify individual patient's blood glucose levels at the onset of counterregulation will permit comparisons and, potentially, identification and discrimination of patients more-versus-less vulnerable to severe hypoglycemia. Additionally, the ability to quantitatively assess individual patient's insulin utilization efficiency for elimination of blood glucose, as well as their counterregulatory response (in terms of both rate and response volume, as well as multimodality), should shed light not only on *which* subjects may be poor candidates for intensive insulin therapy because of vulnerability to severe hypoglycemia, but also *why* they may be so (in functionally mechanistic terms). The availability of an analytical model capable of quantitatively assessing such functional properties of insulin–glucose regulation and counterregulation will also aid in addressing questions such as "What triggers the initial counterregulatory response, and how?" It is therefore believed that applying this quantitative model to an extended set of clinical and experimental data sets will (1) contribute significantly to unraveling and understanding the physiologic mechanisms involved in insulin-dependent regulation of blood glucose dynamics, as well as (2) provide a useful tool for assessing both the propensity for and the functionally mechanistic cause(s) of hypoglycemia in patients with IDDM.

Acknowledgments

This study is supported by the National Institutes of Health grants RO1 DK51562, RO1 DK28288, RROO847 and by the National Science Foundation NSF DIR-8920162. We thank the University of Virginia General Clinical Research Center for the support during the study.

[22] Association of Self-Monitoring Blood Glucose Profiles with Glycosylated Hemoglobin in Patients with Insulin-Dependent Diabetes

By BORIS P. KOVATCHEV, DANIEL J. COX, MARTIN STRAUME, and LEON S. FARHY

Introduction

Researchers have known for more than 20 years that glycosylated hemoglobin (Hb) is a marker for the glycemic control of individuals with insulin-dependent diabetes mellitus (IDDM). Numerous studies have investigated this relationship and found that glycosylated hemoglobin generally reflects the average blood glucose (BG) levels of an IDDM patient over the previous 2 months. Because in the majority of patients with IDDM the BG levels fluctuate considerably over time, it was suggested that the real connection between integrated glucose control and HbA_{1c} can be observed only in patients known to be in stable glucose control over a long period of time.[1] Early studies of such patients derived an almost deterministic relationship between the average BG level in the preceding 5 weeks and HbA_{1c}—a curvilinear association that yielded[1] a correlation of 0.98. In 1993 the Diabetes Control and Complications Trial (DCCT) concluded[2] that HbA_{1c} is

[1] P. Aaby Svendsen, T. Lauritzen, U. Soegard, and J. Nerup, *Diabetologia* **23,** 403 (1982).
[2] J. V. Santiago, *Diabetes* **42,** 1549 (1993).

0076-6879/00 $30.00

"the logical nominee" for a gold-standard glycated hemoglobin assay and established a linear relationship between the preceding mean BG and HbA_{1c}. Guidelines were developed saying that an HbA_{1c} of 7% corresponds to a mean BG of 8.3 m*M* (150 mg/dl), an HbA_{1c} of 9% corresponds to a mean BG of 11.7 m*M* (210 mg/dl), and a 1% increase in HbA_{1c} corresponds to an increase in mean BG of 1.7 m*M* (30 mg/dl, 2). At this time, the DCCT also suggested[2] that because measuring mean BG directly is not practical, one can assess patients' glycemic control with a single simple test, namely HbA_{1c}.

However, the rapid development of home BG monitoring devices somewhat changes this conclusion. Contemporary memory meters store up to several hundred self-monitoring BG (SMBG) readings and can calculate various statistics, including the mean of these BG readings. Such a mean can, to some extent, be used as an estimate of the average BG level of the patient over the observed period of time. The danger here is that a mean BG based on SMBG readings can deviate substantially from the real mean BG value reflected by the HbA_{1c}. For example, if a patient measures at fixed times of day when his or her BG is at its peak, the average of these SMBG readings will be an overestimate of his or her real mean BG. Or, if measurements are taken at times of peak insulin action, then the mean SMBG will be an underestimate of the true average BG. Nevertheless the use of SMBG readings for an assessment of the average BG and HbA_{1c} is tempting, since such readings are routinely taken and readily available for analysis. We previously demonstrated that the Low BG Index, based on mathematically manipulated SMBG data, can accurately assess patients' risk for severe hypoglycemia (SH), predicting more than 40% of upcoming SH episodes.[3,4] High glycosylated hemoglobin is opposite to SH as an extreme manifestation of IDDM, related to long-term complications.[2] An accurate assessment of HbA_{1c}, based on memory meter data would provide an evaluation of the real average BG of a patient over time. This, in combination with the Low BG Index, could be used to assess both low and high BG extremes on the basis of SMBG, thus contributing to on-line optimization of patients' IDDM control.

This study assesses the accuracy of SMBG estimates of average BG on the basis of a large data set containing more than 300,000 SMBG readings of 608 individuals with diabetes, accompanied by HbA_{1c} data. We conclude that SMBG data account for about 50% of the variance of HbA_{1c}. We offer

[3] D. J. Cox, B. P. Kovatchev, D. M. Julian, L. A. Gonder-Frederick, W. Polonsky, D. Schlundt, and W. L. Clarke, *J. Clin. Endocrin. Metab.* **79,** 1659 (1994).

[4] B. P. Kovatchev, D. J. Cox, L. A. Gonder-Frederick, D. Young-Hyman, D. Schlundt, W. L. Clarke, *Diabetes Care* **21,** 1870 (1998).

detailed conversion tables that present the relationship between HbA_{1c} and average SMBG, together with 95% confidence intervals. We conclude that SMBG does not provide a very accurate representation of HbA_{1c}, and should be used cautiously if an accurate estimate of HbA_{1c} is needed. However, SMBG can be clinically useful for a general categorical evaluation of HbA_{1c} through a classification table.

Methods

Subjects

Seven hundred subjects with IDDM participated in a study conducted by Amylin Pharmaceuticals (San Diego, CA). The data of 608 subjects were completed with SMBG and HbA_{1c} records and the SMBG records were considered accurate according to an automated rejection criterion described below.

Procedure

All subjects were instructed to use BG memory meters for 4–6 months and to measure their BG two to four times a day. During the same period of time 5 to 8 HbA_{1c} assays were performed for each subject. The memory meter data were electronically downloaded and stored in a computer for further analysis. This resulted in a database with more than 300,000 SMBG readings and 4180 HbA_{1c} assays taken over 6 months. Given such a large database, we anticipated two things: (1) There might be inaccurate data due to equipment failures (meter calibration, batteries, data transfer, etc.), and (2) such inaccuracies cannot be identified manually by a simple review of the data sets. Thus, we designed an automated data cleaning algorithm to filter all data sets.

Data Cleaning

First, only subjects who had SMBG records and HbA_{1c} assays were selected for analysis. Then the SMBG data were automatically scanned using a data filtering algorithm that is based on comparison of the theoretically derived probability distribution of the last digit of the BG readings and the empirical distribution of the BG data recorded by the meter. A statistical hypothesis for that distribution was formulated and tested, and as a result the data for some patients were partially or completely rejected. This resulted in a rejection of 9.8% of all SMBG readings that later were manually reviewed to reveal various data recording/transfer malfunctions.

TABLE I
DESCRIPTIVE CHARACTERISTICS

Parameter	Mean	Standard deviation	95% CI	20%	40%	60%	80%	Number of subjects
Mean SMBG (m*M*)	10.4	2.2	6.5–16	8.6	9.7	10.6	12.0	608
HbA_{1c} (%)	9.2	1.2	7.1–12	8.3	8.8	9.4	10.0	608

The final count showed 608 subjects, who had on average 436 SMBG readings per subject, accompanied by HbA_{1c} assays.

Results

SMBG and HbA_{1c} Categories

Table I presents the descriptive characteristics of the mean SMBG level per subject and the average HbA_{1c}, together with 95% confidence intervals (CI) and 20th, 40th, 60th, and 80th percentiles.

Table II identifies five subject categories based on the subjects' average SMBG and five subject categories based on the subjects' HbA_{1c}. The Discussion section provides an explanation as to why exactly five categories for each of these two variables were identified. Chi-square test demonstrated that these categories are highly dependent, $p < 0.0001$. To provide a classification correspondence between the average SMBG and HbA_{1c} and to demonstrate that such classification identifies significantly different subgroups of IDDM patients, one-way ANOVA was conducted for HbA_{1c} in the categories defined by the average SMBG, as well as for the average SMBG in the categories defined by HbA_{1c}. Table II presents the results of the first analysis. The five categories were highly significantly different, $F = 111$, $p < 0.00001$. In addition, the average HbA_{1c} was significantly

TABLE II
HbA_{1c} WITHIN CATEGORIES IDENTIFIED BY AVERAGE SMBG

Category	Number	Mean HbA_{1c}	SEM	95% CI
Below 8.6 m*M*	118	8.29	0.06	8.17–8.41
8.6–9.7 m*M*	124	8.70	0.06	8.58–8.82
9.7–10.6 m*M*	119	9.14	0.08	8.99–9.29
10.6–12 m*M*	126	9.50	0.07	9.35–9.64
Above 12 m*M*	121	10.52	0.12	10.29–10.76

TABLE III
AVERAGE SMBG WITHIN CATEGORIES IDENTIFIED BY HbA_{1c}

Category	N	Mean SMBG	SEM	95% CI
Below 8.3%	125	8.58	0.11	8.34–8.81
8.3–8.8%	123	9.54	0.11	9.32–9.76
8.8–9.4%	118	10.28	0.13	10.02–10.53
9.4–10%	116	11.01	0.15	10.7–11.31
Above 10%	126	12.74	0.22	12.3–13.17

different for each pair of categories as demonstrated by Duncan's ranges, $p < 0.01$.

Table III is the "converse" of Table II: the average SMBG is presented within the categories identified by HbA_{1c}. Again, there was a highly significant difference between the five HbA_{1c} categories, $F = 108$, $p < 0.00001$ and the average SMBG was significantly different for each pair of HbA_{1c} categories at $p = 0.01$ as demonstrated by Duncan's ranges.

Linear and Curvilinear Association

We estimated the best linear (as suggested by the DCCT[2]) and the best curvilinear association[1] between the average SMBG and HbA_{1c}. The general parametric form of the curvilinear relationship was $HbA_{1c} = a$

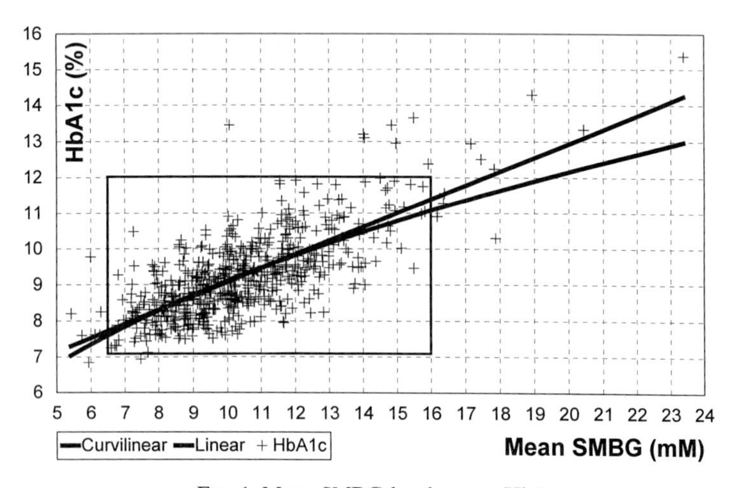

FIG. 1. Mean SMBG level versus HbA_{1c}.

TABLE IV
LINEAR VERSUS CURVILINEAR HbA_{1c}–MEAN SMBG ASSOCIATION

Association	Correlation	Mean square error	Regression
Linear	0.73	0.66	$R^2 = 53\%$, $F = 612$, $p < 0.00001$
Curvilinear	0.71	0.63	$R^2 = 50\%$, $F = 692$, $p < 0.00001$

(mean SMBG)^b. Parameters a and b and the parameters of the linear relationship were estimated using standard least squares minimization. The best linear relationship was HbA_{1c} = 5.21 + 0.39(mean SMBG). The best curvilinear relationship was HbA_{1c} = 3.47(mean SMBG)$^{0.42}$. As with the previous reports,[1] the curvilinear relationship did not differ substantially from its linear counterpart and there is no apparent reason to believe that the association between HbA_{1c} and mean SMBG is nonlinear (Fig. 1). Statistically, we compared the linear and curvilinear approaches to HbA_{1c} using correlations and analysis of the regression residuals. Table IV presents these comparisons.

Figure 1 presents the HbA_{1c}–mean SMBG scatterplot with superimposed straight line and a curve corresponding to the linear and curvilinear estimates of Table IV. Visually, the two lines are not substantially different. The 95% confidence intervals for each variable, mean SMBG and HbA_{1c}, are presented by a black rectangle and are enlarged in Fig. 2. For Fig. 2 the two regression lines were recomputed, which resulted in a reduction

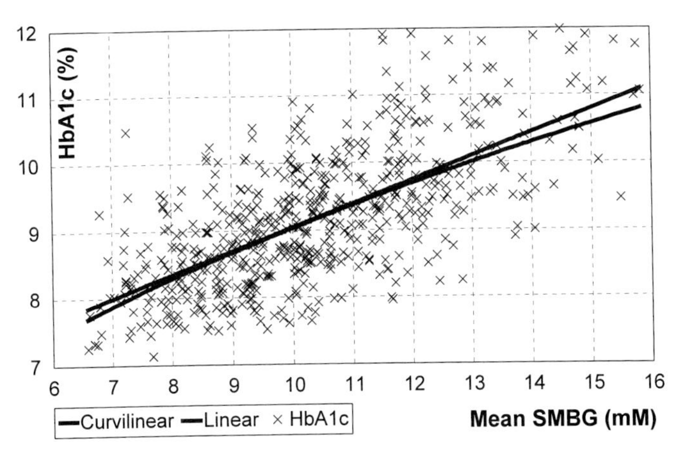

FIG. 2. 95% Cl for mean SMBG and HbA_{1c}.

of the mean square error to 0.55 for the linear relationship and 0.54 for the curvilinear relationship.

Discussion

The relationship between the average BG and HbA_{1c} has been well established and thoroughly investigated during the past two decades. This manuscript investigates this existing relationship from a different angle and with a different purpose. The idea is to find out how well the mean of SMBG data describes the actual mean BG of a patient with IDDM. HbA_{1c} is accepted as an indicator for the actual mean BG over time. Our findings suggest that the mean SMBG data is far from an ideal descriptor of the actual BG level. The correlation between mean SMBG and HbA_{1c} is about 0.7; therefore, only about 50% of the variance of the actual BG is accounted for by mean SMBG. In addition, we found no evidence that a curvilinear relationship between SMBG and HbA_{1c} is superior to a simple linear relationship.

Clinically, the [relatively] poor correlation between HbA_{1c} and mean SMBG might not be that important. Instead, a set of carefully selected categories that classify HbA_{1c} on the basis of mean SMBG (and vice versa) would be of more practical utility. We demonstrated that it is possible to identify nonoverlapping mean SMBG intervals that result in nonoverlapping 95% confidence intervals for HbA_{1c} (Table II). When optimizing these intervals for the mean SMBG we used the following optimization criteria: (1) maximum resolution, a maximum number of mean SMBG categories; (2) approximately equal number of subjects in each category; and (3) nonoverlapping 95% confidence intervals for HbA_{1c}. According to these criteria the optimal number of mean SMBG categories was five; with six categories it was not possible to identify nonoverlapping HbA_{1c} confidence intervals. The inverse problem—to identify nonoverlapping mean SMBG confidence intervals based on a partitioning of HbA_{1c}—was solved in a similar manner. Again, five HbA_{1c} categories were identified (Table III). Because the data set was relatively large—we had 265,102 usable SMBG readings for 608 subjects—we can assume that the classifications in Tables II and III are accurate and generalizable to the general population.

In short, Tables II and III can be used as conversion tables from mean SMBG to HbA_{1c} and vice versa. We believe this provides a useful approach to evaluation of patients' glycemic control on the basis of readily available SMBG readings that could be used between regular HbA_{1c} assessments. At the same time, patients should be encouraged to measure their BG at random times throughout the day in order to achieve a maximally representative characterization of their glycemic control through SMBG.

Acknowledgments

This study is supported by the National Institutes of Health grant RO1 DK51562, by Amylin Pharmaceuticals, San Diego, CA, and by Lifescan Inc., Milpitas, CA.

[23] Outliers and Robust Parameter Estimation

By MICHAEL L. JOHNSON

Introduction

Biomedical researchers are frequently faced with the task of estimating a series of parameters from a set of experimental data. This is commonly done by selecting a set of parameters of an equation such that the equation provides a good description of the data. In most cases the mathematical form of the equation is determined by a theoretical description of the underlying physics, chemistry, physiology, and/or biology. The process of selecting the parameter values is usually done by a least squares parameter estimation procedure.[1-3]

Outliers are those data points that simply did not fall near the central tendency of the majority of the data points. These outlying data points are typically caused by weighing or pipetting errors, but can arise for many reasons. The presence of outliers in a data set forces the investigator to make a series of decisions: Should they be included during the fitting process or should they be eliminated? Which data points are outliers? Are they outliers because the data points are bad or because the mathematical form of the equation (i.e., the theory) is wrong? If a data point is suspected of being an outlier, can an experimental reason (i.e., an experimental error) be found to justify eliminating the particular data point?

Least squares parameter estimation procedures are particularly sensitive to the presence of outliers in a data set. Consider the example shown in Fig. 1. The solid line in the lower panel (Fig. 1) is the least squares "best fit" quadratic polynomial [Eq. (1)].

[1] M. L. Johnson and S. G. Frasier, *Methods Enzymol.* **117,** 301 (1985).
[2] M. L. Johnson and L. M. Faunt, *Methods Enzymol.* **210,** 1 (1992).
[3] M. L. Johnson, *Methods Enzymol.* **240,** 1 (1994).

0076-6879/00 $30.00

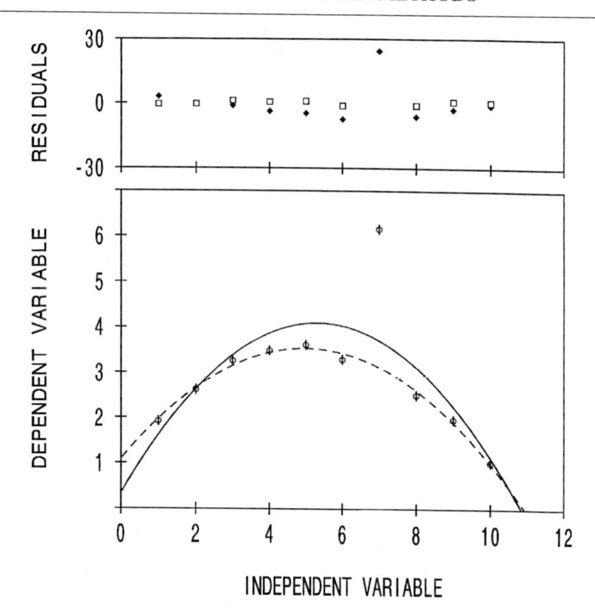

FIG. 1. Least squares analysis of the data by a quadratic polynomial. The solid line in the lower panel is the "best fit" to Eq. (1). The dashed line is the corresponding fit of the data without the data point at an independent variable of 7. The upper panel displays the weighted residuals, solid diamonds correspond to the solid-line fit, and the open squares correspond to the dashed-line fit. The least squares estimated parameters are shown in Table I.

$$G(X, answers) = Y = a + bX + cX^2 \tag{1}$$

The data points are shown in the lower panel (Fig. 1). The dashed line is the corresponding fit of the data without the data point at an independent variable of 7, $X = 7$. The upper panel (Fig. 1) displays the weighted residuals, i.e., the differences between the fitted curve and the data points normalized by the uncertainties of the particular data points (solid diamonds correspond to the solid-line fit and the open squares correspond to the dashed-line fit).

The least squares estimated parameters are shown in Table I. This data set was generated with the values of $a = 1.0$, $b = 1.0$, and $c = -0.1$, and then the data point at $X = 7$ was systematically altered. A small amount (SD = 0.1) of Gaussian distributed random noise was also added to the data set. Clearly, the least squares procedure correctly found the parameter values in the absence of the outlier at $X = 7$, but not when it was present.

From an examination of the fit in the presence of the data point at $X = 7$, it appears that the data point at $X = 7$ is an outlier. The parameter estimation without this data point provides a better description of the data. *However, a better fit is not really a good justification for arbitrarily removing*

TABLE I
PARAMETERS FROM LEAST SQUARES ANALYSIS[a]

Parameter	Solid line		Dashed line	
	Value	Standard error[b]	Value	Standard error[b]
a	0.32	0.87	1.081	0.104
b	1.42	0.37	0.988	0.043
c	−0.13	0.14	−0.100	0.004
s^2	106.52		0.9877	

[a] As shown in Fig. 1.
[b] Standard errors were evaluated by a 3000-cycle bootstrap method.

a data point. The concern is that this particular data point was influential enough to cause systematic errors in the estimated parameters and as a consequence, a major shift in the calculated curve. This systematic shift in the estimated curve makes it difficult to identify a particular data point as an outlier.

The purpose of this manuscript is to discuss one method of fitting data in the presence of outliers. This method is "robust" in that it is less sensitive than least squares methods to the presence of outliers. The reader is, however, cautioned that this method does not have the rigorous statistical basis that least squares methods have.

Least Squares Parameter Estimation

Least squares parameter estimation procedures automatically adjust the parameters of a fitting equation until the weighted sum of the squared residuals [i.e., the sample variance of fit, Eq. (2)] is a minimum:

$$s^2 = \frac{1}{n-m} \sum_{i=1}^{n} \left[\frac{Y_i - G(X_i, answers)}{\sigma_i} \right]^2 \quad (2)$$

where X_i and Y_i are the ith of n data points, σ_i is the standard error of the ith data point, G is the fitting function, and *answers* are the m parameters of the fitting function. For the present example, the *answers* are a, b, and c of Eq. (1) with $m = 3$.

The statistical basis of the least squares procedures is dependent on a series of assumptions about the experimental data and the fitting function. Specifically, these assumptions are made: All of the experimental uncertainties are Gaussian distributed and on the dependent variables (the Y axis), i.e., no errors exist in the independent variables (the X axis) and no systematic (non-Gaussian) uncertainties exist. The fitting equation is the correct equation. The data points are "independent observations." A sufficient number of data points to provide a good sampling of the experimental uncertainties are required; i.e., this implies that the data should not be smoothed to remove the noise components. With these assumptions it can be shown analytically that the least squares procedures will produce the parameter values that have the highest probability of being correct.[1–3] Least squares parameter estimation is a maximum likelihood (highest probability) method. Conversely, if these assumptions are valid and an analysis method produces answers that are different from the values found by the least squares procedure, then the other analysis method is producing answers with less than the maximum likelihood (i.e., they are wrong).

An examination of these assumptions explains why the least squares procedure is significantly biased by the outlier data point in Fig. 1 at $X = 7$. That particular data point contains a significant amount of systematic uncertainty. Therefore, the least squares assumptions are not met and as a consequence the least squares procedure is expected to produce answers that are less than optimal—which is exactly what is seen in Fig. 1 and Table I.

The actual numerical procedure utilized for the parameter estimations (e.g., least squares and others) presented in this work is the Nelder–Mead simplex algorithm.[4] Most available methods for the evaluation of confidence intervals assume that the fitting procedure is a least squares procedure. The bootstrap method[5] was used in this work because it does not assume that the procedure is a least squares procedure.

Robust Parameter Estimation Procedures

From Eq. (2) it is obvious that the least squares procedure weights each data point with the square of the weighted distance between the data point and the fitted curve. This is correct for Gaussian distributed random noise. However, when a few outliers exist with a significant amount of systematic uncertainty, it is clear that the outlier data points have too much influence on the resulting parameter values.

A simple solution is to lower the power in Eq. (2) to something less than two so that these outliers will have less influence. One commonly used method is to minimize the sum of the absolute values, *SAV*, of the weighted residuals, as in Eq. (3).

[4] J. A. Nelder and R. Mead, *Comput. J.* **7,** 308 (1965).

[5] B. Efron and R. J. Tibshirani, "An Introduction to the Bootstrap." Chapman and Hall, New York, 1993.

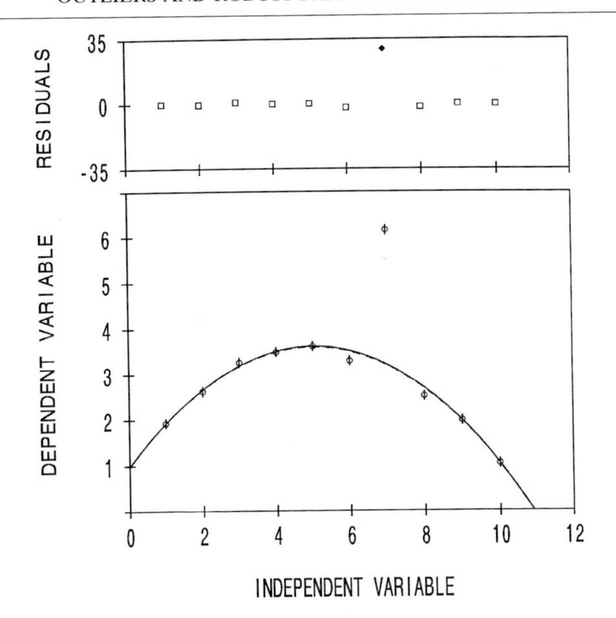

FIG. 2. Robust analysis of the data by a quadratic polynomial. The solid line in the lower panel is the "best fit" to Eq. (1). The dashed line is the corresponding fit of the data without the data point at an independent variable of 7. The upper panel displays the weighted residuals, solid diamonds correspond to the solid-line fit, and the open squares correspond to the dashed-line fit. The least squares estimated parameters are shown in Table II.

$$SAV = \frac{1}{n-m} \sum_{i=1}^{n} \left| \frac{Y_i - G(X_i, answers)}{\sigma_i} \right| \tag{3}$$

This is one of the procedures in the field of robust statistics.[6,7] The reader should be cautioned that while this procedure has some specific applications, such as locating outliers, it is not a maximum likelihood procedure. When applied to data that meet the assumptions of least squares fitting, this procedure will produce results that are not the maximum likelihood results. Under these conditions, this robust procedure will provide answers that are less than optimal.

Robust Analysis of Present Example

Figure 2 and Table II present an analogous robust analysis of the data shown in Fig. 1 and Table I. This analysis was performed by minimizing

[6] P. J. Huber, "Robust Statistics." John Wiley and Sons, New York, 1981.
[7] D. C. Hoaglin, F. Mosteller, and J. W. Tukey, "Understanding Robust and Exploratory Data Analysis." John Wiley and Sons, New York, 1983.

TABLE II
PARAMETERS FROM ROBUST ANALYSIS[a]

Parameter	Solid line		Dashed line	
	Value	Standard error[b]	Value	Standard error[b]
a	0.978	0.089	0.986	0.079
b	1.047	0.040	1.038	0.034
c	−0.104	0.004	−0.103	0.003

[a] As shown in Fig. 2.
[b] Standard errors were evaluated by a 3000-cycle bootstrap method.

the sum of the weighted absolute values of the residuals, SAV in Eq. (3), instead of minimizing the weighted sum of the squares of the residuals, s^2 in Eq. (2). For this analysis the values calculated, while including the data point at $X = 7$ (solid line and sold diamonds), virtually superimpose with the values calculated in the absence of the data point (dashed line and open squares). This indicates that the outlier had little or no contribution to the analysis, thus making it easier to identify since the calculated curve is biased toward the outlier to a lesser degree. However, the answers (i.e., the parameter values) do not agree with the values from the least squares analysis. Figure 3 presents the analysis of the data set when the outlier at an independent variable of 7 is removed. The solid line and solid diamonds correspond to the least squares analysis while the dashed line and open squares correspond to the robust analysis.

Conclusions

This chapter presents an example of robust parameter estimation. The example describes how robust fitting can be employed for data that may contain "outliers." An outlier is simply a data point that for some reason contains a significant amount of systematic (i.e., nonrandom) experimental uncertainty. The example demonstrates how the least squares parameter estimation is sensitive to the presence of these outliers while the robust method is comparatively less sensitive. Thus, it is easier to identify presumptive outliers with the robust method.

Note that the robust method presented here is only one of many possible "robust" methods. This robust method minimized the sum of the first powers of the absolute value of the residuals. In principle, Eq. (3) could be written with a power term, q, as in Eq. (4).

$$SAV_q = \frac{1}{n-m} \sum_{i=1}^{n} \left| \frac{Y_i - G(X_i, answers)}{\sigma_i} \right|^q \tag{4}$$

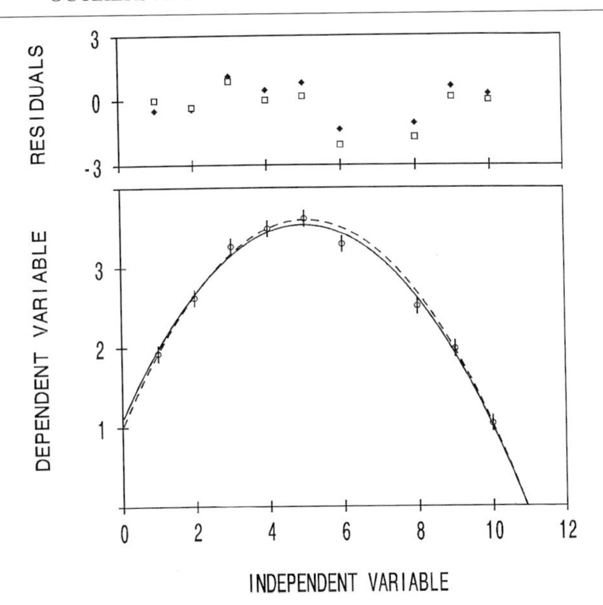

FIG. 3. Least squares and robust analysis of the data in Figs. 1 and 2 with the outlier at an independent variable of 7 removed. The solid line and solid diamonds correspond to the least squares analysis (as shown in Fig. 1 and Table I), whereas the dashed line and open squares correspond to the robust analysis (as shown in Fig. 2 and Table II).

For the present robust example, q is unity. However, any power, q, less than 2 should decrease the sensitivity of the parameter estimation to the presence of outliers. In addition, there are many other "norms" of the data that could be minimized that are less sensitive to outliers than the least squares method.

This is expected since the least squares method is based on the assumption that outliers are not present within the data. Thus, if outliers are present and the least squares method is employed, then it is expected that the results will be less than optimal.

If this is the case, why not always use a robust method? The answer is that the robust methods do not have a rigorous statistical basis. A large body of experimental data is consistent with the assumptions that are outlined above for the least squares method. For data that are consistent with the assumptions, the least squares method provides maximum likelihood parameter values. If outliers are not present (as in Fig. 3), the robust methods do not provide the parameter values with the maximum likelihood of being correct.

If the experimental data are not consistent with the least squares assumptions, then clearly least squares will not provide the optimal parameter values. However, it cannot be demonstrated that the robust method will provide the optimal (i.e., maximum likelihood) parameter values either.

In addition, a large number of techniques have been developed that are based on the concepts of least square parameter estimations. These include goodness-of-fit[8] tests such as autocorrelation, runs test, Kolmogorov–Smirnov test, and Durbin–Watson test. It also includes methods for the evaluation of the precision of the estimated parameters[9,10] (i.e., the parameter confidence intervals). In general, these techniques apply to least squares but not to robust parameter estimation procedures.

So why use robust parameter estimation procedures? Because in many cases they *appear* to provide an analysis of the data that is less sensitive to the presence of outliers. This makes it easier to identify those points that may be outliers.

Acknowledgments

The author acknowledges the support of the National Science Foundation Science and Technology Center for Biological Timing at the University of Virginia (NSF DIR-8920162), the General Clinical Research Center at the University of Virginia (NIH RR-00847), and the University of Maryland at Baltimore Center for Fluorescence Spectroscopy (NIH RR-08119).

[8] M. Straume and M. L. Johnson, *Methods Enzymol.* **210,** 87 (1992).
[9] D. M. Bates and D. G. Watts, "Nonlinear Regression Analysis and Its Applications." John Wiley and Sons, New York, 1988.
[10] D. G. Watts, *Methods Enzymol.* **240,** 23 (1994).

[24] Parameter Correlations while Curve Fitting

By MICHAEL L. JOHNSON

Introduction

While curve fitting experimental data, i.e., simultaneous multiple parameter estimation, an investigator will generally find that the answers are cross-correlated. This cross-correlation is evident when an increase in the variance-of-fit caused by a variation in any parameters can be partially compensated for by a variation in one, or more, of the other parameters. This is generally not a problem if the degree of this compensation, i.e., cross-correlation, is low. However, if the correlation is nearly complete, then

0076-6879/00 $30.00

the parameter estimation procedure (e.g., least squares[1-5] and maximum likelihood[1-5]) will be unable to identify unique values for the parameters being estimated and their associated standard errors. The purpose of this chapter is to review the origin of this parameter cross-correlation and discuss methods to minimize or eliminate it.

When the parameters of a simultaneous least squares analysis are correlated then the estimated uncertainties of these parameters are also correlated. Thus, if the goal is to use the parameter values and their uncertainties to test a hypothesis, it is important that both be well determined. For example, how do we test if a set of data is a straight line? One method is to fit the data to a quadratic polynomial and then ask if the quadratic term, i.e., the curvature, is zero. That test requires accurate values for the quadratic parameter and its associated standard error. However, since the intercept, the linear, and the quadratic terms (and standard errors) are probably all correlated, it is impossible to simply compare the quadratic term with zero. Uncertainties in the evaluation of the intercept and linear terms will be partially compensated by a variation of the quadratic term, but this will not be reflected in an increased uncertainty of the quadratic term. One solution would be to perform the analysis such that the parameters and their associated standard errors are not correlated.[6]

It is important to note that if a parameter estimation procedure reports that two parameters are correlated, it is generally not an indication that there is something correlated about the physiologic or biochemical mechanism underlying the data. It may be simply a consequence of performing the parameter estimation on a limited set of experimental observations. For example, consider a stopped-flow experiment (or a temperature-jump experiment) where the equilibrium position of a chemical reaction is rapidly altered and the time course of the relaxation to the new equilibrium position is monitored. These relaxations can generally be described as a sum of exponential decays. Consider the expected results if the sample consists of a mixture of two independent noninteracting chemical systems each of which is expected to exhibit a single exponential decay as it approaches its new equilibrium position. The mixture will exhibit decay characteristics that are simply the sum of the individual exponential decays. There is nothing correlated about the chemistry, but when the data are analyzed

[1] M. L. Johnson and S. G. Fraiser, *Methods Enzymol.* **117,** 301 (1985).
[2] M. L. Johnson and L. M. Faunt, *Methods Enzymol.* **210,** 1 (1992).
[3] M. L. Johnson, *Methods Enzymol.* **240,** 23 (1994).
[4] D. M. Bates and D. G. Watts, "Nonlinear Regression Analysis and Its Applications." John Wiley and Sons, New York, 1988.
[5] D. G. Watts, *Methods Enzymol.* **240,** 23 (1994).
[6] F. S. Acton, "Analysis of Straight-Line Data," p. 193. Chapman and Hall, New York, 1959.

TABLE I
DATA USED IN FIRST EXAMPLE

Time	Amplitude
0.01	5.77681
0.07	4.74431
0.13	3.91186
0.19	3.22843
0.25	2.68778
0.31	2.22903
0.37	1.86932
0.43	1.58560
0.49	1.32141
0.55	1.11302
0.61	0.960843
0.67	0.788441
0.73	0.687573
0.79	0.584176
0.85	0.495013
0.91	0.435635
0.97	0.372533

by any parameter estimation procedure the resulting decay rates (or half-lives) will be correlated. Examples are shown in Tables I–III and Figs. 1–3.

Examine the relaxation data shown in Table I and Fig. 1. This is an example of the classic ill-posed analysis problem of the numerical analysis literature. These data were simulated for equal amounts of two noninteracting relaxations with rate constants of two and four. These data also contain pseudo-random experimental uncertainties, i.e., noise, with a stan-

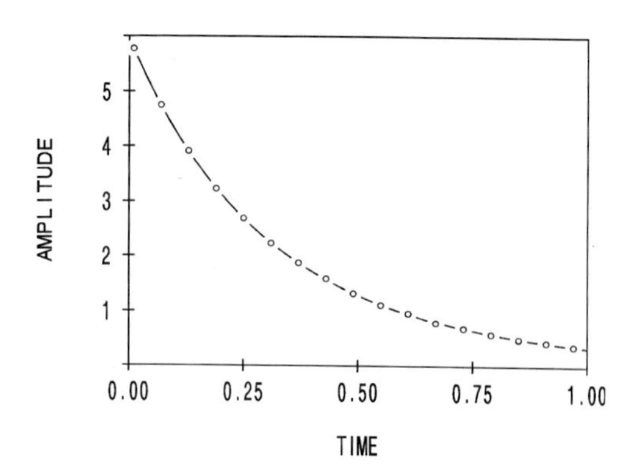

FIG. 1. Two-exponential least squares analysis of the data given in Table I.

TABLE II
PARAMETERS FROM WEIGHTED LEAST SQUARES ANALYSES[a]

Parameter	Value	Confidence region[b]
Single-exponential analysis		
A	5.87	5.81–5.94
K	3.04	2.97–3.08
SSR	398.42	
Double-exponential analysis		
A_1	3.17	2.11–4.41
K_1	4.30	3.81–4.98
A_2	2.80	1.55–3.87
K_2	2.24	1.80–2.49
SSR	9.052	

[a] Of the data shown in Table I and Fig. 1.
[b] A ±2 standard deviation (95.44%) confidence region as evaluated by a bootstrap method.[7]

dard deviation of 0.01 units. The best two-exponential [Eq. (1)] weighted least squares fit of these data is shown as the line in Fig. 1.

$$\text{Amplitude} = A_1 e^{-K_1 time} + A_2 e^{-K_2 time} \tag{1}$$

where the K values are the relaxation decay rates and the A values are the corresponding amplitudes.

The parameter values for one- and two-exponential fits of these data are shown in Table II. Table II also shows the weighted sum of squared residuals, SSR, for these analyses.

$$SSR = \sum_i \left[\frac{Y_i - Curve(X_i)}{\sigma_1} \right]^2 \tag{2}$$

where the summation is over each of the i data points; $Curve(X_i)$ represents the fitting function evaluated at the optimal parameter values and the particular independent variable, X_i; Y_i is the ith dependent variable; and σ_i is the standard error of the mean of Y_i. The residuals are simply the standard deviation weighted differences between the data points and the fitted curve. It is clear from the decrcase in the SSR that the two-exponential fit of these data is substantially better than the one-exponential fit.

Table II also contains estimates of the 95% confidence regions of the fitted parameters as determined by a bootstrap method.[7] Notice that these

[7] B. Efron and R. J. Tibshirani, "An Introduction to the Bootstrap." Chapman and Hall, New York, 1993.

are neither expected nor observed to be symmetrical. The bootstrap procedure is simple. First, a nonlinear regression is done to evaluate the parameters of the fitting equation with the highest probability of being correct. Second, the optimal model parameters are used to simulate a set of experimental data. This simulated set of data must contain observations at exactly the same independent variables, i.e., the X axis. The simulated data set must also contain a realistic amount of random noise (e.g., experimental uncertainties). The "random noise" is generated by randomly selecting from the observed weighted residuals (the weighted differences between the data and the fitted function). For each simulated curve the residuals are randomly selected such that ~37% are used more than once and ~37% are not used. Third, the selected residuals are shuffled to produce a random order. Fourth, the nonlinear regression procedure is repeated on the simulated data to obtain apparent values for the parameters being estimated. Steps 2 to 4 are repeated many times (e.g., 500 to 1000 times) and the apparent parameter values from each cycle saved. These sets of apparent values of the parameters, from the multiple cycle of steps 2 to 4, are used to create distributions of apparent parameter values. These distributions correspond to the complete probability distributions of the estimated parameters.

The correlation matrix (defined below) for the two-exponential analysis is shown in Table III. The correlation matrix is a symmetrical matrix so only the lower diagonal is shown. This matrix presents the cross-correlation coefficients between each pair of the parameters being estimated. For example, the cross-correlation between A_1 and A_2 is -0.999933, an extremely correlated case. The values of the cross-correlation coefficient vary between ± 1, with zero indicating no cross-correlation between the parameters. A cross-correlation above ~0.98 or below ~-0.98 indicated that the parameters are highly correlated. Such high cross-correlation coefficients are also an indication, as will be seen below, that the parameter estimation process is extremely difficult with resulting parameter values and associated standard

TABLE III
CROSS-CORRELATION MATRIX FOR TWO-EXPONENTIAL ANALYSIS[a]

Parameter	Matrix			
A_1	1.00000			
K_1	−0.993626	1.00000		
A_2	−0.999933	0.994455	1.00000	
K_2	−0.997424	0.984443	0.997134	1.00000

[a] Shown in Tables I and II and Fig. 1.

errors that are questionable. The diagonal elements of this matrix correspond to the correlation between each parameter and itself and are thus unity by definition.

The cross-correlation of K_1 and K_2 is shown graphically in Figs. 2 and 3. Figure 2 was generated by specifying a series of values for K_1 between 3.8 and 5.3 (the circles). For each specified value of K_1, the weighted least squares procedure was repeated to determine the apparent values of A_1, A_2, and K_2. The corresponding values of K_2 are plotted with a dashed line connecting them. The procedure was also repeated for specific values of K_2, while the apparent values of A_1, A_2, and K_1 were reevaluated by the weighted least squares procedure (shown as a solid line and diamonds). It is obvious from this example that the values of K_1 and K_2 are correlated. Altering either parameter induces a corresponding change in the apparent value of the other parameter. If K_1 and K_2 were not correlated, then the variation of one would not induce a compensating variation in the other. Also, if K_1 and K_2 were not correlated then Fig. 2 would show two perpendicular lines instead of the observed nearly superimposed lines.

Note that the curvature of the lines in Fig. 2 is a measure of the nonlinear nature of the fitting problem near the solution. If this were a linear least squares fit, then these would be straight lines.

The bootstrap process for the evaluation of the parameter confidence

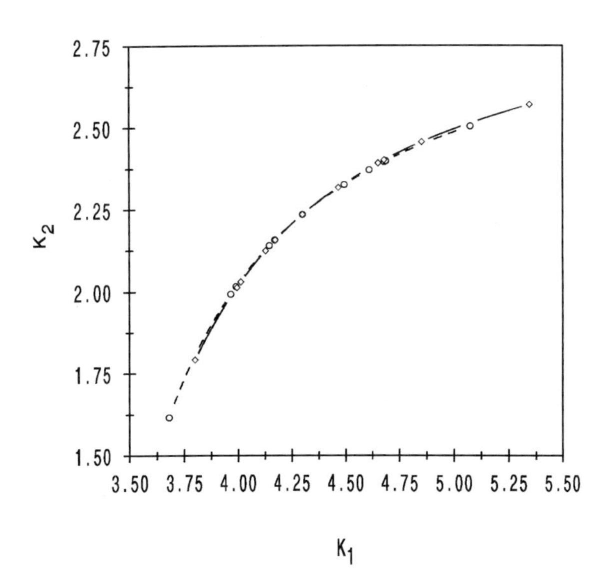

FIG. 2. A plot of the variability of K_2 induced by changes in K_1 (dashed line and circles). Also shown is a plot of the variability of K_1 induced by changes in K_2 (solid line and diamonds).

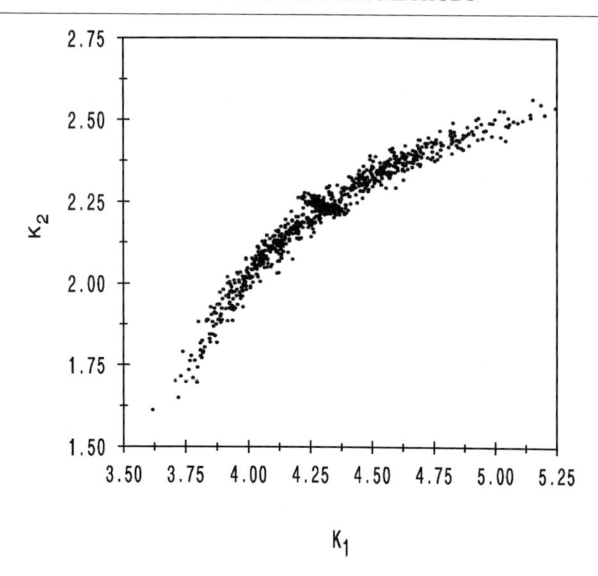

FIG. 3. An analogous plot to Fig. 2 generated from the simulations performed during the evaluation of the confidence intervals of the estimated parameters by a bootstrap procedure.[7]

intervals can be used to create a figure analogous to Fig. 2. Figure 3 presents the 1000 individual pairs of apparent values of K_1 and K_2 determined by the bootstrap process. Figure 3 also demonstrates the nonlinear highly correlated nature of this parameter estimation procedure in a manner analogous to Fig. 2. If this were an uncorrelated linear fitting problem, Fig. 3 would be circular with Gaussian distributions along all cross sections, as in Fig. 6 (shown later).

The cross-correlation coefficients observed in this example are not a result of the chemistry or physiology of the experiment. Parameter cross-correlation coefficients are a function of the following: the form of the fitting equation; the distribution of the data points, particularly the independent variables; the experimental uncertainties; and in nonlinear cases, the specific parameter values. This chapter reviews the mathematical form and origin of parameter cross-correlations and discusses methods to reduce or eliminate them.

Definition of Parameter Cross-Correlation Coefficients

The definition of the parameter cross-correlation coefficients is based on the parameter estimation procedure. In this chapter the mathematical form of the parameter cross-correlation coefficients is derived for the

Gauss–Newton weighted nonlinear least squares algorithm.[1–5] These parameter cross-correlation coefficients are not, however, specific to this algorithm. They are not a consequence of the algorithm. They are a complication with all parameter estimation algorithms.

The Gauss–Newton procedure approximates each of the i data points as a first-order series expansion of the fitting function expanded about an estimation, i.e., guess, of the parameter values, α'.

$$\frac{1}{\sigma_i} Y_i \approx \frac{1}{\sigma_i} G(\alpha, X_i) = \frac{1}{\sigma_i} G(\alpha', X_i) + \frac{1}{\sigma_i} \sum_j \left[\frac{\partial G(\alpha', X_i)}{\partial \alpha'_j} (\alpha_j - \alpha'_j) \right] + \ldots \tag{3}$$

where Y_i is the dependent variable, X_i is the independent variable, σ_i is the standard error of the mean of Y_i, G is the fitting function, α is the vector of parameter values to be determined by the weighted least squares procedure, and α' is a guessed (i.e., the current estimates) vector of parameter values. The j subscript refers to a specific fitting parameter and the i subscript refers to a specific data point. Consider a two-exponential fitting function as an example:

$$G(\alpha, X_i) = A_1 e^{-K_1 X_i} + A_2 e^{-K_2 X_i} \tag{4}$$

where α is the vector of fitting parameters $\{A_1, K_1, A_2, K_2\}$.

The only unknown in Eqs. (3) and (4) is the vector α. The Gauss–Newton procedure solves Eqs. (3) and (4) for the vector α. The easiest presentation of this solution is in matrix notation. Equation (3) can be written in matrix notation as

$$Y^* = J\varepsilon \tag{5}$$

where Y^* is a vector with one element for each of the i data points:

$$Y^* = \begin{pmatrix} \frac{1}{\sigma_1} & [Y_1 - G(\alpha', X_1)] \\ \frac{1}{\sigma_2} & [Y_2 - G(\alpha', X_2)] \\ \frac{1}{\sigma_3} & [Y_3 - G(\alpha', X_3)] \\ & \vdots \end{pmatrix} \tag{6}$$

ε is a vector with one element for each of the j parameters.

$$\varepsilon = \begin{pmatrix} \alpha_1 - \alpha_1' \\ \alpha_2 - \alpha_2' \\ \alpha_3 - \alpha_3' \\ \vdots \end{pmatrix} \tag{7}$$

J is a matrix of partial derivatives

$$J_{ij} = \frac{1}{\sigma_i} \frac{\partial G(\alpha', X_i)}{\partial \alpha_j'} \tag{8}$$

or

$$J = \begin{pmatrix} \frac{1}{\sigma_1}\frac{\partial G(\alpha', X_1)}{\partial \alpha_1'} & \frac{1}{\sigma_1}\frac{\partial G(\alpha', X_1)}{\partial \alpha_2'} & \frac{1}{\sigma_1}\frac{\partial G(\alpha', X_1)}{\partial \alpha_3'} & \cdots \\ \frac{1}{\sigma_2}\frac{\partial G(\alpha', X_2)}{\partial \alpha_1'} & \frac{1}{\sigma_2}\frac{\partial G(\alpha', X_2)}{\partial \alpha_2'} & \frac{1}{\sigma_2}\frac{\partial G(\alpha', X_2)}{\partial \alpha_3'} & \cdots \\ \frac{1}{\sigma_3}\frac{\partial G(\alpha', X_3)}{\partial \alpha_1'} & \frac{1}{\sigma_3}\frac{\partial G(\alpha', X_3)}{\partial \alpha_2'} & \frac{1}{\sigma_3}\frac{\partial G(\alpha', X_3)}{\partial \alpha_3'} & \cdots \\ \frac{1}{\sigma_4}\frac{\partial G(\alpha', X_4)}{\partial \alpha_1'} & \frac{1}{\sigma_4}\frac{\partial G(\alpha', X_4)}{\partial \alpha_2'} & \frac{1}{\sigma_4}\frac{\partial G(\alpha', X_4)}{\partial \alpha_3'} & \cdots \\ \vdots & \vdots & \vdots & \ddots \end{pmatrix} \tag{9}$$

The Gauss–Newton least squares procedure solves Eq. (5) for ε. Matrix J is not a square matrix so it cannot simply be inverted. However, by multiplying both sides of Eq. (5) by the transpose of J the resulting J^TJ matrix is a square symmetric positive definite matrix and can usually be inverted:

$$J^TY^* = (J^TJ)\varepsilon \tag{10}$$

Equation (10) can then be solved for ε. The major requirement for being able to invert the J^TJ is that the number of independent data points is greater than or equal to the number of parameters being estimated.

$$\varepsilon = (J^TJ)^{-1}(J^TY^*) \tag{11}$$

Because ε contains only the parameters being estimated and a known guess of the parameter values, it can easily be solved for the desired parameters.

The algorithm, as stated, is only guaranteed to function properly for linear least squares fitting problems. A linear fitting problem is one where all of the second and higher order partial derivatives of the fitting equation with respect to the parameters being estimated are equal to zero. A quadratic polynomial is an example of a linear fitting equation:

$$G(\alpha, X_i) = a + bX_i + cX_i^2 \tag{12}$$

In this context, linear refers to the series expansion [i.e., Eq. (3)] being linear, not the actual fitting equation being a straight line. The first derivatives of Eq. (12) are

$$\begin{aligned}
\frac{\partial G(\alpha, X_i)}{\partial a} &= 1 \\
\frac{\partial G(\alpha, X_i)}{\partial b} &= X_i \\
\frac{\partial G(\alpha, X_i)}{\partial c} &= X_i^2
\end{aligned} \tag{13}$$

These are not a function of the fitting parameters (i.e., constants) so all of the second and higher partial derivatives are zero.

The two-exponential function [Eq. (4)] is an example of a nonlinear fitting function. The first derivatives of Eq. (4) are

$$\begin{aligned}
\frac{\partial G(\alpha, X_i)}{\partial A_1} &= e^{-K_1 X_i} \\
\frac{\partial G(\alpha, X_i)}{\partial K_1} &= -X_i A_1 e^{-K_1 X_i} \\
\frac{\partial G(\alpha, X_i)}{\partial A_2} &= e^{-K_2 X_i} \\
\frac{\partial G(\alpha, X_i)}{\partial K_2} &= -X_i A_2 e^{-K_2 X_i}
\end{aligned} \tag{14}$$

Clearly all are functions of the parameter values and as a consequence the higher order derivatives are not all equal to zero.

The distinction between linear and nonlinear fitting equations has interesting consequences for the Gauss–Newton procedure. For a linear fitting function the higher order derivatives in Eq. (3) are all zero. Thus, the first-order series expansion in Eq. (3) is exact and Eq. (11) will provide accurate answers for the parameters. The term "linear fitting equation" refers to Eq. (3) being a linear function, not the actual fitting equation. Equation (12) is clearly not a straight line but it is a linear fitting equation.

For nonlinear fitting equations, the series expansion in Eq. (3) will be only approximate. Therefore, the parameter values provided by Eq. (11) will also only be approximate. For such nonlinear problems the parameter values provided by Eq. (11) are used as guesses for Eq. (3) and the algorithm is repeated. Thus, nonlinear least squares algorithms function by successive approximation. The results are used repeatedly as initial guesses and the process repeated until the resulting answers are the same as the current guesses.

This procedure works well if the contributions of the higher order derivatives in Eq. (3) are small. If the contributions are large, then it is possible that the iteration will diverge instead of converging. Several modifications[1–3] (e.g., Marquardt–Levenberg and damped Gauss–Newton) of the basic Gauss–Newton algorithm have been developed to ensure that the algorithm converges.

Cross-Correlation Coefficients

The cross-correlation coefficients are defined in terms of the inverse of the J^TJ matrix in Eqs. (10) and (11):

$$\textit{cross-correlation}_{jk} = \frac{(J^TJ)^{-1}_{jk}}{\sqrt{(J^TJ)^{-1}_{jj}(J^TJ)^{-1}_{kk}}} \tag{15}$$

where the subscripts refer to a particular jk element of the matrix. A necessary and sufficient condition to have no cross-correlation between the parameters is that all of the off-diagonal elements of the inverse of the J^TJ matrix must be equal to zero. This is equivalent to having the off-diagonal elements of the J^TJ matrix equal to zero. If all of the off-diagonal elements are zero, then the parameters are orthogonal to each other. If the parameters are all orthogonal, then the individual parameters can be evaluated without either a previous knowledge of the other parameters or simultaneously evaluating the other parameters.

The elements of the J^TJ matrix are given by:

$$(J^TJ)_{jk} = \sum_i \left[\frac{1}{\sigma_i^2} \frac{\partial G(\alpha', X_i)}{\partial \alpha'_j} \frac{\partial G(\alpha', X_i)}{\partial \alpha'_k} \right] \tag{16}$$

Equation (16) is the key to understanding how to reduce, or eliminate, the parameter cross-correlation. The next sections consider examples of using Eq. (16) to control cross-correlations among estimated parameters.

Polynomials

The simplest application of reducing parameter correlation is with the use of orthogonal polynomials. First, consider the form of the elements of the J^TJ matrix for the quadratic polynomial fitting function [Eq. (12)]. This can be done by inserting Eq. (13) into Eq. (16):

$$(J^TJ)_{ab} = \sum_i \left[\frac{X_i}{\sigma_i^2}\right]$$

$$(J^TJ)_{ac} = \sum_i \left[\frac{X_i^2}{\sigma_i^2}\right] \quad (17)$$

$$(J^TJ)_{bc} = \sum_i \left[\frac{X_i^3}{\sigma_i^2}\right]$$

Clearly, in this form it is unlikely that these terms will be zero for any arbitrary data set.

Consider the simulated data presented in Table IV and Fig. 4. The solid line in Fig. 4 corresponds to the best weighted least squares quadratic polynomial [Eq. (12)] for these data. Figure 5 presents the correlation plot from the bootstrap confidence interval analysis of this example. Figure 5 shows an elliptically shaped region, as is expected for a linear fitting problem. The region is not aligned with the parameter axes, indicating that this was also a correlated analysis. The cross-correlation coefficient for b and c was -0.95, again indicating a correlated analysis problem.

However, if Eq. (12) is written in a slightly different form it is comparatively easy to arrange for the cross-correlations to all be zero:

$$G(\alpha, X_i) = a' + b'(X_i - \beta) + c(X_i - \gamma_1)(X_i - \gamma_2) \quad (18)$$

TABLE IV
DATA USED IN SECOND EXAMPLE

Dependent variable[a]	Independent variable
0.500701	0.0
0.569121	0.1
0.667800	0.2
0.789702	0.3
0.829856	0.4
0.893017	0.5
0.903147	0.6
0.981385	0.7
1.46331	0.8
1.47003	0.9
1.88120	1.0

[a] The weighting factor for the least squares fit was based on a constant 10% coefficient of variation for these observations.

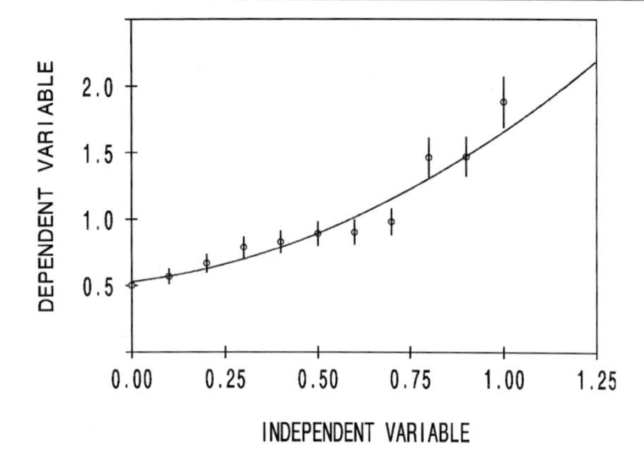

FIG. 4. Quadratic polynomial [Eq. (12)] weighted least squares analysis of the data given in Table IV. The maximum likelihood parameters for this analysis are $a = 0.529$, $b = 0.334$, and $c = 0.794$.

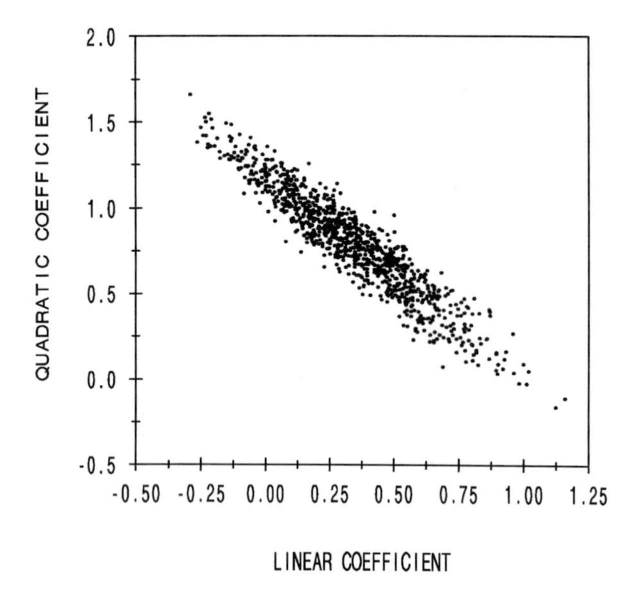

FIG. 5. A correlation plot of the simulations performed during the evaluation of the confidence intervals of the estimated parameters by a bootstrap procedure[7] for the analysis presented in Fig. 4.

where a' is now a different intercept, and b' is the slope at an $X \neq 0$. Terms β, γ_1, and γ_2 are evaluated such that the evaluation of a', b', and c is an orthogonal process, i.e.:

$$(J^TJ)_{a'b'} = \sum_i \left[\frac{(X_i - \beta)}{\sigma_i^2}\right] = 0$$

$$(J^TJ)_{ac} = \sum_i \left[\frac{(X_i - \gamma_1)(X_i - \gamma_2)}{\sigma_i^2}\right] = 0 \tag{19}$$

$$(J^TJ)_{bc} = \sum_i \left[\frac{(X_i - \beta)(X_i - \gamma_1)(X_i - \gamma_2)}{\sigma_i^2}\right] = 0$$

The following summations are over each of the i data points and σ_i is the standard error of the mean for the particular data point.

$$\beta = \frac{\sum_i \left[\frac{X_i}{\sigma_i^2}\right]}{\sum_i \left[\frac{1}{\sigma_i^2}\right]} \tag{20}$$

$$k_1 = \frac{\sum_i \left[\frac{(X_i - \beta)X_i^2}{\sigma_i^2}\right]}{\sum_i \left[\frac{(X_i - \beta)X_i}{\sigma_i^2}\right]} \tag{21}$$

$$k_2 = \frac{\sum_i \left[\frac{X_i^2}{\sigma_i^2}\right] - k_1 \sum_i \left[\frac{X_i}{\sigma_i^2}\right]}{\sum_i \left[\frac{1}{\sigma_i^2}\right]} \tag{22}$$

$$\gamma_1 = \frac{k_1 + \sqrt{k_1^2 - 4k_2}}{2} \tag{23}$$

$$\gamma_2 = \frac{k_1 - \sqrt{k_1^2 - 4k_2}}{2} \tag{24}$$

Figure 6 presents the correlation plot for a weighted least squares analysis of the data in Table IV according to an orthogonal quadratic polynomial [Eq. (18)]. From a comparison of Figs. 5 and 6, it is obvious that the simple transformation of the independent variable eliminated the parameter cross-correlations. The off-diagonal elements of J^TJ, as evaluated by Eq. (15), for this example are all zero.

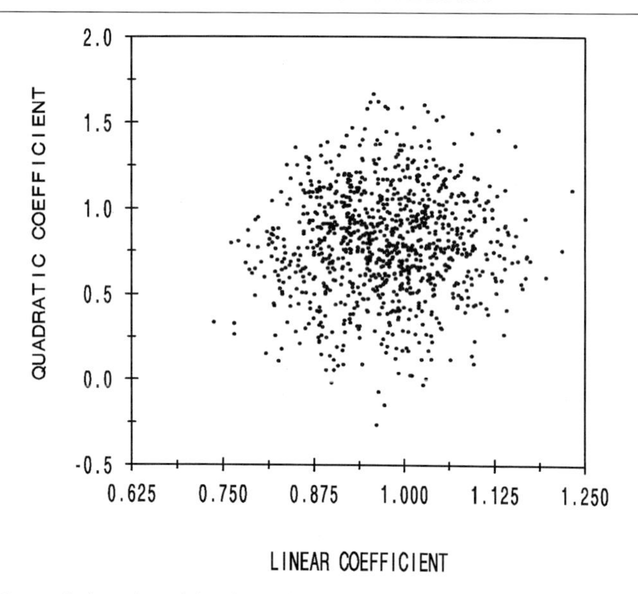

FIG. 6. A correlation plot of the simulations done during the evaluation of the confidence intervals of the estimated parameters by a bootstrap procedure[7] for the data presented in Table IV. Here the fitting function was an orthogonal quadratic polynomial [Eq. (18)]. The parameters of the maximum likelihood orthogonal polynomial are $a' = 0.752$, $b' = 0.962$, $c = 0.794$, $\beta = 0.291$, $\gamma_1 = 0.687$, and $\gamma_2 = 0.104$.

Fourier Series Analysis

The other classic example of eliminating parameter correlation is Fourier time series analysis. The objective of Fourier series analysis is to approximate a series of data points by the sum of a series of sine and/or cosine waves with known periods. Testing for the presence of circadian rhythms is a commonly used application of Fourier time series analysis.

For a continuous series of data, a Fourier series can be written as

$$G(\alpha, X_i) = f(x) = \frac{a_0}{2} + \sum_{n=1}^{\infty} c_n \sin\left(\frac{n\pi X_i}{L} - \phi_n\right) \tag{25}$$

This form of Fourier series is nonlinear and nonorthogonal in the parameters c_n and ϕ_n, and thus requires a complicated nonlinear fitting procedure. For Fourier analysis the fundamental period, $2L$, is generally a known constant.

The linear formulation of the Fourier series is given as the sum of a series of sine and cosine waves.

$$G(\alpha, X_i) = f(x) = \frac{a_0}{2} + \sum_{n=1}^{\infty} \left(a_n \cos \frac{n\pi X_i}{L} + b_n \sin \frac{n\pi X_i}{L} \right) \tag{26}$$

The Fourier series shown in Eq. (26) is orthogonal and linear for a long continuous series of data.

The transformations between Eqs. (25) and (26) are as follows:

$$\begin{aligned} a_n &= c_n \sin \phi_n \\ b_n &= c_n \cos \phi_n \\ c_n &= \sqrt{a_n^2 + b_n^2} \\ \phi_n &= \tan^{-1} \left(\frac{a_n}{b_n} \right) \end{aligned} \tag{27}$$

The coefficients of Eq. (26) can be determined by a linear least squares procedure. However, since this is an orthogonal problem, for some conditions, an exact solution exists for a_n and b_n. It is interesting that the exact solutions can be derived from the least squares formulation, [Eqs. (3)–(11)] for the specific conditions outlined below. This derivation is based on Eq. (26) being linear in the coefficients, which implies that only a single iteration is required, and that the initial estimated values, α', are zero:

$$\begin{aligned} a_n &= \frac{1}{L} \int_{-L}^{L} f(x) \cos \frac{n\pi x}{L}\, dx \\ b_n &= \frac{1}{L} \int_{-L}^{L} f(x) \sin \frac{n\pi x}{L}\, dx \end{aligned} \tag{28}$$

Does this solution apply to a discrete, noncontinuous, case? Consider the form of the off-diagonal elements of the $J^T J$ matrix:

$$\begin{aligned} (J^T J)_{a_n b_m} &= \sum_i \left[\frac{1}{\sigma_i^2} \cos \frac{n\pi X_i}{L} \sin \frac{m\pi X_i}{L} \right], \quad n \neq m \\ (J^T J)_{a_n a_m} &= \sum_i \left[\frac{1}{\sigma_i^2} \cos \frac{n\pi X_i}{L} \cos \frac{m\pi X_i}{L} \right], \quad n \neq m \\ (J^T J)_{b_n b_m} &= \sum_i \left[\frac{1}{\sigma_i^2} \sin \frac{n\pi X_i}{L} \sin \frac{m\pi X_i}{L} \right], \quad n \neq m \end{aligned} \tag{29}$$

Clearly, for the case of a very large number of data points spanning many times the period, $2L$, and constant experimental uncertainties, i.e., constant σ_i, all of the summations in Eq. (29) are equal to zero. Thus, for these conditions the Fourier series analysis, as in Eq. (26), is an orthogonal case.

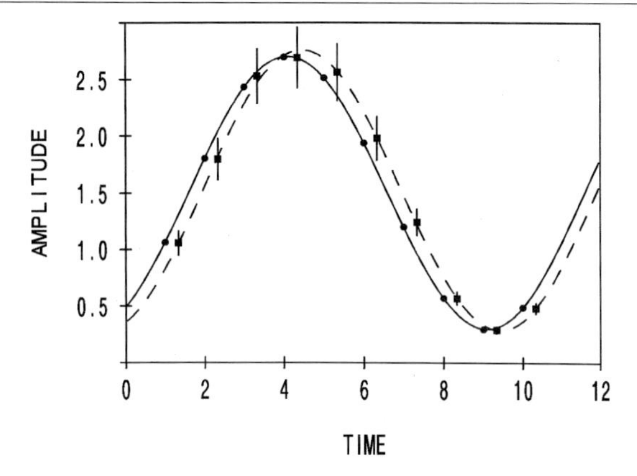

FIG. 7. Two simulated data sets for a Fourier series analysis. The squares contain experimental uncertainties that are proportional to the amplitude, whereas the circles contain uncertainties that are constant at 0.01. The lines are the least squares fits of these data according to Eq. (25).

However, if the data contain variable uncertainties, or variable spaced data points, or a small nonintegral number of periods, then these summations are not expected to be exactly equal to zero. For these conditions the exact solution [Eq. (28)] will yield results with less that the maximum likelihood of being correct, i.e., for these conditions the optimal parameters of the Fourier series cannot be evaluated by Eq. (28).

For example, consider the two least squares fits presented in Fig. 7. The solid line is the least squares fit of Eq. (25) to the solid circles. Note that the circles include constant experimental uncertainties of 0.01 units. When fit to a_0, c_1, and ϕ_1, the analysis is orthogonal and all of the cross-correlations are zero (not shown). The dashed line in Fig. 7 is the least squares fit of Eq. (25) to the solid squares. The squares include experimental uncertainties with a magnitude that is proportional to the amplitude. Table V presents the cross-correlation matrix corresponding to the dashed line. Simply chang-

TABLE V
CROSS-CORRELATION MATRIX FOR DASHED LINE ANALYSIS[a]

Parameter	Matrix		
c_1	1.00000		
ϕ_1	0.671532	1.00000	
a_0	−0.477350	−0.921146	1.00000

[a] In Fig. 7.

TABLE VI
CROSS-CORRELATION MATRIX FOR SOLID LINE ANALYSIS[a]

Parameter	Matrix			
c_1	1.00000			
ϕ_1	0.062116	1.00000		
L	−0.069469	−0.891111	1.00000	
a_0	−0.035931	0.464825	−0.521835	1.00000

[a] In Fig. 7 when the period is also determined.

ing the noise characteristics contained within the data changed one cross-correlation coefficient from 0.0 to 0.92. Clearly, the dashed line analysis is no longer an orthogonal case and thus the exact solution for the Fourier series does not apply to the analysis of these data.

These Fourier series analyses are also not orthogonal for the determination of the period. Table VI presents the cross-correlation for the Fourier series analysis of the closed circles in Fig. 7 when the period is also determined by the least squares procedure. Here the cross-correlation between the phase shift and the period is −0.89. Clearly, the inclusion of the determination of the period into the Fourier series analysis introduced nonorthogonality into the analysis.

Note that the inclusion of the evaluation of the period in the analysis of the variable uncertainty data (squares in Fig. 7) further increased the cross-correlation of all of the parameters. These are shown in Table VII. The effects are cumulative.

Also, although not shown here, the magnitude of the cross-correlation increases if the data points are irregularly spaced or some data points are missing.

TABLE VII
CROSS-CORRELATION MATRIX FOR DASHED LINE ANALYSIS[a]

Parameter	Matrix			
c_1	1.00000			
ϕ_1	0.309562	1.00000		
L	−0.336108	−0.916876	1.00000	
a_0	−0.920338	0.315724	−0.346446	1.00000

[a] In Fig. 7 when the period is also determined.

Exponential Fitting Functions

First, note that exponential fitting functions are neither linear nor orthogonal. They are the quintessential example of an ill-posed numerical analysis problem. There is no trick to eliminating the difficulties caused by the parameter cross-correlation. However, some things can be done to improve the ill-conditioned nature of the analysis of exponential decay data.

Consider a single exponential fitting function with an additive constant. This function can be written in at least four ways:

$$G(\alpha, X_i) = Ae^{-KX_i} + C \tag{30}$$

$$G(\alpha, X_i) = Ae^{-X_i/\tau} + C \tag{31}$$

$$G(\alpha, X_i) = Area\ Ke^{-KX_i} + C \tag{32}$$

$$G(\alpha, X_i) = \frac{Area}{\tau} e^{-X_i/\tau} + C \tag{33}$$

where

$$\begin{aligned} Area &= \frac{A}{K} = A\tau \\ K &= \frac{1}{\tau} \end{aligned} \tag{34}$$

Equations (30) and (32) are formulated with elimination rate constants, whereas Eqs. (31) and (33) are formulated with elimination lifetimes. The preexponential factors in Eqs. (30) and (31) are amplitudes, whereas the preexponential factors in Eqs. (32) and (33) are in terms of the area under the exponential decay curve. Each of these equations has three adjustable parameters. If the data in Table I and Fig. 1 are least squares fit by Eqs. (30) through (33) while $C = 0$, the cross-correlation coefficients are 0.6892, −0.6802, −0.7649, and 0.7672, respectively. For this example it appears that the choice of rate constants versus lifetimes does not affect the cross-correlation coefficients. However, the magnitude of the cross-correlation is increased substantially when the fitting function is formulated with a preexponential factor in terms of the area under the decay curve as compared with an amplitude.

Does the number of data points change the apparent cross-correlation coefficients? The first matrix in Table VIII presents the cross-correlation matrix for the analysis of the data in Table I according to Eq. (30). Here all three parameters are varied. Also shown in Table VIII is an analogous analysis using only the odd numbered data points. Thus, the second matrix in Table VIII corresponds to the same range of data points, just fewer data points. For this example, decreasing the number of data points makes only a small difference in the cross-correlation coefficients.

TABLE VIII
CROSS-CORRELATION MATRIX FOR SUBSETS OF DATA[a]

Parameter	Matrix		
All 17 data points			
A	1.00000		
K	−0.119816	1.00000	
C	−0.454369	0.882387	1.00000
Data points 1–3–7– · · · –17			
A	1.00000		
K	−0.257120	1.00000	
C	−0.569803	0.87336	1.00000
First 9 data points	1.00000		
A	−0.851445		
K	−0.940692	1.00000	
C		0.968304	1.00000

[a] Shown in Table I when fit to Eq. (30).

Does the range of the data points change the apparent cross-correlation coefficients? The third matrix in Table VIII presents the analogous analysis using only the first nine data points. Here the magnitude of the cross-correlation increases significantly, and dramatically, as compared with the complete data set or the data set truncated to every odd-numbered data point. The conclusion drawn is that the wider the range of the independent variable, the lower the cross-correlation coefficients yielded. And as a consequence the fitting procedure will perform better.

The orthogonal polynomials were generated by simply shifting the independent variable axis slightly. Can a similar trick be used to lower the cross-correlation coefficients for the exponential decay case? Consider the modified version of Eq. (1) as shown in Eq. (35):

$$G(\alpha, X_i) = A_1 e^{-K_1(X_i - Q_1)} + A_2 e^{-K_2(X_i - Q_2)} \tag{35}$$

The first derivatives of Eq. (35) with respect to the fitting parameters are

$$\begin{aligned}
\frac{\partial G(\alpha, X_i)}{\partial A_1} &= e^{-K_1(X_i - Q_1)} \\
\frac{\partial G(\alpha, X_i)}{\partial K_1} &= -A_1(X_i - Q_1)e^{-K_1(X_i - Q_1)} \\
\frac{\partial G(\alpha, X_i)}{\partial A_2} &= e^{-K_2(X_i - Q_2)} \\
\frac{\partial G(\alpha, X_i)}{\partial K_2} &= A_2(X_i - Q_2)e^{-K_2(X_i - Q_2)}
\end{aligned} \tag{36}$$

These can be substituted into Eq. (16) to obtain the mathematical form for the individual elements of the J^TJ matrix. It is this matrix that is inverted during the least squares parameter estimation procedures to obtain the values of the cross-correlation coefficients. To obtain an orthogonal system of equations, each of the off-diagonal elements of this matrix must be equal to zero.

$$
\begin{aligned}
(J^TJ)_{A_1K_1} &= \sum_i \left[\frac{1}{\sigma_i^2} (e^{-K_1(X_i-Q_1)})(-A_1(X_i - Q_1)e^{-K_1(X_i-Q_1)}) \right] = 0 \\
(J^TJ)_{A_2K_2} &= \sum_i \left[\frac{1}{\sigma_i^2} (e^{-K_2(X_i-Q_2)})(-A_2(X_i - Q_2)e^{-K_2(X_i-Q_2)}) \right] = 0
\end{aligned}
\tag{37}
$$

Clearly, a pair of Q values can be found for each particular data set such that these two elements of the J^TJ matrix are equal to zero. However, the evaluation of these Q values requires a prior knowledge of the K values. They do not depend on the A values since they can be canceled.

$$
\begin{aligned}
Q_1 &= \frac{\sum_i \frac{1}{\sigma_i^2} X_i e^{-2K_1X_i}}{\sum_i \frac{1}{\sigma_i^2} e^{-2K_1X_i}} \\
Q_2 &= \frac{\sum_i \frac{1}{\sigma_i^2} X_i e^{-2K_2X_i}}{\sum_i \frac{1}{\sigma_i^2} e^{-2K_2X_i}}
\end{aligned}
\tag{38}
$$

The other four off-diagonal elements of the J^TJ matrix cannot be made to be equal to zero if the Q values have been obtained from Eq. (37).

$$
\begin{aligned}
(J^TJ)_{A_1K_2} &= \sum_i \left[\frac{1}{\sigma_i^2} (e^{-K_1(X_i-Q_1)})(-A_2(X_i - Q_2)e^{-K_2(X_i-Q_2)}) \right] \neq 0 \\
(J^TJ)_{A_2K_1} &= \sum_i \left[\frac{1}{\sigma_i^2} (e^{-K_2(X_i-Q_2)})(-A_1(X_i - Q_1)e^{-K_1(X_i-Q_1)}) \right] \neq 0 \\
(J^TJ)_{K_1K_2} &= \sum_i \left[\frac{1}{\sigma_i^2} (-A_1(X_i - Q_1)e^{-K_1(X_i-Q_1)}) \right. \\
&\qquad \left. (-A_2(X_i - Q_2)e^{-K_2(X_i-Q_2)}) \right] \neq 0 \\
(J^TJ)_{A_1A_2} &= \sum_i \left[\frac{1}{\sigma_i^2} (e^{-K_1(X_i-Q_2)})(e^{-K_2(X_i-Q_2)}) \right] \neq 0
\end{aligned}
\tag{39}
$$

Table IX presents the cross-correlation matrix for an analysis of the data in Table I with Eq. (35) as a fitting function. For this analysis the values of Q_1 and Q_2 were specified as 0.098625 and 0.194224. These were evaluated according to Eq. (38). While this procedure improved the cross-correlation, these values are still large and problematic.

It is possible that a different pair of Q values could be derived based on any two of Eq. (39) that would provide a lower parameter cross-correlation coefficient. To test this, a grid search was done to test all possible pairs of Q within the range of ± 5 to a resolution of 0.01. This grid search required a least squares parameter estimation for each of the 10^6 pairs of Q. It was found that for Q values of -0.97 and $+1.63$, the largest cross-correlation coefficient was 0.9845. Consequently, it appears that simply scaling the independent variable, i.e., the X axis, by an additive constant has some small effects, but not enough to warrant the effort.

Conclusion

The objective of this chapter is to acquaint the reader with the origins and consequences of parameter cross-correlations. These commonly are simply a consequence of the fitting process and a limited amount of data. They need not be a consequence of the underlying physical process being studied. It was shown that these parameter cross-correlations can depend on the form of the fitting equation, the distribution of the data points along the independent variables, the distribution of the experimental uncertainties, and for nonlinear cases, the specific parameter values.

When the magnitude of these parameter cross-correlations approaches unity, the parameter estimation process becomes impossible. Consequently, it is important to consider methods to decrease the apparent parameter cross-correlations. Parameter cross-correlations can be modulated by altering the actual form of the fitting equation. They can also be altered by changing the distribution and range of the data points along the independent variable, the X axis.

TABLE IX

CROSS-CORRELATION MATRIX FOR TWO-EXPONENTIAL ANALYSIS OF DATA[a]

Parameter	Matrix			
A_1	1.00000			
K_1	−0.995215	1.00000		
A_2	−0.999967	0.995605	1.00000	
K_2	−0.996475	0.984554	0.996124	1.00000

[a] In Table I, according to Eq. (35), $Q_1 = 0.098625$ and $Q_2 = 0.194224$.

The parameter cross-correlation coefficients are usually not a function of the dependent variable, i.e., the Y axis. Therefore, it is not expected that transformation of the Y axis, such as logarithmic plots, will alter the parameter cross-correlation coefficients. Such transformation may allow alterations of the fitting equation that will improve the parameter cross-correlation, but these are not without a price. One basic assumption of the least squares method is that all of the experimental uncertainties are contained in the dependent variable. Thus, nonlinear transformations of the Y axis will also alter the apparent distribution of the experimental uncertainties contained within the data. Occasionally this can be useful,[8] but usually it is not a good idea.[9]

Acknowledgments

The author acknowledges the support of the National Science Foundation Science and Technology Center for Biological Timing at the University of Virginia (NSF DIR-8920162), the General Clinical Research Center at the University of Virginia (NIH RR-00847), and the University of Maryland at Baltimore Center for Fluorescence Spectroscopy (NIH RR-08119).

[8] R. Abbot and H. P. Gutgesell, *Methods Enzymol.* **240,** 37 (1994).

[9] F. S. Acton, "Analysis of Straight-Line Data," p. 219. Chapman and Hall, New York, 1959.

Author Index

Numbers in parentheses are footnote reference numbers and indicate that an author's work is referred to although the name is not cited in the text.

A

B

C

D

E

F

G

H

I

J

K

L

M

N

O

P

Q

R

S

T

X

Y

Z

Subject Index

A

B

C

F

G

H

I

K

L

M

N

O

P

R

S

ISBN 0-12-182222-2

90038

9 780121 822224